D1559130

Metal Cutting Tool Handbook

Metal Cutting Tool Handbook

Seventh Edition

Published for the
United States Cutting Tool Institute
by
Industrial Press Inc.

Library of Congress Cataloging-in-Publication Data

Metal cutting tool handbook.

  Includes index.
  1. Metal-cutting tools—Handbooks, manuals, etc.
I. United States Cutting Tool Institute.
TJ11 86.M38 1988   671.5'3   87-22825
ISBN 0-8311-1177-1

Industrial Press Inc.
200 Madison Avenue, New York, New York 10016-4078

PREFACE

Advances in technology, tooling, materials, and designs have prompted this revision of the METAL CUTTING TOOL HANDBOOK.

This Seventh Edition presents current data on Twist Drills, Reamers, Counterbores, Taps, Dies, Milling Cutters, Hobs, and Gear Shaving Cutters.

The information contained in each section will assist the users of metal cutting tools.

The text of the 7th edition of the **Metal Cutting Tool Handbook** was prepared by the Metal Cutting Tool Institute, which is now part of the United States Cutting Tool Institute.

United States CUTTING TOOL INSTITUTE

CONTENTS

DRILL SECTION

FOREWORD

Today's highly efficient drill is the result of creative designs, constantly tested in engineering departments, laboratories and production departments of drill manufacturers, machine tool builders and users.

These ideas and tests have amassed knowledge responsible for the "art" of drilling. There is reason to believe that additional improvements will be developed in future years.

This Section offers information on standard designs, application and care, and data useful in economically producing drilled holes.

DRILL SECTION CONTENTS

NOMENCLATURE OF TWIST DRILLS AND COMBINED DRILLS AND COUNTERSINKS

A—TWIST DRILLS

I—DEFINITION

Drill—A rotary end cutting tool having one or more cutting lips, and having one or more helical or straight flutes for the passage of chips and the admission of a cutting fluid.

II—GENERAL CLASSIFICATIONS

A—Classification Based on Construction.

1—*Solid Drills*—Those made of one piece of material such as high speed steel.

2—*Tipped Solid Drills*—Those having a body of one material with cutting lips made of another material brazed or otherwise bonded in place.

3—*Composite Drills*—Those having cutting portions mechanically held in place.

B—Classification Based on Methods of Holding or Driving.

1—*Straight Shank Drills*—Those having cylindrical shanks which may be the same or different diameter than the body of the drill. The shanks may be made with or without driving flats, tang, grooves or threads.

2—*Taper Shank Drills*—Those having conical shanks suitable for direct fitting into tapered holes in machine spindles, driving sleeves or sockets. Tapered shanks generally have a tang.

3—*Taper Square Shank Drills*—Those having tapered shanks with four flat sides for fitting a ratchet or brace.

4—*Shell Core Drills*—Core drills mountable on arbors specifically designed for the purpose. Commonly used with shell reamer arbors.

5—*Threaded Shank Drills*—Those made with threaded shanks generally used in close center multiple spindle applications or portable angle drilling tools.

6—*Beaded Shank Bits*—Drills with flat shanks having raised beads parallel to the axis.

C—Classification Based on Number of Flutes.

1—*Two-Flute Drills*—The conventional type of twist drill used for originating holes.

2—*Single-Flute Drills*—Those having only one flute commonly used for originating holes.

3—*Three-Flute Drills (Core Drills)*—Drills commonly used for enlarging and finishing, drilled, cast, or punched holes. They will not produce original holes.

4—*Four-Flute Drills* (*Core Drills*)—Used interchangeably with three-flute drills. They are of similar construction except for the number of flutes.

D—Classification Based on Hand of Cut.

1—*Right-Hand Cut*—When viewed from the cutting point the counter-clockwise rotation of a drill is required in order to cut. The great majority of drills are made "right hand."

2—*Left-Hand Cut*—When viewed from the cutting point the clockwise rotation of a drill is required in order to cut.

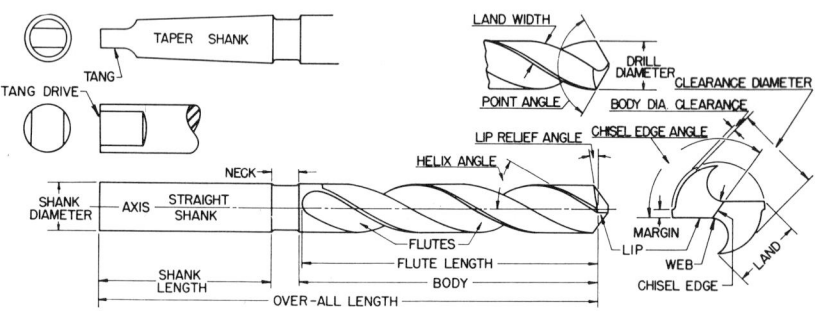

Figure 1—Illustrations of Terms applying to Twist Drills.

III—NOMENCLATURE OF TWIST DRILLS AND OTHER TERMS RELATING TO DRILLING

AXIS—The imaginary straight line which forms the longitudinal center line of the drill.

BACK TAPER—A slight decrease in diameter from front to back in the body of the drill.

BODY—The portion of the drill extending from the shank or neck to the outer corners of the cutting lips.

BODY DIAMETER CLEARANCE—That portion of the land that has been cut away so it will not rub against the walls of the hole.

BUILT-UP EDGE—An adhering deposit of nascent material on the cutting lip or the point of a drill.

CAM RELIEF—The relief from the cutting edge to the back of the land, produced by a cam actuated cutting tool or grinding wheel on a relieving machine.

CHIP BREAKER—Nicks or grooves or special flute shapes designed to reduce the size of chips. They may be steps or grooves in the cutting lip or in the leading face of the land at or adjacent to the cutting lips.

CHIP PACKING—The failure of chips to pass through the flute during the cutting action.

CHIPPING—The breakdown of a cutting lip or margin by loss of fragments broken away during the cutting action.

CHISEL EDGE—The edge at the end of the web that connects the cutting lips.

CHISEL EDGE ANGLE—The angle included between the chisel edge and the cutting lip, as viewed from the end of the drill.

CLEARANCE—The space provided to eliminate undesirable contact between the drill and the work piece.

CLEARANCE DIAMETER—The diameter over the cut away portion of the drill lands.

COMBINED DRILLS AND COUNTERSINKS—See definition on page 9.

CRANKSHAFT OR DEEP HOLE DRILLS—Drills designed for drilling oil holes in crankshafts, connecting rods and similar deep holes. They are generally made with heavy webs and higher helix angles than normal.

CUTTER SWEEP—The section formed by the tool used to generate the flute in leaving the flute.

CUTTING EDGES—See preferred term LIPS.

DOUBLE MARGIN DRILL—A drill whose body diameter clearance is produced to leave two margins on each land and is normally made with margins on the leading edge and on the heel of the land.

DRIFT—A flat tapered bar for forcing a taper shank out of its socket.

DRIFT SLOT—A slot through a socket at the small end of the tapered hole to receive a drift for forcing a taper shank out of the socket.

DRILL DIAMETER—The diameter over the margins of the drill measured at the point.

EXPOSED LENGTH—The distance the large end of a shank projects from the drive socket or large end of the taper ring gage.

EXTERNAL CENTER—The conical point on the shank end of the drill, and the point end on some sizes of core drills.

FLAT DRILL—A drill whose flutes are produced by two parallel or tapered flats.

FLUTES—Helical or straight grooves cut or formed in the body of the drill to provide cutting lips, to permit removal of chips, and to allow cutting fluid to reach the cutting lips.

FLUTE LENGTH—The length from the outer corners of the cutting lips to the extreme back end of the flutes. It includes the sweep of the tool used to generate the flutes and, therefore, does not indicate the usable length of flutes.

GAGE LINE—The axial position on a taper where the diameter is equal to the basic large end diameter of the specified taper.

GALLING—An adhering deposit of nascent work material on the margin adjacent to the leading edge at and near the point of a drill.

GUIDE—A cylindrical portion, following the cutting portion of the flutes, acting as a guide to keep the drill in proper alignment. The guide portion may be fluted, grooved or solid.

GUN DRILL—Special purpose straight flute drills with one or more flutes used for deep hole drilling. They are usually provided with coolant passages through the body. They may be either solid or tipped.

HALF-ROUND DRILL—A drill with a transverse cross-section of approximately half a circle and having one cutting lip.

HAND OF CUT—See general classifications.

HEEL—The trailing edge of the land.

HELICAL FLUTES—Flutes which are formed in a helical path around the axis.

HELIX ANGLE—The angle made by the leading edge of the land with a plane containing the axis of the drill.

LAND—The peripheral portion of the body between adjacent flutes.

LAND CLEARANCE—See preferred term BODY DIAMETER CLEARANCE.

LAND WIDTH—The distance between the leading edge and the heel of the land measured at a right angle to the leading edge.

LEAD—The axial advance of a leading edge of the land in one turn around the circumference.

LEAD OF FLUTE—See preferred term LEAD.

LENGTH OF TWIST—See preferred term FLUTE LENGTH.

LIPS—The cutting edges of a two flute drill extending from the chisel edge to the periphery. [CORE DRILLS]—The cutting edges extending from the bottom of the chamfer to the periphery.

LIP RELIEF—The axial relief on the drill point.

LIP RELIEF ANGLE—The axial relief angle at the outer corner of the lip. It is measured by projection into a plane tangent to the periphery at the outer corner of the lip.

MARGIN—The cylindrical portion of the land which is not cut away to provide clearance.

MULTIPLE-MARGIN DRILL—A drill whose body diameter clearance is produced to leave more than one margin in each land.

NECK—The section of reduced diameter between the body and the shank of a drill.

NOTCHES—See preferred term CHIP BREAKERS.

OIL GROOVES—Longitudinal straight or helical grooves in the shank, or grooves in the lands of a drill to carry cutting fluid to the cutting lips.

OIL HOLES OR TUBES—Holes through the lands or web of a drill for passage of cutting fluid to the cutting lips.

OVERALL LENGTH—The length from the exteme end of the shank to the outer corners of the cutting lips. It does not include the conical shank end often used on straight shank drills, nor does it include the conical cutting point used on both straight and taper shank drills. [CORE DRILLS]—For drills with an external center on the cutting end, same as for two flute drills.

For those with internal centers on the cutting end, the overall length is from the extreme ends of the tool.

PERIPHERY—The outside circumference of a drill.

PERIPHERAL RAKE ANGLE—The angle between the leading edge of the land and an axial plane at the drill point.

PILOT—A cylindrical portion of the drill body preceding the cutting lips. It may be solid, grooved or fluted.

POINT—The cutting end of a drill, made up of the ends of the lands and the web. In form it resembles a cone, but departs from a true cone to furnish clearance behind the cutting lips.

POINT ANGLE—The angle included between the cutting lips projected upon a plane parallel to the drill axis and parallel to the two cutting lips.

RAKE ANGLE—See preferred term PERIPHERAL RAKE ANGLE.

RELATIVE LIP HEIGHT—The difference in indicator reading between the cutting lips of the drill. It is measured at a right angle to the cutting lip at a specific distance from the axis of the tool.

RELIEF—The result of the removal of tool material behind or adjacent to the cutting lip and leading edge of the land to provide clearance and prevent rubbing [heel drag].

SHANK—The part of the drill by which it is held and driven.

SLEEVE—A tapered shell designed to fit into a specified socket and to receive a taper shank smaller than the socket.

SOCKET—The tapered hole in a spindle, adaptor or sleeve, designed to receive, hold, and drive a tapered shank.

SPADE DRILL—A removable cutting drill tip usually attached to a special holder designed for this purpose. Generally used for drilling or enlarging cored holes.

SPIRAL ANGLE—See preferred term HELIX ANGLE.

SPIRAL FLUTES—See preferred term HELICAL FLUTES.

STEP DRILL—A multiple diameter drill with one set of drill lands which are ground to different diameters.

STRAIGHT FLUTES—Flutes which form lands lying in an axial plane.

SUBLAND DRILL—A type of multiple diameter drill which has independent sets of lands in the same body section for each diameter.

TANG—The flattened end of a taper shank, intended to fit into a driving slot in a socket.

TANG DRIVE—Two opposite parallel driving flats on the extreme end of a straight shank.

TAPER DRILL—A drill with part or all of its cutting flute length ground with a specific taper to produce tapered holes. They are used for drilling the original hole or enlarging an existing hole.

TAPER SQUARE SHANK—A taper shank whose cross section is a square.

WEB—The central portion of the body that joins the lands. The extreme end of the web forms the chisel edge on a two-flute drill.

WEB THICKNESS—The thickness of the web at the point, unless another specific location is indicated.

WEB THINNING—The operation of reducing the web thickness at the point to reduce drilling thrust.

B—COMBINED DRILLS AND COUNTERSINKS

I—DEFINITION

Combined Drill and Countersink—Single or double-end cutting tool, having helical or straight flutes, and having a drill portion and an adjacent integral countersink portion, primarily used to produce center holes in work that will be held between machine centers.

II—GENERAL CLASSIFICATIONS

A—Classification Based on Construction.

1—*Solid Combined Drills & Countersinks*—Those made of one piece of material such as high speed steel.

2—*Tipped Solid Combined Drills & Countersinks*—Those having a body or drill portion of one material with cutting edges or lips, or both, made of another material brazed or otherwise bonded in place.

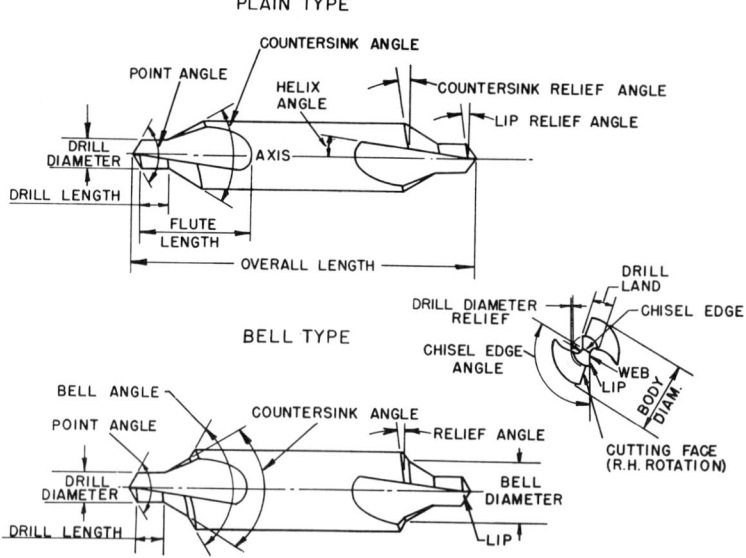

Figure 2—Illustrations of Terms applying to Combined Drills and Countersinks

B—Classification Based on Type.

1—*Plain Type Combined Drills & Countersinks*—Those having a drill portion and a single adjacent integral countersink portion.

2—*Bell Type Combined Drills & Countersinks*—Those having a drill portion and an adjacent integral countersink portion, plus an additional secondary conical section to provide clearance for the bearing surface.

C—Classification Based on Hand of Cut.

1—*Right-Hand Cut*—See definition under Twist Drills.

2—*Left-Hand Cut*—See definition under Twist Drills.

III—NOMENCLATURE OF COMBINED DRILLS AND COUNTERSINKS

AXIS—The imaginary straight line which forms the longitudinal center line of the combined drill and countersink.

BACK TAPER—A slight decrease in diameter from the front to back in the drill length.

BELL ANGLE—The included angle of the secondary conical section providing clearance or protection for the countersink angle conical surface. (It is normally 120°.)

BELL DIAMETER—The diameter at the intersection of the countersink portion and the bell portion at the leading edge of the land.

BODY—The central portion of the tool by which it is held or driven.

COUNTERSINK ANGLE—The included angle of the countersink portion. (It is normally 60°.)

COUNTERSINK RELIEF ANGLE—The angle between a plane at right angles to the axis of the tool, and a line tangent to the surface of the countersink portion at the intersection of the countersink portion and the body and at the leading edge of the land. For the Bell Type, the relief angle is measured at the intersection of the bell portion and the body.

COUNTERSINK PORTION—That part of the tool which produces a conical bearing surface for the work centers. The Bell Type produces an additional secondary conical section to provide clearance or protection for the bearing surface.

DRILL DIAMETER—The diameter of the drill portion measured across the outer corners of the cutting lips.

DRILL DIAMETER RELIEF—The relief provided on the land of the drill portion to reduce contact with the walls of the hole. It is generally of an eccentric form.

DRILL LENGTH—The length of the drill portion from the outer corners of the cutting lips to its intersection with the countersink portion.

DRILL PORTION—That part of the tool extending beyond the countersink portion to produce clearance for the point of a conical center.

FLUTES—The helical or straight grooves cut or formed in the cutting portions and the body to provide cutting edges on both the drill and countersink portions, to permit removal of chips and to allow cutting fluid to reach the cutting lips.

FLUTE LENGTH—The length from the outer corners of the cutting lips to the extreme ends of the flutes. It includes the sweep of the tool used to generate the flutes and, therefore, does not indicate the usable length of flutes.

LAND—The peripheral drill and countersink portions between adjacent flutes.

OVERALL LENGTH—The length between the extreme points of the tools.

CHISEL EDGE
CHISEL EDGE ANGLE
HELIX ANGLE
LIPS
POINT ANGLE   See definition under TWIST DRILL
LIP RELIEF ANGLE
PERIPHERAL RAKE ANGLE
WEB
WEB THICKNESS

CHIP FORMATION

Chip formation in metal cutting has been a subject of considerable research in the United States as well as in other countries concerned with the machining of metals. Studies in chip formation have been made mostly with single point tools in order to simplify the analysis of any given problem. The advanced theories concerning chip formation, supported by laboratory data, have also helped in the analysis of the cutting action of a twist drill.

[1] *The discontinuous chip.*

[2] *The continuous chip without built-up edge.*

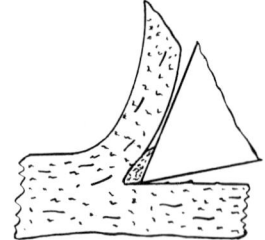

[3] *The continuous chip with a built up edge.*

Metal chips are generally placed in three separate categories or types. These three separate types are diagrammatically illustrated.

While the cutting action of a twist drill is much more complex than that of a single point tool, the type [1] chip creates the least difficulty in drilling. Types [2] and [3] continuous chips present greater problems in drilling, and the type [3] chip decreases the efficiency in drilling as in other machining applications.

Figure 1 illustrates a conventional drill point. It is to be observed that the rake normal to the cutting lip decreases from the periphery toward the center. While the rake angle decreases, the lip relief angle increases. This is illustrated by angles [a] and [b] in Figure 1.

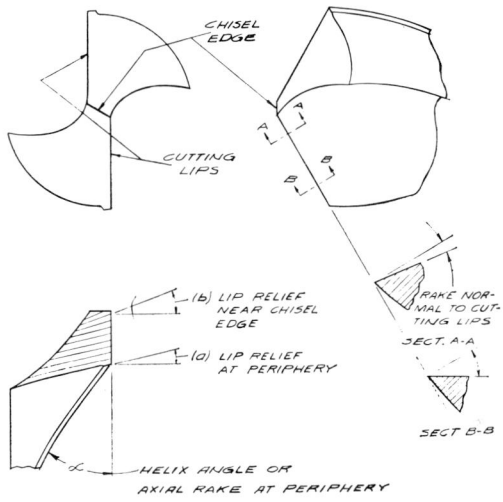

Figure 1.

The chisel edge, considered a cutting edge, has an extremely negative rake. The cutting action of the chisel edge is a very complex one illustrated by the enlarged photographs of chip formations under the chisel edge.

From these illustrations, one can see the cutting action of the chisel edge and the chip forming characteristics of the cutting lips. While in the vast majority of all drilling applications the conventional chisel edge functions very well, providing the simplest of all drill points, there are applications where a modification of the chisel edge is desirable. Different methods have been employed to improve the cutting efficiency of the chisel edge. A few are illustrated in Figures 2 through 6.

In effect, we may consider this reformed chisel edge (Figure 6) as a drill

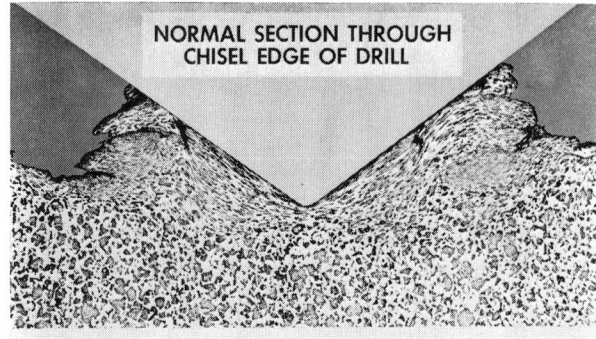

Photomicrograph of section normal to chisel edge, 0.037" radius, ¾" drill.

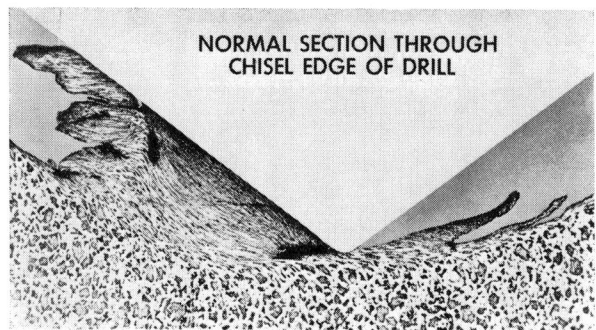

Photomicrograph of section normal to chisel edge, at exact center of hole, ¾" drill.

Photo of chip formed by the chisel edge, at bottom of drilled hole.

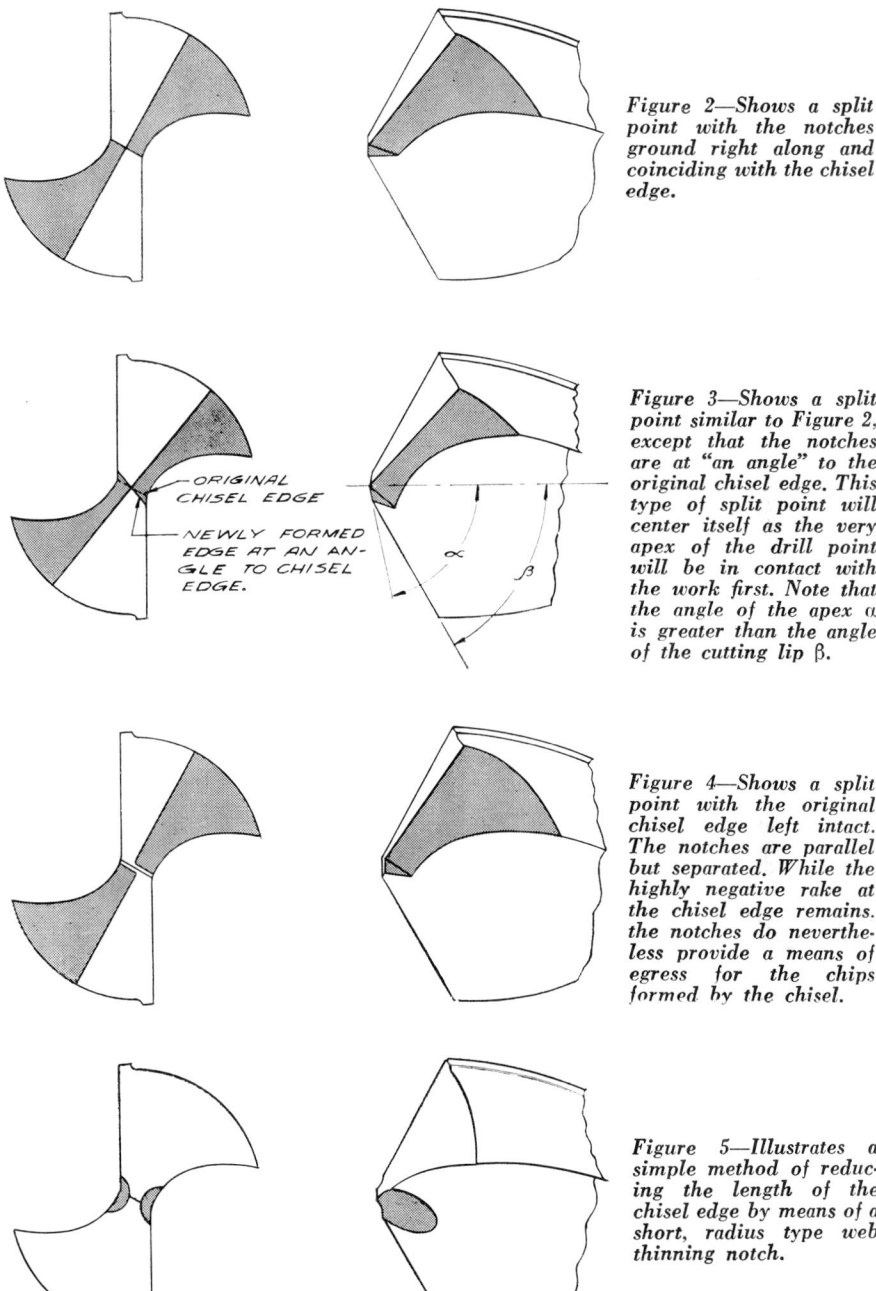

Figure 2—Shows a split point with the notches ground right along and coinciding with the chisel edge.

ORIGINAL CHISEL EDGE

NEWLY FORMED EDGE AT AN ANGLE TO CHISEL EDGE.

Figure 3—Shows a split point similar to Figure 2, except that the notches are at "an angle" to the original chisel edge. This type of split point will center itself as the very apex of the drill point will be in contact with the work first. Note that the angle of the apex α is greater than the angle of the cutting lip β.

Figure 4—Shows a split point with the original chisel edge left intact. The notches are parallel but separated. While the highly negative rake at the chisel edge remains. the notches do nevertheless provide a means of egress for the chips formed by the chisel.

Figure 5—Illustrates a simple method of reducing the length of the chisel edge by means of a short, radius type web thinning notch.

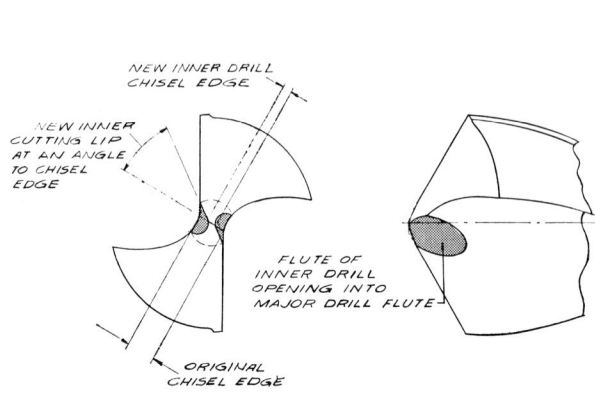

NEW INNER DRILL
CHISEL EDGE

NEW INNER
CUTTING LIP
AT AN ANGLE
TO CHISEL
EDGE

FLUTE OF
INNER DRILL
OPENING INTO
MAJOR DRILL FLUTE

ORIGINAL
CHISEL EDGE

Figure 6—Presents a concept of web thinning that visualizes two new cutting lips being formed at an angle to the chisel edge. The theoretical diameter of this "inner drill" is indicated by the dotted circle.—The radius of the notch provides the flute for the inner drill and an egress into the main flute of the drill for the chips formed by the minor cutting lips.—The newly formed cutting lips are parallel to each other, viewed from the point end of the drill along its axis. These parallel cutting edges should be separated from each other by a distance equalling approximately 10% of the combined length of the two newly formed cutting lips.

within a drill. Having this concept, we can readily see the need to provide adequate chip room and a flute shape that aids in curling and ejecting this secondary chip into the main flutes of the drill. The amount of positive rake that is incorporated, should depend upon the toughness of the material being drilled.

CHIP REMOVAL

When chips have been formed by the cutting lips of the drill point, just one phase of drilling has been completed. The chips must be removed from the hole in such a manner that no compacting of chips will occur and prevent cutting fluid from reaching the cutting lips.

For this reason drills are designed with different helix angles and flute contours. A high helix angle is used for rapid chip removal and where the larger rake angle is also desired. A high helix angle drill is used for aluminum and a low helix drill for brass. An intermediate or general purpose helix has proven best for all around use.

EFFECT OF DRILL STRUCTURE

For hard and tough materials, it is necessary that the drill be made with a heavy cross section because strength and rigidity are of utmost importance. The length of flute should also be reduced and the web increased in thickness from the point to back at a rapid rate. Chip removal under these conditions can be satisfactorily accomplished by withdrawal of the drill at certain increments of depth.

For extremely deep holes, such as oil passage holes in crankshafts, the drill design emphasizes strength at the expense of chip room. Even though this type of drill is generally made with a fast helix for effective chip removal from the point, it is nevertheless necessary to withdraw the entire drill from the hole at frequent intervals to avoid chip congestion.

Parabolic fluted drills have become quite popular in recent years for deep hole drilling using a minimum number of retractions. The drill design features include "open" (parabolic) fluting, high helix, with heavy web, usually parallel, to minimize restrictions to the chip flow. In many deep hole applications they require one-third or less of the number of retractions required by conventionally fluted drills.

EFFECT OF DRILLING POSITION

Chip removal is affected by the position of drilling. In a vertical position the chips must be withdrawn against the force of gravity. For this reason, many drilling applications, especially of deep holes, have been centered around a horizontal application and special machines have been built for inverted drilling where gravity becomes an aid instead of a hindrance. Special provisions must be made for an effective application of cutting fluid under these conditions. Oil hole, oil channel and gun drills are designed to fulfill a need in this respect.

EFFECT OF SPEEDS AND FEEDS

As the type 1 chip involves the least difficulty in chip formation, it also presents the least difficulty in chip removal. The speed and feed functions are therefore used to provide the most efficient stock removal, the criterion of performance being tool life versus hole production and time. The greatest efficiency is usually achieved with a high feed rate per revolution and suitable speed.

CHIP BREAKING

With types 2 and 3 chips, the feed rate that will tend to break up the chips into shorter segments is most desirable. Long continuous coils or stringy chips can cause difficulties by winding around the drill or drill spindle, or whipping around, creating personnel hazards. Well broken chips are the easiest to dispose of in drilling. Heavier feed rates usually help to break the chips into shorter lengths. Some materials however, may require the assistance of chip breakers or curlers ground along the cutting edges of the drill. Reduced rakes at the cutting edges have also been found effective for this purpose.

Figures 1 and 2 illustrate chip breakers designed to assist in breaking the chips into shorter lengths.

Several drill manufacturers offer proprietary "Chip breaker" drills designed for use where feeds high enough to break chips cannot be used. Some restrict the amount of space on the heel side of the flute to provide an obstruction in the normal chip flow path; this can force sufficient extra curling of the chips that they break. Another type provides extra material on the flute face to facilitate grinding a conventional chip breaker groove, illustrated in Figure 3.

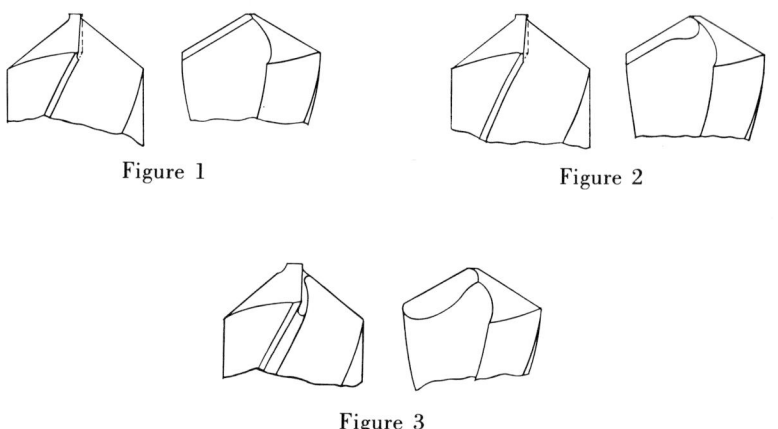

Figure 1   Figure 2

Figure 3

EFFECT OF COOLANT PRESSURE

In many deep hole drilling applications, chip removal is so important that oil hole drills are used. The cutting fluid is forced through holes, tubes or channels in the lands of the drills and the chips are flushed out under pressure. Here again it is usually necessary for the chips to be broken up into smaller segments so that they will readily be ejected along the drill flutes and not become jammed.

Inasmuch as extra energy is required to cause chip breaking, chip breaker drills always require somewhat more torque than conventional drills. Chip breaking drills tend to be less stable than conventional drills and thus may yield out of round or bellmouthed holes, a fact which may be significant if the drilling is to be followed by a precision hole finishing operation.

TYPES OF DRILLS

STRAIGHT SHANK TWIST DRILLS

Are the most widely used class of drills, particularly in the smaller sizes under ½ inch diameter. Such drills have cylindrical shanks, usually of the same nominal diameter as the drill. A few series of straight shank drills have common or uniformly stepped shank diameters for a range of sizes so that they can be driven with specific types of chucks or drivers. Some straight shanks may incorporate driving flats, tangs, grooves, or threads to suit various types of drill drivers.

The most widely used types of straight shank drills and some typical applications are given in the following paragraphs.

STRAIGHT SHANK DRILLS, JOBBERS LENGTH

There is a wide variety of drilling conditions in industry today due to the use of so many different materials and types of equipment. With this in mind, the general purpose drill has been designed to perform satisfactorily under as many different conditions as possible.

This drill may be considered a high production tool for all jobs except those where some element of material or set-up presents a particularly difficult problem. Even under such conditions, this type can be used with some degree of efficiency.

HIGH-HELIX DRILLS

Have been developed especially for drilling deep holes in materials of low tensile strength, such as aluminum, magnesium, copper, die casting materials, wood, some plastic materials, and stacked aluminum sheets. Its wide flutes and high helix help to clear the chips. Ocassional applications in drilling softer steels, free-machining brasses and bronzes, and free-machining stainless steels, have proven successful. In screw machine drilling, some successful applications have been made in drilling of screw stock and similar materials.

This drill can be specially made with heavier cross-section and on many jobs has proved satisfactory in drilling deeper holes in ordinary commercial steels.

Low-helix Drills for Brass

Designed primarily for use in brass, although in many cases successful applications have been made in shallow drilling some of the aluminum and magnesium alloys. The mechanical construction will readily remove the large volume of chips formed by high rates of penetration, particularly in screw machines or turret lathes.

Heavy Duty or Cotter Pin Type Drills

Especially designed for drilling cotter pin holes in bolts, steering knuckles, pins and similar jobs, as well as cross holes in the heads of bolts. Its heavy construction, designed to resist the exceptional strains put on the drill by this kind of work, has given it many successful applications in drilling some of the harder alloy steels, including the harder stainless steels.

When sharpened with a 130° to 140° split point, this drill has been successfully applied to the portable drilling of hard aluminum alloys, aircraft type stainless steels, and titanium alloys.

Drills For Plastics

Developed for use in most rigid plastics, hard rubber, asbestos and some varieties of fiberboard. Their wide flutes and slow helix provide maximum space for chip ejection while their long points (included angle of 90° or less) provide the corner wear resistance required in these often abrasive materials.

LEFT HAND DRILLS, JOBBERS LENGTH

Made regularly of general purpose design only, have the same flute and overall lengths as general purpose jobbers length drills.

They are used considerably in close center multiple spindle gear driven drilling heads, where adjacent spindles operate alternately left and right hand.

They are also used extensively in screw machines, where the sequence of operations makes it advisable to drill while the spindle is revolving in the left hand direction. This quite often occurs in certain automatic screw machines, where a tapping operation is involved and only two spindle speeds are available: a slow right hand speed for tapping and a fast left hand speed for tap reversal and drilling.

STRAIGHT FLUTED DRILLS, JOBBERS LENGTH

Especially adapted for work in brass or other soft materials, as they will not run ahead or "grab." They are also suitable for use in thin sheet material for the same reason.

COBALT H.S.S. HEAVY DUTY DRILLS

These ruggedly constructed heavy duty drills were developed to drill the difficult-to-machine high strength and thermal resistant materials. Cobalt H.S.S. has a harder, stronger matrix which increases abrasion resistance and delays the onset of softening due to the high localized temperatures encountered when machining these tough materials. These drills are usually equipped with split points to reduce the high end thrust required in drilling high strength materials.

The rigid heavy duty construction is necessary to the successful utilization of the superior cutting ability of the cobalt high speed steels because these tool materials, being harder and more highly alloyed, are more susceptible to impact chipping and breakage.

SCREW MACHINE LENGTH DRILLS

Developed primarily for use in screw machines of all types, where conditions require use of short drills.

Considerable savings in time, labor, and tool cost can be made through use of these tools by eliminating the necessity of reworking regular length drills—cutting-off, thinning the webs, and repointing.

The drills are made with short flutes and short overall length to obtain the maximum rigidity without sacrificing any of the cutting ability of these tools. This type of construction gives these drills many applications, not only in screw machine work, but also in portable drilling, sheet metal, and body work. Other applications may be found in drilling tougher and harder steels, such as stainless, high manganese, and others, where short, rigid drills are a basic tool requirement.

CENTER DRILLS

Very short overall and flute length, designed for centering work, where their extreme rigidity reduces runout and eccentricity. For this reason on the larger sizes, they have been quite successful as starting drills in screw-machine drilling.

SPOTTING AND CENTERING DRILLS

Are used to produce accurate and true centers in work in screw machines to obtain a true center for starting follow up drills, and obtaining perfect alignment.

Their short flute and overall lengths and no body clearance permit chucking close to the point so that they will produce a true start or center. Also made with a constant web in order that subsequent repointing of the drills requires no web thinning.

In addition, they are useful in spotting holes for tap drilling and will reduce tap breakage on tapped holes.

TAPER LENGTH DRILLS

Are general purpose straight shank drills having approximately the same overall and flute lengths as taper shank drills. Their shanks are of the same nominal diameter as the drill. They are frequently used in lathes and screw machines, particularly in the larger sizes. When equipped with tangs, they are often used with split sleeve drivers as a substitute for taper shank drills.

HEAVY DUTY TAPER LENGTH DRILLS

Are of heavy duty construction and have flutes somewhat longer than regular taper length drills. The shanks are tanged for use with split sleeve drivers. Above ½ inch diameter, the shanks are smaller than the drill diameter and are arranged in steps to minimize the number of different drivers required.

THREE AND FOUR FLUTE CORE DRILLS

Are designed for enlarging and/or finishing holes previously produced by drilling, casting or punching. They will not produce original holes. Overall and flute lengths and shank diameters are made to the same proportions as taper length drills.

AIRCRAFT DRILLS TO AIA/NAS STANDARDS

The Aerospace Industries Association, in cooperation with drill manufacturers, has standardized on a series of split pointed drills for various drilling operations common to the aerospace industries. The materials to be drilled range from aluminum alloys through tough thermal resistant alloys. They are often used in portable drilling equipment. The accompanying table, abstracted from the current (1986) National Aerospace Standard (NAS) summarizes proportions, construction and intended applications. The general availability of certain types may be limited so this should be checked with the drill supplier before their use is specified.

The AIA also developed a NAS Standard covering short threaded shank drills for use in 45° and 90° angle drives and flexible shaft drives. This permits these drills to be used in close quarters which are inaccessible to regular drilling units.

AIA—NAS 907 Drill Standards (1986)

Type Designation	Length Overall	Flute Lengths	H.S.S.	Application
A	Jobbers	Jobbers	Standard	Limited—Light Duty
B	Jobbers	Jobbers	Standard	General—Medium Duty
C	Screw Machine	Screw Machine	Standard	General—Medium Duty
D	Jobbers	Short Standard Helix	Cobalt	Limited—Heavy Duty
E	Jobbers	Short Low Helix	Cobalt	Limited—Heavy Duty
J	Jobbers	Jobbers	Cobalt	General—Heavy Duty
Extension	6″ & 12″	Jobbers	Standard	For Long Reaches Light—Medium Duty

THREADED SHANK AIRCRAFT DRILLS

1¼″ Overall   *2⅛″ Overall*
                                Hex Shank

These drills have shanks threaded for use in 45° or 90° angle drives and flexible shaft drives with threaded adapters. The combination of angular heads and these short drills permits working in close quarters, where regular drilling units cannot fit. These drills are made to NAS 965–1986.

EXTRA-LENGTH DRILLS

Quite often, because of the inaccessibility or the extreme depth of a hole, it is found that even long series drills are not long enough. The drilling of such holes generally requires the use of an extra-length drill. Such drills are usually made with a general purpose flute design which will perform satisfactorily under the widest range of operating conditions and materials.

In deep holes there is more variation in operating conditions than in shallow holes, and very often it is advantageous to vary the design of extra-length drills in much the same manner as is done with the regular-length drills.

Some of the more important factors which must be taken into account when designing an extra-length drill for a particular operation are:

(1) Material to be drilled
(2) Hardness of material

(3) Position of drill; *i.e.* Vertical, horizontal, or inverted

(4) Type of feed; *i.e.* Intermittent or steady

(5) Depth and size of hole

(6) Is section behind flutes a guide or only an extension?

As an example, crankshaft drills are often used in drilling steel forgings of 300+ Brinell, in a nearly vertical position, to a depth of twenty times the drill diameter, with an intermittent feed. This will require an entirely different design of drill from that best suited for the horizontal drilling of a free-machining brass with a constant feed.

However, the extra-length drill of general purpose design will be found entirely satisfactory for the user who has short runs in a wide range of materials, machines, set-ups, and operating conditions.

DEEP-HOLE OR CRANKSHAFT DRILLS

Designed for drilling the oil holes in forged crankshafts and connecting rods and the like. It has, however, found a great many more applications in the drilling of deep holes in other tough materials. The mechanical construction of this drill is special, as it has a heavy web, a helix angle somewhat higher than normal, and usually a split-point type of web thinning.

PARABOLIC FLUTE DRILLS

These drills are designed with open fluting, heavy parallel web, and high helix. They usually have crankshaft points to reduce the thrust required by the thick web. Because of their heavy web construction they can be used under heavy feed conditions in most materials, in shallow holes, as well as deep holes. They are very popular for deep hole drilling because they work best under heavy feed conditions, with minimal reductions in feed required even in holes 12 or more diameters deep, and usually require one third or less of the number of retractions required by conventionally fluted drills.

COMBINED DRILLS AND COUNTERSINKS

When metal parts are to be machined on centers, as in turning or cylindrical grinding, it is always necessary to Drill and Countersink the center holes to serve as seats or bearings for the machine centers.

Two types of Combined Drills and Countersinks are available; namely, the Plain Type and the so-called Bell Type.

The Plain Type will produce center holes as indicated in Fig. 1. For the ordinary run of work this kind of center hole is quite satisfactory, and for that reason the Plain Type is by far the most commonly used.

If the parts to be machined are to pass through a number of subsequent handlings, where there is danger of marring the edges of the center holes, it is advisable to use the Bell Type and to produce center holes as indicated in Fig. 2. The outer edges are beveled to prevent damage to the center hole itself.

Where a maximum amount of protection of the center hole is required, this can be attained by making the depth greater, as shown in Fig. 3. Tools used in this manner should be relieved on the large diameter. For arbors or other parts, where the center holes are to be used repeatedly, this latter type should always be used. In making this type of center hole, care must be taken to drill only deep enough to give the necessary protection. If the drill penetrates too far, the machine-center may ride on the corner of the straight, large diameter rather than on the 60° surface.

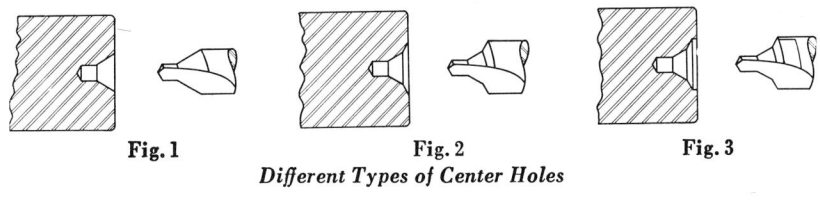

Fig. 1   Fig. 2   Fig. 3
Different Types of Center Holes

SHARPENING

The drill point may be re-sharpened in the conventional manner; but the re-sharpening of the beveled portions is a more difficult procedure. Grinding of the faces of the beveled portions is difficult to perform accurately. It will also tend to widen the flutes, thereby weakening the tool. Grinding on the outside to reproduce the original shape is the only satisfactory way of re-sharpening these beveled parts.

OPERATING PROCEDURE

The importance of properly performing this seemingly simple operation is often not fully realized. It is a fact that the success in general, the quality, and the accuracy of subsequent machining operations depend greatly on good center holes, for these holes control the location and rigidity of the work supports during machining.

SPEEDS AND FEEDS

The permissible surface speeds of these tools must be based on the largest diameter that comes in contact with the work, which means the body diameter rather than the drill-point diameter. This is sometimes overlooked, with disastrous results.

Feeds per revolution, on the other hand, must be based on the diameter of the drill point and should correspond to those given in the table of feeds for drills (see page 77).

Proper Countersinking involves the following:

1. Correct positioning of the center hole.
2. Adequate width of center bearing surface.
3. Correct angle, roundness, and smoothness of the bearing surface.

Positioning of the center hole necessarily is a product of the accuracy and rigidity of work-holding fixtures and of the centering machine itself. It is important, therefore, that both fixtures and machines be maintained in first-class condition at all times.

Cylindrical pieces whose outside surfaces are reasonably round and smooth are best centered by rotating the work in roller rests, holding the center-drill stationary, though this is by no means universal practice today. Work that is not round or smooth must, of course, be held in stationary jaws or clamps, with the center-drill rotating.

Adequate width of center-bearing surfaces is important because this will insure good support for the part to be machined. It will also prevent distortion of the center hole on heavy operations and will avoid scoring of centers.

Widths of center-bearing surfaces should be in reasonable proportion to the size and weight of the part to be machined, but must also take into account the tool pressures that are developed during machining. Thus, it is necessary to have larger centers for milling and heavy-turning operations, than, for example, grinding and light-turning operations.

When Bell Type Drill and Countersinks are used, the width of bearing surface is automatically maintained.

For the majority of centering operations, the standard type of Drill and Countersink is used, and it is here that the width of bearing surface must be watched closely.

The following figures will serve as a guide for checking the proper diameters of center holes:

Size of Drill and Countersink	Diameter of Center Hole	Size of Drill and Countersink	Diameter of Center Hole
No. 1	$3/32''$ to $7/64''$	No. 5	$5/16''$ to $3/8''$
No. 2	$9/64''$ to $11/64''$	No. 6	$3/8''$ to $7/16''$
No. 3	$3/16''$ to $7/32''$	No. 7	$15/32''$ to $17/32''$
No. 4	$15/64''$ to $9/32''$	No. 8	$9/16''$ to $5/8''$

As stated above, these dimensions must be varied to meet special conditions of work weights and tool pressures.

Positive end stops on the work and on the drill spindle should be used at all times, so that uniformity can be maintained. This will be found of value on subsequent operations.

Inasmuch as accuracy and general quality of workmanship on subsequent operations is largely controlled by the quality of center holes, this latter factor is worthy of careful attention.

The first requirement for such quality is the use of Drills and Countersinks having correct bearing surface angles. Secondly, these cutting lips must be maintained true and sharp and the tools removed from use before excessive dulling impairs their accuracy. The operator should be instructed in the correct technique of feeding to produce smooth, undistorted center bearing surfaces on the work.

CENTER DRILLS

These drills, of very short overall and flute length, have been designed for centering work, where their extreme rigidity would reduce runout and eccentricity. For the same reason on the larger sizes, they have been quite successful as starting drills in screw-machine drilling.

Diameter Inches	Length of Flute Inches	Length Overall Inches	Diameter Inches	Length of Flute Inches	Length Overall Inches
1/16	3/4	1 1/4	3/16	1	1 1/2
5/64	3/4	1 1/4	13/64	1	1 1/2
3/32	3/4	1 1/4	7/32	1	1 1/2
7/64	3/4	1 1/4	1/4	1	1 1/2
1/8	3/4	1 1/4	9/32	1	1 1/2
9/64	3/4	1 1/4	5/16	1	1 1/2
5/32	1	1 1/2			
11/64	1	1 1/2			

SPOTTING AND CENTERING DRILLS

Designed to produce accurate, true centers in all types of screw machine work, to obtain a starting spot for other drills. Short flute and overall length, and no body clearance permits chucking close to the point. Constant web thickness requires no thinning when repointing.

Normally furnished with 118° included angle drill point. The best drill point can be determined on each individual operation.

Diameter Inches	Length of Flute Inches	Length Overall Inches
$\frac{3}{8}$	1	2
$\frac{1}{2}$	1	2
$\frac{5}{8}$	$1\frac{1}{8}$	$2\frac{1}{4}$
$\frac{3}{4}$	$1\frac{1}{8}$	$2\frac{1}{4}$
1	$1\frac{1}{4}$	$2\frac{1}{2}$

DRILLS WITH ½ INCH SHANK

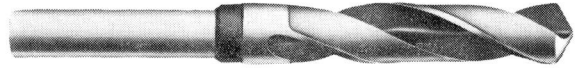

Are of general purpose flute design but all sizes have a ½ inch by 2¼ inch long shank. The maximum overall length is 6 inches and the maximum flute length is 3 inches. This makes them short drills in the larger sizes and thus well suited to use in portable electric drills equipped with ½ inch chucks. They are also used in machines where it is desirable to fit a variety of drill sizes to a single-size driving collet or chuck.

HALF-ROUND DRILLS

Are used mostly in screw machines for horizontal drilling of deep holes in free-machining brass. The hole is usually started with a short, rigid center drill, after which the Half-Round Drill penetrates the balance of the metal.

DRILLS FOR RAILROAD WORK
HIGH SPEED TRACK BITS

Flat Beaded Type Drill

Flat Type Drill with ⅝" (.647) Round Flatted Shank

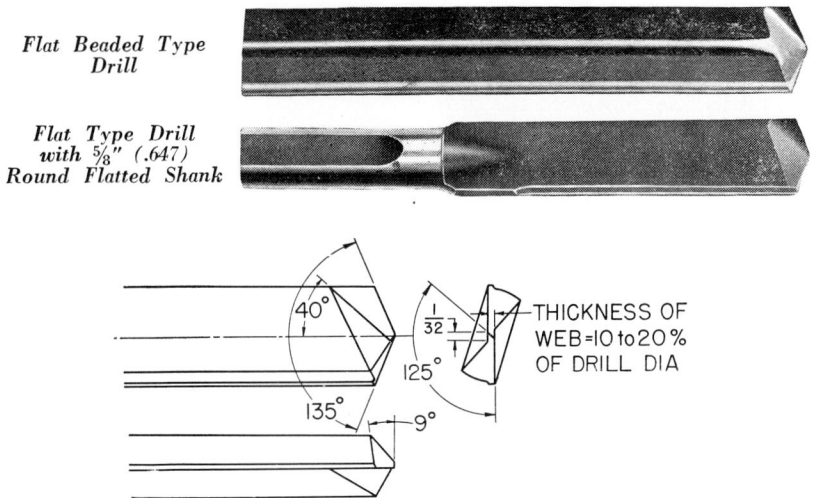

Recommended Point Grinding for Flat Track Drills

These drills have been designed for heavy duty portable track repair work in the field. The above shanks, and those of other types, are designed to fit the various machines in common use.

BONDING DRILLS

These drills are used for drilling holes for bonding wires in track circuit signal work. They are made regularly in two sizes, $\frac{9}{32}$" and $\frac{3}{8}$".

STOVE-BURNER DRILLS

This drill is used to drill the cast iron burners for gas stoves and other heating equipment. It is often furnished with a threaded shank that fits into a threaded holder or driver. Its short flutes and sturdy construction make it ideally suited to overcome the irregularities in the surface of the burners that cause heavy strains to be put on the drills.

THREE AND FOUR FLUTE DRILLS

Three and Four Flute Drills are used for enlarging holes that have previously been drilled, cored, or punched. Because of being used in cored holes, they are often called "Core Drills." Their construction is such that the center portion will not cut. The amount of stock they will remove is limited by the depths of the flutes.

The purposes of using Core Drills instead of ordinary two-fluted drills for enlarging are two-fold:

First, because of greater productivity.

The material to be drilled and its hardness determine and limit the maximum surface speed and the feed per cutting edge. Now, if we use drills with three or four cutting edges instead of two, it is evident that a greater feed per revolution can be used, and more holes can be drilled in a given time. The increase in production should be fairly proportional to the number of cutting lips, though this may be limited in some cases by the strength and rigidity of the setup and by the quality of finish desired. Core Drills, because of their heavy center section, will stand the additional strain imposed by the greater number of cutting lips.

Second, because of the better finish obtained.

Core Drills having three or four cutting lips also have the same number of lands extending along their diameters. This multiple number of lands will support the drills much better while cutting than if they had two lands only. The result is less tendency to wobble, to score up the walls of the drilled hole, and to cut oversize. In fact, the action of Core Drills is much the same as that of Rose Reamers. Often it is found practical to use Core Drills in place of Roughing Reamers.

To obtain the best results, it is highly important that Core Drills be properly sharpened. Care must be taken to see that all cutting lips are exactly the same length, and that they have the same point angle. While this is obvious, it is a detail that is often overlooked. Uneven lengths of the cutting lips will cause a disproportionate strain on the longest ones, resulting in quick dulling and a tendency to crowd the drill to one side. In turn, this will cause holes to come oversize. Further, there is then danger of wearing down the lands at the point, causing the drill to squeal and bind in the holes. It is recommended that a dial indicator or some similar device be used for checking the lengths of cutting lips before putting such drills in operation.

This type of tool is quite often used as a roughing reamer in order to correct the location of a previously formed hole. When used for this purpose, a rigid machine and fixture are required. The fixture should have accurately located drill bushings to guide the tool. When used for this purpose, it is recommended that the drill be pointed with a *very flat* point angle, as shown in the illustration.

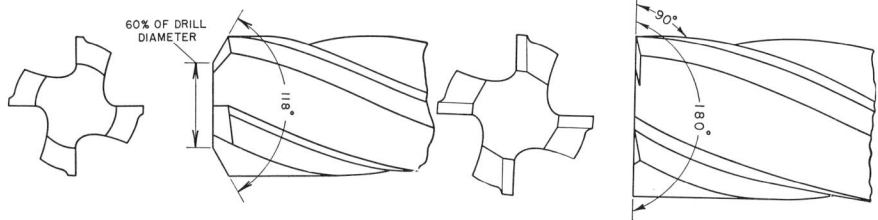

Regular Point for 3- or 4-flute drill   *Point for correcting location of previous hole*

This very flat point angle (often 180°) reduces the tendency of the drill to follow the previously formed hole. It is most important that no chamfer be used on the corners of the point and that the drill be kept quite sharp. As soon as the corners start to wear, the tendency to follow the previous hole is increased.

For most applications there is little choice between a three- and four-fluted drill. A three-fluted drill has slightly more chip clearing ability, but most users find the four-fluted drill a little easier to resharpen and measure.

Core Drills are sometimes made with a replaceable drill tip. The tip is fitted by some mechanical means, such as a taper and locking arrangement, and may be removed when worn or used up.

STRAIGHT SHANK COOLANT HOLE DRILLS

Originally used largely in screw machine and turret lathe production work, recent cutting fluid supply developments have made these coolant hole drills adaptable to a wider variety of drilling operations. In addition to conventional pump cutting fluid systems, these drills are now used in conjunction with air operated systems which supply the fluid in pulsed bursts or as a mist. The cutting fluid is fed through the drill to the cutting region where it improves cooling and lubrication, and assists in ejecting chips from the drilled hole. Properly applied coolant hole drills of optimum design can be useful in drilling difficult-to-machine materials and can improve production rates in ordinary materials. Coolant hole drills for rotating use generally have higher flute helix angles than those for non-rotating application.

Types of Oil Feeding Shanks

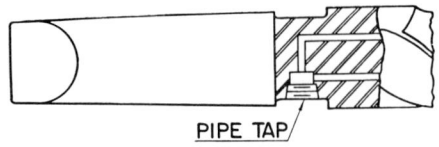

PIPE TAP

"S" SHANK SIZE

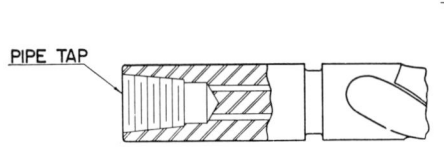

PIPE TAP

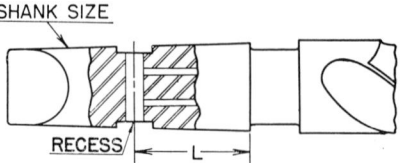

RECESS |←—— L ——→|

Note: Straight-shank oil-hole and oil-tube drills have shanks the same diameter as the body and are furnished with the shank end tapped with the following pipe thread:	

Taper No.	L
1	$1\frac{3}{16}$
2	$1\frac{5}{16}$
3	$1\frac{1}{32}$
4	$1\frac{1}{4}$
5	$1\frac{3}{8}$

$\frac{1}{8}$″ taper pipe tap for drills $\frac{1}{2}$″ through $\frac{3}{4}$″ diameter.
$\frac{1}{4}$″ taper pipe tap for drills $\frac{49}{64}$″ through $1\frac{1}{4}$″ diameter.
$\frac{3}{8}$″ taper pipe tap for drills $1\frac{17}{64}$″ diameter and larger.

TAPER SHANK TWIST DRILLS

Have tanged conical shanks to fit directly into the tapered holes in machine spindles, driving sleeves or sockets. They are available in a variety of designs, generally similar to those used for straight shank drills. Among these are:

1—Regular Taper Shank Drills
2—Heavy Duty Taper Shank Drills
3—Cobalt HSS Taper Shank Drills
4—Three and Four Flute Taper Shank Core Drills

SPECIAL PURPOSE TWIST DRILLS

Are somewhat specialized in their application but are used sufficiently to deserve mention and potential consideration for certain hole producing operations.

GUN DRILLS

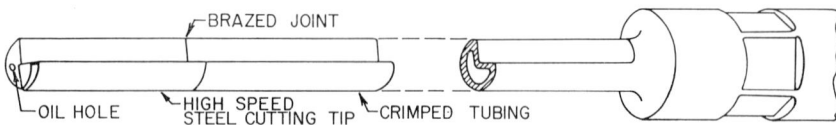

BRAZED JOINT

OIL HOLE — HIGH SPEED STEEL CUTTING TIP — CRIMPED TUBING

Essentially a single-flute straight-fluted drill head, usually cutting to center, mounted in an alloy-steel tubing shank of suitable diameter and length. The shank is crimped or rolled to form a flute and approximate the cross-section of the head, and the two are often joined by welding or brazing. In the larger sizes, the heads may be made so they can be easily detached for greater convenience in resharpening. Both head and shank have provisions for conducting an ample volume of coolant under high pressure to the cutting lip of the drill head. The coolant both cools the tool-work junction and washes the chips out through the flute. The drill head may be made of solid high-speed steel, solid carbide, or high-speed steel with carbide-tipped cutting edges and wear-guide strips.

The single end cutting face of the head is usually sharpened with two intersecting point angles. These point angles, relief angles, and the location of the intersection point are very important. They influence the size of hole produced, and the breaking of the chips into small enough pieces for easy ejection. Exact dimensions of the point details usually are determined for each application.

Properly applied gun drills are capable of producing holes which are straight and accurate as well as of long length. While long length is implied by the word "gun", it is not a necessary requirement. Gun drills can often produce holes which are so accurate and straight that secondary hole finishing operations can be reduced or eliminated.

MULTIPLE-DIAMETER DRILLS

It is often possible to drill a hole of two or more diameters with a single drill of proper construction. Such drills are, in fact, used quite extensively in mass production industries and affect many notable economies.

The simplest form of Multiple-Diameter Drill is the combined Drill and Countersink. Its uses are well known.

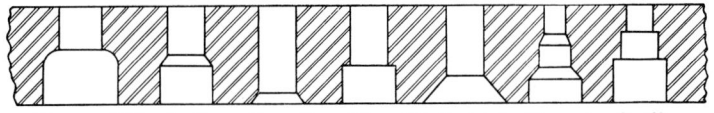

Types of Single Operations Possible with Multiple-Diameter Drills

All of the types of holes shown in the illustration and many others often can be produced in a single operation by Multiple-Diameter Drills.

The following are two examples of tools used for such operations.

STEP DRILL

A Two-diameter Step Drill

A Step Drill may have two or more diameters, produced by grinding various successive steps on the lands of the drill. These steps are usually separated by square or angular cutting edges, as the individual jobs may require. The Step Drill is useful for most of the jobs requiring multiple-diameter drilling. It is extensively used because it can be made by grinding down and stepping an ordinary drill. Some thinning of the web is usually required when so made.

SUBLAND DRILLS

DUAL CUT DRILL

Sub-land Drill

Perform the same functions as the step drill, while its construction is somewhat different. The step drill has its steps or different diameters on the same land, while the Subland Drill has two distinct lands running substantially the entire length of the flutes. This is shown in the illustration.

Subland Drills, like Step Drills, can be made in more than two diameters and are not necessarily restricted to two-fluted tools.

SPADE DRILLS

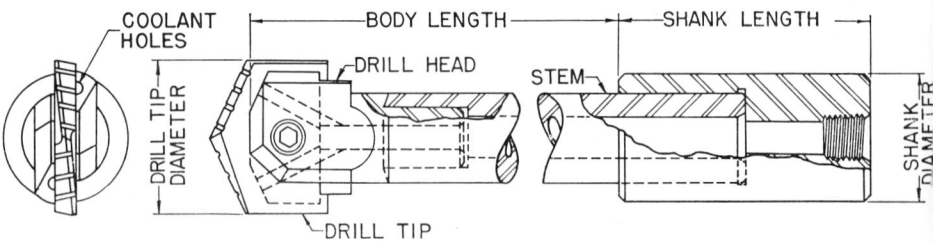

Are an assembly of a cutting blade and a holder or shank. Due to mechanical problems involved in such an assembly, they are principally used in sizes about 1½ inch diameter and larger. The holder may incorporate flutes and usually has a central cutting fluid passage which permits the fluid to reach the cutting blade and to assist in ejecting chips. The cutting blades are readily removable for sharpening which makes it possible to use a single holder with a range of blade sizes.

Spade drills have long been used on turret lathes and similar setups where the drill does not rotate. Recent developments in cutting fluid inducers are permitting their adaptation to drilling and boring machines where the drill must rotate.

Special purpose blades for core drilling and counterboring are available. There is some use of carbide tipped blades in abrasive materials but the use of carbide in these built-up tools requires a good setup and careful tool engineering.

CARBIDE DRILLS

The ability of carbide to retain hardness and a sharp edge at high temperature, plus its excellent abrasion-resistant qualities, makes it occasionally advantageous for use in drilling abrasive materials, such as cast iron, the non-ferrous alloys, hard rubber, and plastics. Carbide-tipped drills [other than gun drills] are not usually recommended for drilling unhardened steel. For hardened steel and glass, drills of special design are required. Carbide drills for concrete, masonry, etc., also are generally of specific design for this application.

CARBIDE TIPPED JOBBERS LENGTH STRAIGHT SHANK DRILLS FOR PLASTICS

May be used for drilling abrasive materials of low tensile strength, such as soft non-ferrous materials, plastics, hard rubber, some masonry, glass, etc.

CARBIDE TIPPED JOBBERS LENGTH STRAIGHT SHANK DRILLS

The drills are primarily for drilling cast iron, cast steel, bronze, hard copper, aluminum and hard non-ferrous materials which have a high degree

of abrasiveness and which require a sturdy drill for strength. They are made with the tips brazed in hardened steel bodies.

CARBIDE TIPPED DRILLS FOR HARD STEEL

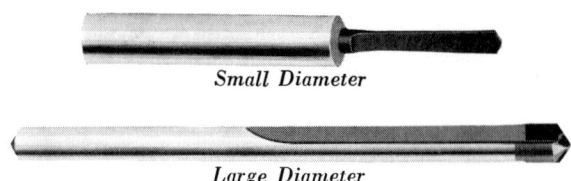

Small Diameter

Large Diameter

Are made for use in drilling hardened steel in the range of 48 to 65 Rockwell C. Holes may be drilled without annealing or appreciably changing the structure of the material.

Smaller sizes are often made flat drill style with the drilling portion being solid carbide. Larger sizes have the web reduced at the point for better and easier feeding characteristics. These drills may be run at speeds from 75 to 100 feet per minute using a good flow of coolant. A steady hand feed should be used. When drilling deep holes, the drill should be withdrawn frequently to clear the chips.

In shallow holes, under good machine conditions with minimum spindle end play, these carbide tipped style hard steel drills may be used to advantage with a light power feed.

The cutting corners, lips and chisel edge should be examined frequently and resharpened when the points begin to dull. This procedure reduces the chance of breakage.

CARBIDE TIPPED TAPER LENGTH STRAIGHT SHANK DRILLS

This drill is the same construction as the Carbide Tipped jobbers length heavy duty drills for cast iron, cast steel, and non-ferrous materials. It is made with a straight shank, not flatted, the same diameter as the fluted section.

CARBIDE TIPPED TAPER SHANK DRILLS

Are of the same construction and application as the Carbide Tipped jobbers length heavy duty drills for cast iron and non-ferrous materials.

Carbide Tipped Four Fluted Taper Shank Drills

Are ideally suited to enlarge holes in cast iron and other similar abrasive materials. If properly handled they are suited for many steel applications as well, particularly where abrasion or high material hardness is a factor.

Solid Carbide Heavy Duty Straight Shank Drills

For use in abrasive materials of low tensile strength such as cast iron, cast aluminum, bronze, copper, rubber and other similar materials.
They offer the advantage of extreme rigidity for greater drilling accuracy.

Solid Carbide Circuit Board Drills

Short solid carbide drills with wide flutes of relatively high helix angle are used to drill the abrasive reinforced plastic-copper laminates used in electronic circuit boards. Frequently used in specialized circuit board drilling machines, all sizes of drills in a series have the same overall length. For this reason, another series of these drills has a common 1/8 inch shank diameter for all sizes.

Carbide Tipped Masonry Drills

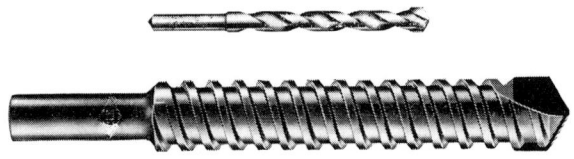

Masonry drills are used in portable electric drills, preferably slow speed. They are applied in concrete, sandstone, cinderblock, and similar materials.

SPECIFICATIONS FOR SPECIAL DRILLS

On many jobs, entirely special drills are needed. In the interest of economy, it is always good practice to exhaust the possibilities of regular tools before deciding on special drills.

If it is decided that special tools are required, information covering the details noted below will be of help to the drill manufacturer in furnishing the most suitable tools.

When specifying special drills, it is always well to make sure that suitable allowance has been made for resharpening and also for clearance for the spindle above the drill bushings.

If a particular style of flute construction is wanted, it should be specified by reference to the regular drill of the required flute style.

When ordering extra-length drills, specify: type of material being drilled, depth of hole, whether drilling in a vertical or horizontal position, and whether feed is intermittent or with only occasional withdrawals.

STRAIGHT SHANK DRILLS

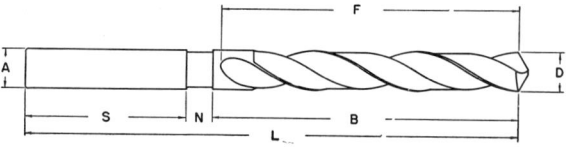

D = Diameter of fluted section. Use decimals instead of fractions.
A = Diameter of Shank.
L = Length Overall.
F = Length of Flute.
B = Length of Body.
S = Length of Shank.
N = Length of Neck.

TAPER SHANK DRILLS

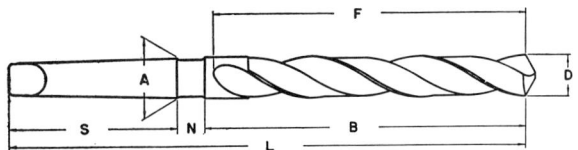

D = Diameter of fluted section. Use decimals instead of fractions.

A }
S } = Size of Shank. If standard shank is ordered, specify as No. 2 Morse, No. 3 Morse, etc.

    For special Taper Shanks, furnish drawing or gauge.

    For dimensions of standard Taper Shanks see page 883.

L = Length Overall.

F = Length of Flute.

B = Length of Body.

MULTIPLE-DIAMETER DRILLS

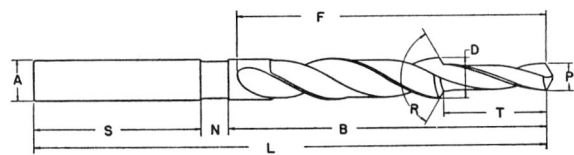

Specify whether construction is to be Step or Subland Type.

D = Diameter of large, fluted section. Use decimals instead of fractions

  P = Diameter of small, fluted section.

  A = Diameter of Shank.

  L = Length Overall.

  F = Length of Flute.

  B = Length of Body.

  T = Length of Small Diameter. (Note that this is measured from the outer corner of the point to the *bottom* or *inner* corner of the cutting shoulder).

  S = Length of Shank.

  N = Length of Neck.

  R = Included angle of cutting shoulder. Note that this is measured as an angle between the two cutting edges (included angle) and *not* as an angle with the center line.

DRILLING CHARACTERISTICS
OF VARIOUS MATERIALS

PLAIN CARBON AND ALLOY STEELS

PLAIN CARBON STEELS

The most common types of steels used in fabricating. They are used in the manufacture of most of the high-volume parts in many industries.

The drillability of plain carbon steels is affected by their carbon content, method of manufacture, hardness level, and possible chemical additions to the steels to make them more machinable. Drilling characteristics cover almost the complete range. They can vary from soft, very gummy materials producing troublesome, long, stringy chips to the hard, highly abrasive, high carbon tool steels. In those with sufficient carbon content to permit heat treatment, partial or full heat treatment to somewhat increase the hardness very often will improve their drillability.

In drilling these materials, the most common mistakes are the use of too high cutting speeds and too light feeds. This will result in more rapid tool wear and in long, stringy chips. These long, stringy chips can cause machine and tool trouble and also create a safety problem. For the most successful drilling of these materials, the lower speeds should be used combined with feeds at or above those normally recommended. See Speeds & Feeds, pages 66–79. In some of the higher carbon content or heat treated varieties, the proper feed may require the use of heavy duty drills instead of the general purpose type.

Where chip problems or excessive wear is experienced, a thorough trial of feed changes and simple drill point alterations should be made. This can result in much more efficient drilling.

Drilling deep holes in plain carbon and alloy steels usually requires special design drills for the particular part or application, and may also require special drilling techniques and setups. Small, deep holes over four to five drill diameters often require a step or woodpecker drilling procedure. Larger holes over four to six diameters deep often require drills designed specifically for the particular application. For additional information on deep hole drilling, refer to pages 77–78.

Drilling small, deep holes is best accomplished by using a low surface speed and a feed which will produce long, continuous, curled chips. In the higher carbon content or hardened material conditions, heavy duty or special drills may be necessary to avoid broken, powdered chips.

ALLOY STEELS [COMMON]

The more common types of alloy steels cover a wide range of drillability, ranging from free machining types to those that are almost impossible to drill efficiently without some previous or extra preparation of the material. Individually, they can vary from an easily drilled material to one that can be very troublesome.

Drilling characteristics of a given material can be changed through selective heat treatment. For example, many of these alloy steels with a low carbon content of .15–.25% and in their normalized condition are very difficult and troublesome to drill.

However, by selectively heat-treating to somewhat increase their hardness level, their machinability can be substantially improved. If trouble is experienced in machining these materials, the possibility of heat treatment to increase machinability should not be overlooked. Many times the cost of an additional or intermediate heat treatment can be readily justified by reductions in tool and machining costs.

Many of these alloys are tough and very gummy in the annealed state. When drilled, they produce long, stringy chips, difficult to curl or break. If the drilling machine has the capabilities and the proper drill is used, a substantial increase in feed along with a slight reduction in speeds can be of great help. Sometimes these conditions will require the use of higher included angle drill points and special chip breakers to curl and break the chip. See pages 14–16. Generally, changes in operating conditions or hardness will be less costly than the use of complex drill pointing.

At the higher hardness levels, these materials increase in toughness and are more abrasive in nature. These characteristics require the use of the heavy duty drills and a reduction in both speed and feed. The feed rate reduction should be only as large as necessary to eliminate chatter of the drill and prevent damage to machine and fixtures. In these cases it may be advantageous to investigate some of the web thinning methods, pages 101–105, which reduce the thrust requirements in drilling.

Some alloy steels also have work-hardening characteristics which present problems. The work-hardening characteristics may result in a hard surface layer created either from fabrication of the part or by the actual drilling operation itself. This condition results in premature and rapid tool wear or failure. Since this characteristic is inherent in the material, the only way to lessen its effects is to use a large enough feed rate, and to check the machine, the setup and the tool for sufficient rigidity to prevent chatter.

When drilling small holes, the overall strength and rigidity are determined by the drill size. Chatter or vibration in the part or the drill can result in magnetism problems. Cold working of some of these materials by manufacturing processes or by chatter or vibration of the cutting tool can result in the tools becoming sufficiently magnetized to prevent proper chip ejection and result in excessive drill breakage.

CAST IRON

All cast iron type materials, including those containing nickel and other additives and malleable iron, are characterized by short, brittle chips and powdery cuttings. These short chips and powder are highly abrasive. They usually cause fairly rapid abrasive wear of the cutting lips and the margins of the drill.

These materials, if of reasonably good grade, do not present any particular drilling problems except in the effort to reduce the rate of wear of the cutting tools. The basic principle involved is to drill the required hole as quickly as possible with the minimum contact between tool and work; i.e., this material should be drilled with the highest possible feed rates at conservative speeds. Since the wear is mostly from abrasion, long drill points, double-angle points, or a long point with corner radii are generally used. See page 111 for details. In addition, web thinning to reduce the length of the chisel edge is also employed.

For optimum results in this class of material, use should be made of cemented carbide tools of either tipped or solid construction. If the quality of the material is reasonably good and the job is carefully set up, the use of carbide types of drills will produce many times the number of holes produced by high speed steel tools.

The most important points to check, whether high speed steel or cemented carbide drills are used [besides proper speeds and feeds], are the spindle end play and the chip disposal means. Because of the high penetration rates used, too much spindle end play will result in damaged and broken drills from the breakthrough. In addition, the high production rates create a large volume of chips which, if not properly disposed of, will result in damage to the tools and setup.

STAINLESS STEEL

Stainless steels are alloys of steel whose prime purposes are to resist corrosion and maintain high strength under elevated temperatures up to approximately 1000°F. Stainless steels fall into three groups:

Group A, Martensitic, which are hardenable and magnetic. Free machining grades contain added sulfur and/or selenium.

Group B, Ferritic, which are non-hardenable, but are magnetic. Free machining grades contain sulfur and/or selenium to improve machinability.

Group C, Austenitic, which are not hardenable by heat treatment and are non-magnetic. They are subject to work-hardening.

The free machining grades of stainless steel are readily drillable because they form short and brittle chips which are easily ejected from the hole. Unless the holes to be drilled are of an unusual nature, general purpose drills are satisfactory.

For high production work, shorter and heavier types of drills will give excellent results. For deep holes, drills with a greater helix angle will be of help in chip removal.

Because of their free machining characteristics, these stainless steels can be machined at speeds and feeds only slightly lower than those used for mild steels.

The austenitic types, because of their work-hardening characteristics, are the most troublesome to drill. Included in this group are precipitation hardened types and manganese types.

To drill this group of materials successfully, these important factors should be remembered: Because of the work-hardening characteristics and the toughness of the materials, both the tools and the setup must be very rigid and the feed per revolution sufficiently high to cut under any work-hardened surface produced by the previous cut. Drills should be pointed with large included point angles, slightly greater lip relief, and the web thinned to reduce the thrust. Any web thinning used should be one that has a positive rake and should reduce the web thickness to an absolute minimum that will not break down under the cut. For additional assistance in drilling, sulfurized and chlorinated cutting oils should be used whenever possible.

Deep holes are not too difficult to drill unless the diameter is very small. Drills of heavy construction and with helix angles greater than normal will be of great help. There must be sufficient rigidity to prevent chatter, undue strain on tool and equipment, and sufficient feed to overcome work-hardening effects and to produce the desired form of chip.

Drilling of stainless steel with portable equipment is similar to other forms of drilling except that the web thinning is more critical so as to obtain maximum thrust reduction and to minimize the possibility of point skidding when starting the hole. In addition, as steady a feed as is possible to attain must be used.

COPPER AND COPPER ALLOYS

Copper and its alloys present two major problems when being drilled: [1] because of their gummy characteristics, long, stringy chips are produced; and [2] because of their fairly high coefficient of expansion, there is a tendency to cause the drills to bind or freeze in the workpiece. The chip problem can be reduced by proper selection of drill designs, drill points, and speeds and feeds. Because copper is known mostly for its soft and plastic characteristics, the tendency in drilling the material is to use a high cutting speed and a light to medium feed rate. Generally, the use of these operating conditions is incorrect and will cause many problems.

To drill copper and its alloys successfully, more conservative speeds than normally recommended should be used and feed rates one and one-half to three times greater should be employed. This will produce a heavy type chip which can more easily be curled and broken. In most cases, a thick, heavy chip will not cling tightly to the drill flutes and will be ejected more easily. In addition, the use of lower speeds and higher feeds plus more than normal web thinning and higher relief angles on the point will reduce the heat generated in drilling.

The other major problem is binding or freezing of the drills in the material. This condition is brought about mainly by two factors: [1] too much heat generated by the cutting operation; and [2] in deep hole drilling, crooked holes, i.e., the holes being produced are not straight. To reduce the heat generated, the drills are quite often altered to reduce the margin width to one-half or less of the original width and to increase the longitudinal back taper. The amount of back taper may vary from .001/inch of flute to .004/inch of flute depending on the size and length of drill used. The straightness of the drilled hole is affected by the type of drill point, the style of drill, and the degree of chip packing encountered. To reduce the possibility of crooked holes, a free-cutting drill and proper chip flow will help eliminate this problem.

ALUMINUM AND ALUMINUM ALLOYS

The drilling of these materials has become quite commonplace and not too many difficulties are experienced. Some of the newer aluminum alloys of high silicon content and some of the cast alloys still present a few problems.

For sheet material of all types, general purpose drills or drills developed for the aircraft industry can be used. For deeper holes, drills of the high helix variety should always be used. The soft alloys will require drills with bright, polished flutes to prevent chip packing and material buildup.

The common problem when drilling deep holes is drill breakage. Most of this breakage is the result of chip packing caused by improper use of the drill, improper cutting fluid, poor setup, or using the wrong type of drill. Generally, deep holes are drilled with high helix drills, while some of the soft, gummy materials will require the low helix type.

In drilling aluminum, high rates of penetration can be used; hence disposal of chips or cuttings is very important. To permit these high penetration rates and still dispose of the chips, drills have to be free cutting to reduce the heat generated and have large flute areas for passage of chips. In addition, the drill points have to be sharpened with high lip relief, and the web at point kept as thin as possible. Drilling some of the aluminum alloys will require altering the drill in the same way as suggested in the section on copper and its alloys.

High silicon aluminum alloys are very abrasive, particularly those alloys with a silicon content above the eutectic, about 11½%. The abrasiveness of the silicon makes the use of carbide tipped, solid carbide, and diamond-tipped drills warranted when any appreciable number of holes is required.

MAGNESIUM AND ZINC ALLOYS

Drilling magnesium and its alloys does not create any difficulties, but requires some changes in general practice to suit its peculiarities. These materials require fairly high drilling speeds and heavy feed rates. Because of the high penetration rates, a large volume of chips is produced. This requires drills with ample chip space and equipment having provisions for efficient chip disposal. These materials should rarely be drilled dry. When drilling, feeds should be heavy enough to produce large, thick chips, thereby reducing the fire hazard.

In deep hole drilling, quite often a helix or spiral effect is produced. This resembles a chatter pattern of spiral or helical grooves which approximate the helix angle of the drill flutes. It results in a hole which appears to be undersize so that a solid rod of equivalent size to the drill cannot be passed through or entered into the hole, yet the drill itself can be reinserted but only by rotating to produce a screw thread motion. The use of smaller included angle drill points can be used to correct this condition.

The drilling of zinc alloys, including the popular die-casting alloys, presents several problems. These materials are soft and gummy, thereby readily clogging drill flutes. Because of their relatively low melting point, the heat generated by the cutting operation can result in the material softening sufficiently to weld in the drill flutes.

To avoid these problems, drills with wide polished flutes should be used. All of the alterations to drills which will reduce the heat generated in drilling, such as high lip relief, longer drill point angles, very thin webs at the point, and narrower margins should be used, particularly on deep holes. Drilling speeds should be conservative and feeds determined by the size and shape of the chips produced and the ease with which they can be ejected. Any alteration made on the drill must be done carefully so as not to impede smooth chip flow.

PLASTICS

The drilling of plastic materials in their pure state, i.e., without any filler material added, is not too difficult. Many of the plastic materials are made with a wide variety of fillers, many of which are highly abrasive. These fillers can change the drilling characteristics of a material considerably, and generally determine the operating conditions.

In their pure state, most plastic materials are relatively soft and are of relatively low temperature compositions. They are usually divided into two classes: thermo plastics and thermo setting types. The thermoplastic type is very soft with a low melting temperature. The heat generated by cutting softens the material causing the chips to pack and stick in the flutes and a tendency to "freeze" the drills in the material. Thermo plastics require drills with wide polished and generally low-helix flutes.

The thermo setting types of plastics are brittle and usually abrasive in nature. Because of these properties they can be readily drilled. The common problem is flaking or break out around the edges of the drilled hole. For these either low-helix or high-helix drills as well as half-round drills can be used depending on the chip forming characteristics of the particular type of plastic. For blind holes and very deep holes, high-helix drills are recommended.

Proper speeds and feeds to be used for any given plastic material are determined by its composition and the type and state of the filler. In addition, job requirements such as no chipout, no fuzzy or shredded hole edges, and smoothness and straightness of holes will also affect the selection of the drill style and operating conditions.

Many of the plastic materials are very abrasive in nature and because of their types of fillers, require sharp tools at all times to produce satisfactory results. Others, such as printed circuit material, also require sharp tools to prevent delamination of the surface material. In such applications, and where production requirements warrant their use, high-helix drills made of solid carbide or tipped with carbide material are used to advantage. Diamond-tipped drills are also becoming increasingly popular.

COMPOSITE MATERIALS

The term "Composite" refers to a material made from two or more separate components as compared to a more homogeneous material such as a metal.

The chief concerns of manufacturing today are not with the older, more easily fabricated glass fiber/epoxy resin type composite. Advanced composite materials made of boron fiber bonded with epoxy resin, and graphite fiber bonded with epoxy resin have been developed and are being increasingly used in structural members of aircraft because of their strength, stiffness, and weight saving characteristics. Kevlar aramid fiber material combined with resins provide another useful material to save weight.

The most common advanced composite material requiring drilling is the graphite/epoxy type. It is relatively abrasive, and the fibers are susceptible to delamination when subjected to the cutting pressure of a dull drill. Carbide drills are used for drilling this material, or sandwiches of graphite/epoxy and aluminum. However, even carbide wears quickly relative to the sharpness needed to minimize delamination. In these cases, diamond-tipped drills may be useful.

Kevlar is difficult to machine in that its fibers are difficult to cut cleanly. Regular drills produce fuzzy holes and surfaces. Drilling into a back-up sacrificial material reduces the effect, where this procedure is practical. Where bushings are used, drills pointed with points similar to type K, p. 62 are useful. Drills made with flatter points and highly hooked lips have also been effective.

HIGH STRENGTH STEELS

These materials are generally machined in the heat-treated stage where the hardness ranges between 40 and 56 Rockwell "C". Extensive experimentation and application show that it is possible to drill these materials at such high hardness levels. The most practical approach is to rough machine these materials at the 25 to 40 Rockwell "C" level, and then finish machine at the high hardness level. When necessary, these materials can be drilled at their high hardness levels by using suitable techniques. Because of their high hardness and strength, high energy is needed to cut them, with resulting high temperatures at the cutting edges. The temperature can be controlled by reducing the cutting speed drastically, but still use a substantial feed. Sufficient feed is required to produce a chip instead of powdery cuttings. To accomplish this, drills, machines, fixtures and workpieces must be made as rigid as possible. In some cases performance can be improved by using drills made of higher alloyed high speed steels or cemented carbide. A further requirement is a positively controlled feed to provide constant thrust and feed rates.

Drill construction must provide maximum rigidity in the tool and maximum strength and support of the cutting edge. It is necessary to use a form of web thinning that will reduce thrust requirements to a minimum and still provide a tool and cutting edge of sufficient strength and cutting ability.

HIGH TEMPERATURE ALLOYS

This class of materials includes those which must retain a large portion of their strength and oxidation resistance at highly elevated temperatures. There are many types of these materials, but because of their common characteristics of high shear strength and extreme tendency to work-harden, their machining characteristics are very similar. Energy requirements for cutting are very high. Because of this, the machine and tooling setups must be rigid and powerful and positive feeds must be used.

By composition, the high temperature alloys can be classified into four categories, and before proceeding with the machining of any material, the proper group classification should be determined.

> 1—Ferritic low alloys—these may be considered an extension of the stainless steel family.
>
> 2—Austenitic alloys—this group covers a wide range of alloys high in chromium, nickel and molybdenum. Some of them have a considerable tungsten and cobalt content.
>
> 3—The nickel base alloys—the principal constituent is nickel, but they contain large percentages of other elements and little or no iron. This group contains many age-hardenable alloys.
>
> 4—The cobalt base alloys—this group has cobalt as the major element, and like the nickel-base alloys, has large percentages of other elements with little or no iron present.

In general, machinability decreases from Group 1 through Group 4; abrasiveness increases from Group 1 through Group 4 and becomes so great that Group 4 requires extremely low cutting speeds.

The metallurgical condition of the alloy at time of machining is established by process consideration. Gains can sometimes be made by machining at the best stage in the metallurgical processing. Precipitation hardenable alloys are usually gummy before heat treatment and abrasive after treatment. Better finish can usually be obtained after treatment. In some cases, machining in the quenched state before precipitation hardening takes place may be economically justified by permitting higher cutting speeds and giving longer tool life.

All of these materials require very heavy web drills with helix angles suitable for the material and job conditions. Because of the high shear strength and abrasiveness of these materials, the double-angle type of point may be helpful to reduce corner wear and increase the strength of the cutting edges. [See Type D, page 52.] Proper web thinning is required to minimize work-hardening and to reduce the thrust requirements without weakening the drill point.

Some of these materials, particularly the cast alloy group, may require the use of solid carbide or carbide tipped drills in order to obtain optimum results. Excellent setup conditions are required. Drills must be kept very sharp, as only slight amounts of wear will result in drill breakage.

All groups require the use of sufficiently rigid tools, good heavy equipment in good condition, and rigid powerful setups. Additional machine requirements are proper speed range increments and fine feed increments. Very small changes in speeds and feeds can make substantial increases or decreases in tool life, often without any further changes in the tools or setups. Better results will be obtained if a coolant is used and substantial increases in results can be achieved through the use of the correct coolant.

TITANIUM AND TITANIUM ALLOYS

Increased use of these materials has broadened the knowledge of working them. Not many of these materials present any difficult problems in drilling unless the job conditions are very unusual in nature.

Titanium and its alloys have unusual thermal, mechanical and chemical properties which, when understood, eliminate actual and potential problems. The low-volume specific heat and low thermal conductivity make them easy to heat at the point of cutting, and difficult to cool because the heat does not readily flow away through the material. When they are hot, titanium chips are plastic and tend to collapse and jam in the drill flutes. Hot titanium is very reactive and susceptible to galling on tool surfaces. It is also very abrasive in its action on cutting tools. Titanium chips are thin, so they move across cutting faces at higher speeds than most other materials. They also curl readily, so the wear area is smaller and concentrated adjacent to the cutting edge.

Thermal problems are best combatted by reducing the rate of heat generation. This can be done by reducing either the speed or the feed. Fortunately, these materials do not work-harden appreciably, so lighter feed can be used successfully. Further reductions of heat generation in drilling can be attained through the use of thinned webs in such a manner as to increase the rake of the cutting edge. Because titanium alloys have a great tendency to weld and gall on the points and margins of drills as temperatures increase, copious supplies of coolant should be used whenever possible.

DRILL POINTS

Drilling today's wide variety of materials efficiently requires a great variety of drill points, types of web thinning, and drill designs. The subject of drill designs and their applications has been covered in a previous section.

The following drill point and web thinning discussion is not confined to specific drill designs, but to the basic principles which can be applied to almost any drill style. It is important to understand the basic principles and purposes in order to modify drills to suit specific applications. The proper use of controlled drill pointing and web thinning can result in substantial savings in drilling costs. Controlled drill pointing includes producing the proper included point angle, cutting lip relief and clearance.

The purpose of using various degrees of included drill point angle with proper lip relief in conjunction with a specific type of web thinning is to achieve the following:

1—Control of the formation of the chip.
2—Control of the size and shape of the chip.
3—Control of the chip flow along the flutes.
4—Increase the strength of the cutting lip.
5—Reduce the rate of cutting lip wear.
6—Reduce the drilling thrust required.
7—Control the drilled hole size, quality and straightness.
8—Control the amount of burr on either or both ends of the drilled hole.
9—Permit greater variations of speeds and feeds for more efficient drilling.
10—Reduce the heat generated.

Most of these factors are also affected materially by the amount and type of lip relief. Too little relief can result in high thrust values, accelerated lip wear, excessive heat generation and undue strain on the drill and equipment. Too much relief can weaken the cutting lip resulting in chipping and breakage of the tool. The amount and type of relief used with any type of drill point is determined by the size of the drill and the material to be drilled. Generally, the amount of relief should be as large as possible without weakening the cutting lip to the point where it breaks down under the thrust and feed loading used.

Since the amount of chip space on any drill is limited by the design requirements, the size and shape of the chip and the ease with which it can be disposed of is very important. It also materially affects the drilling efficiency of any specific application.

Some of the more common types of drill points and web thinning are illustrated. Many of these points and web thinning styles are costly to produce. Before adopting any of the more complex forms of web thinning or drill pointing, various combinations of speeds and feeds and coolants should be tried. These may be more economical overall and require less skill. Close observation of the cutting process and a study of the chips produced can indicate whether or not a drilling operation is efficient. In addition, it can suggest a course of action of further trials.

TYPE A

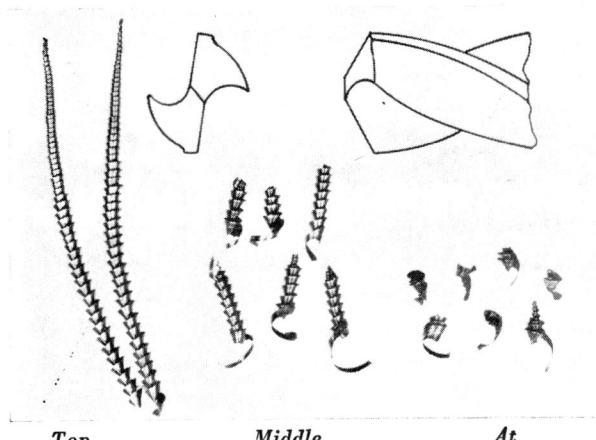

A—CONVENTIONAL 118° POINT—This is the most commonly used drill point. Properly produced, it will give satisfactory drilling results in a wide variety of materials. In use it may also require some form of web thinning when used on drills whose web thickness has increased because of repeated resharpenings, or on drills of heavier web construction.

| *Top Of Hole* | *Middle Of Hole* | *At Breakthrough* |

TYPE B

B—LONG ANGLE POINT —The long angle drill point, generally having an included angle of 60° to 90°, is commonly used on low helix drills in soft plastics and soft non-ferrous metals. It may also be used in soft cast iron and certain types of wood.

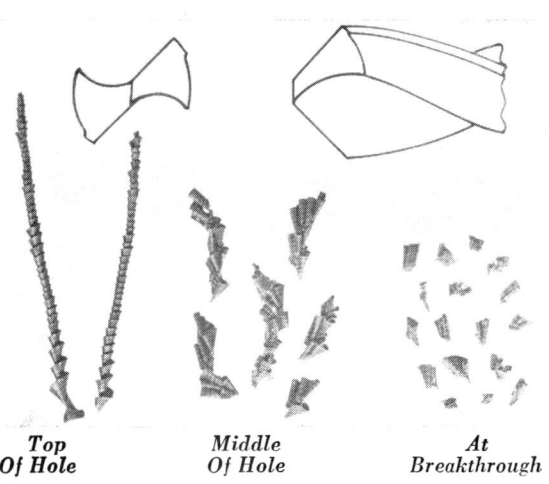

| *Top Of Hole* | *Middle Of Hole* | *At Breakthrough* |

TYPE C

FIGURE A

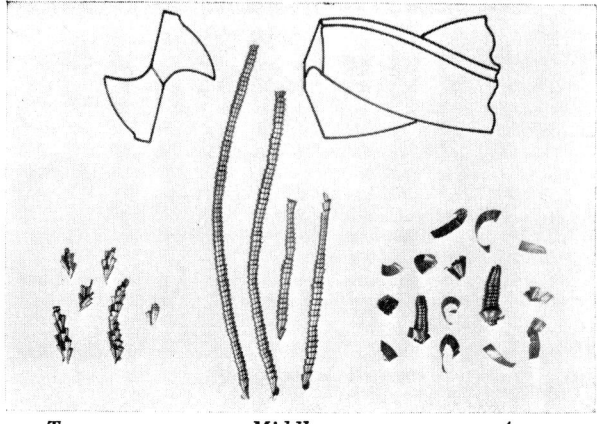

Top Of Hole   Middle Of Hole   At Breakthrough

C—FLAT ANGLE POINT [135° TO 140°] —The flat angle point is generally used for drilling the hard and tough materials. By using this type of point with the reduced feed rates, a thicker chip is produced. It is suggested that guide bushings be used as this point has a tendency to skid or walk on the surface of the work when starting a hole.

The changes in the shape of the chip caused by an increase in feed rate are illustrated. The feed rate used in Figure B was 16% greater than that used in Figure A. Comparison of Figure B with those shown with the Type A point also illustrate the difference in the chip form caused by the difference in included drill point angle.

FIGURE B

Top Of Hole   Middle Of Hole   At Breakthrough

TYPE D

D — DOUBLE-ANGLE POINT —The double-angle point is generally used for the medium and hard cast irons and other very abrasive materials. In effect, the double-angle acts as a chip breaker, and by increasing the length of the cutting lip and increasing the cutting lip corner angle, the rate of wear of the corner is reduced.

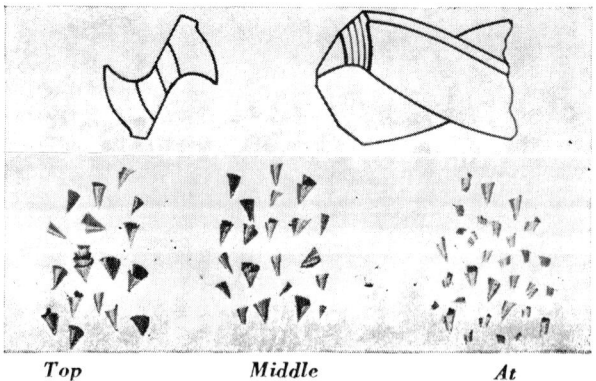

Top Of Hole   Middle Of Hole   At Breakthrough

CRANKSHAFT OR SPLIT POINT

This type of point was originally developed for use on drills designed for drilling deep oil holes in automotive crankshafts. It has gained widespread use on many designs of drills used in a wide variety of hard and soft materials. It can be applied to a variety of drill point angles, perhaps the most common being 135°. The main advantage of this type of point and thinning is the large reduction in thrust and a positive rake cutting edge extending to the center of the drill. In many materials this point will act as a chip breaker to produce small chips which can be readily ejected through the flutes. The split point minimizes skidding or walking of the drill point when starting a hole. This is a distinct advantage when portable drilling is used, or in drillpress work where bushings cannot be used.

These illustrations show the type of chip produced by this style of point when used under various drilling conditions and in different types of material.

Figure A shows the typical chips produced by a crankshaft type of point, when used with an intermittent feed rate, drilling in automotive crankshaft type material.

TYPE E
FIGURE A

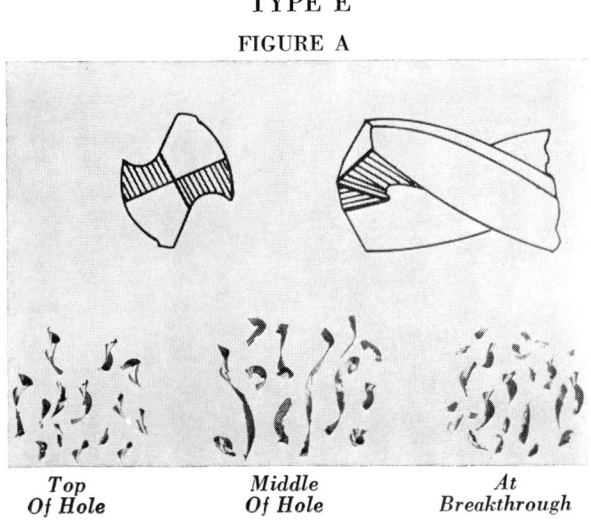

| Top Of Hole | Middle Of Hole | At Breakthrough |

Figures B-1, B-2, B-3 show the type of chips produced by a split point when used in a medium hard material at a given cutting speed, but with an increased feed rate. Figure B-2 chips were produced with a feed rate 100% greater than Figure B-1. Figure B-3 chips were produced with a further increase of 67% over that used in Figure B-2. Of greatest interest are the

FIGURE B-1

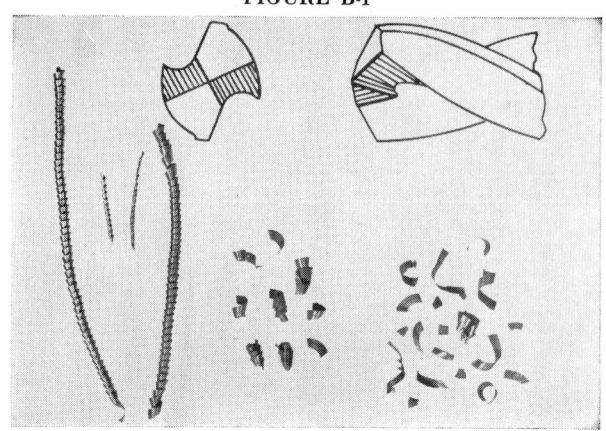

Top	Middle	At
Of Hole	*Of Hole*	*Breakthrough*

FIGURE B-2

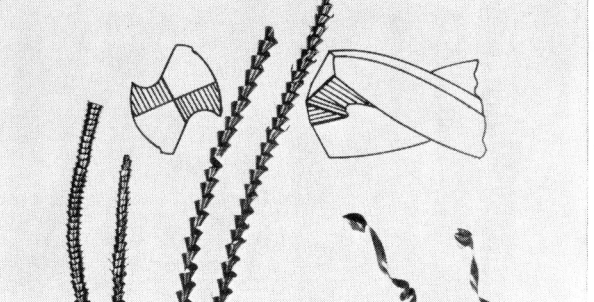

Top	Middle	At
Of Hole	*Of Hole*	*Breakthrough*

FIGURE B-3

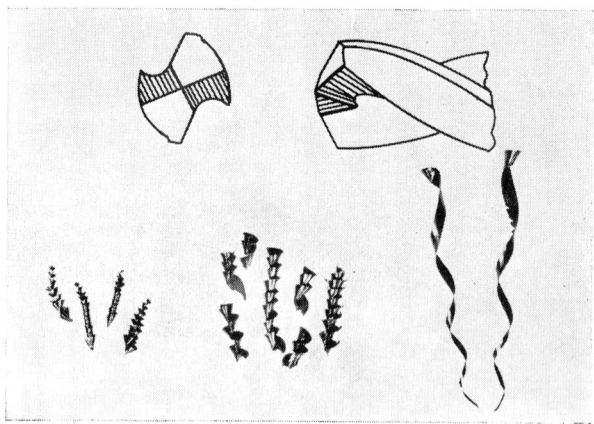

Top	Middle	At
Of Hole	*Of Hole*	*Breakthrough*

typical chips produced by these feed rates illustrated by chips taken from the middle and break-through part of the drilled hole.

Figure C shows the change in the characteristic chip produced by a split point in a different type of alloy steel. Note the ragged edge on the chip. This portion of the chip was produced by the notch.

Figure D shows the two separate and distinct chips produced by the cutting lip and the notch of a split point. These characteristic chips are produced sometimes under certain drilling conditions or in drilling certain types of material.

FIGURE C

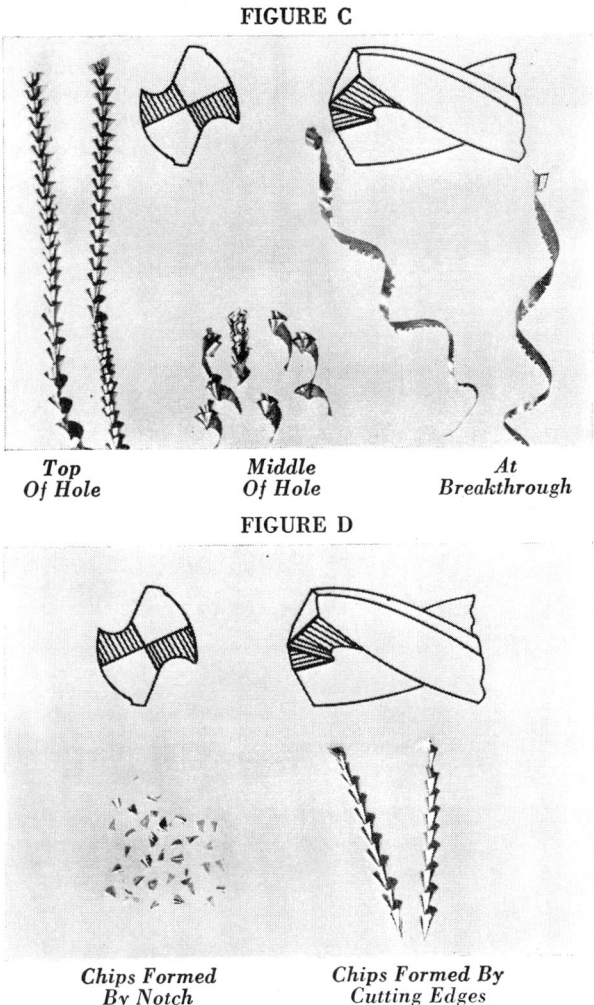

Top
Of Hole   *Middle*
                         Of Hole   *At*
                                                  Breakthrough

FIGURE D

Chips Formed
By Notch   *Chips Formed By*
                              Cutting Edges

HIGH TENSILE NOTCHED POINT

This type of point has been developed for drilling the high tensile and very tough alloys, and the heat resistant alloys. It is generally used with specially designed, low helix, heavy web drills. It will withstand the higher thrust loadings required for drilling these materials.

The changed and distinct types of chip formation caused by small

TYPE F

FIGURE A

| Top Of Hole | Middle Of Hole | At Breakthrough |

FIGURE B

| Top Of Hole | Middle Of Hole | At Breakthrough |

changes in feed rates when drilling in a very hard material are shown. Figure A shows the two separate chips produced by the two parts of each cutting lip. Figure B shows the type of chip produced by a slight increase in feed rate. This change produced a single distorted chip instead of the two chips shown in Figure A.

HEAVY DUTY AIRCRAFT DRILL POINT

This type of drill point in conjunction with a special short flute, heavy duty drill made of high alloy high speed steel was designed primarily for drilling the hard, high strength alloys and the tough, heat resistant alloys presently being used in the aircraft and missile industries. It is similar to Type E except the secondary cutting lip angle formed by the notch thinning is increased so that the notch covers about half of the lip length. The notch rake angle is always positive, averaging about 5°. The drill point angle is usually 135° included. This type of notch thinning also has been applied successfully to other types of heavy duty drills for drilling various tough materials.

Figures A, B & C show typical chips produced by this drill point in different types of high-strength and thermal resistant materials.

Figure D shows the difference in chip formation caused by an increase in feed rate. Drilling conditions and material same as for Figure C.

TYPE G

FIGURE A

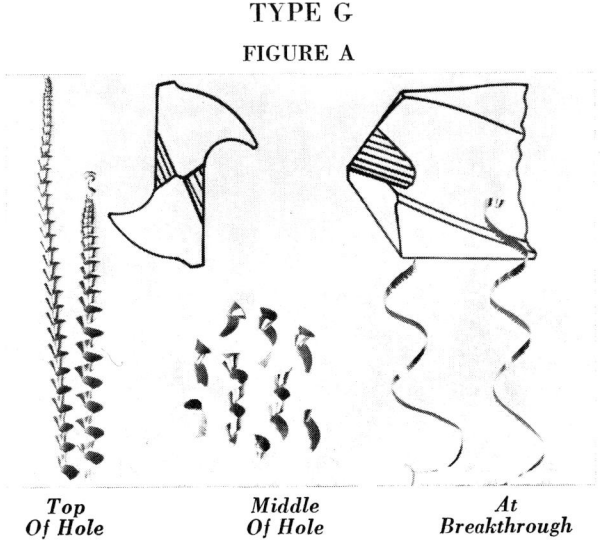

| Top Of Hole | Middle Of Hole | At Breakthrough |

FIGURE B

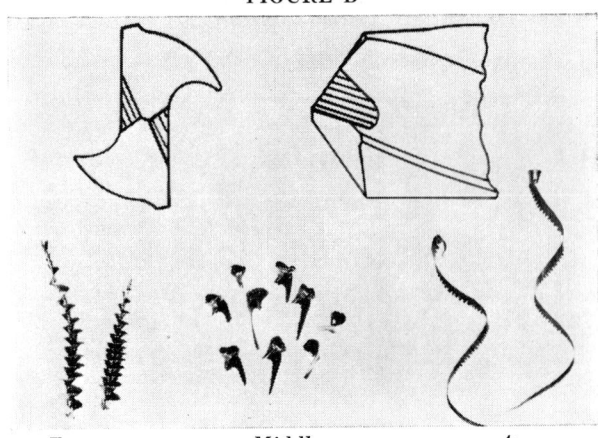

Top Of Hole	Middle Of Hole	At Breakthrough

FIGURE C

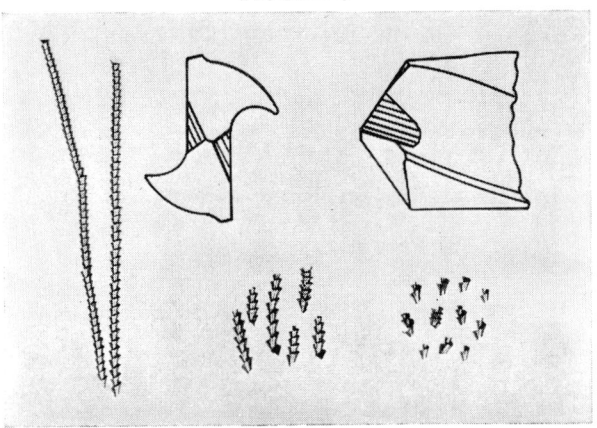

Top Of Hole	Middle Of Hole	At Breakthrough

FIGURE D

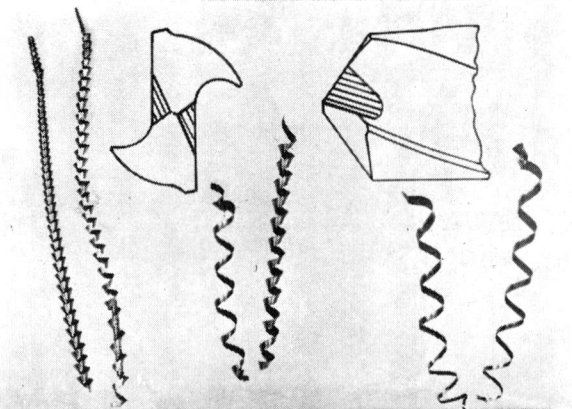

Top Of Hole	Middle Of Hole	At Breakthrough

CONVENTIONAL POINT WITH CONVENTIONAL THINNING

This type of web thinning is the most commonly used and the easiest to apply to a drill point, particularly if the web is not of the very heavy duty type. Refer to page 101 for specific web thinning information.

Figures A, B & C show the change in chip formation, when drilling a commonly used carbon steel, caused by increasing the drill feed only. To permit the use of increased feed rates without large increases in thrust values, this simple form of web thinning can be used.

The preferred types of chips are shown in Figure C.

TYPE H

FIGURE A

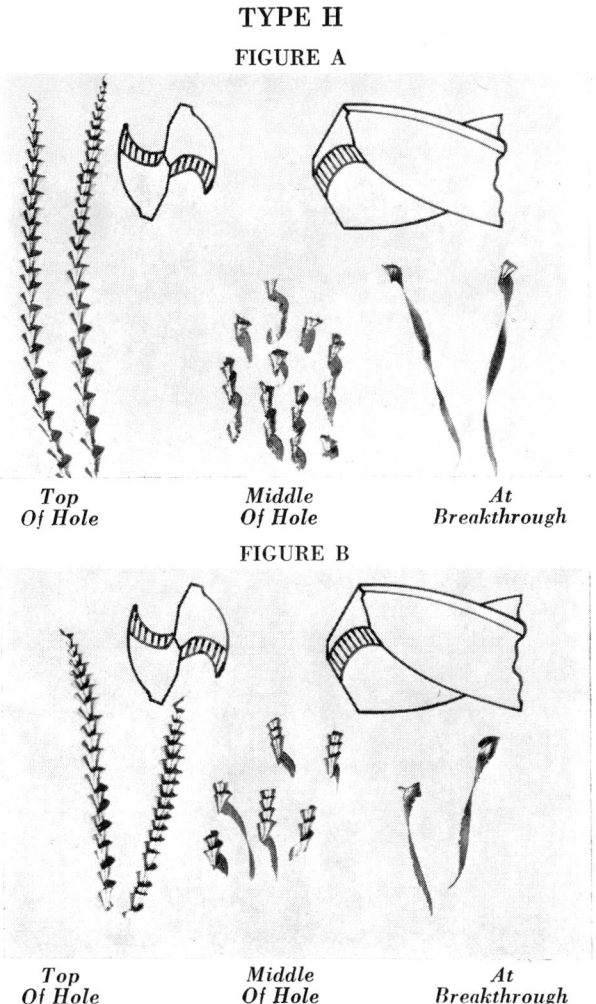

| *Top* | *Middle* | *At* |
| *Of Hole* | *Of Hole* | *Breakthrough* |

FIGURE B

| *Top* | *Middle* | *At* |
| *Of Hole* | *Of Hole* | *Breakthrough* |

FIGURE C

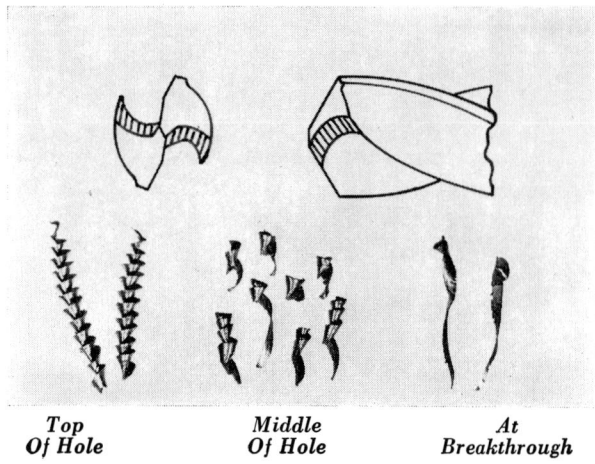

| Top Of Hole | Middle Of Hole | At Breakthrough |

UNDERCUT THINNED POINT

The undercut type of thinning is similar to the conventional thinning except that it usually extends the entire length of the cutting lip and is of such a form that it increases the rake angle of the cutting lip, particularly near the center of the drill. When properly applied, this type of thinning will produce a tightly curled chip.

TYPE I

FIGURE A

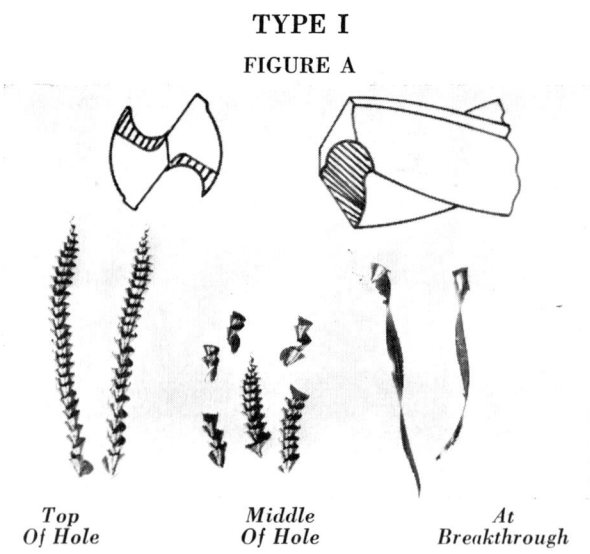

| Top Of Hole | Middle Of Hole | At Breakthrough |

FIGURE B

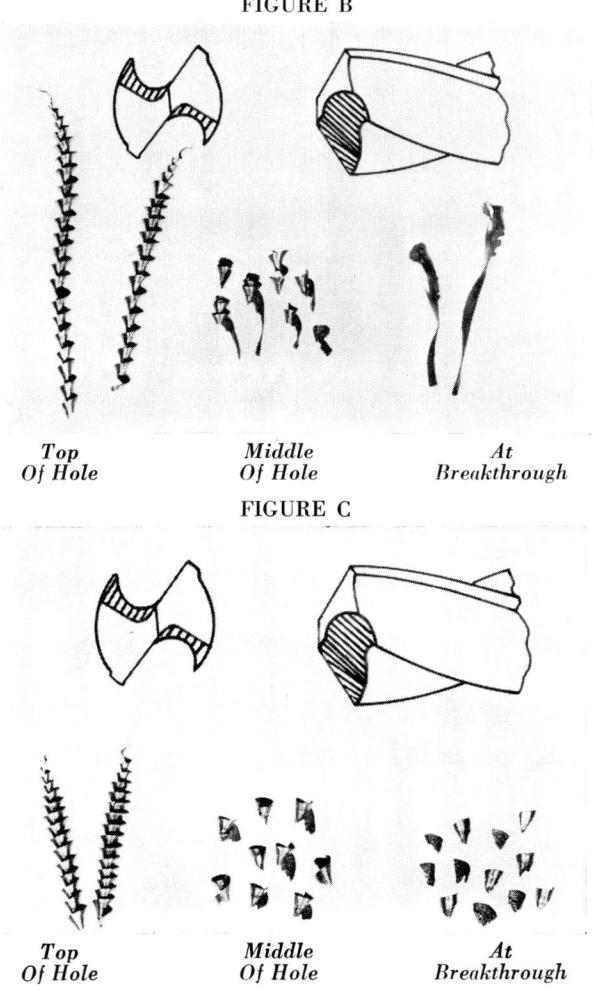

Top
Of Hole

Middle
Of Hole

At
Breakthrough

FIGURE C

Top
Of Hole

Middle
Of Hole

At
Breakthrough

Figures A, B, C illustrate both the chip curling effect of this type of thinning and the chip formation effect of heavier feeds. For materials that are tough and stringy, this type of thinning and variations in drill point angles combined with the proper feed will produce a short, curled, broken chip as shown in Figure C. Chip forms of this type are necessary for ease of chip ejection, particularly when drilling deeper holes.

Figures A & B were produced with the same tool, in the same material, under the same drilling conditions except for rate of feed as in Figure C. Figure C was produced with the heaviest feed.

CONVENTIONAL POINT WITH FLATTED CUTTING LIP
TO REDUCE RAKE

This type of point and thinning reduces the rake angle of the cutting lip to prevent the drill point from "hogging in" in materials such as brass and bronze.

TYPE J

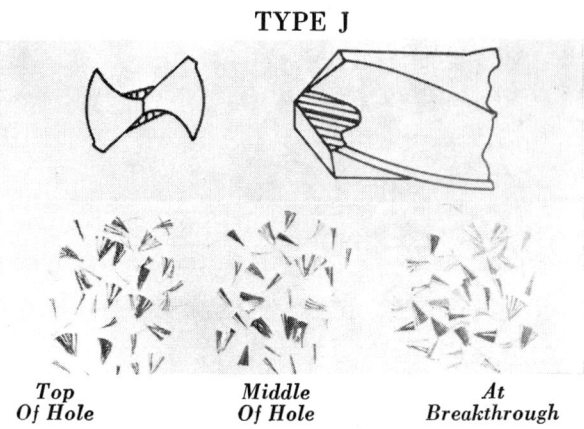

Top Of Hole	Middle Of Hole	At Breakthrough

FISHTAIL POINT

The fishtail point, with small variations, is used to drill holes in thin material where the ratio of hole size to material thickness is very large. A guide bushing is required set close to the material being drilled to center the drill point. The point looks like the tail of a fish and thus the name "fish tail" point.

TYPE K

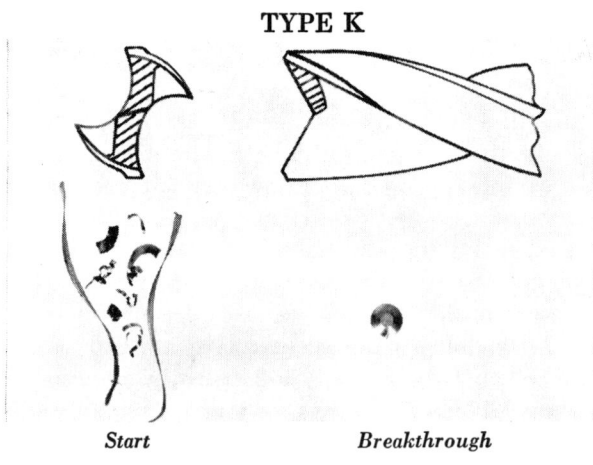

Start   *Breakthrough*

BRAD AND SPUR POINT

The brad and spur point and its many variations are generally used in drilling wood, some of the very soft non-ferrous materials, rubber and leather. Hardwood chips are shown.

In using these types of drill points and web thinnings, the following items are important:

1—The drill point must be uniform and the web thinning must be central.

2—Care must be exercised in web thinning or altering the rake angle of the cutting lip to provide unimpeded chip flow.

3—The drill points themselves must be provided with the proper lip relief. Where secondary clearances are required, they must be uniform and adequate.

TYPE L

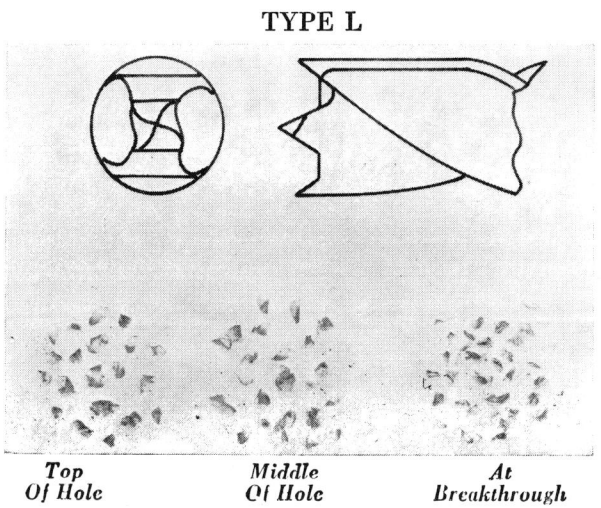

| *Top* | *Middle* | *At* |
| *Of Hole* | *Of Hole* | *Breakthrough* |

SELECTION OF MATERIAL FOR DRILLS

The drilling of holes in the production and assembly of metal parts is performed principally by twist drills made of high speed steel. Carbide drills of several designs are used for drilling certain abrasive ceramic and plastic materials and for some metallic materials of high hardness, but the number of holes drilled with carbide drills is a small percentage of the total. The use of polycrystalline diamond-tipped drills is increasing in applications such as aluminum, composites, plastics, and wood.

HIGH SPEED STEEL FOR DRILLS

Of all types of tools, drills require the highest combination of torsional strength and toughness. The high speed steels of which they are made must be capable of forming cutting edges which will not chip readily and drill bodies which will withstand fluctuating torsional and bending stresses. They must also have good red hardness properties, since the cutting end of a drill buried in the hole it is forming is in contact with hot chips more intimately than is the case with other tools.

The combination of properties required of drills are great strength and toughness plus moderate red hardness. These are exhibited by the common grades of high speed steel, M-1, M-2, M-7, M-10 and T-1. Good evidence that these steels exhibit the optimum combination of such properties lies in the fact that they comprise the greatest percentage of high speed steels used for the manufacture of drills. The molybdenum grades in most cases are a more economical choice than the tungsten type T-1. A substantial majority of all high speed steel drills today are made of the molybdenum types with an ever increasing preference shown for M-7 and M-10.

In the last few years there has been a trend toward machining metal components after heat treatment at increasingly higher hardness levels. At the same time, new thermal-resistant alloys have appeared. These are usually of high alloy content and are considerably more difficult to machine than those formerly used. These trends have necessitated the use of drilling machines and work-supporting fixtures of more rigid construction and have required drills of heavy construction. They also have made it necessary, in some applications, to use drill materials having higher red hardness and abrasion resistance than the standard high speed steels.

Special grades of high speed steel having higher carbon and vanadium contents, and some containing cobalt additions have been used with success in drilling low alloy steels of relatively high hardness as well as many of the thermal resistant alloys. The types of high speed steel containing cobalt have greater red hardness than the standard grades; those having higher carbon and vanadium have greater resistance to abrasion, and those having all three additions exhibit both higher red hardness and greater abrasion resistance. Improvements in abrasion resistance and red hardness are accompanied by some decrease in toughness and this fact must be considered when using drills made of these special grades.

These special grades involve higher unit tool material costs and, because of increased abrasion resistance, higher tool fabrication costs. The economics of using tool materials involving higher material and fabrication costs should always be taken into consideration along with drill performance in reaching a decision to use drills made of the special grades of high speed steel.

The use of molybdenum base, special high speed steels predominates over the use of tungsten base special steels applied to twist drills. M-3, M-33 and M-42 are typical of the special high speed steels that have been in common use for some time. T-15 is the only tungsten base special high speed steel used with any regularity in twist drills. High speed steels are also being fabricated by a process known as particle metallurgy. The PM process can produce steels of the same composition as high speed steels that are produced conventionally. The PM process can also use alloying elements in combinations and quantities not practical by conventional processing.

SURFACE TREATMENTS AND COATINGS

The surfaces of drills have been modified in a number of ways over the years to enhance their performance. This trend is continuing as new processes and techniques are being developed. The chief mechanisms of these treatments have been to increase abrasion resistance, to increase lubricity, and to reduce built-up-edge and related chip welding. The following treatments and coatings are the more common ones in use today:

1. Nitride—A very hard, shallow surface of nitrogen which forms compounds with the metal to provide enhanced abrasion resistance.
2. Steam Oxide—An iron oxide surface formed in a furnace when steam is applied under moderate heat. The oxide acts as a lubricant and reduces welding in ferrous applications. Nitride and steam oxide are often used as complementary treatments.
3. Chromeplate—This surface coating increases lubricity and reduces chip adherence. In drilling, it is most often applied in non-ferrous materials. It is sometimes applied to a nitrided surface.
4. Titanium Nitride—This refractory surface treatment is relatively new. It reduces friction and chip welding, and acts as a thermal insulator between the chip and the tool. The areas of application where it appears to have the most potential are in ferrous materials below Rc 40 in hardness, and in non-ferrous materials.

CARBIDES FOR DRILLS

The drilling of holes in concrete, glass and various ceramic materials is a well-known application for carbide-tipped "masonry" drills. The use of solid carbide drills for drilling glass-laminated plastics is increasing. Carbide-tipped twist drills are now used in production drilling of cast iron, non-ferrous metals, and non-metallic materials. Carbide-tipped "gun" drills with passages for conducting coolant to the drill tip are finding use where hole straightness, close size and parallelism are needed. They have been used effectively on a very wide variety of ferrous and non-ferrous alloys. Very rugged carbide-tipped die drills have been used successfully in drilling low alloy steels of relatively high hardness and for many thermal resistant alloys. A very rigid machine setup is required for this purpose.

The selection of sintered carbide as a material for drills depends on the type of material being drilled and the number of holes to be drilled. Grades of straight tungsten carbide, such as grade classes 1 and 2, are used for the vast majority of carbide drilling applications. Carbide tips must be solidly supported in drill bodies. Hardened high speed steel drill bodies provide solid support for the tips as well as abrasion-resistant margins for wear resistance.

DRILLING TECHNIQUES

SPEEDS AND FEEDS FOR TWIST DRILLS

SPEEDS FOR DRILLING

The speed of a drill is usually measured in terms of the rate at which the outside or periphery of the tool moves in relation to the work being drilled. The common term for this is "Surface Feet Per Minute", abbreviated to sfm. The relation of sfm and Revolutions Per Minute, or rpm is indicated by the following formulas:

$$\text{sfm} = .26 \times \text{rpm} \times \text{Drill Diameter in Inches}$$

$$\text{rpm} = 3.8 \times \frac{\text{sfm}}{\text{Drill Diameter in Inches}}$$

In general when operating a drill at a speed anywhere within its range for the particular material involved, increases in speed result in fewer holes before regrinding becomes necessary, and reductions in speed, more holes before the tool is dulled. On every job there is a problem of choosing a speed which will permit the most economical rate of production which is also determined by drill costs, pieces produced in a given time and down-time for tool changing. The most economical speed for operating a drill will depend on many variables, some of which are:

1—Composition and hardness of material.

2—Depth of hole.

3—Efficiency of cutting fluid.

4—Type and condition of drilling machine.

5—Quality of holes desired.

6—Difficulty of set-up.

SUGGESTED SPEEDS FOR HIGH SPEED STEEL DRILLS

Speeds shown in the following table include a range in feet-per-minute. On most jobs, it is usually better to start with a slower speed and build up to the maximum after trials indicate the job can be run faster.

Material	Speed in feet per minute
Aluminum and its Alloys	200–300
Brass and Bronze (ordinary)	150–300
Bronze (High Tensile)	70–150
Die Castings (Zinc Base)	300–400
High Temperature Alloys	
Cobalt Base: HS25, S816, V36 ↑	10– 20
Iron Base: INCO 800, A286, N155 Solution Treated & Aged	10– 20
Nickel Base: INCONEL 700, U500, René41 ↓	7– 15
Iron-Cast (soft)	75–125
Cast (medium hard)	50–100
Hard Chilled	10– 20
Malleable	80– 90
Magnesium and its Alloys	250–400
Monel Metal or High-Nickel Steel	30– 50
Plastics or Similar Materials (Bakelite)	100–300
Steel-Mild .2 to .3 carbon	80–110
Steel .4 to .5 carbon	70– 80
Tool 1.2 carbon	50– 60
Forgings	40– 50
Alloy-300 to 400 Brinell	20– 30
High Tensile (Heat Treated)	
35 to 40 Rockwell C	30– 40
40 to 45 Rockwell C	25– 35
45 to 50 Rockwell C	15– 25
50 to 55 Rockwell C	7– 15
Maraging (Heat Treated)	7– 20
(Annealed)	40– 55
Stainless Steel	
Free Machining Group: 303, 303SE, 430F, 416F, 420F	30–100
Chromium-Nickel Group (Non-Hardenable):	
(300 Series) (1) (400 Series) (2)	20– 60
Straight Chromium Group (Heat Treated): (400 Series) (3)	10– 30
Titanium Alloys,	
Commercially Pure	50– 60
5Al-2Sn, 8Al-1Mo-1V } Annealed 2Fe-2Cr-Mo	30– 40
6Al-4V, 4Al-4Mn, 7Al-4Mo Annealed	25– 35
6Al-4V, 4Al-4Mn, 7Al-4Mo } Solution Treated & Aged 2Fe-2Cr-2Mo	15– 20
Wood	300–400

(1) Austenitic, (2) Ferritic, (3) Martensitic

Table of Cutting Speeds

Feet per min. dia. inches	10'	20'	30'	40'	50'	60'	70'	80'	90'	100'	150'	200'
						Fractional size drills Revolutions per minute						
1/64	2445	4889	7334	9778	12223	14668	17112	19557	22001	24446	36669	48892
1/32	1222	2445	3667	4889	6112	7334	8556	9778	11001	12223	18335	24446
3/64	815	1630	2445	3259	4074	4889	5704	6519	7334	8149	12224	16298
1/16	611	1222	1833	2445	3056	3667	4278	4889	5500	6112	9168	12224
5/64	489	978	1467	1956	2445	2934	3422	3911	4400	4889	7334	9778
3/32	407	815	1222	1630	2037	2445	2852	3259	3667	4074	6111	8148
7/64	349	698	1048	1397	1746	2095	2445	2794	3143	3492	5238	6984
1/8	306	611	917	1222	1528	1833	2139	2445	2750	3056	4584	6112
9/64	272	543	815	1086	1358	1630	1901	2173	2445	2716	4074	5432
5/32	244	489	733	978	1222	1467	1711	1956	2200	2445	3668	4890
11/64	222	444	667	889	1111	1333	1556	1778	2000	2222	3333	4444
3/16	204	407	611	815	1019	1222	1426	1630	1833	2037	3056	4074
13/64	188	376	564	752	940	1128	1316	1504	1692	1880	2820	3760
7/32	175	349	524	698	873	1048	1222	1397	1572	1746	2619	3492
15/64	163	326	489	652	815	978	1141	1304	1467	1630	2445	3260
1/4	153	306	458	611	764	917	1070	1222	1375	1528	2292	3056
9/32	136	272	407	543	679	815	951	1086	1222	1358	2037	2716
5/16	122	244	367	489	611	733	856	978	1100	1222	1833	2444
11/32	111	222	333	444	556	667	778	889	1000	1111	1667	2222
3/8	102	204	306	407	509	611	713	815	917	1019	1529	2038

13/32	1880	1410	940	846	752	658	564	470	376	282	188	94
7/16	1746	1310	873	786	698	611	524	437	349	262	175	87
15/32	1630	1223	815	733	652	570	489	407	326	244	163	81
1/2	1528	1146	764	688	611	535	458	382	306	229	153	76
9/16	1358	1019	679	611	543	475	407	340	272	204	136	68
5/8	1222	917	611	550	489	428	367	306	244	183	122	61
11/16	1112	834	556	500	444	389	333	278	222	167	111	56
3/4	1018	764	509	458	407	357	306	255	204	153	102	51
13/16	940	705	470	423	376	329	282	235	188	141	94	47
7/8	874	656	437	393	349	306	262	218	175	131	87	44
15/16	814	611	407	367	326	285	244	204	163	122	81	41
1	764	573	382	344	306	267	229	191	153	115	76	38
1 1/8	680	510	340	306	272	238	204	170	136	102	68	34
1 1/4	612	459	306	275	244	214	183	153	122	92	61	31
1 3/8	556	417	278	250	222	194	167	139	111	83	56	28
1 1/2	510	383	255	229	204	178	153	127	102	76	51	25
1 5/8	470	353	235	212	188	165	141	118	94	71	47	24
1 3/4	436	327	218	196	175	153	131	109	87	65	44	22
1 7/8	408	306	204	183	163	143	122	102	81	61	41	20
2	382	287	191	172	153	134	115	95	76	57	38	19
2 1/4	340	255	170	153	136	119	102	85	68	51	34	17
2 1/2	306	230	153	138	122	107	92	76	61	46	31	15
2 3/4	278	209	139	125	111	97	83	69	56	42	28	14
3	254	191	127	115	102	89	76	64	51	38	25	13
3 1/2	218	164	109	98	87	76	65	55	44	33	22	11

Table of Cutting Speeds

Feet per min. no. size					Number size drills Revolutions per minute							
	10'	20'	30'	40'	50'	60'	70'	80'	90'	100'	150'	200'
1	168	335	503	670	838	1005	1173	1340	1508	1675	2513	3350
2	173	346	519	691	864	1037	1210	1382	1555	1728	2592	3456
3	179	359	538	717	897	1076	1255	1434	1614	1793	2690	358
4	183	366	548	731	914	1097	1280	1462	1645	1828	2742	365
5	186	372	558	743	930	1115	1301	1487	1673	1859	2789	371
6	187	374	562	749	936	1123	1310	1498	1685	1872	2808	374
7	190	380	570	760	950	1140	1330	1520	1710	1900	2850	380
8	192	384	576	768	960	1151	1343	1535	1727	1919	2879	383
9	195	390	585	780	975	1169	1364	1559	1754	1949	2924	389
10	197	395	592	790	987	1184	1382	1579	1777	1974	2961	394
11	200	400	600	800	1000	1200	1400	1600	1800	2000	3000	400
12	202	404	606	808	1010	1213	1415	1617	1819	2021	3032	404
13	206	413	619	826	1032	1239	1450	1652	1859	2065	3098	413
14	210	420	630	839	1050	1259	1469	1679	1889	2099	3149	419
15	212	424	637	849	1064	1276	1489	1702	1914	2127	3191	425
16	216	432	647	863	1079	1295	1511	1726	1942	2158	3237	431
17	221	442	662	883	1104	1325	1546	1766	1987	2208	3312	441

18	225	451	676	901	1130	1356	1582	1808	2034	2260	3390	452
19	230	460	690	920	1151	1381	1611	1841	2071	2301	3452	460
20	237	474	712	949	1186	1423	1660	1898	2135	2372	3558	474
21	240	480	721	961	1201	1441	1681	1922	2162	2402	3603	480
22	243	487	730	973	1217	1460	1703	1946	2190	2433	3650	486
23	248	496	744	992	1240	1488	1736	1984	2232	2480	3720	496
24	251	503	754	1005	1257	1508	1759	2010	2262	2513	3770	502
25	255	511	766	1022	1276	1533	1789	2044	2300	2555	3833	511
26	260	520	780	1039	1299	1559	1819	2078	2338	2598	3897	519
27	265	531	796	1061	1327	1592	1857	2122	2388	2653	3980	530
28	272	544	816	1087	1360	1631	1903	2175	2447	2719	4079	543
29	281	562	843	1123	1405	1685	1966	2247	2528	2809	4214	561
30	297	595	892	1189	1487	1784	2081	2378	2676	2973	4460	594
31	318	637	955	1273	1592	1910	2228	2546	2865	3183	4775	636
32	329	659	988	1317	1647	1976	2305	2634	2964	3293	4940	650
33	338	676	1014	1352	1690	2028	2366	2704	3042	3380	5070	676
34	344	688	1032	1376	1721	2065	2409	2753	3097	3442	5163	688
35	347	694	1042	1389	1736	2083	2430	2778	3125	3472	5208	694
36	359	717	1076	1435	1794	2152	2511	2870	3228	3587	5381	717
37	367	735	1102	1469	1837	2204	2571	2938	3306	3673	5510	734
38	376	753	1129	1505	1882	2258	2634	3010	3387	3763	5645	752
39	384	768	1152	1536	1920	2303	2687	3071	3455	3839	5759	767
40	390	780	1169	1559	1949	2339	2729	3118	3508	3898	5847	770

Table of Cutting Speeds

| | | | | | Number size drills | | | | | | | |
no. size	10'	20'	30'	40'	50'	60'	70'	80'	90'	100'	150'	200'
					Revolutions per minute							
41	398	796	1194	1592	1990	2387	2785	3183	3581	3979	5969	7958
42	409	817	1226	1634	2043	2451	2860	3268	3677	4085	6128	8170
43	429	858	1288	1717	2146	2575	3004	3434	3863	4292	6438	8584
44	444	888	1332	1777	2221	2665	3109	3554	3999	4442	6663	8884
45	466	932	1397	1863	2329	2795	3261	3726	4192	4658	6987	9316
46	472	943	1415	1886	2358	2830	3301	3773	4244	4716	7074	9432
47	487	973	1460	1946	2433	2920	3406	3893	4379	4866	7299	9732
48	503	1005	1508	2010	2513	3016	3518	4021	4523	5026	7539	10052
49	523	1046	1570	2093	2617	3140	3663	4186	4710	5233	7850	10466
50	546	1091	1637	2183	2729	3274	3820	4366	4911	5457	8186	10914
51	570	1140	1710	2280	2851	3421	3991	4561	5131	5701	8552	11402
52	602	1203	1805	2406	3008	3609	4211	4812	5414	6015	9023	12030
53	642	1284	1926	2568	3207	3848	4490	5131	5773	6414	9621	12828
54	694	1389	2083	2778	3473	4167	4862	5556	6251	6945	10418	13890
55	735	1469	2204	2938	3673	4408	5142	5877	6611	7346	11019	14692
56	821	1643	2464	3286	4108	4929	5751	6572	7394	8215	12323	16430
57	888	1777	2665	3553	4452	5342	6232	7122	8013	8903	13355	17806
58	909	1819	2728	3638	4547	5456	6367	7275	8186	9095	13643	18190

Feet per min.

59	932	1863	2795	3726	4658	5590	6521	7453	8388	9316	13974	18632
60	955	1910	2865	3820	4775	5729	6684	7639	8594	9549	14324	19098
61	979	1959	2938	3918	4897	5876	6856	7835	8815	9794	14691	19588
62	1005	2010	3016	4021	5025	6030	7035	8040	9045	10050	15075	20100
63	1032	2065	3097	4129	5160	6192	7224	8256	9288	10320	15480	20640
64	1061	2122	3183	4244	5305	6366	7427	8488	9549	10610	15915	21220
65	1091	2183	3274	4365	5455	6546	7637	8728	9819	10910	16365	21820
66	1157	2315	3472	4630	5790	6948	8106	9264	10422	11580	17370	22960
67	1194	2387	3581	4775	5970	7164	8358	9552	10746	11940	17910	23880
68	1232	2464	3696	4929	6160	7392	8624	9856	11088	12320	18480	24640
69	1308	2616	3924	5232	6530	7836	9142	10488	11754	13060	19590	26120
70	1364	2728	4093	5457	6820	8184	9548	10912	12276	13640	20460	27280
71	1469	2938	4407	5876	7365	8838	10311	11784	13257	14730	22095	29460
72	1528	3056	4584	6112	7640	9168	10696	12224	13752	15280	22920	30560
73	1592	3183	4775	6366	7960	9552	11144	12736	14328	15920	23880	31840
74	1698	3395	5093	6791	8510	10212	11914	13616	15318	17020	25530	34040
75	1819	3638	5457	7276	9095	10914	12733	14552	16371	18190	27285	36380
76	1910	3820	5730	7639	9550	11460	13370	15280	17190	19100	28650	38200
77	2122	4244	6366	8488	10610	12732	14854	16976	19098	21220	31830	42440
78	2387	4775	7162	9549	11935	14322	16709	19096	21483	23870	35805	47740
79	2634	5269	7903	10537	13170	15804	18438	21072	23706	26340	39510	52680
80	2829	5659	8488	11318	14150	16980	19810	22640	25470	28300	42450	56600

Table of Cutting Speeds

Feet per min. let. size	10'	20'	30'	40'	50'	Letter size drills 60'	70'	80'	90'	100'	150'	200'
						Revolutions per minute						
A	163	326	490	653	818	982	1145	1309	1472	1636	2454	3272
B	160	321	481	642	803	963	1124	1284	1445	1605	2408	3210
C	158	316	473	631	789	947	1105	1262	1420	1578	2367	3156
D	155	311	466	621	778	934	1089	1245	1400	1556	2334	3112
E	153	306	458	611	764	917	1070	1222	1375	1528	2292	3056
F	149	297	446	595	743	892	1040	1189	1337	1486	2229	2972
G	146	293	439	585	732	878	1024	1170	1317	1463	2195	2926
H	144	287	431	574	718	862	1005	1149	1292	1436	2154	2872
I	140	281	421	562	702	842	983	1123	1264	1404	2106	2808
J	138	276	414	552	690	827	965	1103	1241	1379	2069	2758

K	2718	2039	1359	1223	1087	951	815	680	544	408	272	136
L	2634	1976	1317	1185	1054	922	790	659	527	395	263	132
M	2590	1943	1295	1166	1036	907	777	648	518	383	259	129
N	2530	1898	1265	1139	1012	886	759	633	506	379	253	126
O	2418	1814	1209	1088	967	846	725	605	484	363	242	121
P	2366	1775	1183	1065	946	828	710	592	473	355	237	118
Q	2300	1725	1150	1035	920	805	690	575	460	345	230	115
R	2254	1691	1127	1014	902	789	676	564	451	338	225	113
S	2196	1647	1098	988	878	769	659	549	439	329	220	110
T	2132	1599	1066	959	853	746	640	533	427	320	213	107
U	2076	1557	1038	934	830	727	623	519	415	311	208	104
V	2026	1520	1013	912	810	709	608	507	405	304	203	101
W	1978	1484	989	891	792	693	594	495	396	297	198	99
X	1924	1443	962	865	769	672	576	481	385	289	192	96
Y	1888	1418	945	851	756	662	567	473	378	284	189	95
Z	1850	1388	925	832	740	647	555	462	370	277	185	92

PRODUCTION RATES FOR DRILLED HOLES

Drilling Time in Minutes Required to Penetrate 1″ of Material

R.P.M.	.001	.002	.003	.005	.007	.010	.013	.015	.017	.020	.025	.030
						Feed per Revolution						
6000	.166											
5500	.182											
5000	.200	.100										
4500	.222	.111										
4000	.250	.125	.083									
3500	.286	.143	.095									
3000	.333	.167	.111	.067								
2500	.400	.200	.133	.080								
2000	.500	.250	.166	.100	.071							
1900	.526	.263	.175	.105	.075							
1800	.555	.278	.185	.110	.079							
1700	.588	.294	.196	.118	.084	.059	.045					
1600	.625	.313	.208	.125	.089	.063	.048	.042				
1500	.667	.333	.222	.133	.095	.067	.051	.044				
1400	.714	.357	.238	.143	.102	.071	.055	.048				
1300	.770	.385	.256	.154	.110	.077	.059	.051				
1200	.833	.417	.278	.166	.119	.083	.064	.055				
1100	.909	.454	.303	.182	.130	.091	.070	.061				
1000	1.000	.500	.333	.200	.143	.100	.077	.067	.058			
900		.555	.370	.222	.159	.111	.085	.074	.065	.055		
800		.625	.416	.250	.179	.125	.096	.083	.074	.063	.050	
700		.714	.476	.286	.204	.143	.110	.095	.084	.071	.057	.048
600			.555	.333	.238	.166	.128	.111	.098	.083	.066	.055
500			.666	.400	.286	.200	.154	.133	.117	.100	.080	.067
450				.444	.318	.222	.171	.148	.131	.111	.089	.073
400				.500	.357	.250	.192	.167	.147	.125	.100	.083
350				.571	.408	.285	.220	.190	.168	.143	.114	.095
300				.666	.476	.333	.256	.222	.196	.167	.133	.111
275				.727	.520	.364	.280	.242	.214	.182	.145	.121
250				.800	.572	.400	.308	.267	.236	.200	.160	.133
225				.888	.635	.445	.342	.296	.262	.222	.178	.148
200				1.000	.714	.500	.385	.333	.294	.250	.200	.167
175				1.142	.817	.571	.440	.381	.336	.285	.228	.190
150				1.333	.953	.667	.513	.445	.392	.333	.266	.222
125				1.600	1.142	.800	.615	.533	.470	.400	.320	.266
100				2.000	1.428	1.000	.769	.667	.588	.500	.400	.333
75				2.667	1.905	1.333	1.025	.888	.784	.667	.533	.444
50				4.000	2.858	2.000	1.537	1.333	1.175	1.000	.800	.666

FEEDS FOR DRILLING

Feed rates for drilling are governed by the size of the drill, machinability of the material being drilled and depth of the drilled hole.

Small drills, harder materials and hole depths in excess of 3 to 4 drill diameters require additional consideration in selecting appropriate feeds.

Since the feed partially determines the rate of production and also is a factor in tool life, it should be chosen carefully for each particular job.

In general, the most effective feeds will be found in the following ranges:

Feeds For High Speed Steel Drills

Drill Diameter—Inches	Feed in Inches Per Revolution	
	Conventional Drills	Parabolic Flute Drills
¹⁄₁₆ to ⅛	.001–.003	.001–.004
over ⅛ to ¼	.002–.007	.003–.009
over ¼ to ½	.004–.012	.006–.015
over ½ to 1	.007–.018	.010–.022
over 1	.015–.025	.020–.035

The preceding suggested feeds and speeds for various materials are those for average conditions where the coolant can be efficiently applied, where the strength of the drill is not a critical factor, and where the workpiece can be rigidly supported. When one or more of these conditions vary, the speeds and feeds must also vary.

SPEEDS AND FEEDS FOR DEEP-HOLE DRILLING

Holes which must be drilled three diameters deep or more fall into the "deep-hole" drilling class and some adjustment of feeds and speeds is necessary.

The deeper the hole, the greater the tendency there is for chips to pack and clog the flutes of the drill. This increases the amount of heat generated and prevents the coolant from conducting the heat away from the point. A buildup of heat at the point will eventually result in premature failure.

Step drilling, or the practice of drilling a short distance, then withdrawing the drill, will often reduce the chip packing. The deeper the hole the more frequent the drill must be retracted to be effective.

A reduction in speed and feed to reduce the amount of heat generated is generally required in most deep-hole applications where coolant cannot be effectively applied.

The following table can be used as a guide to proper feed and speed reductions as the depth of hole increases:

Reduction

Depth of Hole	Speed		Feed	
	Conventional	Parabolic	Conventional	Parabolic
3 × drill dia.	0	0	0	0
4	10%	0	0	0
5	20%	5%	10%	0
6–8	30%	10%	20%	0
8–11	40%	20%		0
11–14		30%		0
14–17		40%		0
17–20		50%		0

If excellent cooling conditions are present, such as with oil hole drills with high pressure, the above rates may be increased or even disregarded. However, if no cooling is present, it may be necessary to reduce speeds even lower than indicated above.

Care should be taken with the use of drill bushings. Any bushing set very close to the work becomes, in effect, an extension of the hole itself, and on holes from 2 to 3 diameters deep, complicates the problem of chip disposal.

DRILLING OF SMALL DIAMETER HOLES

Over a period of years of experience and observation, a store of knowledge has been built up regarding the most successful practices to be followed when drilling holes in metals and other materials. Speeds, feeds, and sharpening techniques for various materials, as well as lubricants and other aids, have been fairly well standardized.

It has been found, however, that when holes of very small diameters are to be drilled [for example from .015″ to .040″ in diameter] the standards of operation that are satisfactory for the larger sizes do not hold.

There are several reasons for this:

1—Small drills are very long in proportion to their diameter [from 40 to 60 diameters long]. They are, therefore, subject to much more deflection, both longitudinal and torsional.

2—The webs of small drills are proportionally much heavier than in large drills. This construction increases the required end pressure and decreases the chip space.

3—Small drills are called upon to drill comparatively deep holes, so that almost all small diameter drilling operations come under the classification of deep-hole drilling.

Since the lengths of these small drills must be maintained for utility's sake, and since it is impractical to construct them with thinner webs, it follows that the drilling conditions must be adjusted to give a reasonable amount of service. This involves:

a—Proper feeds and speeds.

b—Guiding of the drill to minimize runout and deflection.

c—Frequent and adequate chip disposal.

d—Careful resharpening, performed often enough to prevent over-dulling.

e—Drills ground with greater lip-relief angles, both at the periphery and at the chisel edge.

FEEDS AND SPEEDS

Perhaps the two most common errors in the operation of very small drills are over-speeding and under-feeding. Feeds should be based on chip formation and not on penetration rate. Speeds should then be adjusted accordingly.

To illustrate:—A No. 70 drill is to operate at a feed of .0005″ per revolution in drilling steel. While the material is soft enough to permit a speed of 80 S.F.M. or 10,912 R.P.M., it is obvious that the drill will not stand a load of .0005″ feed per revolution at this speed, since the rate of penetration would be 5½″ per minute.

The speed should be cut down and the feed per revolution retained until a satisfactory penetration rate is found which will enable the drill to stand the load.

Successful small-hole drilling depends on feeds that will actually produce CHIPS and not POWDER, and on speed adjusted to the strength and loadcarrying capacity of the drill.

DRILL GUIDES

It is always advisable to use a guide bushing when drilling holes with small drills. The guide bushing will prevent the drill point from weaving when the hole is started, and also helps to support the body of the drill so that it will not bend under end pressure. A small space should be provided between the bottom of the bushing and the work, in order to permit the chips to escape at that point.

CHIP REMOVAL

Owing to the limited chip space in small drills, special precautions should be taken to prevent the chips from packing in the flutes which will cause the drill to break. Step drilling, as commonly practiced with deep-hole drills, is therefore recommended.

DRILLING OF THIN METAL

Drilling of sheet metal and other thin sections presents a special problem because of the extraordinary conditions encountered in this type of operation.

Drill breakage is often the main difficulty when drilling thin sections, particularly in the smaller-diameter holes. This is caused by several factors:

First—Drilling is often done with a hand-supported electric or pneumatic drill. Therefore, the drill is not held rigidly in the line of feed, nor is the feeding pressure constant. Further, there is no rigid support when the drill breaks through, so that a distinct shock is produced at this point.

Second—In many cases the feeding pressure causes the metal to be deformed before the drill actually begins to cut. Thus, when the drill does begin to cut, the entire torque load is encountered at once, instead of gradually as with ordinary drilling.

Third—This deformation may also cause the metal to work-harden, and so further increase the load on the drill.

Fourth—In many cases, the drill will skid and rub on the work surface because of a heavy web condition at the point. The web at the point should be thinned to standard thickness or less, to reduce end thrust and permit easier centering of the drill point. Drill points should be sharpened with smaller than standard lip-relief angles to prevent drills from "hogging in" on break-through.

It is not always practical to remedy these conditions by changing the operation or the equipment. They can, however, be minimized by using drills of the proper design for the work at hand.

Generally speaking, the ordinary drill is not sturdy enough for production sheet-metal drilling. A special drill having much shorter flutes and heavy-web construction should be used.

In emergency cases, where short-flute drills are not available, a standard drill may be used by cutting off about one-third of the flute and resharpening. In this way the drill becomes sturdy enough to avoid excessive breakage.

For portable drilling of some of the harder or work-hardening materials the proper selection of a power tool is very important. The power tool should provide, and maintain under load, the correct surface speed for the material being drilled. In addition, the rate of feed is also very critical and should be as uniform as possible.

THRUST AND POWER FOR DRILLING

It is often desirable to know, at least approximately, the thrust or feeding pressure required to feed a drill of a given size. The values given in the following table were developed from the drilling of A.I.S.I. 1112 steel and are the minimum needed to feed a sharp drill with conventional point and web dimensions. No allowance has been made for drill-point wear, which may increase thrust by 100% or more. Improper thinning of the drill point when resharpening can also increase thrust. Therefore, when using the listed values, proper allowance should be made to anticipate maximum conditions.

Thrust in Pounds for Drilling A.I.S.I. 1112

Drill size inches	Feed inches per revolution										
	.001	.002	.004	.006	.008	.010	.013	.016	.020	.025	.030
⅛	23	42	75	108							
¼	45	82	145	215	265	325					
⅜		120	215	305	390	475					
½		150	270	385	490	600	740	890	1100		
⅝			320	450	585	700	880	1050	1300		
¾			375	530	660	820	1025	1200	1500		
⅞			430	610	760	950	1175	1400	1750		
1				700	890	1100	1350	1650	2000		
1¼				880	1100	1400	1750	2100	2550	3100	
1½				1100	1400	1750	2200	2600	3200	3900	
1¾				1350	1750	2150	2700	3250	4000	4750	
2				1650	2100	2600	3250	4000	4800	5800	6800
2¼				2000	2550	3200	4000	4800	5800	7000	8200
2½				2400	3100	3800	4700	5800	7000	8400	9800
2¾				2800	3600	4500	5600	6800	8200	10000	11750
3				3300	4300	5400	6600	8000	9600	11500	13500

Approximate conversion factors for thrust to drill other materials are:	thrust for regular points	thrust for split points
MST 6Al-4VA Titanium Alloy BR.340	2.0	0.8
17-7 PH Stainless Steel BR.400	2.3	1.0
4340 Heat Treated Steel 240,000 to 260,000 P.S.I.	3.5*	1.6
A.I.S.I. 1020	1.4	
A.I.S.I. 1035	1.3	
1.00%C. Tool Steel	1.7	
A.I.S.I. 3150	1.4	
Malleable Iron	0.6	
Gray Cast Iron	0.6	

* Regular points with special thinning are used for drilling this material.

Example: Find the thrust developed in drilling malleable iron with a ⅝ drill at a feed of .013″
From thrust listing, ⅝ drill at .013″ feed = 880 Lbs.
Conversion factor for malleable iron = 0.6
Converting to thrust for M.I. = 880 × 0.6 = 528 Lbs.

The power required to drive twist drills is of interest in planning drilling operations or buying new drilling equipment. The values given in the following listing are the minimum required for the drilling of A.I.S.I. 1112 at a speed of 100 revolutions per minute. No allowance has been made for drill-point wear, for losses in power-transmission to the spindle of the drill

Horsepower Required for Drilling A.I.S.I. 1112 at 100 R.P.M.

Drill size inches	.001	.002	.004	.006	.008	.010	.013	.016	.020	.025	.030
⅛	.002	.004	.006	.009							
¼	.007	.013	.023	.030	.037	.042					
⅜	.016	.027	.048	.68	.80	.94					
½	.027	.042	.081	.115	.135	.163	.20	.25	.28		
¾	.06	.10	.18	.25	.32	.38	.45	.52	.62		
1	.10	.17	.31	.42	.52	.64	.76	.90	1.1	1.3	
1¼	.15	.26	.45	.64	.78	.94	1.15	1.35	1.6	1.9	
1½	.22	.36	.62	.88	1.1	1.3	1.6	1.9	2.2	2.6	
1¾	.30	.50	.82	1.2	1.5	1.7	2.2	2.5	3.0	3.5	
2	.37	.62	1.1	1.5	1.9	2.2	2.7	3.2	3.8	4.5	5.2
2½					2.8	3.3	4	4.8	5.3	6.9	7.8
3								6.9	8	9.1	11

"Feed per revolution" spans the feed columns.

Approximate conversion factors for horsepower required to drill some other material are:

A.I.S.I. 1020	1.6	A.I.S.I. 3150	1.6
A.I.S.I. 1035	1.3	Gray Cast Iron	.5
1.00% C Tool Steel	1.7	Malleable Iron	.6

Stainless Steel

A.I.S.I. No. 416 Free-Machining Martensitic	1.2
A.I.S.I. No. 303 Free-Machining Austenitic	1.6
A.I.S.I. No. 304 Austenitic	1.8
17-7 PH Precipitation Hardened Austenitic	2.0
4340 Steel Heat Treated to 240,000 to 260,000 P.S.I.	2.3

Example: Find the H.P. required to drill a ½" hole in No. 416 Stainless at 75 R.P.M. with a feed of .006
H.P. at 100 R.P.M. for ½" drill at .006 feed (from table) = .115

H.P. at 1 R.P.M. = $\frac{.115}{100}$ = .00115

Material Factor for No. 416 stainless = 1.2

H.P. at 75 R.P.M. = .00115 × 75 × 1.2 = .103

press, or for the small amount of power required to feed the drill into the work. It is recommended, therefore, that a minimum of twice the amount of these values be used in making equipment estimates.

CARBIDE DRILLS

Carbide drills are made from solid carbide or with carbide tips or inserts at the end of a steel drill body. Solid carbide drills are usually made in diameters up to ½" while carbide tipped drills are made as small as ⅛" diameter. The selection of one or the other will depend on the material to be drilled and the economics of the application.

Carbide tipped drills usually are made with a somewhat lower helix angle than general purpose high speed steel drills, while solid carbide drills are designed similar to high speed steel drills. Drills for use in hardened steel are often made with straight flutes, and with negative rake cutting lips.

Deep hole drilling with carbide drills is not generally successful due to lack of rigidity in the drill and consequent breaking down of the cutting edges.

For drilling with carbide, machine conditions must be good and no spindle vibration or end play should be present. Optimum drilling speeds and feeds are governed not only by the material being drilled, but by total job conditions. Remember that vibration and chatter and lack of rigidity are the death of carbides.

The use of guide bushings with carbide drills is highly recommended. If a coolant is used, extreme care in its application must be exercised, otherwise any sudden cooling of the heated tip may cause cracking of the carbide. In the drilling of cast iron or other highly abrasive materials, the use of carbide lined drill bushings is recommended.

Carbide drills should be repointed by machine methods, to assure accuracy and long life. For repointing, a resinoid bond diamond wheel of 180 to 220 grit is suggested. Silicon carbide wheels may be used for repointing, but have a tendency to produce flaked cutting edges. This condition contributes to drill failure in use, and may also affect performance.

In repointing, drills are usually ground with a 118° included angle point with a lip relief of approximately 10° to 12° or greater depending on the size of the drill.

It may be necessary to thin the web, as shown in Figure 1, after repointing to maintain the original web thickness at the point. This thickness should be kept between 10% and 12% of the drill diameter. Quite often in drilling the more ductile materials an additional notch at the chisel edge as illustrated in Figure 2b may be helpful. Sometimes in the more brittle materials breaking of the cutting lip edge with a 30° negative angle by .002 to .005 width may also be advantageous.

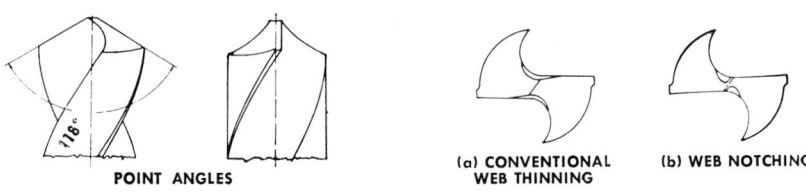

POINT ANGLES

Fig. 1

(a) CONVENTIONAL
WEB THINNING

(b) WEB NOTCHING

Fig. 2

Carbide drills must not be run after the edges dull. Excessive dullness of the edge will cause breakage. The allowable wear land on the carbide cutting lip edge is much less than would be permissible on high speed steel drills.

Carbide drills are generally run at higher speeds than high speed steel drills. Speed is particularly critical in the larger sizes [over ½″ diam.] and should never be under the highest recommended for high speed steel drills. Solid carbide drills often can be run slightly faster than tipped drills, because of their one-piece construction. Generally, small solid drills up to ⅛″ diameter operate at 60 to 100 feet per minute. The intermediate sizes can be operated at up to 125 feet, with the larger sizes running from 110 to 140 feet per minute.

In drilling most non-ferrous materials, drilling speed can be increased still further with excellent results.

Suggested Speeds for Carbide Drills

Material	Speed in feet per minute
Aluminum	150–500
Brass	150–300
Bronze	150–300
Cast Iron [soft]—Drill Sizes ⅛″ thru 1″	90–165
Cast Iron [soft]—Drill Sizes over 1″	125
Cast Iron [chilled]	30–50
Cast Iron [hard]	100–140
Copper	200–300
Non-ferrous Alloys	150–200
Plastic [phenolic, etc.]	100–200
Plastic [Glass-bonded]	50–125
Rubber	200–300
Wood	200
Glass	20
Steel [over 450 Brinell]	75–100

Carbide drills are generally run at lighter feeds and higher speeds than high speed steel drills. Feed can be the same as recommended for high speed steel drills, but a slightly lower feed rate in conjunction with a higher speed gives best results. Machine feeds are recommended for the larger drills in all materials except hardened steel where a hand feed must be used.

Feeds for Carbide Drills

Drill diameter—inches	Feed in inches per revolution
Under 1/16	.0005–.001
1/16 to 1/8	.001 –.003
1/8 to 3/16	.002 –.004
1/4 to 5/16	.003 –.005
3/8 to 7/16	.004 –.008
1/2 to 11/16	.006 –.010
3/4 to 1	.008 –.012

Use the lower feed values for the smaller drills in the range, or for harder materials.
Use hand feed only when drilling hardened steel or glass.

For cast iron feeds of approximately two-thirds those recommended for high speed drills are best. Machine feeds are recommended in all materials on the larger drills except hardened steel where a hand feed must be used.

In some cases it may be advantageous to use a hand feed on the smaller solid carbide drills. In such cases, a sensitive type drill press must be used.

CUTTING FLUIDS

As heat generated in the drilling process can be detrimental to the life of the drill and to the quality of the hole produced, cutting fluids are used for the purpose of lubrication, cooling, and chip removal. The cutting fluid may be in the form of liquid, gas, combination of gas and liquid or solid.

LUBRICATION

Contact between two sliding surfaces is always present during the cutting process. The tool slides along the work before it penetrates; the chip slides along the face of the tool; and the chips often slide along each other. The friction between the surfaces increases the temperature at the cutting edge, and will often cause breakdown of the edge or cratering of the tool face.

A cutting fluid, acting as a lubricant, will reduce the coefficient of friction between the sliding surfaces. This will often keep the cutting temperatures below the point that causes rapid breakdown of the edge or localized cratering.

COOLANT

In production machine work, the function of cooling the tool and the work becomes more important than that of lubricating. The amount of heat generated is usually so great that the life of the tool often depends on the ability to carry this heat away at the same rate as it is generated. If the cutting fluid can do this, there will be no overheating or rapid breakdown. The only dulling action will be that of abrasion.

ANTI-WELD AGENT

A newly formed chip has a very "clean" surface and immediately comes in contact with the tool face at an elevated temperature and pressure. These conditions make it ideal for a weld to occur. Welding will often cause the chips to pack and the drill to fail.

Some cutting fluids contain chemicals which react with the surface of the newly formed chip. The "dirty" surface formed by this reaction will reduce the possibility of welding.

CHIP REMOVAL

Coolant feeding drills take advantage of another function of the cutting fluid in addition to those noted in the preceding paragraphs. If the hole is blind and a liquid cutting fluid is used, the fluid pressure can be used as the agent for carrying the chips out through the flutes. Compressed air may be used to cool the drill point and blow chips up the flute.

In order to function properly, however, the chips must be broken into small pieces at the time they are formed so that they can be flushed out with the fluid. In many cases it is necessary to provide chip breakers at the point of the drill for this purpose.

SELECTING THE CUTTING FLUID

While functions of a cutting fluid may appear to be relatively simple, the non-cutting functions are considerable and it is strongly suggested that cutting fluid problems be referred to a reputable producer of such fluids.

Some of the important considerations in selecting a cutting fluid for drilling are its wetability, its ability to absorb heat and thus minimize the temperature of the cutting tool, its lubricity, its corrosion control, its bacteria control, operator safety, its flammability, and its ability to be economically filtered.

Pure water is one of the best coolants known, but it has little lubricating value and is seldom used without additives. Soluble oils and similar compounds usually have excellent cooling properties and are used to good advantage where hole finish is less important. Rusting of work from the water-soluble compounds is prevented by maintaining enough alkalinity. Sulfurized oils combine good lubricating and cooling qualities, and are used where a better finish is required. They are used extensively in drilling the ordinary steels.

Animal oils should be avoided, if possible, because of their bacterial properties and their tendency to become rancid. Straight mineral oils can be used with fair success on moderately heavy cuts, but they do not compare with the sulfurized oils.

Chemical (synthetic) fluids have increased in use as the costs of oils have increased and OSHA and EPA requirements have been established. They are made to cover a broad range of applications serviced by both cutting oils and soluble oils.

In drilling, cutting fluids are generally more effective when delivered in ample quantity and with sufficient pressure to reach the cutting area. The presence of the cutting fluid in this critical area assists in desirable chip formation and disposal, and increases drill life because of lower workpiece and tool temperatures.

Best results are usually obtained when the cutting fluid temperature is kept below 125° F. Heat transfer is based on the relative differential in temperature, thus cutting fluids of higher temperature fail to properly conduct away the heat. Furthermore, some fluids break down chemically when overheated and can become a hazard from a flammability as well as an operator safety standpoint.

Special drilling operations using highly volatile coolants, carbon tetrachloride, liquid air or compressed carbon dioxide gas should be carefully controlled and ventilated to prevent serious injury from exposure.

Circulated cutting fluids perform best when clean. Chips and dirt should be removed to maintain cutting fluid efficiency. Frequent fluid change, filtering and thorough machine cleaning help to maintain performance, and promote machine and operator well-being.

Cutting Fluids for Drilling

Material Drilled	Suggested Cutting Fluid	Remarks
Plain carbon and low alloy steels	1. Water soluble oils 2. Synthetics 3. Cutting oils—sulfurized and/or chlorinated	
Tool and die steels	1. Cutting oils—sulfurized and/or chlorinated 2. Water soluble oils 3. Synthetics	
Stainless steels	1. Cutting oils—sulfurized and/or chlorinated	
Stainless steels—free machining	1. Water soluble oils 2. Synthetics	
Superalloys (mostly nickel or cobalt base)	1. Cutting oils—sulfurized and/or chlorinated 2. Synthetics	
Cast irons	1. Synthetics 2. Water soluble oils 3. Dry	1. For rust inhibition 2. Good for malleable and ductile only, not for plain gray cast iron—may cause rusting and chip caking
Aluminum and aluminum alloys	1. Synthetics 2. Water soluble oils	
Magnesium and magnesium alloys	1. Mineral oils 2. Dry	Do not use water soluble oils because of reactivity with magnesium
Copper and copper alloys (brasses and bronzes)	1. Water soluble oils 2. Synthetics 3. Mineral oils	Fluids containing sulfur may cause staining
Titanium and titanium alloys	1. Synthetics 2. Water soluble oils 3. Mineral oils	

Notes: In applications involving coated tools, such as TiN, cutting fluid should contain extreme pressure (EP) additives. Fluids without these additives can be detrimental to performance.

EQUIPMENT FOR DRILLING HOLES

With greater emphasis placed on drilling harder and tougher materials, it is noteworthy to mention that strength, rigidity, and power at slow speeds are the factors affecting efficient drill performance.

As chip formation is one of the keys to good performance in drilling, it is very desirable for the drilling equipment to have increments of feed per revolution that can be established and retained under load. Backlash in a drilling machine may result in heavy torque loads at the drill point, both at the start of drilling as well as at the break-through. The machine and the drill must be properly mated for the most satisfactory results.

DRILLING EQUIPMENT

The drilling machine [drill press] is probably the most widely used of all machine tools. Its basic purpose is to produce holes. The machines vary widely in design and construction, depending on the purpose for which they are to be used.

They range in size from small, sensitive bench types for producing holes for parts of a watch or a small instrument to huge units with multiple heads having a number of spindles in each head for drilling large forgings or castings such as engine blocks for large Marine engines.

There are numerous types of power operated drilling machines. Some of the more common are:

1—SENSITIVE DRILLING MACHINES

The sensitive drill press may be either bench or floor type, having a box or round column. Feeds are manual or power.

2—UPRIGHT DRILLING MACHINES

This type of machine is often called a plain drilling machine. It has power as well as manual feed and is built to perform heavier work. The column may be of the round or box type. The upright drilling machine is the most common type found in the average shop. It usually has available a number of spindle feeds and speeds.

3—RADIAL DRILLING MACHINES

On radial drilling machines the drill head is mounted on a radial arm which is connected to and swings about a vertical column. They range in size from light-weight machines, having short arms to giants with arms 12 feet or more in length. The size range of radial machines enables the drilling of holes from the smallest diameter to holes four inches in diameter and larger.

4—GANG DRILLING MACHINES

Machines having two or more drilling spindles mounted on the same table are called gang drilling machines. Often an operator can run all the spindles at one time.

5—MULTIPLE SPINDLE DRILLING MACHINES

This is a variation of the gang drill. These machines may have any number of spindles driven from the spindle drive gear in the head. They are used to drill a number of holes simultaneously.

6—TRANSFER MACHINES

Transfer machines have been adapted for "in-line" production work. They are built up of any number of unit heads, having single or multiple spindles placed in a horizontal, vertical, or angular position and perform specific machining operations at each station.

7—NUMERICAL CONTROL MACHINING CENTERS

Numerical-control units have been developed into machining center concepts. The machining center may be single or multiple station and contain banks of tools from which specific tools are selected to perform a variety of operations.

8—DEEP HOLE DRILLING MACHINES

Deep hole drilling machines are also referred to as gun drilling machines. They are used for drilling holes to 12" or more in depth. The size tolerance can generally be held closer and a finer finish is generally produced with a gun drilling machine due to the special design of the machine and the tool used for producing the hole.

Power operated drilling equipment offers a wide choice of capacities, table and spindle sizes, speeds, feeds, etc. Some drilling machines provide for mounting tools at various stations. The tools are selected from each station in the proper sequence by having a turret rotate. The proper tools are selected and mounted in the operating spindle. Tape or numerically controlled machines are another recent development for drilling. They can be set to perform a repetitive cycle, produce holes in any set pattern, and automatically cease to operate when the pattern is completed.

With today's exotic materials and complex structures, portable power feed drilling equipment is often required right at the job site. This type of equipment is available from many sources. The advantage of portable power feed drilling equipment over hand operated drilling is increased drill life and hole quality. Usage of the equipment should be considered when thick or hard materials must be drilled on site.

DRILL BUSHINGS

The primary function of a drill bushing is to locate a designated position and to guide the drill in producing this hole. This bushing must be resistant to wear and securely fastened in the drill jig. Good practice in designing drill bushings is to make the length of the bushing approximately twice the diameter of the drill to support and guide the drill. Bushings that are too short will not keep the drill in line. If the bushing is longer than necessary it will cut the effective length of the drill and sometimes necessitate using special length drills.

The hole in a bushing should be very close to the diameter of the drill, but, under no circumstances, be so tight as to prevent the drill from turning freely. Drill bushings with insufficient clearance will cause the drill to bind, whereas bushings too loose in the fixture, or having too much clearance between the drill and the bushing, will result in chipped margins on the drill.

Judgment must be used in determining the proper space between the bushing and the work. Enough space should be provided between the work and the bushing so that chips can be ejected without flowing through the bushing. This improves the performance of the drill, keeping the flutes free of chips and admitting more coolant to the cutting edges. If the bushing is placed too far away from the work, the drill is no longer properly guided, sometimes resulting in oversize holes and drill breakage. At times it is desirable to use a split bushing to guide the drill when starting the hole. The bushing is ejected by the formation and flow of chips.

DRILL JIGS AND FIXTURES

Much drilling is done with jigs and fixtures. It is essential that drill jigs be constructed to hold the work securely. The jig must be of sturdy construction to minimize deflection of the jig or the workpiece during the drilling operation.

TWIST DRILL DRIVING APPLIANCES

Between the power-driven spindle of a drilling machine and the drill itself there must be a secure and satisfactory connection. As only a comparatively few drills will fit directly into the taper sockets with which drilling machines are ordinarily provided, this connection must be made by means of some appliance, such as a taper sleeve, a chuck, or similar arrangement.

Drill-driving appliances should have certain definite characteristics in order to produce satisfactory results. These characteristics may be enumerated as follows:

1. Positive gripping of the drill, to prevent rotary or longitudinal slippage under maximum loads.
2. Concentricity of the drill holder with the drilling machine spindle, to eliminate danger of misalignment or run-out.
3. Adaptability to take a wide range of sizes of standard drills as furnished from manufacturers' stocks.
4. Easy removal of drill for sharpening or changing to different size, and equally easy placing of another drill in the holder.
5. Durability, so as to maintain accuracy and gripping ability over a long period and under adverse conditions.
6. Longitudinal adjustment in the case of multiple-spindle operation.

Several kinds of drill driving appliances are in more or less common use in the metal-working industries. Each one of these appliances has certain advantages as well as certain disadvantages.

The guide in selecting any one of the above systems for a given job must be the quantity of work to be produced, the machine equipment available, and the first cost of drills and driving appliances. Generally speaking, it is more economical to use standard straight shank or taper shank drills than drills prepared for special drives, because the former can be obtained at regular prices and from manufacturers' stocks. Special drives should be limited to cases where standard drills will not serve on account of individual conditions surrounding the job.

DRIVING MEANS FOR TAPER SHANK TOOLS

DRILL SOCKETS

Taper shank drills fit directly into the spindle of a drill press having the same size taper as the drills. When the hole in the spindle is larger, a reducing socket or sleeve may be used.

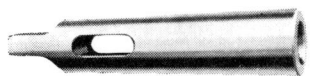

Sleeve or Shell Socket

Fitted Socket

The shorter sleeve is preferred rather than the fitted socket because of the added rigidity. The use of a fitted socket to permit the use of a larger drill than can be fitted directly into the spindle is not good practice except for occasional holes.

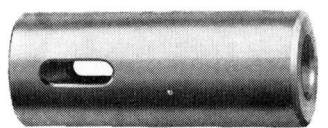

Screw Machine Socket

Rough Socket

Screw machine sockets are made in various diameters to fit into turrets or tool-holding fixtures on automatic screw machines.

Rough sockets have the shanks unfinished and may be finished by the user to fit any holder.

IMPORTANT FACT TO CONSIDER

Taper shank drives are designed to absorb the bulk of the torque in the fit of the taper. Worn, scratched, marred, or dirty sleeves or shanks destroy this fit and throw most of the burden on the tang. This often results in broken tangs. Another result of dirty or worn sleeves is the loss of accurate alignment. Sleeves and sockets should be discarded as soon as they have become worn and battered.

To insure the best fit, the insides of sleeves and the outsides of sockets should be wiped before they are assembled.

DRIFT OR KEY

Drifts are used for the removal of the taper shanks from the sockets. They are designed to eliminate injury to the tang of the drill and thus keep the driving action of the shank at its maximum.

The use of the tang of a file to drive a shank from a socket is definitely bad practice. The possible breakage of the file endangers the operator. Also, the action of the file tang on the drill tang is injurious in that the drill tang becomes flattened out and will not fit properly in the socket. This cuts down the driving action of the shank, causing slippage, breaking of the tang, and chipping of the drill point.

MODIFICATIONS OF TAPER SHANKS

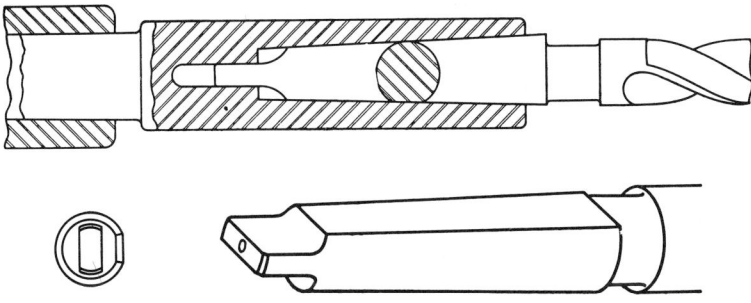

Fig. 1. *Use-Em-Up Flats on Drill Shanks*

Variations of the taper shank drive are occasionally used for heavy-duty work. The most common of these is the Use-Em-Up or flatted shank, Fig. 1. Here the taper shank has a flat on the side of the taper, and the socket has a corresponding flat on the inside, thus providing a positive drive the full length of the shank. This system calls for special, flatted shanks and special sockets.

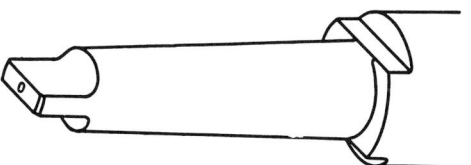

Fig. 2. *Two Flats at Large End of Shank*

Another variation of the taper shank drive is often used on large diameter drills. This type of drive shown in Fig. 2 uses a pair of flats at the large end of the taper shank which fits into a slot in the spindle nose.

Other variations of shanks that have been used occasionally for special applications in the manufacturing field are shown below. These shanks are entirely special and are shown for their reference value only.

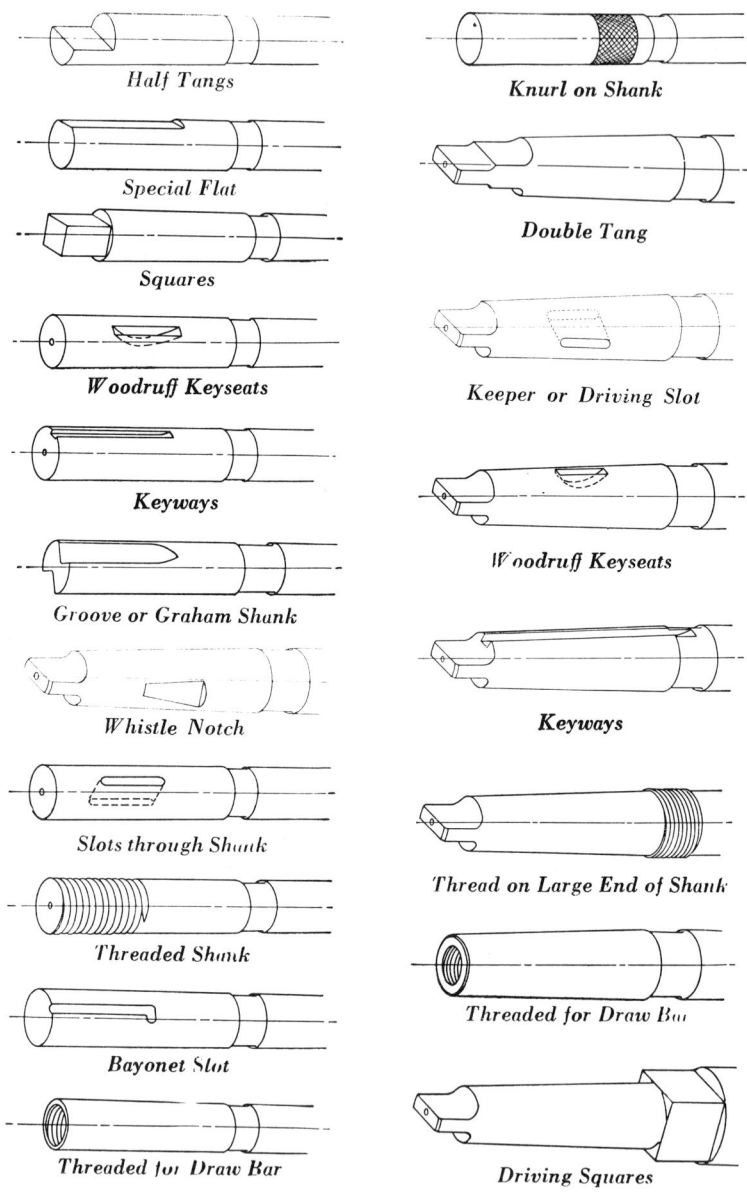

Half Tangs

Knurl on Shank

Special Flat

Double Tang

Squares

Woodruff Keyseats

Keeper or Driving Slot

Keyways

Woodruff Keyseats

Groove or Graham Shank

Whistle Notch

Keyways

Slots through Shank

Threaded Shank

Thread on Large End of Shank

Bayonet Slot

Threaded for Draw Bar

Threaded for Draw Bar

Driving Squares

DRIVING MEANS FOR STRAIGHT SHANK TOOLS

The Three-Jawed Drill Chuck, Fig. 1, is probably the most commonly used driver for small, straight shank drills. It may be of the wrench-tightening or of the wrenchless variety. Either kind has the advantage of being able to handle a fairly wide range of sizes of drills. These chucks can be easily adjusted from one size to another, within their range.

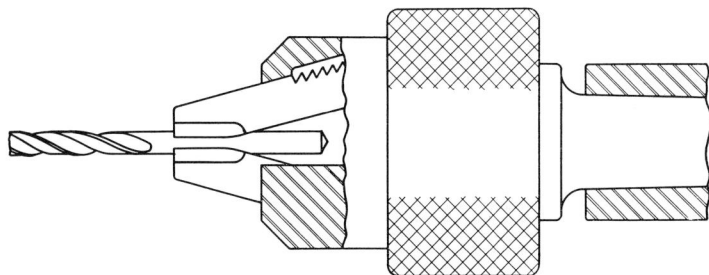

Fig. 1. *Three-Jawed Drill Chuck*

The Three-Jawed Drill Chuck depends entirely on the friction grip of the jaws against the drill shank. Care must be exercised to make sure that the jaws are tightened sufficiently to prevent slippage of the drill in the chuck; otherwise injury to the drill shank may result. Further, if this condition persists, the wear on the jaws of the chuck can cause misalignment of the drills for successive applications.

On some production jobs, tapered, split sleeves are used for driving small, straight shank drills. Such sleeves have a taper on the outside to fit the spindle socket and a straight hole on the inside to fit the drill shank. The drill sometimes is provided with flats or a square at the end to fit into a driving slot as shown in Fig. 2. This system is of advantage in multiple-spindle setups with close centers, because these sleeves are of smaller diameters than the conventional drill chucks.

The disadvantages of this system are the necessity of a separate sleeve **for** each diameter of drill to be used and the drifting of the sleeves each time drills are to be changed.

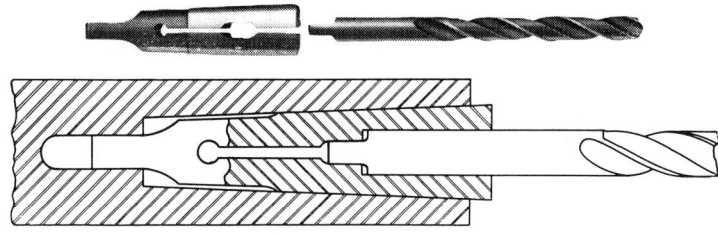

Fig. 2. *Split Sleeves for Straight Shank Tools*

STRAIGHT SHANK DRILLS WITH TANG DRIVE

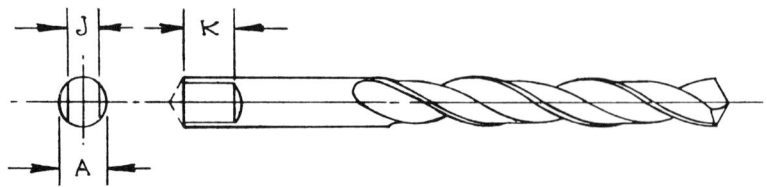

GENERAL DIMENSIONS OF TANGS TO FIT DRILL DRIVERS

Nominal Diameter of Shank			Thickness of Tang		Length of Tang
			J		K
A			Max.	Min.	
⅛	thru	³⁄₁₆	0.094	0.090	⁹⁄₃₂
Over ³⁄₁₆	"	¼	0.122	0.118	⁵⁄₁₆
Over ¼	"	⁵⁄₁₆	0.162	0.158	¹¹⁄₃₂
Over ⁵⁄₁₆	"	⅜	0.203	0.199	⅜
Over ⅜	"	¹⁵⁄₃₂	0.243	0.239	⁷⁄₁₆
Over ¹⁵⁄₃₂	"	⁹⁄₁₆	0.303	0.297	½
Over ⁹⁄₁₆	"	²¹⁄₃₂	0.373	0.367	⁹⁄₁₆
Over ²¹⁄₃₂	"	¾	0.443	0.437	⅝
Over ¾	"	⅞	0.514	0.508	¹¹⁄₁₆
Over ⅞	"	1"	0.609	0.601	¾
Over 1"	"	1³⁄₁₆	0.700	0.692	¹³⁄₁₆
Over 1³⁄₁₆	"	1⅜	0.817	0.809	⅞

All dimensions are given in inches.
The tangs in this table are for use with split-sleeve-type drill-driving collets.

Concentricity of Tang

From 0.090 to 0.243 thickness incl. = 0.006 total indicator variation.
Over 0.243 to 0.443 thickness incl. = 0.007 total indicator variation.
Over 0.443 to 0.609 thickness incl. = 0.008 total indicator variation.
Over 0.609 to 0.817 thickness incl. = 0.009 total indicator variation.

The larger sizes of straight shank drills are generally used in turret lathes or similar machines and are then held in a straight sleeve as indicated in Fig. 3. Set screws are most often used for clamping the drill in place.

A different size sleeve is, of course, necessary for each different diameter of shank to be handled. With such setups the sleeve must fit the shank of the drill very closely, so as to avoid misalignment when clamping. Set screws also have a tendency to mar up the drill shanks.

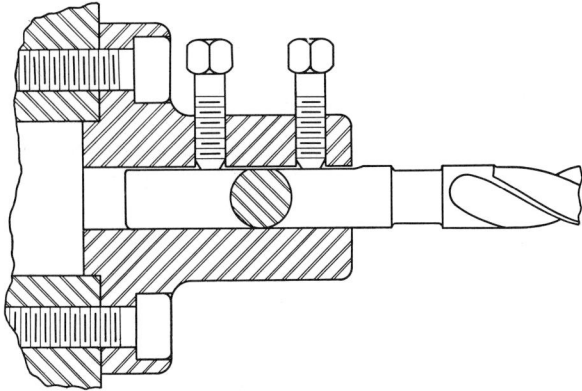

Fig. 3. *Straight Sleeve for Larger Sizes of Straight Shank Drills*

OTHER TYPES OF DRIVES

A split bushing is often used as a holding means in turret lathes and automatic screw machines. Two types of these split bushings are shown in Figs. 4 and 5.

 Style A

Style B

Fig. 4. *A short, split bushing for adapting drills, reamers, counterbores, taps, etc.*

Fig. 5. *A long, split bushing for use in turret holes, multiple turning heads, flanged tool holders, and other tools having deep holes*

In multiple-spindle drill press operation, it is usually necessary to have some means of adjusting the drills longitudinally in each spindle. This permits accurate setting of the depth of each hole. In through holes it permits setting the drills so that they all break through at about the same time.

The adapter shown in Figs. 6 and 6A can be moved in and out of the spindle by adjusting the knurled nut, which is then locked into one of the four slots with a set screw. The drive is through a Woodruff key, and the adapter is held in the spindle by another set screw.

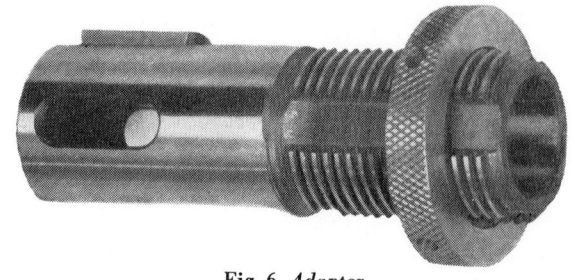

Fig. 6. *Adapter*

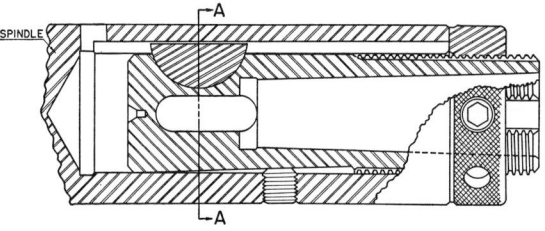

SECTION A-A

Fig. 6A. *Adjustable Adapter Assembly*

DRILL RECONDITIONING

Of all the factors which affect overall drill economy, proper drill reconditioning is one of the most important and one of the most underestimated. In order to lay out an effective procedure for the handling of this problem, it is necessary that there be a thorough understanding of the factors involved. There are three separate and distinct steps in the conditioning of dull or worn drills.

REMOVAL OF WORN SECTION

As soon as a new drill is placed in operation, it starts to wear. This wear is of several types, each of which has its effect upon drill life. Drill wear usually starts at the corners of the drill, as shown by area A in Figure 1.

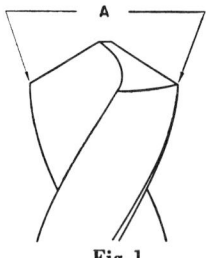

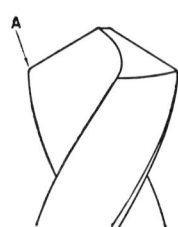

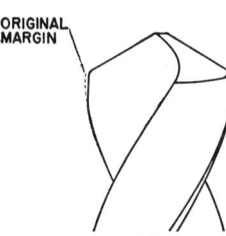

Fig. 1   Fig. 2   Fig. 3

Wear of Drill Points

This wear starts as a slight rounding, as shown by area B in Figure 2. At the same time, the cutting edges or lips, as well as the chisel-edge, start to wear away and form a truly conical surface of narrow width adjacent to these edges. This conical surface has no clearance and tends to rub in the hole rather than to cut. The power and thrust required to force this slightly dulled edge into the work increase, resulting in greater heat generation at the cutting edges and a faster rate of wear.

This increase in wear at the corners travels back along the margins, resulting in a loss in size and a negative back-taper in the tool, as shown by area C in Figure 3. This wear can easily be measured with micrometers. The first principle of proper reconditioning is to remove all of this worn section. Failure to do so results in a drill which will not cut properly and also in a very short life on all subsequent grinds.

The wear on a drill, like that on any other cutting tool, is not proportional to the holes or pieces produced between grinds. This is a basic fact which must be considered if reasonable economy is the user's goal. Wear occurs at an accelerated rate. As explained, as the drill wears, it cuts harder, more heat is generated, and it wears faster. In other words, there is more wear on the twentieth hole than on the tenth, still more on the thirtieth, and this continues until the last hole is drilled, at which time the rate of wear is greatest.

As an example: A 1″ drill under certain conditions may be able to drill 100 holes before it is completely dull and refuses to cut. At this time the margins may be worn undersize for a distance of 1 inch back from the point, and the drill must be shortened 1 inch before it can be reused. If the drill had been removed from the machine at the end of 75 holes, this wear might have been only ⅓ inch, and the drill would only need shortening by this amount. In the second case, the user would get 225 holes per inch of drill used instead of only 100 when running the drill to absolute dullness. The above example is not at all extreme and shows the advantage of a program which will promote the removal of drills from machines before the point of complete dulling.

In any event, regardless of the amount of wear on the margins near the point, the drill should be shortened to remove this excessively worn portion.

WEB-THINNING

Most drills today are made with webs which increase in thickness towards the shank of the drill, as shown in Figures 4 and 5.

After shortening the drill to remove the worn portion, the drill will appear as shown in Figure 6. The chisel edge of a drill does no actual cutting but

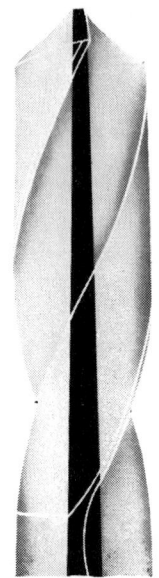

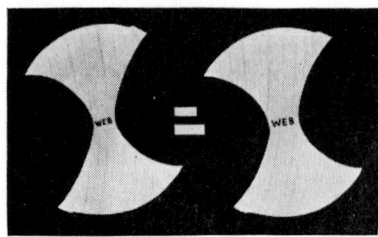

Fig. 5. *The section on the left was cut from a drill near the point while the section on the right was cut near the shank. The difference in the thickness of the web at these two points is shown by the length of the white lines between the two sections in the illustration*

Fig. 4. *The web is the metal column which separates the flutes*

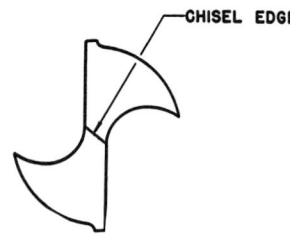

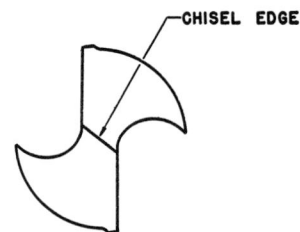

Point of drill when drill is new

Point of drill after drill has been cut back in use and repointed

Fig. 6

pushes the metal out of the way. The long chisel-edge of the shortened drill in Figure 6 will require a great deal of power and will generate much heat, resulting in short drill-life. In order to recondition the drill properly, it is necessary to reduce this web, so that the chisel edge is restored to its normal length. This operation is called web-thinning. The second principle of proper reconditioning is to restore approximately the original web-thickness of all drills which have been shortened.

The best web-thickness for any size of drill may vary slightly for different jobs, but unless unusual conditions prevail, the following approximate web-thicknesses expressed as a percentage of drill-diameter will be found to be satisfactory:

Diameter of drill, inches	Web-thickness expressed as a percentage of drill-diameter
⅛	20
¼	17
½	14
1	12
Over 1	11

The thinning of webs can be done in a special web-thinning grinder, or it can be done as a free-hand grinding operation. Machine grinding is to be preferred to hand grinding because of its greater accuracy and reliability. However, the absence of a machine for this operation should never be considered a sufficient reason for neglecting this important step in drill-reconditioning. The operation is not at all difficult to learn and can be performed by any competent tool-grinder.

Several different types of web-thinning are in common use. In Figure 7 is shown a type that is perhaps the most common. The length A is usually ½ to ¾ the length of the cutting lip. In this type of thinning, as well as in all others, it is important that the thinning cut extend far enough up the flute so that an abrupt wedge is not formed at the extreme point. The distance up the flute to which the thinning extends will vary with the amount of thinning required, but an average of ¼ to ½ of the drill diameter will usually insure that the thinning is not too abrupt. The general relationship of the drill and grinding wheel is shown in Figure 8.

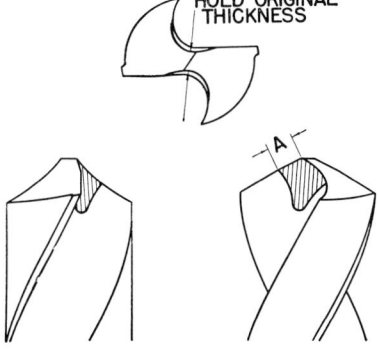

HOLD ORIGINAL THICKNESS

A

Fig. 7. *Usual method of thinning the point of a drill when the web has become too thick because of repeated re-pointing*

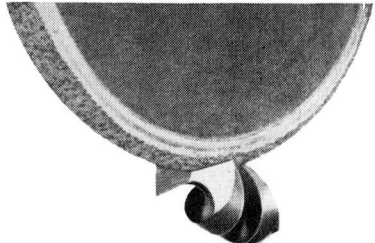

Fig. 8. *Relation of drill and grinding wheel when thinning the web*

Sometimes it is advisable to extend the thinning out to the extreme edge in order to change the shape of the chip. In this type of thinning, shown in Figure 9, a positive, effective rake is maintained the full length of the cutting-edge.

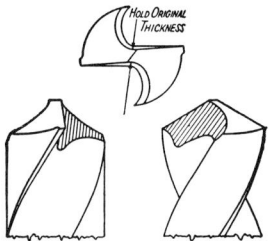

Fig. 9. *Undercut Thinned Point — Another common method of web thinning. If properly done this type of thinning will produce a fine curled chip*

A third type of thinning quite often used is the offset, notched, split, or crankshaft point, as it is variously called. This type of thinning is shown in Figure 10.

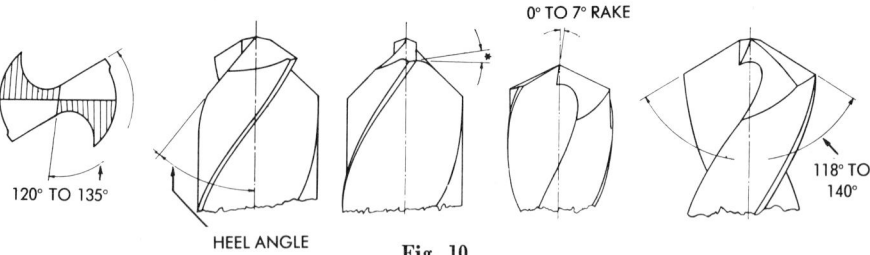

Fig. 10

* See suggested values of lip relief angles, p. 107.

The included angle of the point can be adjusted to the individual job, an angle of about 135° being perhaps most common. In grinding the original point-surfaces, it is usually found best to have a chisel-edge angle of 90° to 120°, as shown in Figure 11. The two notching cuts are then made to meet at the center, forming a new chisel-edge angle of about 120° to 135°.

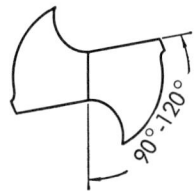

Fig. 11. *Point of crankshaft drill before notching*

There are several important features of this type of web thinning. The angle of the notch is shown in Figure 10. This angle is subject to some variation, but if increased too much, the result is a shallow notch which is not big enough to accommodate the chips formed by the short, center-cutting edges.

In certain applications, it is advantageous to grind the notch in such a way as to produce a positive rake angle in the face of the notch as illustrated in Figure 10. This type of notching will produce a freer cutting edge on the secondary cutting lip.

GENERAL WEB-THINNING INFORMATION

Regardless of the type of web-thinning employed—whether by machine or by hand—it is important that the web-thinning be done evenly with the same amount of stock removed from each cutting-edge.

The grinding wheel should be soft enough to remove this stock without danger of burning the cutting edges.

It should always be remembered that chips must form on the drill lips and flow into the flutes, and the shape of the thinning should be such that it does not interfere in any way with the chip-flow.

REGRINDING THE SURFACE OF THE POINT

Besides removing the worn portion of the drill and thinning the web, it is necessary that the surfaces of the point be reground. These two conical surfaces intersect with the faces of the flutes to form the cutting lips. They also intersect with each other to form the chisel-edge as shown in Figure 1.

As in the case of any other cutting tool, the surface behind these cutting lips must not rub on the work, but must be relieved in order to permit the

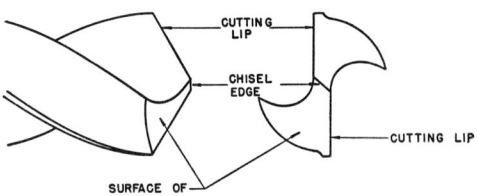

Fig. 1. *Nomenclature of Drill Point*

edge to penetrate. Without such a relief a drill would appear like that shown in Figure 2. This drill could not penetrate the metal, but would only rub around and around.

Grinding clearance on the surface back of the cutting edge, illustrated in Figure 4, makes the heel lower than the cutting edge and permits the cutting

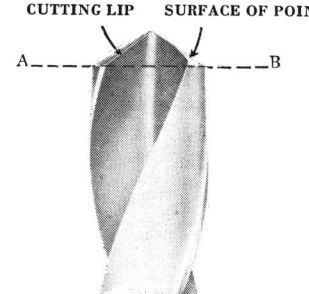

Fig. 2. *A drill point without any lip relief. Note that corners of cutting lip A and of heel B are in approximately the same plane*

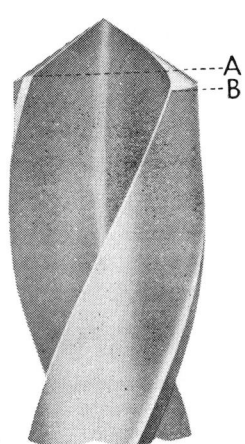

Fig. 4. *Proper lip relief. Note how much lower the heel line B is than the cutting lip line A. This difference indicates clearance on the drill point*

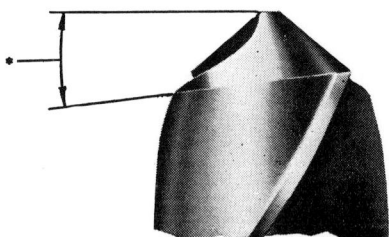

Fig. 3. *Proper way to grind lip relief. The angle indicated is the angle, measured across the margin, at the periphery of the drill*

edge to penetrate while drilling. This clearance, called lip-relief, is measured at the periphery of the drill across the margin portion of the land, illustrated in Figure 3. The amount may be varied to suit the application.

The degree of relief increases toward the center and the relieved surfaces of each edge intersect at center forming a chisel-edge angle as illustrated in Figure 5. This angle bears a relationship to the lip-relief angle. Since the lip-relief is greater on small drills than on large drills, the chisel-edge angle will also be greater for small drills. For example, the chisel-edge angle may be 125° for lip-reliefs of 8°–10°, 135° for lip reliefs of 12°–15°, and 145° for a lip-reliefs of 25°.

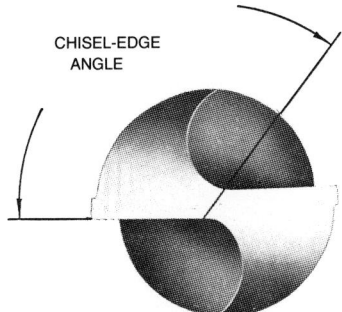

CHISEL-EDGE
ANGLE

Fig. 5. *Chisel-Edge Angle*

The suggested relief angles on drill points for general purpose use, shown below, are to serve as a guide for proper repointing and lip-relief angles.

Suggested Lip Relief Angles

Drill diameter range	Lip relief angle
# 80 to # 61 (0.0135 to 0.0390)	24°
# 60 to # 41 (0.0400 to 0.0960)	21°
# 40 to # 31 (0.0980 to 0.1200)	18°
⅛ to ¼ (0.1250 to 0.2500)	16°
F to ¹¹/₃₂ (0.2570 to 0.3438)	14°
S to ½ (0.3480 to 0.5000)	12°
³³/₆₄ to ¾ (0.5156 to 0.7500)	10°
⁴⁹/₆₄ and larger (0.7656 to ——)	8°

For harder materials it is suggested to reduce these values by approximately 2°, and for soft and non-ferrous materials that they be increased by approximately 2°.

ACCURACY OF CUTTING LIPS

The two cutting lips of a drill should be accurately ground. Regardless of the point angle, the angles of the two cutting lips A1 and A2, Figure 6, must be equal. Similarly, the length of the two lips L1 and L2 should be equal. The point should be symmetrical and the difference in relative lip height, as

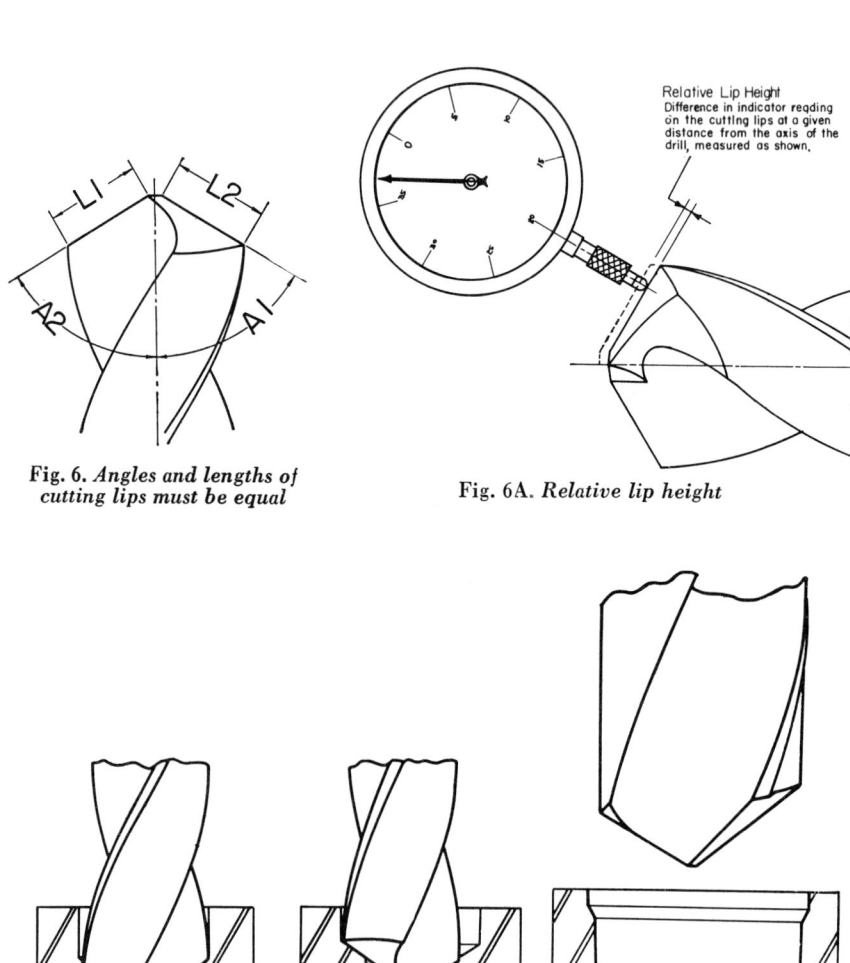

Relative Lip Height
Difference in indicator reading on the cutting lips at a given distance from the axis of the drill, measured as shown.

Fig. 6. *Angles and lengths of cutting lips must be equal*

Fig. 6A. *Relative lip height*

Fig. 7. *Incorrect Point. Lips of equal length but at unequal angles*

Fig. 8. *Incorrect Point. Lips of unequal length but at equal angles*

Fig. 9. *Incorrect Point. Lips of unequal length and at unequal angles*

shown in Figure 6A, should be kept to an absolute minimum. There are many drill-point gages available which can be used to measure these angles and lengths.

In Figure 7 is shown a drill with a point having lips of equal length, but which make unequal angles with the axis of the drill. One cutting edge does most of the cutting. This type of point will cause an oversize hole. In addition, unnecessarily severe strains are thrown on the drill and spindle. Excessive wear and short tool-life will usually be another result.

Much the same result is obtained if the angles are equal, but the lips are

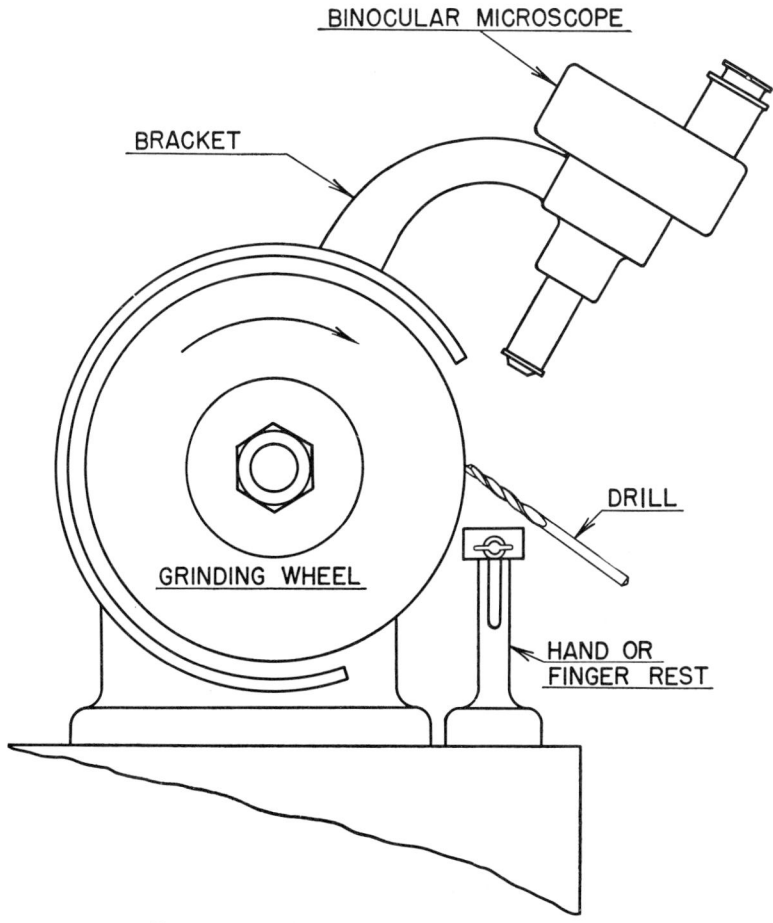

Fig. 10. *Use of binocular microscope in grinding very small drills*

of unequal length, as shown in Figure 8. A combination of both unequal lengths and angles is shown in Figure 9.

In order to maintain the necessary accuracy of point-angles, lip-lengths, lip-relief angle, and chisel-edge angle, the use of machine point-grinding is recommended. There are many commercial drill-point grinders available today, the use of which will be of great help in the accurate repointing of drills.

However, the lack of a drill-point grinding machine should never be considered a sufficient reason to excuse poor points. Drills can be pointed accurately by hand, although the operation takes skill and practice. On very small drills, the use of a binocular microscope [see Figure 10] focused on the wheel at the point of grinding is quite helpful. Magnifications ordinarily used range from 7 to 10.

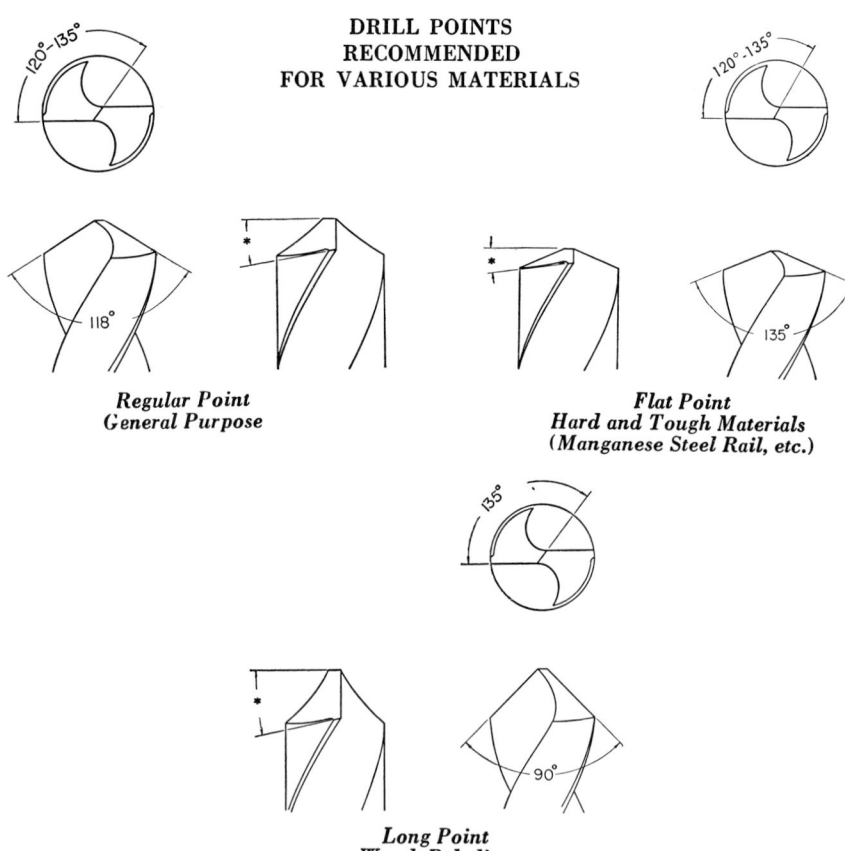

**DRILL POINTS
RECOMMENDED
FOR VARIOUS MATERIALS**

*Regular Point
General Purpose*

*Flat Point
Hard and Tough Materials
(Manganese Steel Rail, etc.)*

*Long Point
Wood, Bakelite,
Hard Rubber and Fibers*

DIFFERENT TYPES OF DRILL-POINTS

It has been generally noted by those engaged in production-drilling of steels, especially of deep holes in the harder steels, that better results can be obtained by increasing the included point-angle from the average 118° to 135° or 140°. Similarly, it has been found that cast iron and related materials can often be drilled to better advantage with a drill having a smaller included point-angle, such as 90° to 100°.

Many people have followed the theory that the effect of changing point

  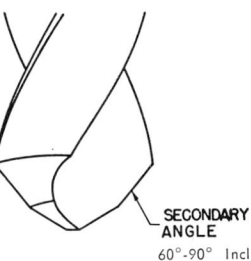

Fig. 3   Fig. 4   Fig. 5

Drill Points for Cast Iron

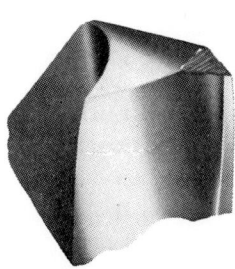 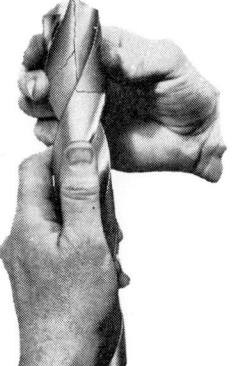 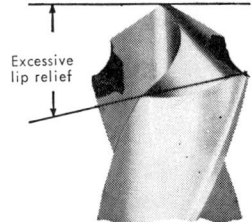

The Indication of Too Great Speed. The outer corners of the drill have worn away rapidly because excessive speed has drawn the temper

Here the drill was given insufficient lip-relief. As a result there ceased to be any effective cutting lips whatsoever, and, as the feed-pressure was applied, the drill could not enter the work—as a result it "splits up the center"

Showing results of giving a drill too great lip-relief—the edges of the cutting lips have broken down because of insufficient support

angles upon drill performance was due to changes in the effective rake angle of the cutting edge. However, recent investigations have shown that this effect is negligible except at the inner ends of the drill lips. Relatively little drilling energy is expended in this region of the drill.

It now appears that the major benefit derived from changing drill point angles is related to the thickness and width of the chips produced by the drill lips.

Small included point angles give drill lips which are relatively long so

COMMON DRILL TROUBLES AND CAUSES

Indications	Causes
Outer Corners Break Down	Cutting speed too high—Hard spots in material—No cutting compound at drill point—Flutes clogged with chips.
Cutting Lips Chip	Too much feed—Lip relief too great.
Checks or Cracks in Cutting Lips	Overheated or too quickly cooled while sharpening or drilling.
Margin Chips	Oversize jig bushing.
Drill Breaks	Point improperly ground—Feed too heavy—Spring or back lash in drill press, fixture or work—Drill is dull—Flutes clogged with chips.
Tang Breaks	Imperfect fit between taper shank and socket caused by dirt or chips, or burred or badly worn sockets.
Drill Breaks when Drilling Brass or Wood	Flutes clogged with chips—Improper type drill.
Drill Splits up Center	Lip relief too small—Too much feed.
Drill Will Not Enter Work	Drill is dull—Lip relief too small—Too heavy a web.
Hole Rough	Point improperly ground or dull—No cutting compound at drill point—Improper cutting compound—Feed too great—Fixture not rigid.
Hole Oversize	Unequal angle or length of the cutting edges—or both—Loose spindle.
Chip Shape changes while Drilling	Drill becomes dull or cutting lips chipped.
Large Chip coming out of one Flute, Small Chip out of other Flute	Point improperly ground, one lip doing all the cutting.

the chips are wide and thin. This distributes the cutting load over the long cutting edge in such a manner that the load per unit length of lip is reduced. Such reduced loading enables the lips to better resist wear caused by the abrasive action of cast iron and certain plastics.

A large included point angle is helpful with hard-to-drill materials because the chips produced are thicker. It has been shown that most materials are more efficiently cut in the form of thick chips. Torque requirements are definitely reduced when steel is drilled with a drill having a large included point angle. Further, thicker chips tend to minimize strain hardening of the workpiece. Strain hardening can make materials difficult to drill and may reduce drill life.

If a drill is sharpened with a large included point-angle, as in Figure 3, the zone of abrasion is comparatively small, as at **M**. Dulling will, therefore, be proportionately hastened. By sharpening with a small included angle, as in Figure 4, the area of the zone of abrasion **N** is greatly increased and with it the life per grind. To further increase this abrasive area, and with it the grinding life, it is sometimes advisable to grind a secondary angle at the corners of the drill, as shown in Figure 5.

USE AND CARE OF CARBIDE DRILLS

Carbide drills are made from solid carbide or with carbide tips or inserts at the end of a steel drill body. Solid carbide drills are usually made in diameters up to ½″ while carbide tipped drills are made as small as ⅛″ diameter. The selection of one or the other will depend on the material to be drilled and the economics of the application.

Carbide tipped drills usually are made with a somewhat lower helix angle than general pupose high speed steel drills, while solid carbide drills are designed similar to high speed steel drills. Drills for use in hardened steel are often made with straight flutes, and with negative rake cutting lips.

Carbide drills must not be run after the edges dull. Excessive dullness of the edge will cause breakage. The allowable wear land on the carbide cutting lip edge is much less than would be permissible on high speed steel drills.

Carbide drills should be repointed by machine methods, to assure accuracy and long life. For repointing, a resinoid bond diamond wheel of 180 to 220 grit is suggested. Silicon-carbide wheels may be used for repointing, but have a tendency to produce flaked cutting edges. This condition contributes to drill failure in use, and may also affect performance.

In repointing, carbide drills both tipped and solid are usually ground with a 118° point, illustrated in Figures 1 and 3. The lip-relief should be approximately that listed for H.S.S. on page 107 for conical ground points. A secondary clearance of a few degrees more than the primary relief is sometimes used on tipped drills to remove the steel back-up portion prior to grinding the carbide tip with a diamond wheel. A secondary clearance is sometimes required when the point grind equipment used produces a flat or profile type of relief. When this type of grind is used, the primary relief is usually less than that used for the conventional conical relief.

It may be necessary to thin the web, as shown in Figure 2a, after repointing to maintain the original web thickness at the point.

The thickness should be kept between 10% and 12% of the drill diameter. Quite often in drilling the more ductile materials an additional notch at the chisel edge as illustrated in Figure 2b may be helpful. Sometimes in the more brittle materials breaking of the cutting lip edge with a 30° to 45° negative angle by .002 to .005 width may also be advantageous.

Drills for special application may require special point geometry depending upon the application requirements or drill construction.

For most applications the conventional 118° point is used. On some materials, such as plastics, a 90° point is required. Solid carbide drills can usually be modified without a problem. However, if applied to the conventional tipped drill point shown in Figure 3, a smaller included angle will reduce the usable length of carbide considerably.

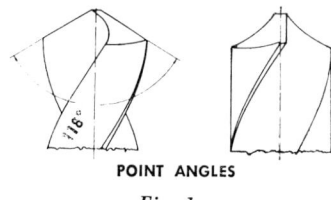

POINT ANGLES

Fig. 1

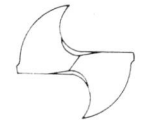

(a) CONVENTIONAL
WEB THINNING

(b) WEB NOTCHING

Fig. 2

Drills for drilling hard steels require a point similar to that shown in Figure 4. The exact configuration will vary according to the size and individual manufacturer's design. The heavy drill web and lip thickness together with the severity of the application requires an accurately located thinning notch with negative angles for strength. On some designs the lip is ground with a negative angle extending the length of the cutting lip.

Circuit board drills, which have the greatest usage in sizes smaller than 1/16″ diameter, are usually ground with a "four-facet" type point shown in Figure 5. The geometry is adaptable to the point grind equipment generally used in the circuit board industry. There is a primary and secondary relief. The secondary extends to center, parallel with the cutting edge, removing the primary chisel edge. The primary relief is approximately 10°–15° in the larger sizes and 15°–20° in the smaller. These drills are used in high speed automatic equipment and accuracy of the point is essential to the application.

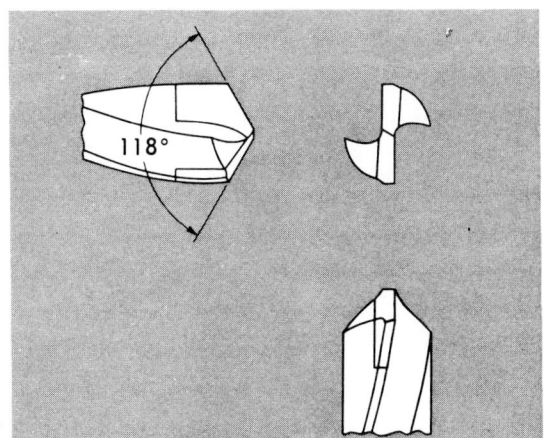

Fig. 3

CIRCUIT BOARD DRILL POINT

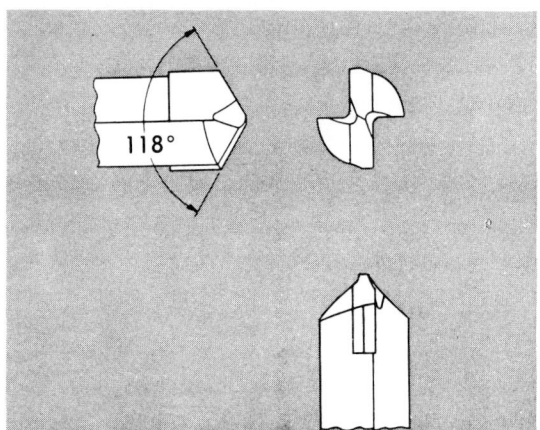

Fig. 4

DRILL POINT FOR HARD STEEL

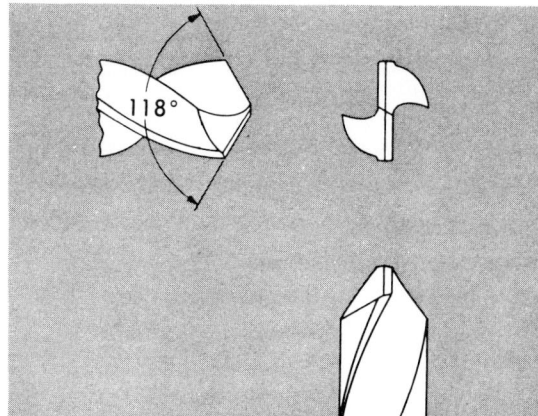

Fig. 5

STANDARD TIPPED DRILL POINT

DRILL TESTING

Drill performance tests are undertaken for a variety of reasons but the underlying theme of all these is economic—the production of drilled holes at the lowest possible cost. The actual tests may be run to determine the best drilling technique, to compare different drill designs, to compare competitive brands of tools or to check the quality of incoming tools.

Drill tests may be run under carefully controlled laboratory conditions or on actual production machinery in the shop. Laboratory tests are usually conducted to develop basic drilling techniques and to develop and evaluate new drill designs and tool materials. Here, a considerable effort is made to measure drill performance alone, and to eliminate the effect of the machine and fixtures with which the drill is used. This requires special attention to all the conditions surrounding the test, especially the machine rigidity and an adequate smooth flow of power to the drill. Production tests, on the other hand, are aimed at securing the best possible drill performance in combination with existing drilling machines and fixtures.

Accelerated tests are often misleading and should be avoided. Type of wear or failure may change under accelerated conditions and lead to erroneous conclusions.

In either laboratory or shop testing it will be quickly discovered that a casual approach to the problem will yield poor and often misleading results. Considerable variation in performance will be found between individual drills and between successive resharpenings of any single drill. A small part of this variation may be due to manufacturing tolerance variations between individual drills, but most of it is due to nonuniformity in work material, machine condition, drill sharpening and to the difficulty in establishing a positive failure criterion which indicates the end of useful drill life. Also involved is the fact that a drill is a complex helical shape having an extreme length to diameter ratio. It is subject to columnar bending and torsional vibration. Slight inaccuracies in sharpening magnify wear on lips and margins. Scatter in test results is a normal part of drill testing and an adequate number of drills is required to compensate for this characteristic.

The most important factor in successful drill testing is the maintenance of uniform operating conditions throughout the test. Factors which must be controlled include:

1—Drilling machine and fixture condition.
2—Speeds and feeds.
3—Cutting fluids.
4—Work material.
5—Drill sharpening.
6—Drill failure criterion.

A loss of control of the uniformity of any of the factors can lead to poor results and a worthless test. Needless to say, before any test is started it is prudent to examine each of the above factors to insure that the best possible conditions prevail.

Because of the forementioned unavoidable variations encountered, it is necessary that a sufficiently large group of drills be used in the tests. If performance differences between the test groups are relatively large, of the order of 30% or more, a test of as few as six drills of a type run through several resharpenings may be adequate. Small differences, of the order of 15% or less, are very difficult to detect reliably so much larger test groups must be used. Simple statistical analysis procedures are available which will indicate whether small differences in observed performance of different groups are actually significant.

In shop tests it is sometimes impossible to control all of the machine and workpiece variations. Testing under such conditions is still possible but very large groups of drills and a great deal of time is required. The drills must be systematically rotated between various spindle and workpiece groups to minimize the effect of the variations, and as unplanned variations appear they should be noted in a log.

It is important that an accurate log of the performance of each drill is maintained. Any unusual machine conditions, workpiece changes, accidental breakage, etc. must be noted in the log. Such notations are of great assistance in analyzing the test results. They sometimes permit useful data to be obtained from an otherwise useless test. They should also be permanently filed for future reference.

The determination of the end of useful drill life can be quite difficult. Some of the criteria which are used include:

1—Drill noise [scream or cry].
2—Wear land development.
3—Refusal of the drill to cut.
4—Total destructive failure of the drill.
5—Hole size.
6—Hole finish or the development of burrs.
7—Drill margin wear.
8—Drill breakage [especially on small drills].
9—A predetermined increase in torque, thrust or power.

Destructive failure of the drill may not be a very good criterion for drills which are normally resharpened. Here, consideration must be given to the amount of drill which must be removed to restore it to good condition. Workpiece material, cutting fluid, and other operating conditions influence the type of failure which is obtained. A certain amount of preliminary testing is often necessary to determine the most reasonable and representative indicator of drill life, and all operators on the test should agree to the same criterion.

Not only should there be a sufficient number of drills but the drills should be resharpened enough times to represent fairly the practice in the shop. It goes without saying that the sharpening should duplicate that originally used on the new drills. Resharpening should include cutting off that part of the drill which has metal pickup on the margin or has reverse back taper. Thinning the web to the original values is occasionally neglected by the tool resharpener. The resharpened drill should have accurate symmetry and be free from burn and checks.

In most test work the arithmetic average life of the various runs is used as an index of performance. With drills, occasional extremely high or extremely low runs are often obtained. Unless the test group is very large, the inclusion of such runs can unduly influence the average life. A much better index of performance is the median drill life. This is the drill life which is exceeded by half the drills in each test group. It is found by arranging the drills in order of increasing number of holes produced in the test. If the median and average drill lives differ by more than about 10%, the validity of the test is questionable and statistical analysis procedures should be used.

A test is not finished when merely tool costs per unit of product is determined. Analysis of the whole machining operation should follow.

DECIMAL EQUIVALENTS

Fractional, Wire, Letter and Metric Sizes

Decimal	Fract. Wire Letter	mm.	Decimal	Fract. Wire Letter	mm.	Decimal	Fract. Wire Letter	mm.	Decimal	Fract. Wire Letter	mm.
.0059	97		.0413	58	1.05	.1065	36		.1960	9	
.0063	96		.0420	57		.1083		2.75	.1969		5.
.0067	95		.0430			.1094	7/64″		.1990	8	
.0071	94		.0433		1.1	.1100	35		.2008		5.1
.0075	93		.0453	56	1.15	.1102		2.8	.2010	7	
.0079	92	.2	.0465			.1110	34		.2031	13/64″	
.0083	91		.0469	3/64″		.1130	33		.2040	6	
.0087	90	.22	.0472		1.2	.1142		2.9	.2047		5.2
.0091	89		.0492		1.25	.1160	32		.2055	5	
.0095	88		.0512	55	1.3	.1181		3.	.2067		5.25
.0098		.25	.0520			.1200	31		.2087		5.3
.0100	87		.0531		1.35	.1220		3.1	.2090	4	
.0105	86		.0550	54		.1250	1/8″		.2126		5.4
.0110	85	.28	.0551		1.4	.1260		3.2	.2130	3	
.0115	84		.0571		1.45	.1280		3.25	.2165		5.5
.0118		.3	.0591		1.5	.1285	30		.2188	7/32″	
.0120	83		.0595	53		.1299		3.3	.2205		5.6
.0125	82		.0610		1.55	.1339		3.4	.2210	2	
.0126		.32	.0625	1/16″		.1360	29		.2244		5.7
.0130	81		.0630		1.6	.1378		3.5	.2264		5.75

(Continued on following page)

(Continued from preceding page)

DECIMAL EQUIVALENTS (Continued)
Fractional, Wire, Letter and Metric Sizes

Decimal	Fract. Wire Letter	mm.	Decimal	Fract. Wire Letter	mm.	Decimal	Fract. Wire Letter	mm.	Decimal	Fract. Wire Letter	mm.
.0135	80		.0635	52		.1405	28		.2280	1	
.0138		.35	.0650		1.65	.1406	9/64"		.2283		5.8
.0145	79		.0669		1.7	.1417		3.6	.2323		5.9
.0156	1/64"		.0670	51		.1440	27		.2340	A	
.0157		.4	.0689		1.75	.1457		3.7	.2344	15/64"	
.0160	78		.0700	50		.1470	26		.2362		6.
.0177		.45	.0709		1.8	.1476		3.75	.2380	B	
.0180	77		.0728		1.85	.1495	25		.2402		6.1
.0197		.5	.0730	49		.1496		3.8	.2420	C	
.0200	76		.0748		1.9	.1520	24		.2441		6.2
.0210	75		.0760	48		.1535		3.9	.2460	D	
.0217		.55	.0768		1.95	.1540	23		.2461		6.25
.0225	74		.0781	5/64"		.1562	5/32"		.2480		6.3
.0236		.6	.0785	47		.1570	22		.2500	1/4" E	
.0240	73		.0787		2.	.1575		4.	.2520		6.4
.0250	72		.0807		2.05	.1590	21		.2559		6.5
.0256		.65	.0810	46		.1610	20		.2570	F	
.0260	71		.0820	45		.1614		4.1	.2598		6.6
.0276		.7	.0827		2.1	.1654		4.2	.2610	G	
.0280	70		.0846		2.15	.1660	19		.2638		6.7

Dec.	Size	mm	Dec.	Size	mm	Dec.	Size	mm	Dec.	Size	mm
.0292	69		.0860	44		.1673		4.25	.2656	17/64"	
.0295		.75	.0866		2.2	.1693		4.3	.2657		6.75
.0310	68		.0886		2.25	.1695	18		.2660	H	
.0312	1/32"		.0890	43		.1719	11/64"		.2677		6.8
.0315		.8	.0906		2.3	.1730	17		.2717		6.9
.0320	67		.0925		2.35	.1732		4.4	.2720	I	
.0330	66		.0935	42		.1770	16		.2756		7.
.0335		.85	.0938	3/32"		.1772		4.5	.2770	J	
.0350	65		.0945		2.4	.1800	15		.2795		7.1
.0354		.9	.0960	41		.1811		4.6	.2810	K	
.0360	64		.0965		2.45	.1820	14		.2812	9/32"	
.0370	63		.0980	40		.1850	13	4.7	.2835		7.2
.0374		.95	.0984		2.5	.1870		4.75	.2854		7.25
.0380	62		.0995	39		.1875	3/16"		.2874		7.3
.0390	61		.1015	38		.1890	12	4.8	.2900	L	
.0394		1.	.1024		2.6	.1910	11		.2913		7.4
.0400	60		.1040	37		.1929		4.9			
.0410	59		.1063		2.7	.1935	10				
.2950	M		.4409			.8465		21.5	1.3594	123/64"	
.2953		7.5	.4528		11.5	.8594	55/64"		1.3750	13/8"	
.2969	19/64"		.4531	29/64"		.8661		22.	1.3780		35.
.2992		7.6	.4646		11.8	.8750	7/8"		1.3906	125/64"	

(Continued on following page)

(Continued from preceding page)

DECIMAL EQUIVALENTS (*Continued*)
Fractional, Wire, Letter and Metric Sizes

Decimal	Fract. Wire Letter	mm.	Decimal	Fract. Wire Letter	mm.	Decimal	Fract. Wire Letter	mm.	Decimal	Fract. Wire Letter	mm.
.3020	N		.4688	15/32"		.8858		22.5	1.3976		35.5
.3031		7.7	.4724		12.	.8906	57/64"		1.4062	1 13/32"	
.3051		7.75	.4803		12.2	.9055		23.	1.4173		36.
.3071		7.8	.4844	31/64"		.9062	29/32"		1.4219	1 27/64"	
.3110		7.9	.4921		12.5	.9219	59/64"		1.4370		36.5
.3125	5/16"		.5000	1/2"		.9252		23.5	1.4375	1 7/16"	
.3150		8.	.5039		12.8	.9375	15/16"		1.4531	1 29/64"	
.3160	O		.5118		13.	.9449		24.	1.4567		37.
.3189		8.1	.5156	33/64"		.9531	61/64"		1.4688	1 15/32"	
.3228		8.2	.5197		13.2	.9646		24.5	1.4764		37.5
.3230	P		.5312	17/32"		.9688	31/32"		1.4844	1 31/64"	
.3248		8.25	.5315		13.5	.9843		25.	1.4961		38.
.3268		8.3	.5433		13.8	.9844	63/64"		1.5000	1 1/2"	
.3281	21/64"		.5469	35/64"		1.0000	1"		1.5156	1 33/64"	
.3307		8.4	.5512		14.	1.0039		25.5	1.5157		38.5
.3320	Q		.5610		14.25	1.0156	1 1/64"		1.5312	1 17/32"	

Size	Decimal	Size	Decimal	Size	Decimal	Size	Decimal
8.5	.3346	9/16″	.5625	26.	1.0236	39.	1.5354
8.6	.3386	14.5	.5709	1 1/32″	1.0312	1 35/64″	1.5469
R	.3390	37/64″	.5781	26.5	1.0433	39.5	1.5551
8.7	.3425	14.75	.5807	1 3/64″	1.0469	1 9/16″	1.5625
11/32″	.3438	15.	.5906	1 1/16″	1.0625	40.	1.5748
8.75	.3445	19/32″	.5938	27.	1.0630	1 37/64″	1.5781
8.8	.3465	15.25	.6004	1 5/64″	1.0781	1 19/32″	1.5938
S	.3480	39/64″	.6094	27.5	1.0827	40.5	1.5945
8.9	.3504	15.5	.6102	1 3/32″	1.0938	1 39/64″	1.6094
9.	.3543	15.75	.6201	28.	1.1024	41.	1.6142
T	.3580	5/8″	.6250	1 7/64″	1.1094	1 5/8″	1.6250
9.1	.3583	16.	.6299	28.5	1.1220	41.5	1.6339
23/64″	.3594	16.25	.6398	1 1/8″	1.1250	1 41/64″	1.6406
9.2	.3622	41/64″	.6406	1 9/64″	1.1406	42.	1.6535
9.25	.3642	16.5	.6496	29.	1.1417	1 21/32″	1.6562
9.3	.3661	21/32″	.6562	1 5/32″	1.1562	1 43/64″	1.6719
U	.3680	16.75	.6598	29.5	1.1614	42.5	1.6732
9.4	.3701	17.	.6693	1 11/64″	1.1719	1 11/16″	1.6875
9.5	.3740	43/64″	.6719	30.	1.1811	43.	1.6929
3/8″	.3750	17.25	.6791	1 3/16″	1.1875	1 45/64″	1.7031

(Continued on following page)

(Continued from preceding page)

DECIMAL EQUIVALENTS (*Continued*)
Fractional, Wire, Letter and Metric Sizes

Decimal	Fract. Wire Letter	mm.	Decimal	Fract. Wire Letter	mm.	Decimal	Fract. Wire Letter	mm.	Decimal	Fract. Wire Letter	mm.
.3770	V		.6875	11/16"		1.2008		30.5	1.7126		43.5
.3780		9.6	.6890		17.5	1.2031	1 13/64"		1.7188	1 23/32"	
.3819		9.7	.7031	45/64"		1.2188	1 7/32"		1.7323		44.
.3839		9.75	.7087		18.	1.2205		31.	1.7344	1 47/64"	
.3858		9.8	.7188	23/32"		1.2344	1 15/64"		1.7500	1 3/4"	
.3860	W		.7283		18.5	1.2402		31.5	1.7520		44.5
.3898		9.9	.7344	47/64"		1.2500	1 1/4"		1.7656	1 49/64"	
.3906	25/64"		.7480		19.	1.2598		32.	1.7717		45.
.3937		10.	.7500	3/4"		1.2656	1 17/64"		1.7812	1 25/32"	
.3970	X		.7656	49/64"		1.2795		32.5	1.7913		45.5
.4016		10.2	.7677		19.5	1.2812	1 9/32"		1.7969	1 51/64"	
.4040	Y		.7812	25/32"		1.2969	1 19/64"		1.8110		46.
.4062	13/32"		.7874		20.	1.2992		33.	1.8125	1 13/16"	
.4130	Z		.7969	51/64"		1.3125	1 5/16"		1.8281	1 53/64"	
.4134		10.5	.8071		20.5	1.3189		33.5	1.8307		46.5
.4219	27/64"		.8125	13/16"		1.3281	1 21/64"		1.8438	1 27/32"	
.4252		10.8	.8268		21.	1.3386		34.	1.8504		47.
.4331		11.	.8281	53/64"		1.3438	1 11/32"		1.8594	1 55/64"	
.4375	7/16"		.8438	27/32"		1.3583		34.5	1.8701		47.5
.4409		11.2									

Decimal	Fraction	mm
1.8750	$1\frac{7}{8}''$	
1.8898		48.
1.8906	$1\frac{57}{64}''$	
1.9062	$1\frac{29}{32}''$	
1.9094		48.5
1.9219	$1\frac{59}{64}''$	
1.9291		49.
1.9375	$1\frac{15}{16}''$	
1.9488		49.5
1.9531	$1\frac{61}{64}''$	
1.9685		50.
1.9688	$1\frac{31}{32}''$	
1.9844	$1\frac{63}{64}''$	
1.9882		50.5
2.0000	$2''$	
2.0079		51.
2.0156	$2\frac{1}{64}''$	
2.0276		51.5
2.0312	$2\frac{1}{32}''$	
2.0469	$2\frac{3}{64}''$	
2.0472		52.
2.0625	$2\frac{1}{16}''$	
2.0669		52.5
2.0781	$2\frac{5}{64}''$	

Decimal	Fraction	mm
2.1562	$2\frac{5}{32}''$	
2.1654		55.
2.1719	$2\frac{11}{64}''$	
2.1850		55.5
2.1875	$2\frac{3}{16}''$	
2.2031	$2\frac{13}{64}''$	
2.2047		56.
2.2188	$2\frac{7}{32}''$	
2.2244		56.5
2.2344	$2\frac{15}{64}''$	
2.2441		57.
2.2500	$2\frac{1}{4}''$	
2.2638		57.5
2.2656	$2\frac{17}{64}''$	
2.2812	$2\frac{9}{32}''$	
2.2835		58.
2.2969	$2\frac{19}{64}''$	
2.3031		58.5
2.3125	$2\frac{5}{16}''$	
2.3228		59.
2.3281	$2\frac{21}{64}''$	
2.3425		59.5
2.3438	$2\frac{11}{32}''$	
2.3594	$2\frac{23}{64}''$	

Decimal	Fraction	mm
2.4375	$2\frac{7}{16}''$	
2.4409		62.
2.4531	$2\frac{29}{64}''$	
2.4606		62.5
2.4688	$2\frac{15}{32}''$	
2.4803		63.
2.4844	$2\frac{31}{64}''$	
2.5000	$2\frac{1}{2}''$	
2.5156	$2\frac{33}{64}''$	
2.5197		64.
2.5312	$2\frac{17}{32}''$	
2.5394		64.5
2.5469	$2\frac{35}{64}''$	
2.5591		65.
2.5625	$2\frac{9}{16}''$	
2.5781	$2\frac{37}{64}''$	
2.5787		65.5
2.5938	$2\frac{19}{32}''$	
2.5984		66.
2.6094	$2\frac{39}{64}''$	
2.6181		66.5
2.6250	$2\frac{5}{8}''$	
2.6378		67.
2.6406	$2\frac{41}{64}''$	

Decimal	Fraction	mm
2.7188	$2\frac{23}{32}''$	
2.7344	$2\frac{47}{64}''$	
2.7362		69.5
2.7500	$2\frac{3}{4}''$	
2.7559		70.
2.7656	$2\frac{49}{64}''$	
2.7756		70.5
2.7812	$2\frac{25}{32}''$	
2.7953		71.
2.7969	$2\frac{51}{64}''$	
2.8125	$2\frac{13}{16}''$	
2.8150		71.5
2.8281	$2\frac{53}{64}''$	
2.8346		72.
2.8438	$2\frac{27}{32}''$	
2.8543		72.5
2.8594	$2\frac{55}{64}''$	
2.8740		73.
2.8750	$2\frac{7}{8}''$	
2.8906	$2\frac{57}{64}''$	
2.8937		73.5
2.9062	$2\frac{29}{32}''$	
2.9134		74.
2.9219	$2\frac{59}{64}''$	

(Continued on following page)

(Continued from preceding page)

DECIMAL EQUIVALENTS (Continued)
Fractional, Wire, Letter and Metric Sizes

Decimal	Fract. Wire Letter	mm.	Decimal	Fract. Wire Letter	mm.	Decimal	Fract. Wire Letter	mm.	Decimal	Fract. Wire Letter	mm.
2.0866		53.	2.3622		60.	2.6562	$2\frac{21}{32}''$		2.9331		74.5
2.0938	$2\frac{3}{32}''$		2.3750	$2\frac{3}{8}''$		2.6575		67.5	2.9375	$2\frac{15}{16}''$	
2.1063		53.5	2.3819		60.5	2.6719	$2\frac{43}{64}''$		2.9528		75.
2.1094	$2\frac{7}{64}''$		2.3906	$2\frac{25}{64}''$		2.6772		68.	2.9531	$2\frac{61}{64}''$	
2.1250	$2\frac{1}{8}''$		2.4016		61.	2.6875	$2\frac{11}{16}''$		2.9688	$2\frac{31}{32}''$	
2.1260		54.	2.4062	$2\frac{13}{32}''$		2.6968		68.5	2.9724		75.5
2.1406	$2\frac{9}{64}''$		2.4213		61.5	2.7031	$2\frac{45}{64}''$		2.9844	$2\frac{63}{64}''$	
2.1457		54.5	2.4219	$2\frac{27}{64}''$		2.7165		69.	2.9921		76.
...	...	...	...	...	...	...	...	...	3.0000	3	...

(Concluded)

Table 804

Dimensional Tolerances

For High Speed Steel

General Purpose, Two, Three, and Four Flute Drills

DRILL DIAMETER AT POINT

	#97 to #81 inclusive	Plus 0.0002 to Minus 0.0002
	0.15 mm to 0.33 mm inclusive	Plus 0.005 mm to Minus 0.005 mm
Over	#81 to 1/8" inclusive	Plus 0.0000 to Minus 0.0005
Over	0.33 mm to 3.18 mm inclusive	Plus 0.000 mm to Minus 0.013 mm
Over	1/8" to 1/4" inclusive	Plus 0.0000 to Minus 0.0007
Over	3.18 mm to 6.35 mm inclusive	Plus 0.000 mm to Minus 0.018 mm
Over	1/4" to 1/2" inclusive	Plus 0.0000 to Minus 0.0010
Over	6.35 mm to 12.70 mm inclusive	Plus 0.000 mm to Minus 0.025 mm
Over	1/2" to 1" inclusive	Plus 0.0000 to Minus 0.0012
Over	12.70 mm to 25.40 mm inclusive	Plus 0.000 mm to Minus 0.030 mm
Over	1" to 2" inclusive	Plus 0.0000 to Minus 0.0015
Over	25.40 mm to 50.80 mm inclusive	Plus 0.000 mm to Minus 0.038 mm
Over	2" to 3-1/2" inclusive	Plus 0.0000 to Minus 0.0020
Over	50.80 mm to 88.90 mm inclusive	Plus 0.000 mm to Minus 0.051 mm

SHANK DIAMETER (STRAIGHT SHANK DRILLS)

	#97 to #81 inclusive	Plus 0.0002 to Minus 0.0002
	0.15 mm to 0.33 mm inclusive	Plus 0.005 mm to Minus 0.005 mm
Over	#81 to 1/8" inclusive	Minus 0.0000 to Minus 0.0025
Over	0.33 mm to 3.18 mm inclusive	Minus 0.000 mm to Minus 0.064 mm
Over	1/8" to 1/4" inclusive	Minus 0.0005 to Minus 0.0030
Over	3.18 mm to 6.35 mm inclusive	Minus 0.013 mm to Minus 0.076 mm
Over	1/4" to 1/2" inclusive	Minus 0.0005 to to Minus 0.0045
Over	6.35 mm to 12.70 mm inclusive	Minus 0.013 mm to Minus 0.114 mm
Over	1/2" to 2" inclusive	Minus 0.0005 to Minus 0.0030
Over	12.70 mm to 50.80 mm inclusive	Minus 0.013 mm to Minus 0.076 mm

BACK TAPER

	#97 to #81 inclusive	None
	0.15 mm to 0.33 mm inclusive	None
Over	#81 to 1/8" inclusive	0.0000 to 0.0008 per unit of length
Over	0.33 mm to 3.18 mm inclusive	
Over	1/8" to 1/4" inclusive	0.0002 to 0.0008 per unit of length
Over	3.18 mm to 6.35 mm inclusive	

(Continued on following page)

Table 804

Dimensional Tolerances

For High Speed Steel

General Purpose, Two, Three, and Four Flute Drills

(Continued)

BACK TAPER

Over 1/4" to 1/2" inclusive Over 6.35 mm to 12.70 mm inclusive	0.0002 to 0.0009 per unit of length	
Over 1/2" to 1" inclusive Over 12.70 mm to 25.40 mm inclusive	0.0003 to 0.0011 per unit of length	
Over 1" to 3-1/2" inclusive Over 25.40 mm to 88.90 mm inclusive	0.0004 to 0.0015 per unit of length	

FLUTE LENGTH

#97 to #81 inclusive 0.15 mm to 0.33 mm inclusive	Plus 1/64 to Minus 1/64 Plus 0.4 mm to Minus 0.4 mm
Over #81 to 1/8" inclusive Over 0.33 mm to 3.18 mm inclusive	Plus 1/8 to Minus 1/16 Plus 3.2 mm to Minus 1.6 mm
Over 1/8" to 1/2" inclusive Over 3.18 mm to 12.70 mm inclusive	Plus 1/8 to Minus 1/8 Plus 3.2 mm to Minus 3.2 mm
Over 1/2" to 1" inclusive Over 12.70 mm to 25.40 mm inclusive	Plus 1/4 to Minus 1/8 Plus 6.4 mm to Minus 3.2 mm
Over 1" to 2" inclusive Over 25.40 mm to 50.80 mm inclusive	Plus 1/4 to Minus 1/4 Plus 6.4 mm to Minus 6.4 mm
Over 2" to 3-1/2" inclusive Over 50.80 mm to 88.90 mm inclusive	Plus 3/8 to Minus 3/8 Plus 8.5 mm to Minus 9.5 mm

OVERALL LENGTH

#97 to #81 inclusive 0.15 mm to 0.33 mm inclusive	Plus 1/32 to Minus 1/32 Plus 0.8 mm to Minus 0.8 mm
Over #81 to 1/8" inclusive Over 0.33 mm to 3.18 mm inclusive	Plus 1/8 to Minus 1/16 Plus 3.2 mm to Minus 1.6 mm
Over 1/8" to 1/2" inclusive Over 3.18 mm to 12.70 mm inclusive	Plus 1/8 to Minus 1/8 Plus 3.2 mm to Minus 3.2 mm
Over 1/2" to 1" inclusive Over 12.70 mm to 25.40 mm inclusive	Plus 1/4 to Minus 1/8 Plus 6.4 mm to Minus 3.2 mm
Over 1" to 2" inclusive Over 25.40 mm to 50.80 mm inclusive	Plus 1/4 to Minus 1/4 Plus 6.4 mm to Minus 6.4 mm
Over 2" to 3-1/2" inclusive Over 50.80 mm to 88.90 mm inclusive	Plus 3/8 to Minus 3/8 Plus 9.5 mm to Minus 9.5 mm

Metric dimensions and tolerances are translations of inch values.

Table 805

Element Tolerances

For High Speed Steel

General Purpose, Two Flute Twist Drills

Included Angle of Point

Drill Diameter Range	Included Angle	Tolerance
1/16" to 1/2" inclusive 1.59 mm to 12.70 mm inclusive	118° 118°	±5° ±5°
Over 1/2" to 1-1/2" inclusive Over 12.70 mm to 38.10 mm inclusive	118° 118°	±3o ±3°
Over 1-1/2" to 3-1/2" inclusive Over 38.10 mm to 88.90 mm inclusive	118° 118°	±2° ±2°

Lip Height

Drill Diameter Range	Tolerance (Total Indicator Variation)
1/16" to 1/8" inclusive 1.59 mm to 3.18 mm inclusive	0.0020 0.051 mm
Over 1/8" to 1/4" inclusive Over 3.18 mm to 6.35 mm inclusive	0.0030 0.076 mm
Over 1/4" to 1/2" inclusive Over 6.35 mm to 12.70 mm inclusive	0.0040 0.102 mm
Over 1/2" to 1" inclusive Over 12.70 mm to 25.40 mm inclusive	0.0050 0.127 mm
Over 1" to 3-1/2" inclusive Over 25.40 mm to 88.90 mm inclusive	0.0060 0.152 mm

95% of drills in any one lot to fall within above tolerance.
Method of Measurement: Rotate the drill in a V-Block against a back end stop. Measure the cutting lip height variation on a comparator, or with an indicator set at a location approximately 75% of the distance from the center to the periphery of the drill.

(Continued on Following page)

Table 805

Element Tolerances

For High Speed Steel

General Purpose, Two Flute Twist Drills

(Continued)

Centrality of Web	
Drill Diameter Range	**Tolerance (Total Indicator Variation)**
1/16" to 1/8" inclusive 1.59 mm to 3.18 mm inclusive	0.0030 0.076 mm
Over 1/8" to 1/4" inclusive Over 3.18 mm to 6.35 mm inclusive	0.0040 0.102 mm
Over 1/4" to 1/2" inclusive Over 6.35 mm to 12.70 mm inclusive	0.0050 0.127 mm
Over 1/2" to 1" inclusive Over 12.70 mm to 25.40 mm inclusive	0.0070 0.178 mm
Over 1" to 3-1/2" inclusive Over 25.40 mm to 88.90 mm inclusive	0.0100 0.254 mm

95% of drills in any one lot to fall within above tolerance.
Method of Measurement: Rotate the drill in a close fitting bushing. Record the difference in indicator readings of the web at the point as the drill is indexed 180°.

Flute Spacing		
	Tolerance	
Drill Diameter Range	**(tiv)**	**Actual Deviation**
1/16" to 1/8" inclusive 1.59 mm to 3.18 mm inclusive	0.0030 0.076 mm	0.0015 0.038 mm
Over 1/8" to 1/4" inclusive Over 3.18 mm to 6.35 mm inclusive	0.0060 0.152 mm	0.0030 0.076 mm
Over 1/4" to 1/2" inclusive Over 6.35 mm to 12.70 mm inclusive	0.0100 0.254 mm	0.0050 0.127 mm
Over 1/2" to 1" inclusive Over 12.70 mm to 25.40 mm inclusive	0.0140 0.356 mm	0.0070 0.178 mm
Over 1" to 3-1/2" inclusive Over 25.40 mm to 88.90 mm inclusive	0.0260 0.660 mm	0.0130 0.330 mm

95% of drills in any one lot to fall within above tolerance.
Method of Measurement: Place the drill in a V-Block against a back end stop, and rotate it against a radial finger stop. Take an indicator reading at the leading edge of the margin on the opposite flute. Repeat for the other flute and note the difference between the two readings. The deviation in flute spacing is equal to one-half the difference between the two readings.

Metric dimensions and tolerances are translations of inch values.

Runout Tolerances for High Speed Steel Two Flute Twist Drills
Drill Runout (Straightness)

Jobbers Lengths: For drills smaller than .03125″, runout tolerances to be to manufacturer's discretion. For drills .03125″ and larger, runout tolerances are to be calculated per the following formula:

$$\text{Maximum TIR} = \left(.0001316 \times \frac{\text{overall length}}{\text{drill diameter}}\right) + .00368''$$

Examples

Diameter	Maximum TIR
1/16″	.008″
1/8″	.007″
1/4″	.006″
3/8″	.005″
1/2″	.005″

Taper Lengths: For drills smaller than .03125″, runout tolerances to be to manufacturer's discretion. For drills .03125″ and larger, runout tolerances are to be calculated per the following formula:

$$\text{Maximum TIR} = \left(.000375 \times \frac{\text{overall length}}{\text{drill diameter}}\right) + .0027''$$

The maximum allowable value, however, is .020″ TIR.

Examples

Diameter	Maximum TIR
1/16″	.021″ (use .020″)
1/8″	.018″
1/4″	.012″
3/8″	.009″
1/2″	.009″
3/4″	.008″
1″	.007″
1½″	.006″

(Continued on following page)

Runout Tolerances for High Speed Steel Two Flute Twist Drills
Drill Runout (Straightness)

(Continued)

Screw Machine Lengths: For drills smaller than .0400″, runout tolerances to be to manufacturer's discretion. For drills .0400″ and larger, runout tolerances are to be calculated per the following formula:

$$\text{Maximum TIR} = \left(.0001316 \times \frac{\text{overall length}}{\text{drill diameter}}\right) + .00368″$$

The maximum allowable value, however, is .007″ TIR.

Examples

Diameter	Maximum TIR
1/16″	.007″
1/8″	.006″
1/4″	.005″
3/8″	.005″
1/2″	.005″
3/4″	.005″
1″	.004″
1 1/2″	.004″

6″ Aircraft Drills

Diameter Range	Maximum TIR
.0400″ – .0624″	.020″
.0625″ – .1249″	.015″
.1250″ – .2499″	.012″
.2500″ – .3749″	.010″
.3750″ – .4999″	.007″
.5000″	.005″

(Continued on following page)

Runout Tolerances for High Speed Steel Two Flute Twist Drills
Drill Runout (Straightness)

(Continued)

12″ Aircraft Drills

Diameter Range	Maximum TIR
.0400″−.0624″	.040″
.0625″−.1934″	.035″
.1935″−.5000″	.025″

Taper Shank

Diameter Range	Maximum TIR
.1250″−.3749″	.015″
.3750″−1.4999″	.010″
1.5000″ & larger	.007″

1. 95% of drills in any one lot to fall within above tolerance.

2. Method of Measurement:

   Straight shank drills are to be checked with 1″ of shank in a V-block with measurements taken on the margins at the point. If shank length is shorter than 1″, use the full length of shank.

   Taper shank drills are to be checked with the shank in a tapered V-block or a precision socket.

3. Calculated values are to be rounded to the nearer thousandth of an inch.

4. Metric sizes, .15 (.0059″) to 17.50 (.6890″).

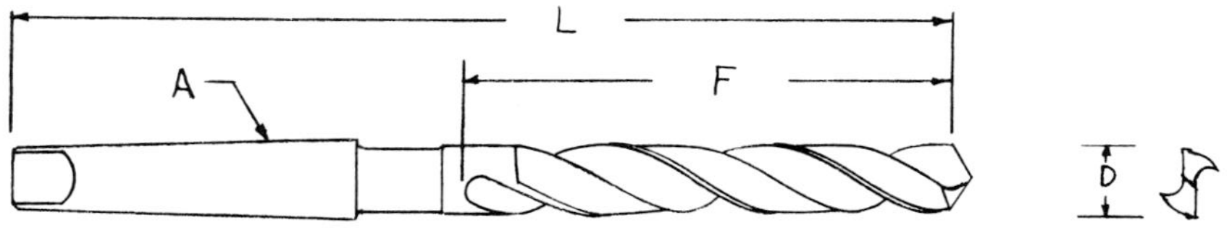

TWIST DRILLS
with Regular Taper Shank
High Speed Steel
Standard Sizes and Dimensions

Diameter of Drill				Regular Shank				
D		Decimal Inch Equivalent	Millimeter Equivalent	Morse Taper Number	Flute Length F		Overall Length L	
Fraction	Millimeters			A	Inch	Millimeters	Inch	Millimeters
	3.00	0.1181	3.000	1	1⅞	48	5⅛	130
⅛		0.1250	3.175	1	1⅞	48	5⅛	130
	3.20	0.1260	3.200	1	2⅛	54	5⅜	137
	3.50	0.1378	3.500	1	2⅛	54	5⅜	137
9/64		0.1406	3.571	1	2⅛	54	5⅜	137
	3.80	0.1496	3.800	1	2⅛	54	5⅜	137
5/32		0.1562	3.967	1	2⅛	54	5⅜	137
	4.00	0.1575	4.000	1	2½	64	5¾	146
	4.20	0.1654	4.200	1	2½	64	5¾	146
11/64		0.1719	4.366	1	2½	64	5¾	146
	4.50	0.1772	4.500	1	2½	64	5¾	146
3/16		0.1875	4.762	1	2½	64	5¾	146
	4.80	0.1890	4.800	1	2¾	70	6	152
	5.00	0.1969	5.000	1	2¾	70	6	152
13/64		0.2031	5.159	1	2¾	70	6	152
	5.20	0.2047	5.200	1	2¾	70	6	152
	5.50	0.2165	5.500	1	2¾	70	6	152

(Continued on following page)

TWIST DRILLS (continued)
with Regular Taper Shank
High Speed Steel
Standard Sizes and Dimensions

Diameter of Drill				Regular Shank				
D		Decimal Inch Equivalent	Millimeter Equivalent	Morse Taper Number	Flute Length F		Overall Length L	
Fraction	Millimeters			A	Inch	Millimeters	Inch	Millimeters
7/32		0.2183	5.558	1	2 3/4	70	6	152
	5.80	0.2223	5.800	1	2 7/8	73	6 1/8	156
15/64		0.2344	5.954	1	2 7/8	73	6 1/8	156
	6.00	0.2362	6.000	1	2 7/8	73	6 1/8	156
	6.20	0.2441	6.200	1	2 7/8	73	6 1/8	156
1/4		0.2500	6.350	1	2 7/8	73	6 1/8	156
	6.50	0.2559	6.500	1	3	76	6 1/4	159
17/64		0.2656	6.746	1	3	76	6 1/4	159
	6.80	0.2677	6.800	1	3	76	6 1/4	159
	7.00	0.2756	7.000	1	3	76	6 1/4	159
9/32		0.2812	7.142	1	3	76	6 1/4	159
	7.20	0.2835	7.200	1	3 1/8	79	6 3/8	162
	7.50	0.2953	7.500	1	3 1/8	79	6 3/8	162
19/64		0.2969	7.541	1	3 1/8	79	6 3/8	162
	7.80	0.3071	7.800	1	3 1/8	79	6 3/8	162
5/16		0.3125	7.938	1	3 1/8	79	6 3/8	162
	8.00	0.3150	8.000	1	3 1/4	83	6 1/2	165
	8.20	0.3228	8.200	1	3 1/4	83	6 1/2	165
21/64		0.3281	8.334	1	3 1/4	83	6 1/2	165
	8.50	0.3346	8.500	1	3 1/4	83	6 1/2	165
11/32		0.3438	8.733	1	3 1/4	83	6 1/2	165
	8.80	0.3465	8.800	1	3 1/2	89	6 3/4	171
	9.00	0.3543	9.000	1	3 1/2	89	6 3/4	171

(Continued on following page)

TWIST DRILLS (continued)
with Regular Taper Shank
High Speed Steel
Standard Sizes and Dimensions

Diameter of Drill				Morse Taper Number	Regular Shank			
D		Decimal Inch Equivalent	Millimeter Equivalent		Flute Length F		Overall Length L	
Fraction	Millimeters			A	Inch	Millimeters	Inch	Millimeters
23/64		0.3594	9.129	1	3 1/2	89	6 3/4	171
	9.20	0.3622	9.200	1	3 1/2	89	6 3/4	171
	9.50	0.3740	9.500	1	3 1/2	89	6 3/4	171
3/8		0.3750	9.525	1	3 1/2	89	6 3/4	171
	9.80	0.3858	9.800	1	3 5/8	92	7	178
25/64		0.3906	9.921	1	3 5/8	92	7	178
	10.00	0.3937	10.000	1	3 5/8	92	7	178
	10.20	0.4016	10.200	1	3 5/8	92	7	178
13/32		0.4062	10.320	1	3 5/8	92	7	178
	10.50	0.4134	10.500	1	3 7/8	98	7 1/4	184
27/64		0.4219	10.716	1	3 7/8	98	7 1/4	184
	10.80	0.4252	10.800	1	3 7/8	98	7 1/4	184
	11.00	0.4331	11.000	1	3 7/8	98	7 1/4	184
7/16		0.4375	11.112	1	3 7/8	98	7 1/4	184
	11.20	0.4409	11.200	1	4 1/8	105	7 1/2	190
	11.50	0.4528	11.500	1	4 1/8	105	7 1/2	190
29/64		0.4531	11.509	1	4 1/8	105	7 1/2	190
	11.80	0.4646	11.800	1	4 1/8	105	7 1/2	190
15/32		0.4688	11.900	1	4 1/8	105	7 1/2	190
	12.00	0.4724	12.000	2	4 3/8	111	8 1/4	210
	12.20	0.4803	12.200	2	4 3/8	111	8 1/4	210
31/64		0.4844	12.304	2	4 3/8	111	8 1/4	210
	12.50	0.4921	12.500	2	4 3/8	111	8 1/4	210

(Continued on following page)

TWIST DRILLS (continued)
with Regular Taper Shank
High Speed Steel
Standard Sizes and Dimensions

Diameter of Drill			Morse Taper Number	Regular Shank			
D		Decimal Inch Equivalent		Flute Length F		Overall Length L	
Fraction	Millimeters		Millimeter Equivalent	A	Inch	Millimeters	Inch	Millimeters

Fraction	Millimeters	Decimal Inch Equivalent	Millimeter Equivalent	A	Flute Inch	Flute mm	Overall Inch	Overall mm
1/2		0.5000	12.700	2	4 3/8	111	8 1/4	210
	12.80	0.5034	12.800	2	4 5/8	117	8 1/2	216
	13.00	0.5118	13.000	2	4 5/8	117	8 1/2	216
33/64		0.5156	13.096	2	4 5/8	117	8 1/2	216
	13.20	0.5197	13.200	2	4 5/8	117	8 1/2	216
17/32		0.5312	13.492	2	4 5/8	117	8 1/2	216
	13.50	0.5315	13.500	2	4 5/8	117	8 1/2	216
	13.80	0.5433	13.800	2	4 7/8	124	8 3/4	222
35/64		0.5469	13.891	2	4 7/8	124	8 3/4	222
	14.00	0.5572	14.000	2	4 7/8	124	8 3/4	222
	14.25	0.5610	14.250	2	4 7/8	124	8 3/4	222
9/16		0.5625	14.288	2	4 7/8	124	8 3/4	222
	14.50	0.5709	14.500	2	4 7/8	124	8 3/4	222
37/64		0.5781	14.684	2	4 7/8	124	8 3/4	222
	14.75	0.5807	14.750	2	4 7/8	124	8 3/4	222
	15.00	0.5906	15.000	2	4 7/8	124	8 3/4	222
19/32		0.5938	15.083	2	4 7/8	124	8 3/4	222
	15.25	0.6004	15.250	2	4 7/8	124	8 3/4	222
39/64		0.6094	15.479	2	4 7/8	124	8 3/4	222
	15.50	0.6102	15.500	2	4 7/8	124	8 3/4	222
	15.75	0.6201	15.750	2	4 7/8	124	8 3/4	222
5/8		0.6250	15.815	2	4 7/8	124	8 3/4	222
	16.00	0.6299	16.000	2	5 1/8	130	9	229

(Continued on following page)

TWIST DRILLS (continued)
with Regular Taper Shank
High Speed Steel
Standard Sizes and Dimensions

Diameter of Drill				Morse Taper Number	Regular Shank			
					Flute Length F		Overall Length L	
Fraction	D Millimeters	Decimal Inch Equivalent	Millimeter Equivalent	A	Inch	Millimeters	Inch	Millimeters
	16.25	0.6398	16.250	2	5⅛	130	9	229
41/64		0.6406	16.271	2	5⅛	130	9	229
	16.50	0.6496	16.500	2	5⅛	130	9	229
21/32		0.6562	16.667	2	5⅛	130	9	229
	16.75	0.6594	16.750	2	5⅜	137	9¼	235
	17.00	0.6693	17.000	2	5⅜	137	9¼	235
43/64		0.6719	17.066	2	5⅜	137	9¼	235
	17.25	0.6791	17.250	2	5⅜	137	9¼	235
11/16		0.6875	17.462	2	5⅜	137	9¼	235
	17.50	0.6880	17.500	2	5⅝	143	9½	241
45/64		0.7031	17.859	2	5⅝	143	9½	241
	18.00	0.7087	18.000	2	5⅝	143	9½	241
23/32		0.7188	18.258	2	5⅝	143	9½	241
	18.50	0.7283	18.500	2	5⅞	149	9¾	248
47/64		0.7344	18.654	2	5⅞	149	9¾	248
	19.00	0.7480	19.000	2	5⅞	149	9¾	248
3/4		0.7500	19.050	2	5⅞	149	9¾	248
49/64		0.7656	19.446	2	6	152	9⅞	251
	19.50	0.7677	19.500	2	6	152	9⅞	251
25/32		0.7812	19.843	2	6	152	9⅞	251
	20.00	0.7821	20.000	3	6⅛	156	10¾	273
51/64		0.7969	20.241	3	6⅛	156	10¾	273
	20.50	0.8071	20.500	3	6⅛	156	10¾	273

(Continued on following page)

TWIST DRILLS (continued)
with Regular Taper Shank
High Speed Steel
Standard Sizes and Dimensions

Diameter of Drill (D)				Morse Taper Number	Regular Shank			
					Flute Length (F)		Overall Length (L)	
Fraction	Millimeters	Decimal Inch Equivalent	Millimeter Equivalent	A	Inch	Millimeters	Inch	Millimeters
$\frac{13}{16}$		0.8125	20.638	3	$6\frac{1}{8}$	156	$10\frac{3}{4}$	273
	21.00	0.8268	21.000	3	$6\frac{1}{8}$	156	$10\frac{3}{4}$	273
$\frac{53}{64}$		0.8281	21.034	3	$6\frac{1}{8}$	156	$10\frac{3}{4}$	273
$\frac{27}{32}$		0.8438	21.433	3	$6\frac{1}{8}$	156	$10\frac{3}{4}$	273
	21.50	0.8465	21.500	3	$6\frac{1}{8}$	156	$10\frac{3}{4}$	273
$\frac{55}{64}$		0.8594	21.829	3	$6\frac{1}{8}$	156	$10\frac{3}{4}$	273
	22.00	0.8661	22.000	3	$6\frac{1}{8}$	156	$10\frac{3}{4}$	273
$\frac{7}{8}$		0.8750	22.225	3	$6\frac{1}{8}$	156	$10\frac{3}{4}$	273
	22.50	0.8858	22.500	3	$6\frac{1}{8}$	156	$10\frac{3}{4}$	273
$\frac{57}{64}$		0.8906	22.621	3	$6\frac{1}{8}$	156	$10\frac{3}{4}$	273
	23.00	0.9055	23.000	3	$6\frac{1}{8}$	156	$10\frac{3}{4}$	273
$\frac{29}{32}$		0.9062	23.017	3	$6\frac{1}{8}$	156	$10\frac{3}{4}$	273
$\frac{59}{64}$		0.9219	23.416	3	$6\frac{1}{8}$	156	$10\frac{3}{4}$	273
	23.50	0.9252	23.500	3	$6\frac{1}{8}$	156	$10\frac{3}{4}$	273
$\frac{15}{16}$		0.9375	23.862	3	$6\frac{1}{8}$	156	$10\frac{3}{4}$	273
	24.00	0.9449	24.000	3	$6\frac{3}{8}$	162	11	279
$\frac{61}{64}$		0.9531	24.209	3	$6\frac{3}{8}$	162	11	279
	24.50	0.9646	24.500	3	$6\frac{3}{8}$	162	11	279
$\frac{31}{32}$		0.9688	24.608	3	$6\frac{3}{8}$	162	11	279
	25.00	0.9843	25.000	3	$6\frac{3}{8}$	162	11	279
$\frac{63}{64}$		0.9844	25.004	3	$6\frac{3}{8}$	162	11	279
1		1.0000	25.400	3	$6\frac{3}{8}$	162	11	279
	25.50	1.0039	25.500	3	$6\frac{1}{2}$	165	$11\frac{1}{8}$	283

(Continued on following page)

TWIST DRILLS (continued)
with Regular Taper Shank
High Speed Steel
Standard Sizes and Dimensions

Diameter of Drill		Decimal Inch Equivalent	Millimeter Equivalent	Morse Taper Number	Regular Shank			
					Flute Length F		Overall Length L	
Fraction (D)	Millimeters			A	Inch	Millimeters	Inch	Millimeters
1 1/64		1.0156	25.796	3	6½	165	11⅛	283
	26.00	1.0236	26.000	3	6½	165	11⅛	283
1 1/32		1.0312	26.192	3	6½	165	11⅛	283
1 3/64		1.0433	26.500	3	6⅝	168	11¼	286
	26.50	1.0469	26.591	3	6⅝	168	11¼	286
1 1/16		1.0625	26.988	3	6⅝	168	11¼	286
	27.00	1.0630	27.000	3	6⅝	168	11¼	286
1 5/64		1.0781	27.384	4	6⅞	175	12½	318
	27.50	1.0827	27.500	4	6⅞	175	12½	318
1 3/32		1.0938	27.783	4	6⅞	175	12½	318
1 7/64		1.1024	28.000	4	7⅛	181	12¾	324
	28.00	1.1094	28.179	4	7⅛	181	12¾	324
1 1/8		1.1220	28.500	4	7⅛	181	12¾	324
	28.50	1.1250	28.575	4	7⅛	181	12¾	324
1 9/64		1.1406	28.971	4	7¼	184	12⅞	327
	29.00	1.1417	29.000	4	7¼	184	12⅞	327
1 5/32		1.1562	29.367	4	7¼	184	12⅞	327
	29.50	1.1614	29.500	4	7⅜	187	13	330
1 11/64		1.1719	29.766	4	7⅜	187	13	330
	30.00	1.1811	30.000	4	7⅜	187	13	330
1 3/16		1.1875	30.162	4	7⅜	187	13	330
	30.50	1.2008	30.500	4	7½	190	13⅛	333
1 13/64		1.2031	30.559	4	7½	190	13⅛	333

(Continued on following page)

TWIST DRILLS (continued) with Regular Taper Shank High Speed Steel Standard Sizes and Dimensions

Diameter of Drill				Regular Shank				
D		Decimal Inch Equivalent	Millimeter Equivalent	Morse Taper Number	Flute Length F		Overall Length L	
Fraction	Millimeters			A	Inch	Millimeters	Inch	Millimeters
1 7/32		1.2188	30.958	4	7 1/2	190	13 1/8	333
	31.00	1.2205	31.000	4	7 7/8	200	13 1/2	343
1 15/64		1.2344	31.354	4	7 7/8	200	13 1/2	343
	31.50	1.2402	31.500	4	7 7/8	200	13 1/2	343
1 1/4		1.2500	31.750	4	7 7/8	200	13 1/2	343
	32.00	1.2598	32.000	4	8 1/2	216	14 1/8	359
1 17/64		1.2656	32.146	4	8 1/2	216	14 1/8	359
	32.50	1.2795	32.500	4	8 1/2	216	14 1/8	359
1 9/32		1.2812	32.542	4	8 1/2	216	14 1/8	359
1 19/64		1.2969	32.941	4	8 5/8	219	14 1/4	362
	33.00	1.2992	33.000	4	8 5/8	219	14 1/4	362
1 5/16		1.3125	33.338	4	8 5/8	219	14 1/4	362
	33.50	1.3189	33.500	4	8 3/4	222	14 3/8	365
1 21/64		1.3281	33.734	4	8 3/4	222	14 3/8	365
	34.00	1.3386	34.000	4	8 3/4	222	14 3/8	365
1 11/32		1.3438	34.133	4	8 3/4	222	14 3/8	365
	34.50	1.3583	34.500	4	8 7/8	225	14 1/2	368
1 23/64		1.3594	34.529	4	8 7/8	225	14 1/2	368
1 3/8		1.3750	34.925	4	8 7/8	225	14 1/2	368
	35.00	1.3780	35.000	4	9	229	14 5/8	371
1 25/64		1.3906	35.321	4	9	229	14 5/8	371
	35.50	1.3976	35.500	4	9	229	14 5/8	371
1 13/32		1.4062	35.717	4	9	229	14 5/8	371

(Continued on following page)

TWIST DRILLS (continued)
with Regular Taper Shank
High Speed Steel
Standard Sizes and Dimensions

Diameter of Drill D				Morse Taper Number	Regular Shank			
					Flute Length F		Overall Length L	
Fraction	Millimeters	Decimal Inch Equivalent	Millimeter Equivalent	A	Inch	Millimeters	Inch	Millimeters
	36.00	1.4173	36.000	4	$9\frac{1}{8}$	232	$14\frac{3}{4}$	375
$1\frac{27}{64}$		1.4219	36.116	4	$9\frac{1}{8}$	232	$14\frac{3}{4}$	375
	36.50	1.4370	36.500	4	$9\frac{1}{8}$	232	$14\frac{3}{4}$	375
$1\frac{7}{16}$		1.4375	36.512	4	$9\frac{1}{8}$	232	$14\frac{3}{4}$	375
$1\frac{29}{64}$		1.4531	36.909	4	$9\frac{1}{4}$	235	$14\frac{7}{8}$	378
	37.00	1.4567	37.000	4	$9\frac{1}{4}$	235	$14\frac{7}{8}$	378
$1\frac{15}{32}$		1.4688	37.308	4	$9\frac{1}{4}$	235	$14\frac{7}{8}$	378
	37.50	1.4764	37.500	4	$9\frac{3}{8}$	238	15	381
$1\frac{31}{64}$		1.4844	37.704	4	$9\frac{3}{8}$	238	15	381
	38.00	1.4961	38.000	4	$9\frac{3}{8}$	238	15	381
$1\frac{1}{2}$		1.5000	38.100	4	$9\frac{3}{8}$	238	15	381
$1\frac{33}{64}$		1.5156	38.496	…	…	…	…	…
$1\frac{17}{32}$		1.5312	38.892	5	$9\frac{3}{8}$	238	$16\frac{3}{8}$	416
	39.00	1.5354	39.000	5	$9\frac{5}{8}$	244	$16\frac{5}{8}$	422
$1\frac{35}{64}$		1.5469	39.291	…	…	…	…	…
$1\frac{9}{16}$		1.5625	39.688	5	$9\frac{5}{8}$	244	$16\frac{5}{8}$	422
	40.00	1.5748	40.000	5	$9\frac{7}{8}$	251	$16\frac{7}{8}$	429
$1\frac{37}{64}$		1.5781	40.084	…	…	…	…	…
$1\frac{19}{32}$		1.5938	40.483	5	$9\frac{7}{8}$	251	$16\frac{7}{8}$	429
$1\frac{39}{64}$		1.6094	40.879	…	…	…	…	…
	41.00	1.6142	41.000	5	10	254	17	432
$1\frac{5}{8}$		1.6250	41.275	5	10	254	17	432
$1\frac{41}{64}$		1.6406	41.671	…	…	…	…	…

(Continued on following page)

TWIST DRILLS (continued)
with Regular Taper Shank
High Speed Steel
Standard Sizes and Dimensions

Diameter of Drill — D				Morse Taper Number A	Regular Shank — Flute Length F		Overall Length L	
Fraction	Millimeters	Decimal Inch Equivalent	Millimeter Equivalent		Inch	Millimeters	Inch	Millimeters
	42.00	1.6535	42.000	5	10 1/8	257	17 1/8	435
1 21/32		1.6562	42.067	5	10 1/8	257	17 1/8	435
1 43/64		1.6719	42.466					
1 11/16		1.6875	42.862	5	10 1/8	257	17 1/8	435
	43.00	1.6929	43.000	5	10 1/8	257	17 1/8	435
1 45/64		1.7031	43.259					
1 23/32		1.7188	43.658	5	10 1/8	257	17 1/8	435
	44.00	1.7323	44.000	5	10 1/8	257	17 1/8	435
1 47/64		1.7344	44.054					
1 3/4		1.7500	44.450	5	10 1/8	257	17 1/8	435
	45.00	1.7717	45.000	5	10 1/8	257	17 1/8	435
1 25/32		1.7812	45.242	5	10 1/8	257	17 1/8	435
	46.00	1.8110	46.000	5	10 1/8	257	17 1/8	435
1 13/16		1.8125	46.038	5	10 1/8	257	17 1/8	435
1 27/32		1.8438	46.833	5	10 1/8	257	17 1/8	435
	47.00	1.8504	47.000	5	10 3/8	264	17 3/8	441
1 7/8		1.8750	47.625	5	10 3/8	264	17 3/8	441
	48.00	1.8898	48.000	5	10 3/8	264	17 3/8	441
1 29/32		1.9062	48.417	5	10 3/8	264	17 3/8	441
	49.00	1.9291	49.000	5	10 3/8	264	17 3/8	441
1 15/16		1.9375	49.212	5	10 3/8	264	17 3/8	441
	50.00	1.9625	50.000	5	10 3/8	264	17 3/8	441
1 31/32		1.9688	50.008	5	10 3/8	264	17 3/8	441

(Continued on following page)

TWIST DRILLS (continued)
with Regular Taper Shank
High Speed Steel
Standard Sizes and Dimensions

Diameter of Drill (D)				Morse Taper Number	Regular Shank			
					Flute Length (F)		Overall Length (L)	
Fraction	Millimeters	Decimal Inch Equivalent	Millimeter Equivalent	A	Inch	Millimeters	Inch	Millimeters
2		2.0000	50.800	5	10 3/8	264	17 3/8	441
	51.00	2.0079	51.000	5	10 3/8	264	17 3/8	441
2 1/32		2.0312	51.592	5	10 3/8	264	17 3/8	441
	52.00	2.0472	52.000	5	10 1/4	260	17 3/8	441
2 1/16		2.0625	52.388	5	10 1/4	260	17 3/8	441
	53.00	2.0866	53.000	5	10 1/4	260	17 3/8	441
2 3/32		2.0938	53.183	5	10 1/4	260	17 3/8	441
2 1/8		2.1250	53.975	5	10 1/4	260	17 3/8	441
	54.00	2.1260	54.000	5	10 1/4	260	17 3/8	441
2 5/32		2.1562	54.767	5	10 1/4	260	17 3/8	441
	55.00	2.1654	55.000	5	10 1/4	260	17 3/8	441
2 3/16		2.1875	55.512	5	10 1/4	260	17 3/8	441
	56.00	2.2000	56.000	5	10 1/8	257	17 3/8	441
2 7/32		2.2188	56.358	5	10 1/8	257	17 3/8	441
	57.00	2.2441	57.000	5	10 1/8	257	17 3/8	441
2 1/4		2.2500	57.150	5	10 1/8	257	17 3/8	441
	58.00	2.2835	58.000	5	10 1/8	257	17 3/8	441
2 5/16		2.3125	58.738	5	10 1/8	257	17 3/8	441
	59.00	2.3228	59.000	5	10 1/8	257	17 3/8	441
	60.00	2.3622	60.000	5	10 1/8	257	17 3/8	441
2 3/8		2.3750	60.325	5	10 1/8	257	17 3/8	441
	61.00	2.4016	61.000	5	11 1/4	286	18 3/4	476
2 7/16		2.4375	61.912	5	11 1/4	286	18 3/4	476

(Continued on following page)

TWIST DRILLS (continued)
with Regular Taper Shank
High Speed Steel
Standard Sizes and Dimensions

Diameter of Drill				Regular Shank				
D		Decimal Inch Equivalent	Millimeter Equivalent	Morse Taper Number	Flute Length F		Overall Length L	
Fraction	Millimeters			A	Inch	Millimeters	Inch	Millimeters
	62.00	2.4409	62.000	5	11¼	286	18¾	476
	63.00	2.4803	63.000	5	11¼	286	18¾	476
2½		2.5000	63.500	5	11¼	286	18¾	476
	64.00	2.5197	64.000	5	11⅞	302	19½	495
	65.00	2.5591	65.000	5	11⅞	302	19½	495
2 9/16		2.5625	65.088	5	11⅞	302	19½	495
	66.00	2.5984	66.000	5	11⅞	302	19½	495
2⅝		2.6250	66.675	5	11⅞	302	19½	495
	67.00	2.6378	67.000	5	12¾	324	20⅜	518
	68.00	2.6772	68.000	5	12¾	324	20⅜	518
2 11/16		2.6875	68.262	5	12¾	324	20⅜	518
	69.00	2.7165	69.000	5	12¾	324	20⅜	518
2¾		2.7500	69.850	5	12¾	324	20⅜	518
	70.00	2.7559	70.000	5	13⅜	340	21⅛	537
	71.00	2.7953	71.000	5	13⅜	340	21⅛	537
2 13/16		2.8125	71.438	5	13⅜	340	21⅛	537
	72.00	2.8346	72.000	5	13⅜	340	21⅛	537
	73.00	2.8740	73.000	5	13⅜	340	21⅛	537
2⅞		2.8750	73.025	5	13⅜	340	21⅛	537
	74.00	2.9134	74.000	5	14	356	21¾	552
2 15/16		2.9375	74.612	5	14	356	21¾	552
	75.00	2.9528	75.000	5	14	356	21¾	552
	76.00	2.9921	76.000	5	14	356	21¾	552

(Concluded on following page)

TWIST DRILLS (concluded)
with Regular Taper Shank
High Speed Steel
Standard Sizes and Dimensions

Diameter of Drill				Morse Taper Number	Regular Shank			
D		Decimal Inch Equivalent	Millimeter Equivalent	A	Flute Length F		Overall Length L	
Fraction	Millimeters				Inch	Millimeters	Inch	Millimeters
3		3.0000	76.200	5	14	356	21¾	552
	77.00	3.0315	77.000	6	14⅝	371	24½	622
	78.00	3.0709	78.000	6	14⅝	371	24½	622
3⅛		3.1250	79.375	6	14⅝	371	24½	622
3¼		3.2500	82.550	6	15½	394	25½	648
3½		3.5000	88.900	...	...	...	...	...

See T-804 for Dimensional Tolerances
See T-805 for Element Tolerances

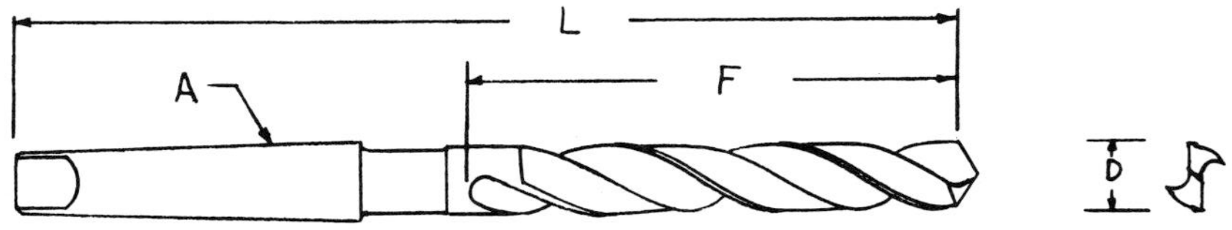

TWIST DRILLS
with Taper Shank Larger than Regular High Speed Steel—Fraction and Metric Standard Sizes and Dimensions

Diameter of Drill (D)				Morse Taper Number	Regular Shank			
					Flute Length (F)		Overall Length (L)	
Fraction	Millimeters	Decimal Inch Equivalent	Millimeter Equivalent	A	Inch	Millimeters	Inch	Millimeters
3/8		0.3750	9.525	2	3½	89	7⅜	187
25/64		0.3906	9.921	2	3⅝	92	7½	190
13/32		0.4062	10.317	2	3⅝	92	7½	190
27/64		0.4219	10.716	2	3⅞	98	7¾	197
7/16		0.4375	11.112	2	3⅞	98	7¾	197
29/64		0.4531	11.509	2	4⅛	105	8	203
15/32		0.4688	11.908	2	4⅛	105	8	203
41/64		0.6406	16.271	3	5⅛	130	9¾	248
21/32		0.6562	16.667	3	5⅛	130	9¾	248
43/64		0.6719	17.066	3	5⅜	137	10	254
11/16		0.6875	17.462	3	5⅜	137	10	254
45/64		0.7031	17.859	3	5⅝	143	10¼	260
23/32		0.7188	18.258	3	5⅝	143	10¼	260
47/64		0.7344	18.654	3	5⅞	149	10½	267
3/4		0.7500	19.050	3	5⅞	149	10½	267
49/64		0.7656	19.446	3	6	152	10⅝	270
25/32		0.7812	19.843	3	6	152	10⅝	270

(Concluded on following page)

TWIST DRILLS (concluded)
with Taper Shank Larger than Regular
High Speed Steel—Fraction and Metric
Standard Sizes and Dimensions

Diameter of Drill				Morse Taper Number	Regular Shank			
D					Flute Length		Overall Length	
					F		L	
Fraction	Millimeters	Decimal Inch Equivalent	Millimeter Equivalent	A	Inch	Millimeters	Inch	Millimeters
1		1.0000	25.400	4	6⅜	162	12	305
1¹⁄₃₂		1.0312	26.192	4	6½	165	12⅛	308
1¹⁄₁₆		1.0625	26.988	4	6⅝	168	12¼	311

See T-804 for Dimensional Tolerances
See T-805 for Element Tolerances

Twist Drills
Taper Shank, Heavy Duty
High Speed Steel
Fractional Sizes

Standard Sizes and Dimensions

Diameter Inches	Length Flute Inches	Length Overall Inches	No. of Taper Shank	Diameter Inches	Length Flute Inches	Length Overall Inches	No. of Taper Shank
$1/4$	$2\frac{7}{8}$	$6\frac{1}{8}$	1	$23/32$	$5\frac{5}{8}$	$9\frac{1}{2}$	2
$17/64$	3	$6\frac{1}{4}$	1	$47/64$	$5\frac{7}{8}$	$9\frac{3}{4}$	2
$9/22$	3	$6\frac{1}{4}$	1	$3/4$	$5\frac{7}{8}$	$9\frac{3}{4}$	2
$19/34$	$3\frac{1}{8}$	$6\frac{3}{8}$	1	$49/64$	6	$9\frac{7}{8}$	2
$5/16$	$3\frac{1}{8}$	$6\frac{3}{8}$	1	$25/32$	6	$9\frac{7}{8}$	2
$31/64$	$3\frac{1}{4}$	$6\frac{1}{2}$	1	$13/16$	$6\frac{1}{8}$	$10\frac{3}{4}$	3
$11/32$	$3\frac{1}{4}$	$6\frac{1}{2}$	1	$27/32$	$6\frac{1}{8}$	$10\frac{3}{4}$	3
$23/64$	$3\frac{1}{2}$	$6\frac{3}{4}$	1	$7/8$	$6\frac{1}{8}$	$10\frac{3}{4}$	3
$3/8$	$3\frac{1}{2}$	$6\frac{3}{4}$	1	$29/32$	$6\frac{1}{4}$	$10\frac{3}{4}$	3
$25/64$	$3\frac{5}{8}$	7	1	$15/16$	$6\frac{1}{4}$	$10\frac{3}{4}$	3
$13/32$	$3\frac{5}{8}$	7	1	$31/32$	$6\frac{3}{8}$	11	3
$27/64$	$3\frac{7}{8}$	$7\frac{1}{4}$	1	1	$6\frac{3}{8}$	11	3
$7/16$	$3\frac{7}{8}$	$7\frac{1}{4}$	1	$1\frac{1}{16}$	$6\frac{5}{8}$	$11\frac{1}{4}$	3
$29/64$	$4\frac{1}{8}$	$7\frac{1}{2}$	1	$1\frac{1}{8}$	$7\frac{1}{8}$	$12\frac{3}{4}$	4
$15/32$	$4\frac{1}{8}$	$7\frac{1}{2}$	1	$1\frac{3}{16}$	$7\frac{3}{8}$	13	4
$31/64$	$4\frac{3}{8}$	$8\frac{1}{4}$	2	$1\frac{1}{4}$	$7\frac{7}{8}$	$13\frac{1}{2}$	4
$1/2$	$4\frac{3}{8}$	$8\frac{1}{4}$	2				
$33/64$	$4\frac{5}{8}$	$8\frac{1}{2}$	2				
$17/32$	$4\frac{5}{8}$	$8\frac{1}{2}$	2				
$35/64$	$4\frac{7}{8}$	$8\frac{3}{4}$	2				
$9/16$	$4\frac{7}{8}$	$8\frac{3}{4}$	2				
$37/64$	$4\frac{7}{8}$	$8\frac{3}{4}$	2				
$19/32$	$4\frac{7}{8}$	$8\frac{3}{4}$	2				
$39/64$	$4\frac{7}{8}$	$8\frac{3}{4}$	2				
$5/8$	$4\frac{7}{8}$	$8\frac{3}{4}$	2				
$41/64$	$5\frac{1}{8}$	9	2				
$21/32$	$5\frac{1}{8}$	9	2				
$43/64$	$5\frac{3}{8}$	$9\frac{1}{4}$	2				
$11/16$	$5\frac{3}{8}$	$9\frac{1}{4}$	2				
$45/64$	$5\frac{5}{8}$	$9\frac{1}{2}$	2				

See T-804 for Dimensional Tolerances

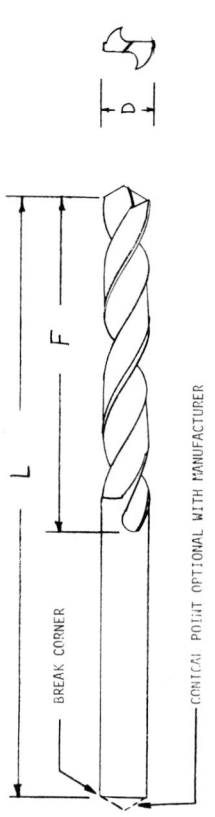

BREAK CORNER

CONICAL POINT OPTIONAL WITH MANUFACTURER

TWIST DRILLS
Straight Shank Taper Length
Thru ½ inch (12.7 mm) Diameter
Fraction, Wire, Letter, Metric
High Speed Steel—Standard Sizes and Dimensions

Fraction	Diameter of Drill				Flute Length F		Overall Length L	
	No.	mm	Decimal Inch Equivalent	Millimeter Equivalent	Inch	mm	Inch	mm
	60	1.00	0.0394	1.000	1⅛	29	2¼	57
	59		0.0400	1.016	1⅛	29	2¼	57
	58	1.05	0.0410	1.041	1⅛	29	2¼	57
	57		0.0413	1.050	1⅛	29	2¼	57
			0.0420	1.067	1⅛	29	2¼	57
			0.0430	1.092	1⅛	29	2¼	57
		1.10	0.0433	1.100	1⅛	29	2¼	57
		1.15	0.0453	1.150	1⅛	29	2¼	57
	56		0.0465	1.181	1⅛	29	2¼	57
3/64			0.0469	1.191	1⅛	29	2¼	57
		1.20	0.0472	1.200	1¾	44	3	76
		1.25	0.0492	1.250	1¾	44	3	76
		1.30	0.0512	1.300	1¾	44	3	76
	55		0.0520	1.321	1¾	44	3	76
		1.35	0.0531	1.350	1¾	44	3	76
	54		0.0550	1.397	1¾	44	3	76
		1.40	0.0551	1.400	1¾	44	3	76

(Continued on following page)

TWIST DRILLS (continued)
Straight Shank Taper Length
Thru ½ inch (12.7 mm) Diameter
Fraction, Wire, Letter, Metric
High Speed Steel—Standard Sizes and Dimensions

Diameter of Drill					Flute Length F		Overall Length L	
Fraction	No. (D)	mm	Decimal Inch Equivalent	Millimeter Equivalent	Inch	mm	Inch	mm
		1.45	0.0571	1.450	1¾	44	3	76
		1.50	0.0591	1.500	1¾	44	3	76
	53		0.0595	1.511	1¾	44	3	76
		1.55	0.0610	1.550	1¾	44	3	76
1/16			0.0625	1.588	1¾	44	3	76
		1.60	0.0630	1.600	2	51	3¾	95
	52		0.0635	1.613	2	51	3¾	95
		1.65	0.0650	1.650	2	51	3¾	95
		1.70	0.0669	1.700	2	51	3¾	95
	51		0.0670	1.702	2	51	3¾	95
		1.75	0.0689	1.750	2	51	3¾	95
	50		0.0700	1.778	2	51	3¾	95
		1.80	0.0709	1.800	2	51	3¾	95
		1.85	0.0728	1.850	2	51	3¾	95
	49		0.0730	1.854	2	51	3¾	95
		1.90	0.0748	1.900	2	51	3¾	95
	48		0.0760	1.930	2	51	3¾	95
		1.95	0.0768	1.950	2	51	3¾	95
5/64			0.0781	1.984	2	51	3¾	95
	47		0.0785	1.994	2¼	57	4¼	108
		2.00	0.0787	2.000	2¼	57	4¼	108
		2.05	0.0807	2.050	2¼	57	4¼	108
	46		0.0810	2.057	2¼	57	4¼	108

(Continued on following page)

TWIST DRILLS (continued)
Straight Shank Taper Length
Thru ½ inch (12.7 mm) Diameter
Fraction, Wire, Letter, Metric
High Speed Steel—Standard Sizes and Dimensions

Diameter of Drill (D)					Flute Length (F)		Overall Length (L)	
Fraction	No.	mm	Decimal Inch Equivalent	Millimeter Equivalent	Inch	mm	Inch	mm
	45		0.0820	2.083	2¼	57	4¼	108
		2.10	0.0827	2.100	2¼	57	4¼	108
		2.15	0.0846	2.150	2¼	57	4¼	108
	44		0.0860	2.184	2¼	57	4¼	108
		2.20	0.0866	2.200	2¼	57	4¼	108
		2.25	0.0886	2.250	2¼	57	4¼	108
	43		0.0890	2.261	2¼	57	4¼	108
		2.30	0.0906	2.300	2¼	57	4¼	108
		2.35	0.0925	2.350	2¼	57	4¼	108
	42		0.0935	2.375	2¼	57	4¼	108
3/32			0.0938	2.383	2¼	57	4¼	108
		2.40	0.0945	2.400	2½	64	4⅝	117
	41		0.0960	2.438	2½	64	4⅝	117
		2.45	0.0965	2.450	2½	64	4⅝	117
	40		0.0980	2.489	2½	64	4⅝	117
		2.50	0.0984	2.500	2½	64	4⅝	117
	39		0.0995	2.527	2½	64	4⅝	117
	38		0.1015	2.578	2½	64	4⅝	117
		2.60	0.1024	2.600	2½	64	4⅝	117
	37		0.1040	2.642	2½	64	4⅝	117
		2.70	0.1063	2.700	2½	64	4⅝	117
	36		0.1065	2.705	2½	64	4⅝	117
7/64			0.1094	2.779	2½	64	4⅝	117

(Continued on following page)

TWIST DRILLS (continued)
Straight Shank Taper Length
Thru ½ inch (12.7 mm) Diameter
Fraction, Wire, Letter, Metric
High Speed Steel—Standard Sizes and Dimensions

Diameter of Drill					Flute Length		Overall Length	
	D		Decimal Inch Equivalent	Millimeter Equivalent	F		L	
Fraction	No.	mm			Inch	mm	Inch	mm
	35		0.1100	2.794	2¾	70	5⅛	130
		2.80	0.1102	2.800	2¾	70	5⅛	130
	34		0.1110	2.819	2¾	70	5⅛	130
	33		0.1130	2.870	2¾	70	5⅛	130
		2.90	0.1142	2.900	2¾	70	5⅛	130
	32		0.1160	2.946	2¾	70	5⅛	130
		3.00	0.1181	3.000	2¾	70	5⅛	130
	31		0.1200	3.048	2¾	70	5⅛	130
		3.10	0.1220	3.100	2¾	70	5⅛	130
⅛			0.1250	3.175	2¾	70	5⅛	130
		3.20	0.1260	3.200	3	76	5⅜	137
	30		0.1285	3.264	3	76	5⅜	137
		3.30	0.1299	3.300	3	76	5⅜	137
		3.40	0.1339	3.400	3	76	5⅜	137
	29		0.1360	3.464	3	76	5⅜	137
		3.50	0.1378	3.500	3	76	5⅜	137
	28		0.1405	3.569	3	76	5⅜	137
9/64			0.1406	3.571	3	76	5⅜	137
		3.60	0.1417	3.600	3	76	5⅜	137
	27		0.1440	3.658	3	76	5⅜	137
		3.70	0.1457	3.700	3	76	5⅜	137
	26		0.1470	3.734	3	76	5⅜	137
	25		0.1495	3.797	3	76	5⅜	137

(Continued on following page)

TWIST DRILLS (continued)
Straight Shank Taper Length
Thru ½ inch (12.7 mm) Diameter
Fraction, Wire, Letter, Metric
High Speed Steel—Standard Sizes and Dimensions

Diameter of Drill			Decimal Inch Equivalent	Millimeter Equivalent	Flute Length F		Overall Length L	
Fraction	D No.	mm			Inch	mm	Inch	mm
	24	3.80	0.1496	3.800	3	76	5⅜	137
			0.1520	3.861	3	76	5⅜	137
	23	3.90	0.1535	3.900	3	76	5⅜	137
			0.1540	3.912	3	76	5⅜	137
5/32			0.1562	3.967	3	76	5⅜	137
	22		0.1570	3.988	3⅜	86	5¾	146
		4.00	0.1575	4.000	3⅜	86	5¾	146
	21		0.1590	4.039	3⅜	86	5¾	146
	20		0.1610	4.089	3⅜	86	5¾	146
		4.10	0.1614	4.100	3⅜	86	5¾	146
		4.20	0.1654	4.200	3⅜	86	5¾	146
	19		0.1660	4.216	3⅜	86	5¾	146
		4.30	0.1693	4.300	3⅜	86	5¾	146
	18		0.1695	4.305	3⅜	86	5¾	146
11/64			0.1719	4.366	3⅜	86	5¾	146
	17		0.1730	4.394	3⅜	86	5¾	146
		4.40	0.1732	4.400	3⅜	86	5¾	146
	16		0.1770	4.496	3⅜	86	5¾	146
		4.50	0.1772	4.500	3⅜	86	5¾	146
	15		0.1800	4.572	3⅜	86	5¾	146
		4.60	0.1811	4.600	3⅜	86	5¾	146
	14		0.1820	4.623	3⅜	86	5¾	146
	13	4.70	0.1850	4.700	3⅜	86	5¾	146

(Continued on following page)

TWIST DRILLS (continued)
Straight Shank Taper Length
Thru ½ inch (12.7 mm) Diameter
Fraction, Wire, Letter, Metric
High Speed Steel—Standard Sizes and Dimensions

Diameter of Drill					Flute Length F		Overall Length L	
Fraction	No. (D)	mm	Decimal Inch Equivalent	Millimeter Equivalent	Inch	mm	Inch	mm
3/16	12		0.1875	4.762	3⅜	86	5¾	146
	11	4.80	0.1890	4.800	3⅝	92	6	152
	11		0.1910	4.851	3⅝	92	6	152
		4.90	0.1929	4.900	3⅝	92	6	152
	10		0.1935	4.915	3⅝	92	6	152
	9		0.1960	4.978	3⅝	92	6	152
		5.00	0.1969	5.000	3⅝	92	6	152
	8		0.1990	5.054	3⅝	92	6	152
		5.10	0.2008	5.100	3⅝	92	6	152
	7		0.2010	5.105	3⅝	92	6	152
13/64			0.2031	5.159	3⅝	92	6	152
	6		0.2040	5.182	3⅝	92	6	152
		5.20	0.2047	5.200	3⅝	92	6	152
	5		0.2055	5.220	3⅝	92	6	152
		5.30	0.2087	5.300	3⅝	92	6	152
	4		0.2090	5.309	3⅝	92	6	152
		5.40	0.2126	5.400	3⅝	92	6	152
	3		0.2130	5.410	3⅝	92	6	152
		5.50	0.2165	5.500	3⅝	92	6	152
7/32			0.2188	5.558	3⅝	92	6	152
		5.60	0.2205	5.600	3¾	95	6⅛	156
	2		0.2210	5.613	3¾	95	6⅛	156
		5.70	0.2244	5.700	3¾	95	6⅛	156

(Continued on following page)

TWIST DRILLS (continued)
Straight Shank Taper Length
Thru ½ inch (12.7 mm) Diameter
Fraction, Wire, Letter, Metric
High Speed Steel—Standard Sizes and Dimensions

Diameter of Drill					Flute Length		Overall Length	
	D				F		L	
Fraction	No.	mm	Decimal Inch Equivalent	Millimeter Equivalent	Inch	mm	Inch	mm
	1		0.2280	5.791	3¾	95	6⅛	156
		5.80	0.2283	5.800	3¾	95	6⅛	156
		5.90	0.2323	5.900	3¾	95	6⅛	156
15/64			0.2344	5.954	3¾	95	6⅛	156
		6.00	0.2362	6.000	3¾	95	6⅛	156
		6.10	0.2402	6.100	3¾	95	6⅛	156
		6.20	0.2441	6.200	3¾	95	6⅛	156
		6.30	0.2480	6.300	3¾	95	6⅛	156
¼			0.2500	6.350	3¾	95	6⅛	156
		6.40	0.2520	6.400	3⅞	98	6¼	159
		6.50	0.2559	6.500	3⅞	98	6¼	159
17/64			0.2656	6.746	3⅞	98	6¼	159
		6.80	0.2677	6.800	3⅞	98	6¼	159
		7.00	0.2756	7.000	3⅞	98	6¼	159
9/32			0.2812	7.142	3⅞	98	6¼	159
		7.20	0.2835	7.200	4	102	6⅜	162
		7.50	0.2953	7.500	4	102	6⅜	162
19/64			0.2969	7.541	4	102	6⅜	162
		7.80	0.3071	7.800	4	102	6⅜	162
5/16			0.3125	7.938	4	102	6⅜	162
		8.00	0.3150	8.000	4⅛	105	6½	165
		8.20	0.3228	8.200	4⅛	105	6½	165
21/64			0.3281	8.334	4⅛	105	6½	165

(Continued on following page)

TWIST DRILLS (continued)
Straight Shank Taper Length
Thru ½ inch (12.7 mm) Diameter
Fraction, Wire, Letter, Metric
High Speed Steel—Standard Sizes and Dimensions

\	Diameter of Drill				Flute Length F		Overall Length L	
D No.	Fraction	mm	Decimal Inch Equivalent	Millimeter Equivalent	Inch	mm	Inch	mm
		8.50	0.3346	8.500	4⅛	105	6½	165
	11/32		0.3438	8.733	4⅛	105	6½	165
		8.80	0.3465	8.800	4¼	108	6¾	171
		9.00	0.3543	9.000	4¼	108	6¾	171
	23/64		0.3594	9.129	4¼	108	6¾	171
		9.20	0.3622	9.200	4¼	108	6¾	171
		9.50	0.3740	9.500	4¼	108	6¾	171
	3/8		0.3750	9.525	4¼	108	6¾	171
		9.80	0.3858	9.800	4⅜	111	7	178
	25/64		0.3906	9.921	4⅜	111	7	178
		10.00	0.3937	10.000	4⅜	111	7	178
		10.20	0.4016	10.200	4⅜	111	7	178
	13/32		0.4062	10.317	4⅜	111	7	178
		10.50	0.4134	10.500	4⅝	117	7¼	184
	27/64		0.4219	10.716	4⅝	117	7¼	184
		10.80	0.4252	10.800	4⅝	117	7¼	184
		11.00	0.4331	11.000	4⅝	117	7¼	184
	7/16		0.4375	11.112	4⅝	117	7¼	184
		11.20	0.4409	11.200	4¾	121	7½	190
		11.50	0.4528	11.500	4¾	121	7½	190
	29/64		0.4531	11.509	4¾	121	7½	190
		11.80	0.4646	11.800	4¾	121	7½	190
	15/32		0.4688	11.908	4¾	121	7½	190

(Concluded on following page)

TWIST DRILLS (concluded)
Straight Shank Taper Length
Thru ½ inch (12.7 mm) Diameter
Fraction, Wire, Letter, Metric
High Speed Steel—Standard Sizes and Dimensions

Diameter of Drill					Flute Length		Overall Length	
D					F		L	
Fraction	No.	mm	Decimal Inch Equivalent	Millimeter Equivalent	Inch	mm	Inch	mm
		12.00	0.4724	12.000	4¾	121	7¾	197
		12.20	0.4803	12.200	4¾	121	7¾	197
31/64			0.4844	12.304	4¾	121	7¾	197
		12.50	0.4921	12.500	4¾	121	7¾	197
½			0.5000	12.700	4¾	121	7¾	197

See T-804 for Dimensional Tolerances
See T-805 for Element Tolerances

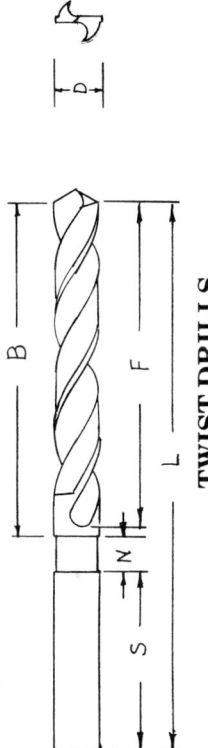

TWIST DRILLS
Straight Shank Taper Length
Over ½ inch (12.7 mm) Diameter
Fraction and Metric—High Speed Steel
Standard Sizes and Dimensions

Diameter of Drill Frac.	Diameter of Drill mm	Decimal Inch Equiv.	Millimeter Equiv.	Flute Length F Inch	Flute Length F mm	Overall Length L Inch	Overall Length L mm	Length of Body B Inch	Length of Body B mm	Minimum Length of Shk. S Inch	Minimum Length of Shk. S mm	Maximum Length of Neck N Inch	Maximum Length of Neck N mm
	12.80	0.5039	12.800	4¾	121	8	203	4⅞	124	2⅝	66	½	13
	13.00	0.5117	13.000	4¾	121	8	203	4⅞	124	2⅝	66	½	13
33/64		0.5156	13.096	4¾	121	8	203	4⅞	124	2⅝	66	½	13
	13.20	0.5197	13.200	4¾	121	8	203	4⅞	124	2⅝	66	½	13
17/32		0.5312	13.492	4¾	121	8	203	4⅞	124	2⅝	66	½	13
	13.50	0.5315	13.500	4¾	121	8	203	4⅞	124	2⅝	66	½	13
	13.80	0.5433	13.800	4⅞	124	8¼	210	5	127	2¾	70	½	13
35/64		0.5469	13.891	4⅞	124	8¼	210	5	127	2¾	70	½	13
	14.00	0.5512	14.000	4⅞	124	8¼	210	5	127	2¾	70	½	13
	14.25	0.5610	14.250	4⅞	124	8¼	210	5	127	2¾	70	½	13
9/16		0.5625	14.288	4⅞	124	8¼	210	5	127	2¾	70	½	13
	14.50	0.5709	14.500	4⅞	124	8¾	222	5	127	3⅛	79	⅝	16
37/64		0.5781	14.684	4⅞	124	8¾	222	5	127	3⅛	79	⅝	16
	14.75	0.5807	14.750	4⅞	124	8¾	222	5	127	3⅛	79	⅝	16
	15.00	0.5906	15.000	4⅞	124	8¾	222	5	127	3⅛	79	⅝	16
19/32		0.5938	15.083	4⅞	124	8¾	222	5	127	3⅛	79	⅝	16

(Continued on following page)

TWIST DRILLS (continued)
Straight Shank Taper Length
Over ½ inch (12.7 mm) Diameter
Fraction and Metric—High Speed Steel
Standard Sizes and Dimensions

Diameter of Drill				Flute Length F		Overall Length L		Length of Body B		Minimum Length of Shk. S		Maximum Length of Neck N	
Frac.	mm (D)	Decimal Inch Equiv.	Millimeter Equiv.	Inch	mm	Inch	mm	Inch	mm	Inch	mm	Inch	mm
	15.25	0.6004	15.250	4⁷⁄₈	124	8¾	222	5	127	3⅛	79	⅝	16
39⁄64		0.6094	15.479	4⁷⁄₈	124	8¾	222	5	127	3⅛	79	⅝	16
	15.50	0.6102	15.500	4⁷⁄₈	124	8¾	222	5	127	3⅛	79	⅝	16
	15.75	0.6201	15.750	4⁷⁄₈	124	8¾	222	5	127	3⅛	79	⅝	16
5⁄8		0.6250	15.875	4⁷⁄₈	124	8¾	222	5	127	3⅛	79	⅝	16
	16.00	0.6299	16.000	5⅛	130	9	228	5¼	133	3⅛	79	⅝	16
	16.25	0.6398	16.250	5⅛	130	9	228	5¼	133	3⅛	79	⅝	16
41⁄64		0.6406	16.271	5⅛	130	9	228	5¼	133	3⅛	79	⅝	16
	16.50	0.6496	16.500	5⅛	130	9	228	5¼	133	3⅛	79	⅝	16
21⁄32		0.6562	16.667	5⅛	130	9	228	5¼	133	3⅛	79	⅝	16
	16.75	0.6594	16.750	5⅜	137	9¼	235	5½	140	3⅛	79	⅝	16
	17.00	0.6693	17.000	5⅜	137	9¼	235	5½	140	3⅛	79	⅝	16
43⁄64		0.6719	17.066	5⅜	137	9¼	235	5½	140	3⅛	79	⅝	16
	17.25	0.6791	17.250	5⅜	137	9¼	235	5½	140	3⅛	79	⅝	16
11⁄16		0.6875	17.462	5⅜	137	9¼	235	5½	140	3⅛	79	⅝	16
	17.50	0.6890	17.500	5⅝	143	9½	241	5¾	146	3⅛	79	⅝	16
45⁄64		0.7031	17.859	5⅝	143	9½	241	5¾	146	3⅛	79	⅝	16
	18.00	0.7087	18.000	5⅝	143	9½	241	5¾	146	3⅛	79	⅝	16
23⁄32		0.7188	18.258	5⅝	143	9½	241	5¾	146	3⅛	79	⅝	16
	18.50	0.7283	18.500	5⅞	149	9¾	247	6	152	3⅛	79	⅝	16
47⁄64		0.7344	18.654	5⅞	149	9¾	247	6	152	3⅛	79	⅝	16
	19.00	0.7480	19.000	5⅞	149	9¾	247	6	152	3⅛	79	⅝	16

(Continued on following page)

TWIST DRILLS (continued)
Straight Shank Taper Length
Over ½ inch (12.7 mm) Diameter
Fraction and Metric—High Speed Steel
Standard Sizes and Dimensions

Diameter of Drill D		Decimal Inch Equiv.	Milli-meter Equiv.	Flute Length F		Overall Length L		Length of Body B		Minimum Length of Shk. S		Maximum Length of Neck N	
Frac.	mm			Inch	mm	Inch	mm	Inch	mm	Inch	mm	Inch	mm
3/4		0.7500	19.050	5⅞	149	9¾	247	6	152	3⅛	79	⅝	16
49/64		0.7656	19.446	6	152	9⅞	251	6⅛	156	3⅛	79	⅝	16
	19.50	0.7677	19.500	6	152	9⅞	251	6⅛	156	3⅛	79	⅝	16
25/32		0.7812	19.842	6	152	9⅞	251	6⅛	156	3⅛	79	⅝	16
	20.00	0.7874	20.000	6⅛	156	10	254	6¼	159	3⅛	79	⅝	16
51/64		0.7969	20.241	6⅛	156	10	254	6¼	159	3⅛	79	⅝	16
	20.50	0.8071	20.500	6⅛	156	10	254	6¼	159	3⅛	79	⅝	16
13/16		0.8125	20.638	6⅛	156	10	254	6¼	159	3⅛	79	⅝	16
	21.00	0.8268	21.000	6⅛	156	10	254	6¼	159	3⅛	79	⅝	16
53/64		0.8281	21.034	6⅛	156	10	254	6¼	159	3⅛	79	⅝	16
27/32		0.8438	21.433	6⅛	156	10	254	6¼	159	3⅛	79	⅝	16
	21.50	0.8465	21.500	6⅛	156	10	254	6¼	159	3⅛	79	⅝	16
55/64		0.8594	21.829	6⅛	156	10	254	6¼	159	3⅛	79	⅝	16
	22.00	0.8661	22.000	6⅛	156	10	254	6¼	159	3⅛	79	⅝	16
7/8		0.8750	22.225	6⅛	156	10	254	6¼	159	3⅛	79	⅝	16
57/64		0.8858	22.500	6⅛	156	10	254	6¼	159	3⅛	79	⅝	16
	22.50	0.8906	22.621	6⅛	156	10	254	6¼	159	3⅛	79	⅝	16
29/32		0.9055	23.000	6⅛	156	10	254	6¼	159	3⅛	79	⅝	16
	23.00	0.9062	23.017	6⅛	156	10	254	6¼	159	3⅛	79	⅝	16
59/64		0.9219	23.416	6⅛	156	10¾	273	6¼	159	3⅞	98	⅝	16
15/16		0.9252	23.500	6⅛	156	10¾	273	6¼	159	3⅞	98	⅝	16
	23.50	0.9375	23.812	6⅛	156	10¾	273	6¼	159	3⅞	98	⅝	16

(Continued on following page)

TWIST DRILLS (continued)
Straight Shank Taper Length
Over ½ inch (12.7 mm) Diameter
Fraction and Metric—High Speed Steel
Standard Sizes and Dimensions

Diameter of Drill				Flute Length (F)		Overall Length (L)		Length of Body (B)		Minimum Length of Shk. (S)		Maximum Length of Neck (N)	
Frac. (D)	mm	Decimal Inch Equiv.	Millimeter Equiv.	Inch	mm	Inch	mm	Inch	mm	Inch	mm	Inch	mm
	24.00	0.9449	24.000	6 3/8	162	11	279	6 1/2	165	3 7/8	98	5/8	16
61/64		0.9531	24.209	6 3/8	162	11	279	6 1/2	165	3 7/8	98	5/8	16
	24.50	0.9646	24.500	6 3/8	162	11	279	6 1/2	165	3 7/8	98	5/8	16
31/32		0.9688	24.608	6 3/8	162	11	279	6 1/2	165	3 7/8	98	5/8	16
	25.00	0.9843	25.000	6 3/8	162	11	279	6 1/2	165	3 7/8	98	5/8	16
63/64		0.9844	25.004	6 3/8	162	11	279	6 1/2	165	3 7/8	98	5/8	16
1		1.0000	25.400	6 3/8	162	11	279	6 1/2	165	3 3/4	98	5/8	16
	25.50	1.0039	25.500	6 1/2	165	11 1/8	282	6 5/8	168	3 7/8	98	5/8	16
1 1/64		1.0156	25.796	6 1/2	165	11 1/8	282	6 5/8	168	3 7/8	98	5/8	16
	26.00	1.0236	26.000	6 1/2	165	11 1/8	282	6 5/8	168	3 7/8	98	5/8	16
1 1/32		1.0312	26.192	6 1/2	165	11 1/8	282	6 5/8	168	3 7/8	98	5/8	16
	26.50	1.0433	26.560	6 5/8	168	11 1/4	286	6 3/4	172	3 7/8	98	5/8	16
1 3/64		1.0469	26.591	6 5/8	168	11 1/4	286	6 3/4	172	3 7/8	98	5/8	16
1 1/16		1.0625	26.988	6 5/8	168	11 1/4	286	6 3/4	172	3 7/8	98	5/8	16
	27.00	1.0630	27.000	6 5/8	168	11 1/4	286	6 3/4	172	3 7/8	98	5/8	16
1 5/64		1.0781	27.384	6 7/8	175	11 1/2	292	7	178	3 7/8	98	5/8	16
	27.50	1.0827	27.500	6 7/8	175	11 1/2	292	7	178	3 7/8	98	5/8	16
1 3/32		1.0938	27.783	6 7/8	175	11 1/2	292	7	178	3 7/8	98	5/8	16
	28.00	1.1024	28.000	7 1/8	181	11 3/4	298	7 1/4	184	3 7/8	98	5/8	16
1 7/64		1.1094	28.179	7 1/8	181	11 3/4	298	7 1/4	184	3 7/8	98	5/8	16
	28.50	1.1220	28.500	7 1/8	181	11 3/4	298	7 1/4	184	3 7/8	98	5/8	16
1 1/8		1.1250	28.575	7 1/8	181	11 3/4	298	7 1/4	184	3 7/8	98	5/8	16

(Continued on following page)

TWIST DRILLS (continued)
Straight Shank Taper Length
Over ½ inch (12.7 mm) Diameter
Fraction and Metric—High Speed Steel
Standard Sizes and Dimensions

Diameter of Drill				Flute Length F		Overall Length L		Length of Body B		Minimum Length of Shk. S		Maximum Length of Neck N	
Frac.	mm	Decimal Inch Equiv.	Millimeter Equiv.	Inch	mm	Inch	mm	Inch	mm	Inch	mm	Inch	mm
1 9/64		1.1406	28.971	7¼	184	11⅞	301	7⅜	187	3⅞	98	⅝	16
	29.00	1.1417	29.000	7¼	184	11⅞	301	7⅜	187	3⅞	98	⅝	16
1 5/32		1.1562	29.367	7¼	184	11⅞	301	7⅜	187	3⅞	98	⅝	16
	29.50	1.1614	29.500	7⅜	187	12	305	7½	191	3⅞	98	⅝	16
1 11/64		1.1719	29.766	7⅜	187	12	305	7½	191	3⅞	98	⅝	16
	30.00	1.1811	30.000	7⅜	187	12	305	7½	191	3⅞	98	⅝	16
1 3/16		1.1875	30.162	7⅜	187	12	305	7½	191	3⅞	98	⅝	16
	30.50	1.2008	30.500	7½	190	12⅛	308	7⅝	194	3⅞	98	⅝	16
1 13/64		1.2031	30.559	7½	190	12⅛	308	7⅝	194	3⅞	98	⅝	16
1 7/32		1.2188	30.958	7½	190	12⅛	308	7⅝	194	3⅞	98	⅝	16
	31.00	1.2205	31.000	7⅞	200	12½	317	8	203	3⅞	98	⅝	16
1 15/64		1.2344	31.354	7⅞	200	12½	317	8	203	3⅞	98	⅝	16
	31.50	1.2402	31.500	7⅞	200	12½	317	8	203	3⅞	98	⅝	16
1 1/4		1.2500	31.750	7⅞	200	12½	317	8	203	3⅞	98	⅝	16
	32.00	1.2598	32.000	8½	216	14⅛	359	8⅝	219	4⅞	124	⅝	16
	32.50	1.2795	32.500	8½	216	14⅛	359	8⅝	219	4⅞	124	⅝	16
1 9/32		1.2812	32.542	8½	216	14⅛	359	8⅝	219	4⅞	124	⅝	16
	33.00	1.2992	33.000	8⅝	219	14¼	362	8¾	222	4⅞	124	⅝	16
1 5/16		1.3125	33.338	8⅝	219	14¼	362	8¾	222	4⅞	124	⅝	16
	33.50	1.3189	33.500	8¾	222	14⅜	365	8⅞	225	4⅞	124	⅝	16
	34.00	1.3386	34.000	8¾	222	14⅜	365	8⅞	225	4⅞	124	⅝	16
1 11/32		1.3438	34.133	8¾	222	14⅜	365	8⅞	225	4⅞	124	⅝	16

(Concluded on following page)

TWIST DRILLS (concluded)
Straight Shank Taper Length
Over ½ inch (12.7 mm) Diameter
Fraction and Metric—High Speed Steel
Standard Sizes and Dimensions

Diameter of Drill				Flute Length F		Overall Length L		Length of Body B		Minimum Length of Shk. S		Maximum Length of Neck N	
Frac.	D mm	Decimal Inch Equiv.	Millimeter Equiv.	Inch	mm	Inch	mm	Inch	mm	Inch	mm	Inch	mm
1 3/8	34.50	1.3583	34.500	8 7/8	225	14 1/2	368	9	229	4 7/8	124	5/8	16
		1.3750	34.925	8 7/8	225	14 1/2	368	9	229	4 7/8	124	5/8	16
	35.00	1.3780	35.000	9	229	14 5/8	372	9 1/8	232	4 7/8	124	5/8	16
	35.50	1.3976	35.500	9	229	14 5/8	372	9 1/8	232	4 7/8	124	5/8	16
1 13/32		1.4062	35.717	9	229	14 5/8	372	9 1/8	232	4 7/8	124	5/8	16
	36.00	1.4173	36.000	9 1/8	232	14 3/4	375	9 1/4	235	4 7/8	124	5/8	16
	36.50	1.4370	36.500	9 1/8	232	14 3/4	375	9 1/4	235	4 7/8	124	5/8	16
1 7/16		1.4375	36.512	9 1/8	232	14 3/4	375	9 1/4	235	4 7/8	124	5/8	16
	37.00	1.4567	37.000	9 1/4	235	14 7/8	378	9 3/8	238	4 7/8	124	5/8	16
1 15/32		1.4688	37.308	9 1/4	235	14 7/8	378	9 3/8	238	4 7/8	124	5/8	16
	37.50	1.4764	37.500	9 3/8	238	15	381	9 1/2	241	4 7/8	124	5/8	16
	38.00	1.4961	38.000	9 3/8	238	15	381	9 1/2	241	4 7/8	124	5/8	16
1 1/2		1.5000	38.100	9 3/8	238	15	381	9 1/2	241	4 7/8	124	5/8	16
1 9/16		1.5625	39.688	9 5/8	244	15 1/4	387	9 3/4	247	4 7/8	124	5/8	16
1 5/8		1.6250	41.275	9 7/8	251	15 5/8	397	10	254	4 7/8	124	3/4	19
1 3/4		1.7500	44.450	10 1/2	267	16 1/4	413	10 5/8	270	4 7/8	124	3/4	19

See T-804 for Dimensional Tolerances
See T-805 for Element Tolerances

TWIST DRILLS

Straight Shank, Taper Length
Heavy Duty, Long Flute
Tang Drive, High Speed Steel

Designed for strength and rigidity, adaptable for severe drilling conditions. Rugged design suitable for drilling tough alloy steels and steel forgings. Straight shanks, tanged for use with split sleeve type drivers.

Diameter Inches	Length Flute Inches	Length Overall Inches	Shank Diameter Inches	Diameter Inches	Length Flute Inches	Length Overall Inches	Shank Diameter Inches
1/8	3 3/8	5 1/8	1/8	19/32	6 1/2	8 3/4	1/2
9/64	3 5/8	5 3/8	9/64	39/64	6 1/2	8 3/4	1/2
5/32	3 3/4	5 3/8	5/32	5/8	6 1/2	8 3/4	1/2
11/64	4 1/8	5 3/4	11/64	41/64	6 3/4	9	5/8
3/16	4 1/8	5 3/4	3/16	21/32	6 3/4	9	5/8
13/64	4 3/8	6	13/64	11/16	6 7/8	9 1/4	5/8
7/32	4 3/8	6	7/32	23/32	7 1/8	9 1/2	5/8
15/64	4 13/16	6 1/8	15/64	3/4	7 3/8	9 3/4	3/4
1/4	4 13/16	6 1/8	1/4				
17/64	5	6 1/4	17/64				
9/32	5	6 1/4	9/32				
19/64	5 1/8	6 3/8	19/64				
5/16	5 1/8	6 3/8	5/16				
21/64	5 1/4	6 1/2	21/64				
11/32	5 1/4	6 1/2	11/32				
23/64	5 3/8	6 3/4	23/64				
3/8	5 3/8	6 3/4	3/8				
25/64	5 5/8	7	25/64				
13/32	5 5/8	7	13/32				
27/64	5 11/16	7 1/4	27/64				
7/16	5 11/16	7 1/4	7/16				
29/64	5 3/4	7 1/2	29/64				
15/32	5 3/4	7 1/2	15/32				
31/64	5 3/4	7 3/4	31/64				
1/2	5 3/4	7 3/4	1/2				
33/64	6	8	1/2				
17/32	6	8	1/2				
35/64	6 1/4	8 1/4	1/2				
9/16	6 1/4	8 1/4	1/2				
37/64	6 1/2	8 3/4	1/2				

Four Flute Core Drills

High Speed Steel
With Regular Taper Shank

Standard Sizes and Dimensions

Diameter Inches	Flute Length Inches	Overall Length Inches	Morse Taper Shank	Diameter Inches	Flute Length Inches	Overall Length Inches	Morse Taper Shank
1/2	4 3/8	8 1/4	2	1 11/32	8 3/4	14 3/8	4
17/32	4 5/8	8 1/2	2	1 3/8	8 7/8	14 1/2	4
9/16	4 7/8	8 3/4	2	1 13/32	9	14 5/8	4
19/32	4 7/8	8 3/4	2	1 7/16	9 1/8	14 3/4	4
5/8	4 7/8	8 3/4	2	1 15/32	9 1/4	14 7/8	4
21/32	5 1/8	9	2	1 1/2	9 3/8	15	4
11/16	5 3/8	9 1/4	2	1 17/32	9 3/8	16 3/8	5
23/32	5 5/8	9 1/2	2	1 9/16	9 5/8	16 5/8	5
3/4	5 7/8	9 3/4	2	1 19/32	9 7/8	16 7/8	5
25/32	6	9 7/8	2	1 5/8	10	17	5
13/16	6 1/8	10 3/4	3	1 21/32	10 1/8	17 1/8	5
27/32	6 1/8	10 3/4	3	1 11/16	10 1/8	17 1/8	5
7/8	6 1/8	10 3/4	3	1 23/32	10 1/8	17 1/8	5
29/32	6 1/8	10 3/4	3	1 3/4	10 1/8	17 1/8	5
15/16	6 1/8	10 3/4	3	1 25/32	10 1/8	17 1/8	
31/32	6 3/8	11	3	1 13/16	10 1/8	17 1/8	5
1	6 3/8	11	3	1 27/32	10 1/8	17 1/8	5
1 1/32	6 1/2	11 1/8	3	1 7/8	10 3/8	17 3/8	5
1 1/16	6 5/8	11 1/4	3	1 29/32	10 3/8	17 3/8	5
1 3/32	6 7/8	12 1/2	4	1 15/16	10 3/8	17 3/8	5
1 1/8	7 1/8	12 3/4	4	1 31/32	10 3/8	17 3/8	5
1 5/32	7 1/4	12 7/8	4	2	10 3/8	17 3/8	5
1 3/16	7 3/8	13	4	2 1/8	10 1/4	17 3/8	5
1 7/32	7 1/2	13 1/8	4	2 1/4	10 1/8	17 3/8	5
1 1/4	7 7/8	13 1/2	4	2 3/8	10 1/8	17 3/8	5
1 9/32	8 1/8	14 1/8	4	2 1/2	11 1/4	18 3/4	5
1 5/16	8 5/8	14 1/4	4				

Three Flute Core Drills

High Speed Steel
With Regular Taper Shank

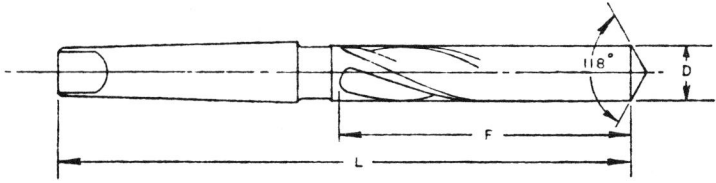

Drill Diameter $^{11}/_{32}''$ & Smaller

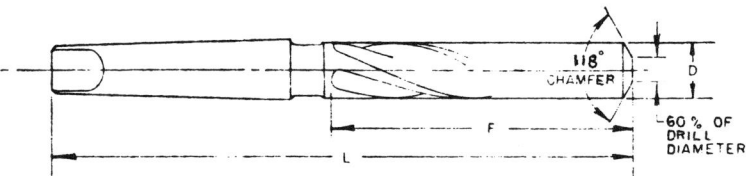

Drill Diameter $^3/_8''$ and Larger

For standard shank dimensions, see Table 801.

Standard Sizes and Dimensions

Diameter Inches	Flute Length Inches	Overall Length Inches	No. of Taper Shank	Diameter Inches	Flute Length Inches	Overall Length Inches	No. of Taper Shank
D	F	L		D	F	L	
$^1/_4$	$2^7/_8$	$6^1/_8$	1	$^{27}/_{32}$	$6^1/_8$	$10^3/_4$	3
$^9/_{32}$	3	$6^1/_4$	1	$^7/_8$	$6^1/_8$	$10^3/_4$	3
$^5/_{16}$	$3^1/_8$	$6^3/_8$	1	$^{29}/_{32}$	$6^1/_8$	$10^3/_4$	3
$^{11}/_{32}$	$3^1/_4$	$6^1/_2$	1	$^{15}/_{16}$	$6^1/_8$	$10^3/_4$	3
$^3/_8$	$3^1/_2$	$6^3/_4$	1	$^{31}/_{32}$	$6^3/_8$	11	3
$^{13}/_{32}$	$3^5/_8$	7	1	1	$6^3/_8$	11	3
$^7/_{16}$	$3^7/_8$	$7^1/_4$	1	$1\,^1/_{16}$	$6^5/_8$	$11^1/_4$	3
$^{15}/_{32}$	$4^1/_8$	$7^1/_2$	1	$1\,^1/_8$	$7^1/_8$	$12^3/_4$	4
$^1/_2$	$4^3/_8$	$8^1/_4$	2	$1\,^3/_{16}$	$7^3/_8$	13	4
$^{17}/_{32}$	$4^5/_8$	$8^1/_2$	2	$1\,^1/_4$	$7^7/_8$	$13^1/_2$	4
$^9/_{16}$	$4^7/_8$	$8^3/_4$	2				
$^{19}/_{32}$	$4^7/_8$	$8^3/_4$	2				
$^5/_8$	$4^7/_8$	$8^3/_4$	2				
$^{21}/_{32}$	$5^1/_8$	9	2				
$^{11}/_{16}$	$5^3/_8$	$9^1/_4$	2				
$^{23}/_{32}$	$5^5/_8$	$9^1/_2$	2				
$^3/_4$	$5^7/_8$	$9^3/_4$	2				
$^{25}/_{32}$	6	$9^7/_8$	2				
$^{13}/_{16}$	$6^1/_8$	$10^3/_4$	3				

Three Flute Core Drills

High Speed Steel
With Straight Shank

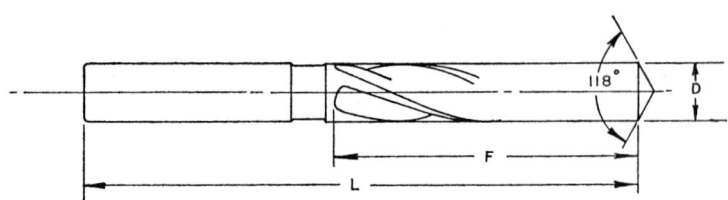

Drill Diameter 11/32″ & Smaller

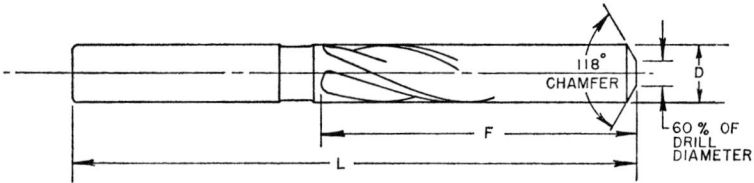

Drill Diameter 3/8″ & Larger

All sizes have shank same diameter as body.

Standard Sizes and Dimensions

Diam- eter Inches	Flute Length Inches	Overall Length Inches	Diam- eter Inches	Flute Length Inches	Overall Length Inches
D	F	L	D	F	L
1/4	3 3/4	6 1/8	1/2	4 3/4	7 3/4
9/32	3 7/8	6 1/4	17/32	4 3/4	8
5/16	4	6 3/8	9/16	4 7/8	8 1/4
11/32	4 1/8	6 1/2	19/32	4 7/8	8 3/8
3/8	4 1/4	6 3/4	5/8	4 7/8	8 3/4
13/32	4 3/8	7	21/32	5 1/8	9
7/16	4 5/8	7 1/4	11/16	5 3/8	9 1/4
15/32	4 3/4	7 1/2	3/4	5 7/8	9 3/4

Four Flute Core Drills

High Speed Steel
With Straight Shank

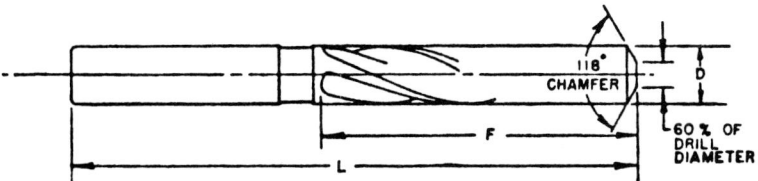

Standard Sizes and Dimensions

Diameter Inches	Flute Length Inches	Overall Length Inches
D	F	L
1/2	4¾	7¾
17/32	4¾	8
9/16	4⅞	8¼
19/32	4⅞	8¾
5/8	4⅞	8¾
21/32	5⅓	9
11/16	5⅜	9¼
23/32	5⅝	9½
3/4	5⅞	9¾
25/32	6	9⅞
13/16	6⅛	10
27/32	6⅛	10
7/8	6⅛	10
29/32	6⅛	10
15/16	6⅛	10¾
31/32	6⅜	11
1	6⅜	11

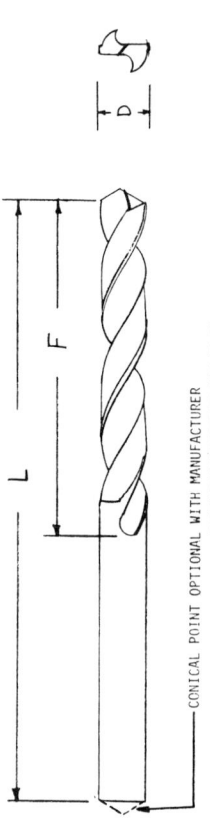

CONICAL POINT OPTIONAL WITH MANUFACTURER

TWIST DRILLS
Straight Shank, Jobbers Length
Fraction, Wire, Letter, Metric
High Speed Steel
Standard Sizes and Dimensions

Diameter of Drill					Flute Length		Overall Length	
		D			F		L	
Fraction	No.	Ltr.	Millimeter	Decimal Inch Equivalent	Millimeter Equivalent	Inch	mm	Inch	mm
	97		0.15	0.0059	0.150	1/16	1.6	3/4	19
	96		0.16	0.0063	0.160	1/16	1.6	3/4	19
	95		0.17	0.0067	0.170	1/16	1.6	3/4	19
	94		0.18	0.0071	0.180	1/16	1.6	3/4	19
	93		0.19	0.0075	0.190	1/16	1.6	3/4	19
	92		0.20	0.0079	0.201	1/16	1.6	3/4	19
	91			0.0083	0.211	5/64	2.0	3/4	19
	90		0.22	0.0087	0.221	5/64	2.0	3/4	19
	89			0.0091	0.231	5/64	2.0	3/4	19
	88			0.0095	0.241	5/64	2.0	3/4	19
	87		0.25	0.0098	0.250	5/64	2.0	3/4	19
	86			0.0100	0.254	5/64	2.0	3/4	19
	85			0.0105	0.267	3/32	2.4	3/4	19
	84		0.28	0.0110	0.280	3/32	2.4	3/4	19
				0.0115	0.292	3/32	2.4	3/4	19
	83		0.30	0.0118	0.300	3/32	2.4	3/4	19
				0.0120	0.305	3/32	2.4	3/4	19

(Continued on following page)

TWIST DRILLS (continued)
Straight Shank, Jobbers Length
Fraction, Wire, Letter, Metric
High Speed Steel
Standard Sizes and Dimensions

Diameter of Drill (D)						Flute Length (F)		Overall Length (L)	
Fraction	No.	Ltr.	Millimeter	Decimal Inch Equivalent	Millimeter Equivalent	Inch	mm	Inch	mm
	82			0.0125	0.318	3/32	2.4	3/4	19
			0.32	0.0126	0.320	3/32	2.4	3/4	19
	81			0.0130	0.330	3/32	2.4	3/4	19
	80			0.0135	0.343	1/8	3	3/4	19
			0.35	0.0138	0.350	1/8	3	3/4	19
	79			0.0145	0.368	1/8	3	3/4	19
			0.38	0.0150	0.380	3/16	5	3/4	19
1/64				0.0156	0.396	3/16	5	3/4	19
			0.40	0.0157	0.400	3/16	5	3/4	19
	78			0.0160	0.406	3/16	5	7/8	22
			0.42	0.0165	0.420	3/16	5	7/8	22
			0.45	0.0177	0.450	3/16	5	7/8	22
	77			0.0180	0.457	3/16	5	7/8	22
			0.48	0.0189	0.480	3/16	5	7/8	22
			0.50	0.0197	0.500	3/16	5	7/8	22
	76			0.0200	0.508	3/16	5	7/8	22
	75			0.0210	0.533	1/4	6	1	25
			0.55	0.0217	0.550	1/4	6	1	25
	74			0.0225	0.572	1/4	6	1	25
			0.60	0.0236	0.600	5/16	8	1 1/8	29
	73			0.0240	0.610	5/16	8	1 1/8	29
	72			0.0250	0.635	5/16	8	1 1/8	29
			0.65	0.0256	0.650	3/8	10	1 1/4	32

(Continued on following page)

TWIST DRILLS (continued)
Straight Shank, Jobbers Length
Fraction, Wire, Letter, Metric
High Speed Steel
Standard Sizes and Dimensions

Diameter of Drill						Flute Length		Overall Length	
D				Decimal Inch Equivalent	Millimeter Equivalent	*F*		*L*	
Fraction	No.	Ltr.	Millimeter			Inch	mm	Inch	mm
	71			0.0260	0.660	3/8	10	1 1/4	32
			0.70	0.0276	0.700	3/8	10	1 1/4	32
	70			0.0280	0.711	3/8	10	1 1/4	32
	69			0.0292	0.742	1/2	13	1 3/8	35
			0.75	0.0295	0.750	1/2	13	1 3/8	35
	68			0.0310	0.787	1/2	13	1 3/8	35
1/32				0.0312	0.792	1/2	13	1 3/8	35
			0.80	0.0315	0.800	1/2	13	1 3/8	35
	67			0.0320	0.813	1/2	13	1 3/8	35
	66			0.0330	0.838	1/2	13	1 3/8	35
			0.85	0.0335	0.850	5/8	16	1 1/2	38
	65			0.0350	0.889	5/8	16	1 1/2	38
			0.90	0.0354	0.899	5/8	16	1 1/2	38
	64			0.0360	0.914	5/8	16	1 1/2	38
	63			0.0370	0.940	5/8	16	1 1/2	38
			0.95	0.0374	0.950	5/8	16	1 1/2	38
	62			0.0380	0.965	5/8	16	1 1/2	38
	61			0.0390	0.991	11/16	17	1 5/8	41
			1.00	0.0394	1.000	11/16	17	1 5/8	41
	60			0.0400	1.016	11/16	17	1 5/8	41
	59			0.0410	1.041	11/16	17	1 5/8	41
			1.05	0.0413	1.050	11/16	17	1 5/8	41
	58			0.0420	1.067	11/16	17	1 5/8	41

(Continued on following page)

TWIST DRILLS (continued)
Straight Shank, Jobbers Length
Fraction, Wire, Letter, Metric
High Speed Steel
Standard Sizes and Dimensions

Diameter of Drill (D)						Flute Length (F)		Overall Length (L)	
Fraction	No.	Ltr.	Millimeter	Decimal Inch Equivalent	Millimeter Equivalent	Inch	mm	Inch	mm
	57			0.0430	1.092	3/4	19	1 3/4	44
			1.10	0.0433	1.100	3/4	19	1 3/4	44
			1.15	0.0453	1.150	3/4	19	1 3/4	44
	56			0.0465	1.181	3/4	19	1 3/4	44
3/64				0.0469	1.191	3/4	19	1 3/4	44
			1.20	0.0472	1.200	7/8	22	1 7/8	48
			1.25	0.0492	1.250	7/8	22	1 7/8	48
			1.30	0.0512	1.300	7/8	22	1 7/8	48
	55			0.0520	1.321	7/8	22	1 7/8	48
			1.35	0.0531	1.350	7/8	22	1 7/8	48
	54			0.0550	1.397	7/8	22	1 7/8	48
			1.40	0.0551	1.400	7/8	22	1 7/8	48
			1.45	0.0571	1.450	7/8	22	1 7/8	48
			1.50	0.0591	1.500	7/8	22	1 7/8	48
	53			0.0595	1.511	7/8	22	1 7/8	48
			1.55	0.0610	1.550	7/8	22	1 7/8	48
1/16				0.0625	1.588	7/8	22	1 7/8	48
			1.60	0.0630	1.600	7/8	22	1 7/8	48
	52			0.0635	1.613	7/8	22	1 7/8	48
			1.65	0.0650	1.650	1	25	2	51
			1.70	0.0669	1.700	1	25	2	51
	51			0.0670	1.702	1	25	2	51
			1.75	0.0689	1.750	1	25	2	51

(Continued on following page)

TWIST DRILLS (continued)
Straight Shank, Jobbers Length
Fraction, Wire, Letter, Metric
High Speed Steel
Standard Sizes and Dimensions

Diameter of Drill (D)						Flute Length (F)		Overall Length (L)	
Fraction	No.	Ltr.	Millimeter	Decimal Inch Equivalent	Millimeter Equivalent	Inch	mm	Inch	mm
	50			0.0700	1.778	1	25	2	51
			1.80	0.0709	1.800	1	25	2	51
			1.85	0.0728	1.850	1	25	2	51
	49			0.0730	1.854	1	25	2	51
			1.90	0.0748	1.900	1	25	2	51
	48			0.0760	1.930	1	25	2	51
			1.95	00.768	1.950	1	25	2	51
5/64				0.0781	1.984	1	25	2	51
	47			0.0785	1.994	1	25	2	51
			2.00	0.0787	2.000	1	25	2	51
			2.05	0.0807	2.050	1⅛	29	2⅛	54
	46			0.0810	2.057	1⅛	29	2⅛	54
	45			0.0820	2.083	1⅛	29	2⅛	54
			2.10	0.0827	2.100	1⅛	29	2⅛	54
			2.15	0.0846	2.150	1⅛	29	2⅛	54
	44			0.0860	2.184	1⅛	29	2⅛	54
			2.20	0.0866	2.200	1¼	32	2¼	57
			2.25	0.0886	2.250	1¼	32	2¼	57
	43			0.0890	2.261	1¼	32	2¼	57
			2.30	0.0906	2.300	1¼	32	2¼	57
			2.35	0.0925	2.350	1¼	32	2¼	57
	42			0.0935	2.375	1¼	32	2¼	57
3/32				0.0938	2.383	1¼	32	2¼	57

(Continued on following page)

TWIST DRILLS (continued)
Straight Shank, Jobbers Length
Fraction, Wire, Letter, Metric
High Speed Steel
Standard Sizes and Dimensions

Diameter of Drill (D)				Decimal Inch Equivalent	Millimeter Equivalent	Flute Length (F)		Overall Length (L)	
Fraction	No.	Ltr.	Millimeter			Inch	mm	Inch	mm
			2.40	0.0945	2.400	1 3/8	35	2 3/8	60
	41			0.0960	2.438	1 3/8	35	2 3/8	60
			2.45	0.0965	2.450	1 3/8	35	2 3/8	60
	40			0.0980	2.489	1 3/8	35	2 3/8	60
			2.50	0.0984	2.500	1 3/8	35	2 3/8	60
	39			0.0995	2.527	1 3/8	35	2 3/8	60
	38			0.1015	2.578	1 7/16	37	2 1/2	64
			2.60	0.1024	2.600	1 7/16	37	2 1/2	64
	37			0.1040	2.642	1 7/16	37	2 1/2	64
			2.70	0.1063	2.700	1 7/16	37	2 1/2	64
	36			0.1065	2.705	1 7/16	37	2 1/2	64
7/64				0.1094	2.779	1 1/2	38	2 5/8	67
	35			0.1100	2.794	1 1/2	38	2 5/8	67
			2.80	0.1102	2.800	1 1/2	38	2 5/8	67
	34			0.1110	2.819	1 1/2	38	2 5/8	67
	33			0.1130	2.870	1 1/2	38	2 5/8	67
			2.90	0.1142	2.900	1 5/8	41	2 3/4	70
	32			0.1160	2.946	1 5/8	41	2 3/4	70
			3.00	0.1181	3.000	1 5/8	41	2 3/4	70
	31			0.1200	3.048	1 5/8	41	2 3/4	70
			3.10	0.1220	3.100	1 5/8	41	2 3/4	70
1/8				0.1250	3.175	1 5/8	41	2 3/4	70
			3.20	0.1260	3.200	1 5/8	41	2 3/4	70

(Continued on following page)

TWIST DRILLS (continued)
Straight Shank, Jobbers Length
Fraction, Wire, Letter, Metric
High Speed Steel
Standard Sizes and Dimensions

Fraction	Diameter of Drill (D)					Flute Length (F)		Overall Length (L)	
	No.	Ltr.	Millimeter	Decimal Inch Equivalent	Millimeter Equivalent	Inch	mm	Inch	mm
	30			0.1285	3.264	1⅝	41	2¾	70
			3.30	0.1299	3.300	1¾	44	2⅞	73
			3.40	0.1339	3.400	1¾	44	2⅞	73
	29			0.1360	3.454	1¾	44	2⅞	73
			3.50	0.1378	3.500	1¾	44	2⅞	73
	28			0.1405	3.569	1¾	44	2⅞	73
9/64				0.1406	3.571	1¾	44	2⅞	73
			3.60	0.1417	3.600	1⅞	48	3	76
	27			0.1440	3.658	1⅞	48	3	76
			3.70	0.1457	3.700	1⅞	48	3	76
	26			0.1470	3.734	1⅞	48	3	76
	25			0.1495	3.797	1⅞	48	3	76
			3.80	0.1496	3.800	1⅞	48	3	76
	24			0.1520	3.861	2	51	3⅛	79
			3.90	0.1535	3.900	2	51	3⅛	79
	23			0.1540	3.912	2	51	3⅛	79
5/32				0.1562	3.967	2	51	3⅛	79
	22			0.1570	3.988	2	51	3⅛	79
			4.00	0.1575	4.000	2⅛	54	3¼	83
	21			0.1590	4.039	2⅛	54	3¼	83
	20			0.1610	4.089	2⅛	54	3¼	83
			4.10	0.1614	4.100	2⅛	54	3¼	83
			4.20	0.1654	4.200	2⅛	54	3¼	83

(Continued on following page)

TWIST DRILLS (continued)
Straight Shank, Jobbers Length
Fraction, Wire, Letter, Metric
High Speed Steel
Standard Sizes and Dimensions

Diameter of Drill (D)						Flute Length (F)		Overall Length (L)	
Fraction	No.	Ltr.	Millimeter	Decimal Inch Equivalent	Millimeter Equivalent	Inch	mm	Inch	mm
	19			0.1660	4.216	2⅛	54	3¼	83
			4.30	0.1693	4.300	2⅛	54	3¼	83
	18			0.1695	4.305	2⅛	54	3¼	83
11/64				0.1719	4.366	2⅛	54	3¼	83
	17			0.1730	4.394	2³⁄₁₆	56	3⅜	86
			4.40	0.1732	4.400	2³⁄₁₆	56	3⅜	86
	16			0.1770	4.496	2³⁄₁₆	56	3⅜	86
			4.50	0.1772	4.500	2³⁄₁₆	56	3⅜	86
	15			0.1800	4.572	2³⁄₁₆	56	3⅜	86
			4.60	0.1811	4.600	2³⁄₁₆	56	3⅜	86
	14			0.1820	4.623	2³⁄₁₆	56	3⅜	86
	13		4.70	0.1850	4.700	2⁵⁄₁₆	59	3½	89
3/16				0.1875	4.762	2⁵⁄₁₆	59	3½	89
	12		4.80	0.1890	4.800	2⁵⁄₁₆	59	3½	89
	11			0.1910	4.851	2⁵⁄₁₆	59	3½	89
			4.90	0.1929	4.900	2⁷⁄₁₆	62	3⅝	92
	10			0.1935	4.915	2⁷⁄₁₆	62	3⅝	92
	9			0.1960	4.978	2⁷⁄₁₆	62	3⅝	92
			5.00	0.1969	5.000	2⁷⁄₁₆	62	3⅝	92
	8			0.1990	5.055	2⁷⁄₁₆	62	3⅝	92
			5.10	0.2008	5.100	2⁷⁄₁₆	62	3⅝	92
	7			0.2010	5.105	2⁷⁄₁₆	62	3⅝	92
13/64				0.2031	5.159	2⁷⁄₁₆	62	3⅝	92

(Continued on following page)

TWIST DRILLS (continued)
Straight Shank, Jobbers Length
Fraction, Wire, Letter, Metric
High Speed Steel
Standard Sizes and Dimensions

Diameter of Drill				Decimal Inch Equivalent	Millimeter Equivalent	Flute Length F		Overall Length L	
Fraction	No.	Ltr.	Millimeter D			Inch	mm	Inch	mm
	6			0.2040	5.182	2½	64	3¾	95
	5		5.20	0.2047	5.200	2½	64	3¾	95
	4			0.2055	5.220	2½	64	3¾	95
			5.30	0.2087	5.300	2½	64	3¾	95
				0.2090	5.309	2½	64	3¾	95
	3		5.40	0.2126	5.400	2½	64	3¾	95
				0.2130	5.410	2½	64	3¾	95
			5.50	0.2165	5.500	2½	64	3¾	95
7/32				0.2188	5.558	2½	64	3¾	95
			5.60	0.2205	5.600	2⅝	67	3⅞	98
	2			0.2210	5.613	2⅝	67	3⅞	98
			5.70	0.2244	5.700	2⅝	67	3⅞	98
				0.2280	5.791	2⅝	67	3⅞	98
	1		5.80	0.2283	5.800	2⅝	67	3⅞	98
			5.90	0.2323	5.900	2⅝	67	3⅞	98
		A		0.2340	5.944	2⅝	67	3⅞	98
15/64				0.2344	5.954	2⅝	67	3⅞	98
		B	6.00	0.2362	6.000	2¾	70	4	102
				0.2380	6.045	2¾	70	4	102
			6.10	0.2402	6.100	2¾	70	4	102
		C		0.2420	6.147	2¾	70	4	102
			6.20	0.2441	6.200	2¾	70	4	102
		D		0.2460	6.248	2¾	70	4	102

(Continued on following page)

TWIST DRILLS (continued)
Straight Shank, Jobbers Length
Fraction, Wire, Letter, Metric
High Speed Steel
Standard Sizes and Dimensions

Diameter of Drill (D)						Flute Length (F)		Overall Length (L)	
Fraction	No.	Ltr.	Millimeter	Decimal Inch Equivalent	Millimeter Equivalent	Inch	mm	Inch	mm
			6.30	0.2480	6.300	2 3/4	70	4	102
1/4		E		0.2500	6.350	2 3/4	70	4	102
			6.40	0.2520	6.400	2 7/8	73	4 1/8	105
			6.50	0.2559	6.500	2 7/8	73	4 1/8	105
		F		0.2570	6.528	2 7/8	73	4 1/8	105
			6.60	0.2598	6.600	2 7/8	73	4 1/8	105
		G		0.2610	6.629	2 7/8	73	4 1/8	105
			6.70	0.2638	6.700	2 7/8	73	4 1/8	105
17/64				0.2656	6.746	2 7/8	73	4 1/8	105
		H		0.2660	6.756	2 7/8	73	4 1/8	105
			6.80	0.2677	6.800	2 7/8	73	4 1/8	105
			6.90	0.2717	6.900	2 7/8	73	4 1/8	105
		I		0.2720	6.909	2 7/8	73	4 1/8	105
			7.00	0.2756	7.000	2 7/8	73	4 1/8	105
		J		0.2770	7.036	2 7/8	73	4 1/8	105
			7.10	0.2795	7.100	2 15/16	75	4 1/4	108
		K		0.2810	7.137	2 15/16	75	4 1/4	108
9/32				0.2812	7.142	2 15/16	75	4 1/4	108
			7.20	0.2835	7.200	2 15/16	75	4 1/4	108
			7.30	0.2874	7.300	2 15/16	75	4 1/4	108
		L		0.2900	7.366	2 15/16	75	4 1/4	108
			7.40	0.2913	7.400	3 1/16	78	4 3/8	111
		M		0.2950	7.493	3 1/16	78	4 3/8	111

(Continued on following page)

TWIST DRILLS (continued)
Straight Shank, Jobbers Length
Fraction, Wire, Letter, Metric
High Speed Steel
Standard Sizes and Dimensions

Diameter of Drill				Decimal Inch Equivalent	Millimeter Equivalent	Flute Length F		Overall Length L	
Fraction	No.	Ltr.	Millimeter			Inch	mm	Inch	mm
			7.50	0.2953	7.500	3 1/16	78	4 3/8	111
19/64				0.2969	7.541	3 1/16	78	4 3/8	111
			7.60	0.2992	7.600	3 1/16	78	4 3/8	111
		N		0.3020	7.671	3 1/16	78	4 3/8	111
			7.70	0.3031	7.700	3 3/16	81	4 1/2	114
			7.80	0.3071	7.800	3 3/16	81	4 1/2	114
			7.90	0.3110	7.900	3 3/16	81	4 1/2	114
5/16				0.3125	7.938	3 3/16	81	4 1/2	114
			8.00	0.3150	8.000	3 3/16	81	4 1/2	114
		O		0.3160	8.026	3 3/16	81	4 1/2	114
			8.10	0.3189	8.100	3 5/16	84	4 5/8	117
			8.20	0.3228	8.200	3 5/16	84	4 5/8	117
		P		0.3230	8.204	3 5/16	84	4 5/8	117
			8.30	0.3268	8.300	3 5/16	84	4 5/8	117
21/64				0.3281	8.334	3 5/16	84	4 5/8	117
			8.40	0.3307	8.400	3 7/16	87	4 3/4	121
		Q		0.3320	8.433	3 7/16	87	4 3/4	121
			8.50	0.3346	8.500	3 7/16	87	4 3/4	121
			8.60	0.3386	8.600	3 7/16	87	4 3/4	121
		R		0.3390	8.611	3 7/16	87	4 3/4	121
			8.70	0.3425	8.700	3 7/16	87	4 3/4	121
11/32				0.3438	8.733	3 7/16	87	4 3/4	121
			8.80	0.3465	8.800	3 1/2	89	4 7/8	124

(Continued on following page)

TWIST DRILLS (continued)
Straight Shank, Jobbers Length
Fraction, Wire, Letter, Metric
High Speed Steel
Standard Sizes and Dimensions

| Diameter of Drill (D) | | | | | | Flute Length (F) | | Overall Length (L) | |
Fraction	No.	Ltr.	Millimeter	Decimal Inch Equivalent	Millimeter Equivalent	Inch	mm	Inch	mm
		S		0.3480	8.839	3½	89	4⅞	124
			8.90	0.3504	8.900	3½	89	4⅞	124
			9.00	0.3543	9.000	3½	89	4⅞	124
		T		0.3580	9.093	3½	89	4⅞	124
			9.10	0.3583	9.100	3½	89	4⅞	124
23/64				0.3594	9.129	3½	89	4⅞	124
			9.20	0.3622	9.200	3⅝	92	5	127
			9.30	0.3661	9.300	3⅝	92	5	127
		U		0.3680	9.347	3⅝	92	5	127
			9.40	0.3701	9.400	3⅝	92	5	127
			9.50	0.3740	9.500	3⅝	92	5	127
3/8				0.3750	9.525	3⅝	92	5	127
		V		0.3770	9.576	3⅝	92	5	127
			9.60	0.3780	9.600	3¾	95	5⅛	130
			9.70	0.3819	9.700	3¾	95	5⅛	130
			9.80	0.3858	9.800	3¾	95	5⅛	130
		W		0.3860	9.804	3¾	95	5⅛	130
			9.90	0.3898	9.900	3¾	95	5⅛	130
25/64				0.3906	9.921	3¾	95	5⅛	130
			10.00	0.3937	10.000	3¾	95	5⅛	130
		X		0.3970	10.084	3¾	95	5⅛	130
			10.20	0.4016	10.200	3⅞	98	5¼	133
		Y		0.4040	10.262	3⅞	98	5¼	133

(Continued on following page)

TWIST DRILLS (continued)
Straight Shank, Jobbers Length
Fraction, Wire, Letter, Metric
High Speed Steel
Standard Sizes and Dimensions

| Diameter of Drill (D) | | | | | | Flute Length (F) | | Overall Length (L) | |
Fraction	No.	Ltr.	Millimeter	Decimal Inch Equivalent	Millimeter Equivalent	Inch	mm	Inch	mm
13/32				0.4062	10.317	3 7/8	98	5 1/4	133
		Z		0.4130	10.490	3 7/8	98	5 1/4	133
			10.50	0.4134	10.500	3 7/8	98	5 1/4	133
27/64				0.4219	10.716	3 15/16	100	5 3/8	137
			10.80	0.4252	10.800	4 1/16	103	5 1/2	140
			11.00	0.4331	11.000	4 1/16	103	5 1/2	140
7/16				0.4375	11.112	4 1/16	103	5 1/2	140
			11.20	0.4409	11.200	4 3/16	106	5 5/8	143
			11.50	0.4528	11.500	4 3/16	106	5 5/8	143
29/64				0.4531	11.509	4 3/16	106	5 5/8	143
			11.80	0.4646	11.800	4 5/16	110	5 3/4	146
15/32				0.4688	11.908	4 5/16	110	5 3/4	146
			12.00	0.4724	12.000	4 3/8	111	5 7/8	149
			12.20	0.4803	12.200	4 3/8	111	5 7/8	149
31/64				0.4844	12.304	4 3/8	111	5 7/8	149
			12.50	0.4921	12.500	4 1/2	114	6	152
1/2				0.5000	12.700	4 1/2	114	6	152
			12.80	0.5039	12.800	4 1/2	114	6	152
			13.00	0.5118	13.000	4 1/2	114	6	152
33/64				0.5156	13.096	4 13/16	122	6 5/8	168
			13.20	0.5197	13.200	4 13/16	122	6 5/8	168
17/32				0.5312	13.492	4 13/16	122	6 5/8	168
			13.50	0.5315	13.500	4 13/16	122	6 5/8	168

(Continued on following page)

TWIST DRILLS (continued)
Straight Shank, Jobbers Length
Fraction, Wire, Letter, Metric
High Speed Steel
Standard Sizes and Dimensions

Diameter of Drill (D)						Flute Length (F)		Overall Length (L)	
Fraction	No.	Ltr.	Millimeter	Decimal Inch Equivalent	Millimeter Equivalent	Inch	mm	Inch	mm
			13.80	0.5433	13.800	4 13/16	122	6 5/8	168
35/64				0.5469	13.891	4 13/16	122	6 5/8	168
			14.00	0.5512	14.000	4 13/16	122	6 5/8	168
			14.25	0.5610	14.250	4 13/16	122	6 5/8	168
9/16				0.5625	14.288	4 13/16	122	6 5/8	168
			14.50	0.5709	14.500	4 13/16	122	6 5/8	168
37/64				0.5781	14.684	4 13/16	122	6 5/8	168
			14.75	0.5807	14.750	5 3/16	132	7 1/8	181
			15.00	0.5906	15.000	5 3/16	132	7 1/8	181
19/32				0.5938	15.083	5 3/16	132	7 1/8	181
			15.25	0.6004	15.250	5 3/16	132	7 1/8	181
39/64				0.6094	15.479	5 3/16	132	7 1/8	181
			15.50	0.6102	15.500	5 3/16	132	7 1/8	181
			15.75	0.6201	15.750	5 3/16	132	7 1/8	181
5/8				0.6250	15.875	5 3/16	132	7 1/8	181
			16.00	0.6299	16.000	5 3/16	132	7 1/8	181
			16.25	0.6398	16.250	5 3/16	132	7 1/8	181
41/64				0.6406	16.271	5 3/16	132	7 1/8	181
			16.50	0.6496	16.500	5 3/16	132	7 1/8	181
21/32				0.6562	16.667	5 3/16	132	7 1/8	181
			16.75	0.6594	16.750	5 5/8	143	7 5/8	194
			17.00	0.6693	17.000	5 5/8	143	7 5/8	194
43/64				0.6719	17.066	5 5/8	143	7 5/8	194

(Concluded on following page)

TWIST DRILLS (concluded)
Straight Shank, Jobbers Length
Fraction, Wire, Letter, Metric
High Speed Steel
Standard Sizes and Dimensions

Diameter of Drill D					Flute Length F		Overall Length L	
Fraction	No.	Ltr.	Millimeter	Decimal Inch Equivalent	Millimeter Equivalent	Inch	mm	Inch	mm
11/16			17.25	0.6791	17.250	5⁵⁄₈	143	7⁵⁄₈	194
				0.6875	17.462	5⁵⁄₈	143	7⁵⁄₈	194
			17.50	0.6890	17.500	5⁵⁄₈	143	7⁵⁄₈	194

See Table 804 for Dimensional Tolerances
See Table 805 for Element Tolerances

TWIST DRILLS
Straight Shank Screw Machine Length
Thru 1 inch (25.4 mm) Diameter
Fraction, Wire, Letter, Metric
High Speed Steel—Standard Sizes and Dimensions

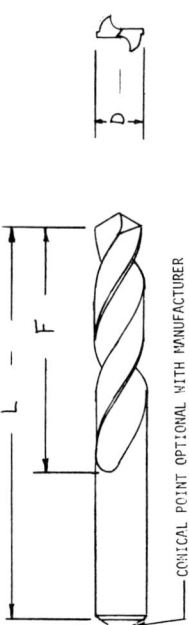

CONICAL POINT OPTIONAL WITH MANUFACTURER

Diameter of Drill					Flute Length F		Overall Length L	
Frac.	No.	Ltr.	mm	Decimal Inch Equivalent	Millimeter Equivalent	Inch	mm	Inch	mm
	60		1.00	0.0394	1.000	½	13	1⅜	35
	59			0.0400	1.016	½	13	1⅜	35
	58			0.0410	1.041	½	13	1⅜	35
			1.05	0.0413	1.050	½	13	1⅜	35
	57			0.0420	1.067	½	13	1⅜	35
			1.10	0.0430	1.092	½	13	1⅜	35
			1.15	0.0433	1.100	½	13	1⅜	35
	56			0.0453	1.150	½	13	1⅜	35
				0.0465	1.181	½	13	1⅜	35
3/64				0.0469	1.191	½	13	1⅜	35
			1.20	0.0472	1.200	⅝	16	1⅝	41
			1.25	0.0492	1.250	⅝	16	1⅝	41
	55		1.30	0.0512	1.300	⅝	16	1⅝	41
				0.0520	1.321	⅝	16	1⅝	41
			1.35	0.0531	1.350	⅝	16	1⅝	41
	54			0.0550	1.397	⅝	16	1⅝	41
			1.40	0.0551	1.400	⅝	16	1⅝	41

(Continued on following page)

TWIST DRILLS (continued)
Straight Shank Screw Machine Length
Thru 1 inch (25.4 mm) Diameter
Fraction, Wire, Letter, Metric
High Speed Steel—Standard Sizes and Dimensions

Diameter of Drill						Flute Length F		Overall Length L	
Frac.	No.	Ltr.	mm	Decimal Inch Equivalent	Millimeter Equivalent	Inch	mm	Inch	mm
			1.45	0.0571	1.450	5/8	16	1 5/8	41
			1.50	0.0591	1.500	5/8	16	1 5/8	41
	53			0.0595	1.511	5/8	16	1 5/8	41
			1.55	0.0610	1.550	5/8	16	1 5/8	41
1/16				0.0625	1.588	5/8	16	1 5/8	41
			1.60	0.0630	1.600	11/16	17	1 11/16	43
	52			0.0635	1.613	11/16	17	1 11/16	43
			1.65	0.0650	1.650	11/16	17	1 11/16	43
			1.70	0.0669	1.700	11/16	17	1 11/16	43
	51			0.0670	1.702	11/16	17	1 11/16	43
			1.75	0.0689	1.750	11/16	17	1 11/16	43
	50			0.0700	1.778	11/16	17	1 11/16	43
			1.80	0.0709	1.800	11/16	17	1 11/16	43
			1.85	0.0728	1.850	11/16	17	1 11/16	43
	49			0.0730	1.854	11/16	17	1 11/16	43
			1.90	0.0748	1.900	11/16	17	1 11/16	43
	48			0.0760	1.930	11/16	17	1 11/16	43
			1.95	0.0768	1.950	11/16	17	1 11/16	43
5/64				0.0781	1.984	11/16	17	1 11/16	43
	47			0.0785	1.994	11/16	17	1 11/16	43
			2.00	0.0787	2.000	11/16	17	1 11/16	43
			2.05	0.0807	2.050	3/4	19	1 3/4	44
	46			0.0810	2.057	3/4	19	1 3/4	44

(Continued on following page)

TWIST DRILLS (continued)
Straight Shank Screw Machine Length
Thru 1 inch (25.4 mm) Diameter
Fraction, Wire, Letter, Metric
High Speed Steel—Standard Sizes and Dimensions

Diameter of Drill D				Decimal Inch Equivalent	Millimeter Equivalent	Flute Length F		Overall Length L	
Frac.	No.	Ltr.	mm			Inch	mm	Inch	mm
	45			0.0820	2.083	3/4	19	1 3/4	44
			2.10	0.0827	2.100	3/4	19	1 3/4	44
			2.15	0.0846	2.150	3/4	19	1 3/4	44
	44			0.0860	2.184	3/4	19	1 3/4	44
			2.20	0.0866	2.200	3/4	19	1 3/4	44
			2.25	0.0886	2.250	3/4	19	1 3/4	44
	43			0.0890	2.261	3/4	19	1 3/4	44
			2.30	0.0906	2.300	3/4	19	1 3/4	44
			2.35	0.0925	2.350	3/4	19	1 3/4	44
	42			0.0935	2.375	3/4	19	1 3/4	44
3/32				0.0938	2.383	3/4	19	1 3/4	44
			2.40	0.0945	2.400	13/16	21	1 13/16	46
	41			0.0960	2.438	13/16	21	1 13/16	46
			2.45	0.0965	2.450	13/16	21	1 13/16	46
	40			0.0980	2.489	13/16	21	1 13/16	46
			2.50	0.0984	2.500	13/16	21	1 13/16	46
	39			0.0995	2.527	13/16	21	1 13/16	46
	38			0.1015	2.578	13/16	21	1 13/16	46
			2.60	0.1024	2.600	13/16	21	1 13/16	46
	37			0.1040	2.642	13/16	21	1 13/16	46
			2.70	0.1063	2.700	13/16	21	1 13/16	46
	36			0.1065	2.705	13/16	21	1 13/16	46
7/64				0.1094	2.779	13/16	21	1 13/16	46

(Continued on following page)

TWIST DRILLS (continued)
Straight Shank Screw Machine Length
Thru 1 inch (25.4 mm) Diameter
Fraction, Wire, Letter, Metric
High Speed Steel—Standard Sizes and Dimensions

Diameter of Drill (D)				Decimal Inch Equivalent	Millimeter Equivalent	Flute Length (F)		Overall Length (L)	
Frac.	No.	Ltr.	mm			Inch	mm	Inch	mm
	35			0.1100	2.794	7/8	22	1 7/8	48
			2.80	0.1102	2.800	7/8	22	1 7/8	48
	34			0.1110	2.819	7/8	22	1 7/8	48
	33			0.1130	2.870	7/8	22	1 7/8	48
			2.90	0.1142	2.900	7/8	22	1 7/8	48
	32			0.1160	2.946	7/8	22	1 7/8	48
			3.00	0.1181	3.000	7/8	22	1 7/8	48
	31			0.1200	3.048	7/8	22	1 7/8	48
			3.10	0.1220	3.100	7/8	22	1 7/8	48
1/8				0.1250	3.175	7/8	22	1 7/8	48
			3.20	0.1260	3.200	15/16	24	1 15/16	49
	30			0.1285	3.264	15/16	24	1 15/16	49
			3.30	0.1299	3.300	15/16	24	1 15/16	49
			3.40	0.1339	3.400	15/16	24	1 15/16	49
	29			0.1360	3.464	15/16	24	1 15/16	49
			3.50	0.1378	3.500	15/16	24	1 15/16	49
	28			0.1405	3.569	15/16	24	1 15/16	49
9/64				0.1406	3.571	15/16	24	1 15/16	49
			3.60	0.1417	3.600	1	25	2 1/16	52
	27			0.1440	3.658	1	25	2 1/16	52
			3.70	0.1457	3.700	1	25	2 1/16	52
	26			0.1470	3.734	1	25	2 1/16	52
	25			0.1495	3.797	1	25	2 1/16	52

(Continued on following page)

TWIST DRILLS (continued)
Straight Shank Screw Machine Length
Thru 1 inch (25.4 mm) Diameter
Fraction, Wire, Letter, Metric
High Speed Steel—Standard Sizes and Dimensions

Diameter of Drill						Flute Length		Overall Length	
D				Decimal Inch Equivalent	Millimeter Equivalent	*F*		*L*	
Frac.	No.	Ltr.	mm			Inch	mm	Inch	mm
			3.80	0.1496	3.800	1	25	2 1/16	52
	24			0.1520	3.861	1	25	2 1/16	52
			3.90	0.1535	3.900	1	25	2 1/16	52
	23			0.1540	3.912	1	25	2 1/16	52
5/32				0.1562	3.967	1	25	2 1/16	52
	22			0.1570	3.988	1 1/16	27	2 1/8	54
			4.00	0.1575	4.000	1 1/16	27	2 1/8	54
	21			0.1590	4.039	1 1/16	27	2 1/8	54
	20			0.1610	4.089	1 1/16	27	2 1/8	54
			4.10	0.1614	4.100	1 1/16	27	2 1/8	54
			4.20	0.1654	4.200	1 1/16	27	2 1/8	54
	19			0.1660	4.216	1 1/16	27	2 1/8	54
			4.30	0.1693	4.300	1 1/16	27	2 1/8	54
	18			0.1695	4.305	1 1/16	27	2 1/8	54
11/64				0.1719	4.366	1 1/16	27	2 1/8	54
	17			0.1730	4.394	1 1/8	29	2 3/16	56
			4.40	0.1732	4.400	1 1/8	29	2 3/16	56
	16			0.1770	4.496	1 1/8	29	2 3/16	56
			4.50	0.1772	4.500	1 1/8	29	2 3/16	56
	15			0.1800	4.572	1 1/8	29	2 3/16	56
			4.60	0.1811	4.600	1 1/8	29	2 3/16	56
	14			0.1820	4.623	1 1/8	29	2 3/16	56
	13		4.70	0.1850	4.700	1 1/8	29	2 3/16	56

(Continued on following page)

TWIST DRILLS (continued)
Straight Shank Screw Machine Length
Thru 1 inch (25.4 mm) Diameter
Fraction, Wire, Letter, Metric
High Speed Steel—Standard Sizes and Dimensions

Diameter of Drill (D)				Decimal Inch Equivalent	Millimeter Equivalent	Flute Length (F)		Overall Length (L)	
Frac.	No.	Ltr.	mm			Inch	mm	Inch	mm
3/16				0.1875	4.762	1 1/8	29	2 3/16	56
	12		4.80	0.1890	4.800	1 3/16	30	2 1/4	57
	11			0.1910	4.851	1 3/16	30	2 1/4	57
			4.90	0.1929	4.900	1 3/16	30	2 1/4	57
	10			0.1935	4.915	1 3/16	30	2 1/4	57
	9			0.1960	4.978	1 3/16	30	2 1/4	57
			5.00	0.1969	5.000	1 3/16	30	2 1/4	57
	8			0.1990	5.054	1 3/16	30	2 1/4	57
			5.10	0.2008	5.100	1 3/16	30	2 1/4	57
	7			0.2010	5.105	1 3/16	30	2 1/4	57
13/64				0.2031	5.159	1 3/16	30	2 1/4	57
	6			0.2040	5.182	1 1/4	32	2 3/8	60
			5.20	0.2047	5.200	1 1/4	32	2 3/8	60
	5			0.2055	5.220	1 1/4	32	2 3/8	60
			5.30	0.2087	5.300	1 1/4	32	2 3/8	60
	4			0.2090	5.309	1 1/4	32	2 3/8	60
			5.40	0.2126	5.400	1 1/4	32	2 3/8	60
	3			0.2130	5.410	1 1/4	32	2 3/8	60
			5.50	0.2165	5.500	1 1/4	32	2 3/8	60
7/32				0.2188	5.558	1 1/4	32	2 3/8	60
			5.60	0.2205	5.600	1 5/16	33	2 7/16	62
	2			0.2210	5.613	1 5/16	33	2 7/16	62
			5.70	0.2244	5.700	1 5/16	33	2 7/16	62

(Continued on following page)

TWIST DRILLS (continued)
Straight Shank Screw Machine Length
Thru 1 inch (25.4 mm) Diameter
Fraction, Wire, Letter, Metric
High Speed Steel—Standard Sizes and Dimensions

Diameter of Drill (D)				Decimal Inch Equivalent	Millimeter Equivalent	Flute Length (F)		Overall Length (L)	
Frac.	No.	Ltr.	mm			Inch	mm	Inch	mm
	1			0.2280	5.791	15/16	33	2 7/16	62
			5.80	0.2283	5.800	15/16	33	2 7/16	62
			5.90	0.2323	5.900	15/16	33	2 7/16	62
		A		0.2340	5.944	15/16	33	2 7/16	62
15/64				0.2344	5.954	15/16	33	2 7/16	62
			6.00	0.2362	6.000	1 3/8	35	2 1/2	64
		B		0.2380	6.045	1 3/8	35	2 1/2	64
			6.10	0.2402	6.100	1 3/8	35	2 1/2	64
		C		0.2420	6.147	1 3/8	35	2 1/2	64
			6.20	0.2441	6.200	1 3/8	35	2 1/2	64
		D		0.2460	6.248	1 3/8	35	2 1/2	64
			6.30	0.2480	6.300	1 3/8	35	2 1/2	64
1/4		E		0.2500	6.350	1 3/8	35	2 1/2	64
			6.40	0.2520	6.400	1 7/16	37	2 5/8	67
			6.50	0.2559	6.500	1 7/16	37	2 5/8	67
		F		0.2570	6.528	1 7/16	37	2 5/8	67
			6.60	0.2598	6.600	1 7/16	37	2 5/8	67
		G		0.2610	6.629	1 7/16	37	2 5/8	67
			6.70	0.2638	6.700	1 7/16	37	2 5/8	67
17/64				0.2656	6.746	1 7/16	37	2 5/8	67
		H		0.2660	6.756	1 1/2	38	2 11/16	68
			6.80	0.2677	6.800	1 1/2	38	2 11/16	68
			6.90	0.2717	6.900	1 1/2	38	2 11/16	68

(Continued on following page)

TWIST DRILLS (continued)
Straight Shank Screw Machine Length
Thru 1 inch (25.4 mm) Diameter
Fraction, Wire, Letter, Metric
High Speed Steel—Standard Sizes and Dimensions

Diameter of Drill (D)				Decimal Inch Equivalent	Millimeter Equivalent	Flute Length (F)		Overall Length (L)	
Frac.	No.	Ltr.	mm			Inch	mm	Inch	mm
		I		0.2720	6.909	1½	38	2 11/16	68
			7.00	0.2756	7.000	1½	38	2 11/16	68
		J		0.2770	7.036	1½	38	2 11/16	68
			7.10	0.2795	7.100	1½	38	2 11/16	68
		K		0.2810	7.137	1½	38	2 11/16	68
9/32				0.2812	7.142	1½	38	2 11/16	68
			7.20	0.2835	7.200	1 9/16	40	2 3/4	70
			7.30	0.2874	7.300	1 9/16	40	2 3/4	70
		L		0.2900	7.366	1 9/16	40	2 3/4	70
			7.40	0.2913	7.400	1 9/16	40	2 3/4	70
		M		0.2950	7.493	1 9/16	40	2 3/4	70
			7.50	0.2953	7.500	1 9/16	40	2 3/4	70
19/64				0.2969	7.541	1 9/16	40	2 3/4	70
			7.60	0.2992	7.600	1 5/8	41	2 13/16	71
		N		0.3020	7.671	1 5/8	41	2 13/16	71
			7.70	0.3031	7.700	1 5/8	41	2 13/16	71
			7.80	0.3071	7.800	1 5/8	41	2 13/16	71
			7.90	0.3110	7.900	1 5/8	41	2 13/16	71
5/16				0.3125	7.938	1 5/8	41	2 13/16	71
			8.00	0.3150	8.000	1 11/16	43	2 15/16	75
		O		0.3160	8.026	1 11/16	43	2 15/16	75
			8.10	0.3189	8.100	1 11/16	43	2 15/16	75
			8.20	0.3228	8.200	1 11/16	43	2 15/16	75

(Continued on following page)

TWIST DRILLS (continued)
Straight Shank Screw Machine Length
Thru 1 inch (25.4 mm) Diameter
Fraction, Wire, Letter, Metric
High Speed Steel—Standard Sizes and Dimensions

Diameter of Drill (D)				Decimal Inch Equivalent	Millimeter Equivalent	Flute Length (F)		Overall Length (L)	
Frac.	No.	Ltr.	mm			Inch	mm	Inch	mm
		P		0.3230	8.204	1 11/16	43	2 15/16	75
			8.30	0.3268	8.300	1 11/16	43	2 15/16	75
21/64				0.3281	8.334	1 11/16	43	2 15/16	75
			8.40	0.3307	8.400	1 11/16	43	3	76
		Q		0.3320	8.433	1 11/16	43	3	76
			8.50	0.3346	8.500	1 11/16	43	3	76
			8.60	0.3386	8.600	1 11/16	43	3	76
		R		0.3390	8.611	1 11/16	43	3	76
			8.70	0.3425	8.700	1 11/16	43	3	76
11/32				0.3438	8.733	1 11/16	43	3	76
			8.80	0.3465	8.800	1 3/4	44	3 1/16	78
		S		0.3480	8.839	1 3/4	44	3 1/16	78
			8.90	0.3504	8.900	1 3/4	44	3 1/16	78
			9.00	0.3543	9.000	1 3/4	44	3 1/16	78
		T		0.3580	9.093	1 3/4	44	3 1/16	78
			9.10	0.3583	9.100	1 3/4	44	3 1/16	78
23/64				0.3594	9.129	1 3/4	44	3 1/16	78
			9.20	0.3622	9.200	1 13/16	46	3 1/8	79
			9.30	0.3661	9.300	1 13/16	46	3 1/8	79
		U		0.3680	9.347	1 13/16	46	3 1/8	79
			9.40	0.3701	9.400	1 13/16	46	3 1/8	79
			9.50	0.3740	9.500	1 13/16	46	3 1/8	79
3/8				0.3750	9.525	1 13/16	46	3 1/8	79

(Continued on following page)

TWIST DRILLS (continued)
Straight Shank Screw Machine Length
Thru 1 inch (25.4 mm) Diameter
Fraction, Wire, Letter, Metric
High Speed Steel—Standard Sizes and Dimensions

Diameter of Drill				Decimal Inch Equivalent	Millimeter Equivalent	Flute Length		Overall Length	
Frac.	No.	Ltr.	mm			Inch	mm	Inch	mm
		V	9.60	0.3770	9.576	1⅞	48	3¼	83
			9.70	0.3780	9.600	1⅞	48	3¼	83
				0.3819	9.700	1⅞	48	3¼	83
		W	9.80	0.3858	9.800	1⅞	48	3¼	83
				0.3860	9.804	1⅞	48	3¼	83
			9.90	0.3898	9.900	1⅞	48	3¼	83
25/64				0.3906	9.921	1⅞	48	3¼	83
			10.00	0.3937	10.000	1 15/16	49	3 5/16	84
		X		0.3970	10.084	1 15/16	49	3 5/16	84
			10.20	0.4016	10.200	1 15/16	49	3 5/16	84
		Y		0.4040	10.262	1 15/16	49	3 5/16	84
13/32				0.4062	10.320	1 15/16	49	3 5/16	84
		Z		0.4130	10.490	2	51	3⅜	86
			10.50	0.4134	10.500	2	51	3⅜	86
27/64				0.4219	10.716	2	51	3⅜	86
			10.80	0.4252	10.800	2 1/16	52	3 7/16	87
			11.00	0.4331	11.000	2 1/16	52	3 7/16	87
7/16				0.4375	11.112	2 1/16	52	3 7/16	87
			11.20	0.4409	11.200	2⅛	54	3 9/16	90
			11.50	0.4528	11.500	2⅛	54	3 9/16	90
29/64				0.4531	11.509	2⅛	54	3 9/16	90
			11.80	0.4646	11.800	2⅛	54	3⅝	92
15/32				0.4688	11.908	2⅛	54	3⅝	92

(Continued on following page)

TWIST DRILLS (continued)
Straight Shank Screw Machine Length
Thru 1 inch (25.4 mm) Diameter
Fraction, Wire, Letter, Metric
High Speed Steel—Standard Sizes and Dimensions

Diameter of Drill D				Decimal Inch Equivalent	Millimeter Equivalent	Flute Length F		Overall Length L	
Frac.	No.	Ltr.	mm			Inch	mm	Inch	mm
			12.00	0.4724	12.000	2 3/16	56	3 11/16	94
			12.20	0.4803	12.200	2 3/16	56	3 11/16	94
31/64				0.4844	12.304	2 3/16	56	3 11/16	94
			12.50	0.4921	12.500	2 1/4	57	3 3/4	95
1/2				0.5000	12.700	2 1/4	57	3 3/4	95
			12.80	0.5039	12.800	2 3/8	60	3 7/8	98
			13.00	0.5118	13.000	2 3/8	60	3 7/8	98
33/64				0.5156	13.096	2 3/8	60	3 7/8	98
			13.20	0.5197	13.200	2 3/8	60	3 7/8	98
17/32				0.5312	13.492	2 3/8	60	3 7/8	98
			13.50	0.5315	13.500	2 3/8	60	3 7/8	98
			13.80	0.5433	13.800	2 1/2	64	4	102
35/64				0.5469	13.891	2 1/2	64	4	102
			14.00	0.5512	14.000	2 1/2	64	4	102
			14.25	0.5610	14.250	2 1/2	64	4	102
9/16				0.5625	14.288	2 1/2	64	4	102
			14.50	0.5709	14.500	2 5/8	67	4 1/8	105
37/64				0.5781	14.684	2 5/8	67	4 1/8	105
			14.75	0.5807	14.750	2 5/8	67	4 1/8	105
			15.00	0.5906	15.000	2 5/8	67	4 1/8	105
19/32				0.5938	15.083	2 5/8	67	4 1/8	105
			15.25	0.6004	15.250	2 3/4	70	4 1/4	108
39/64				0.6094	15.479	2 3/4	70	4 1/4	108

(Continued on following page)

TWIST DRILLS (continued)
Straight Shank Screw Machine Length
Thru 1 inch (25.4 mm) Diameter
Fraction, Wire, Letter, Metric
High Speed Steel—Standard Sizes and Dimensions

Diameter of Drill *D*						Flute Length *F*		Overall Length *L*	
Frac.	No.	Ltr.	mm	Decimal Inch Equivalent	Millimeter Equivalent	Inch	mm	Inch	mm
			15.50	0.6102	15.500	2 3/4	70	4 1/4	108
			15.75	0.6201	15.750	2 3/4	70	4 1/4	108
5/8				0.6250	15.815	2 3/4	70	4 1/4	108
			16.00	0.6299	16.000	2 7/8	73	4 1/2	114
			16.25	0.6398	16.250	2 7/8	73	4 1/2	114
41/64				0.6406	16.271	2 7/8	73	4 1/2	114
			16.50	0.6496	16.500	2 7/8	73	4 1/2	114
21/32				0.6562	16.667	2 7/8	73	4 1/2	114
			16.75	0.6594	16.750	2 7/8	73	4 1/2	114
			17.00	0.6693	17.000	2 7/8	73	4 1/2	114
43/64				0.6719	17.066	2 7/8	73	4 1/2	114
			17.25	0.6791	17.250	2 7/8	73	4 1/2	114
11/16				0.6875	17.462	2 7/8	73	4 1/2	114
			17.50	0.6890	17.500	3	76	4 3/4	121
45/64				0.7031	17.859	3	76	4 3/4	121
			18.00	0.7087	18.000	3	76	4 3/4	121
23/32				0.7188	18.258	3	76	4 3/4	121
			18.50	0.7283	18.500	3 1/8	79	5	127
47/64				0.7344	18.654	3 1/8	79	5	127
			19.00	0.7480	19.000	3 1/8	79	5	127
3/4				0.7500	19.050	3 1/8	79	5	127
49/64				0.7656	19.446	3 1/4	83	5 1/8	130
			19.50	0.7677	19.500	3 1/4	83	5 1/8	130

(Continued on following page)

TWIST DRILLS (continued)
Straight Shank Screw Machine Length
Thru 1 inch (25.4 mm) Diameter
Fraction, Wire, Letter, Metric
High Speed Steel—Standard Sizes and Dimensions

| Diameter of Drill (D) | | | | | | Flute Length (F) | | Overall Length (L) | |
Frac.	No.	Ltr.	mm	Decimal Inch Equivalent	Millimeter Equivalent	Inch	mm	Inch	mm
25/32				0.7812	19.845	3 1/4	83	5 1/8	130
51/64			20.00	0.7879	20.000	3 3/8	86	5 1/4	133
				0.7969	20.241	3 3/8	86	5 1/4	133
13/16			20.50	0.8071	20.500	3 3/8	86	5 1/4	133
				0.8125	20.638	3 3/8	86	5 1/4	133
			21.00	0.8268	21.000	3 1/2	89	5 3/8	137
53/64				0.8281	21.034	3 1/2	89	5 3/8	137
27/32				0.8438	21.433	3 1/2	89	5 3/8	137
			21.50	0.8465	21.500	3 1/2	89	5 3/8	137
55/64				0.8594	21.829	3 1/2	89	5 3/8	137
			22.00	0.8661	22.000	3 1/2	89	5 3/8	137
7/8				0.8750	22.225	3 1/2	89	5 3/8	137
			22.50	0.8858	22.500	3 5/8	92	5 5/8	143
57/64				0.8906	22.621	3 5/8	92	5 5/8	143
			23.00	0.9055	23.000	3 5/8	92	5 5/8	143
29/32				0.9062	23.017	3 5/8	92	5 5/8	143
59/64				0.9219	23.416	3 3/4	95	5 3/4	146
			23.50	0.9252	23.500	3 3/4	95	5 3/4	146
15/16				0.9375	23.812	3 3/4	95	5 3/4	146
			24.00	0.9449	24.000	3 7/8	98	5 7/8	149
61/64				0.9531	24.209	3 7/8	98	5 7/8	149
			24.50	0.9646	24.500	3 7/8	98	5 7/8	149
31/32				0.9688	24.608	3 7/8	98	5 7/8	149

(Concluded on following page)

TWIST DRILLS (concluded)
Straight Shank Screw Machine Length
Thru 1 inch (25.4 mm) Diameter
Fraction, Wire, Letter, Metric
High Speed Steel—Standard Sizes and Dimensions

Diameter of Drill					Flute Length		Overall Length	
D			Decimal Inch Equivalent	Millimeter Equivalent	F		L	
Frac.	No.	Ltr.	mm			Inch	mm	Inch	mm
63/64				0.9843	25.000	4	102	6	152
				0.9844	25.004	4	102	6	152
1			25.00	1.0000	25.400	4	102	6	152

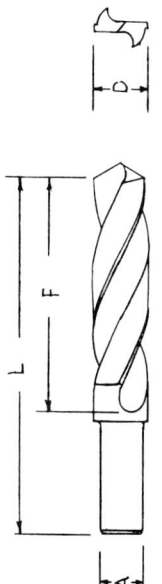

TWIST DRILLS
Straight Shank Screw Machine Length
Over 1 inch (25.4 mm) Diameter
Fraction and Metric Sizes
High Speed Steel—Standard Sizes and Dimensions

| Diameter of Drill | | | Flute Length | | Overall Length | | Shank Diameter | |
| D | | | F | | L | | A | |
Frac.	mm	Decimal Inch Equivalent	Millimeter Equivalent	Inch	mm	Inch	mm	Decimal Inch	mm
	25.50	1.0039	25.500	4	102	6	152	0.9843	25.00
	26.00	1.0236	26.000	4	102	6	152	0.9843	25.00
1 1/16		1.0625	26.988	4	102	6	152	1.0000	25.40
	28.00	1.1024	28.000	4	102	6	152	0.9843	25.00
1 1/8		1.1250	28.575	4	102	6	152	1.0000	25.40
	30.00	1.1811	30.000	4 1/4	108	6 5/8	168	0.9843	25.00
1 3/16		1.1875	30.162	4 1/4	108	6 5/8	168	1.0000	25.40
1 1/4		1.2500	31.750	4 3/8	111	6 3/4	171	1.0000	25.40
	32.00	1.2598	32.000	4 3/8	111	7	178	1.2402	31.50
1 5/16		1.3125	33.338	4 3/8	111	7	178	1.2500	31.75
	34.00	1.3386	34.000	4 1/2	114	7 1/8	181	1.2402	31.50
1 3/8		1.3750	34.925	4 1/2	114	7 1/8	181	1.2500	31.75
	36.00	1.4173	36.000	4 3/4	121	7 3/8	187	1.2402	31.50
1 7/16		1.4375	36.512	4 3/4	121	7 3/8	187	1.2500	31.75
	38.00	1.4961	38.000	4 7/8	124	7 1/2	190	1.2402	31.50
1 1/2		1.5000	38.100	4 7/8	124	7 1/2	190	1.2500	31.75
1 9/16		1.5625	39.688	4 7/8	124	7 3/4	197	1.5000	38.10

(Concluded on following page)

TWIST DRILLS (concluded)
Straight Shank Screw Machine Length
Over 1 inch (25.4 mm) Diameter
Fraction and Metric Sizes
High Speed Steel—Standard Sizes and Dimensions

Diameter of Drill (D)				Flute Length (F)		Overall Length (L)		Shank Diameter (A)	
Frac.	mm	Decimal Inch Equivalent	Millimeter Equivalent	Inch	mm	Inch	mm	Decimal Inch	mm
	40.00	1.5748	40.000	4 7/8	124	7 3/4	197	1.4961	38.00
1 5/8		1.6250	41.275	4 7/8	124	7 3/4	197	1.5000	38.10
	42.00	1.6535	42.000	5 1/8	130	8	203	1.4961	38.00
1 11/16		1.6875	42.862	5 1/8	130	8	203	1.5000	38.10
	44.00	1.7323	44.000	5 1/8	130	8	203	1.4961	38.00
1 3/4		1.7500	44.450	5 1/8	130	8	203	1.5000	38.10
	46.00	1.8110	46.000	5 3/8	137	8 1/4	210	1.4961	38.00
1 13/16		1.8125	46.038	5 3/8	137	8 1/4	210	1.5000	38.10
1 7/8		1.8750	47.625	5 3/8	137	8 1/4	210	1.5000	38.10
	48.00	1.8898	48.000	5 5/8	143	8 1/2	216	1.4961	38.00
1 15/16		1.9375	49.212	5 5/8	143	8 1/2	216	1.5000	38.10
	50.00	1.9685	50.000	5 5/8	143	8 1/2	216	1.4961	38.00
2		2.0000	50.800	5 5/8	143	8 1/2	216	1.5000	38.10

COMBINED DRILL AND COUNTERSINK

High Speed Steel

Plain Type

Bell Type

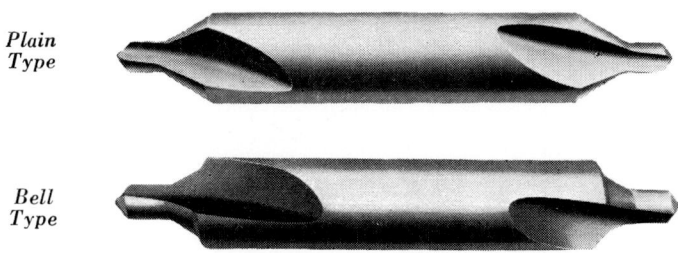

Plain Type

Size Designation	Dimensions—Inches			
	Body Diam.	Drill Diam.	Drill Length	Overall Length
	A	D	C	L
00	⅛	.025	.040	1⁷⁄₃₂
0	⅛	¹⁄₃₂	³⁄₆₄	1⁷⁄₃₂
1	⅛	³⁄₆₄	³⁄₆₄	1¼
2	³⁄₁₆	⁵⁄₆₄	⁵⁄₆₄	1⅞
3	¼	⁷⁄₆₄	⁷⁄₆₄	2
4	⁵⁄₁₆	⅛	⅛	2⅛
5	⁷⁄₁₆	³⁄₁₆	³⁄₁₆	2¾
6	½	⁷⁄₃₂	⁷⁄₃₂	3
7	⅝	¼	¼	3¼
8	¾	⁵⁄₁₆	⁵⁄₁₆	3½

Bell Type

Size Designation	Dimensions — Inches				
	Body Diam.	Drill Diam.	Drill Length	Overall Length	Bell Diam.
	A	D	C	L	E
11	⅛	³⁄₆₄	³⁄₆₄	1¼	.100
12	³⁄₁₆	¹⁄₁₆	¹⁄₁₆	1⅞	.150
13	¼	³⁄₃₂	³⁄₃₂	2	.200
14	⁵⁄₁₆	⁷⁄₆₄	⁷⁄₆₄	2⅛	.250
15	⁷⁄₁₆	⁵⁄₃₂	⁵⁄₃₂	2¾	.350
16	½	³⁄₁₆	³⁄₁₆	3	.400
17	⅝	⁷⁄₃₂	⁷⁄₃₂	3¼	.500
18	¾	¼	¼	3½	.600

CARBIDE TIPPED JOBBERS LENGTH STRAIGHT SHANK DRILLS

FRACTION SIZES

Diam. Inches	Decimal Equivalents	Length of Flute Inches	Length Overall Inches
1/8	.1250	1 5/8	2 3/4
9/64	.1406	1 3/4	2 7/8
5/32	.1562	2	3 1/8
11/64	.1719	2 1/8	3 1/4
3/16	.1875	2 5/16	3 1/2
13/64	.2031	2 7/16	3 5/8
7/32	.2188	2 1/2	3 3/4
15/64	.2344	2 5/8	3 7/8
1/4	.2500	2 3/4	4
17/64	.2656	2 7/8	4 1/8
9/32	.2812	2 15/16	4 1/4
19/64	.2969	3 1/16	4 3/8
5/16	.3125	3 3/16	4 1/2
21/64	.3281	3 5/16	4 5/8
11/32	.3438	3 7/16	4 3/4
23/64	.3594	3 1/2	4 7/8
3/8	.3750	3 5/8	5
25/64	.3906	3 3/4	5 1/8
13/32	.4062	3 7/8	5 1/4
27/64	.4219	3 15/16	5 3/8
7/16	.4375	4 1/16	5 1/2
29/64	.4531	4 3/16	5 5/8
15/32	.4688	4 5/16	5 3/4
31/64	.4844	4 3/8	5 7/8
1/2	.5000	4 1/2	6

WIRE GAUGE SIZES

Wire Gauge No.	Decimal Diam. Inches	Length of Flute Inches	Length Overall Inches
1	.2280	2 5/8	3 7/8
2	.2210	2 5/8	3 7/8
3	.2130	2 1/2	3 3/4
4	.2090	2 1/2	3 3/4
5	.2055	2 1/2	3 3/4
6	.2040	2 1/2	3 3/4
7	.2010	2 7/16	3 5/8
8	.1990	2 7/16	3 5/8
9	.1960	2 7/16	3 5/8
10	.1935	2 7/16	3 5/8
11	.1910	2 5/16	3 1/2
12	.1890	2 5/16	3 1/2
13	.1850	2 5/16	3 1/2
14	.1820	2 3/16	3 3/8
15	.1800	2 3/16	3 3/8
16	.1770	2 3/16	3 3/8
17	.1730	2 3/16	3 3/8
18	.1695	2 1/8	3 1/4
19	.1660	2 1/8	3 1/4
20	.1610	2 1/8	3 1/4
21	.1590	2 1/8	3 1/4
22	.1570	2	3 1/8
23	.1540	2	3 1/8
24	.1520	2	3 1/8
25	.1495	1 7/8	3
26	.1470	1 7/8	3
27	.1440	1 7/8	3
28	.1405	1 3/4	2 7/8
29	.1360	1 3/4	2 7/8
30	.1285	1 5/8	2 3/4
31	.1200	1 5/8	2 3/4
32	.1160	1 5/8	2 3/4

LETTER SIZES

Diam.	Decimal Diam. Inches	Length of Flute Inches	Length Overall Inches
A	.234	2 5/8	3 7/8
B	.238	2 3/4	4
C	.242	2 3/4	4
D	.246	2 3/4	4
E	.250	2 3/4	4
F	.257	2 7/8	4 1/8
G	.261	2 7/8	4 1/8
H	.266	2 7/8	4 1/8
I	.272	2 7/8	4 1/8
J	.277	2 7/8	4 1/8
K	.281	2 15/16	4 1/4
L	.290	2 15/16	4 1/4
M	.295	3 1/16	4 3/8
N	.302	3 1/16	4 3/8
O	.316	3 3/16	4 1/2
P	.323	3 5/16	4 5/8
Q	.332	3 7/16	4 3/4
R	.339	3 7/16	4 3/4
S	.348	3 1/2	4 7/8
T	.358	3 1/2	4 7/8
U	.368	3 5/8	5
V	.377	3 5/8	5
W	.386	3 3/4	5 1/8
X	.397	3 3/4	5 1/8
Y	.404	3 7/8	5 1/4
Z	.413	3 7/8	5 1/4

CARBIDE TIPPED TAPER SHANK DRILLS

Diameter Inches	Decimal Equivalents	Length of Flute Inches	Length Overall Inches	Morse Taper Shank	Diameter Inches	Decimal Equivalents	Length of Flute Inches	Length Overall Inches	Morse Taper Shunk
1/8	.1250	1 7/8	5 1/8	1	3/4	.7500	5 7/8	9 3/4	2
3/16	.1875	2 1/2	5 3/4	1	25/32	.7812	6	9 7/8	2
1/4	.2500	2 7/8	6 1/8	1	13/16	.8215	6 1/8	10 3/4	3
5/16	.3125	3 1/8	6 3/8	1	27/32	.8438	6 1/8	10 3/4	3
3/8	.3750	3 1/2	6 3/4	1	7/8	.8750	6 1/8	10 3/4	3
7/16	.4375	3 7/8	7 1/4	1	29/32	.9062	6 1/8	10 3/4	3
1/2	.5000	4 3/8	8 1/4	2	15/16	.9375	6 1/8	10 3/4	3
17/32	.5312	4 5/8	8 1/2	2	31/32	.9688	6 3/8	11	3
9/16	.5625	4 7/8	8 3/4	2	1	1.0000	6 3/8	11	3
19/32	.5938	4 7/8	8 3/4	2	1 1/16	1.0625	6 5/8	11 1/4	3
5/8	.6250	4 7/8	8 3/4	2	1 1/8	1.1250	7 1/8	12 3/4	4
21/32	.6562	5 1/8	9	2	1 3/16	1.1875	7 3/8	13	4
11/16	.6875	5 3/8	9 1/4	2	1 1/4	1.2500	7 7/8	13 1/2	4
23/32	.7188	5 5/8	9 1/2	2					

CARBIDE TIPPED TAPER LENGTH STRAIGHT SHANK DRILLS

Diam- eter	Decimal Equiva- lent Inches	Length Flute Inches	Length Overall Inches	Diam- eter	Decimal Equiva- lent Inches	Length Flute Inches	Length Overall Inches
1/8	.1250	2¾	5⅛	37/64	.5781	4⅞	8¾
9/64	.1406	3	5⅜	19/32	.5938	4⅞	8¾
5/32	.1562	3	5⅜	39/64	.6094	4⅞	8¾
11/64	.1719	3⅜	5¾	5/8	.6250	4⅞	8¾
3/16	.1875	3⅜	5¾	41/64	.6406	5⅛	9
13/64	.2031	3⅝	6	21/32	.6562	5⅛	9
7/32	.2188	3⅝	6	43/64	.6719	5⅜	9¼
15/64	.2344	3¾	6⅛	11/16	.6875	5⅜	9¼
1/4	.2500	3¾	6⅛	45/64	.7031	5⅝	9½
17/64	.2656	3⅞	6¼	23/32	.7188	5⅝	9½
9/32	.2812	3⅞	6¼	47/64	.7344	5⅞	9¾
19/64	.2969	4	6⅜	3/4	.7500	5⅞	9¾
5/16	.3125	4	6⅜	49/64	.7656	6	9⅞
21/64	.3281	4⅛	6½	25/32	.7812	6	9⅞
11/32	.3438	4⅛	6½	51/64	.7969	6⅛	10
23/64	.3594	4¼	6¾	13/16	.8125	6⅛	10
3/8	.3750	4¼	6¾	53/64	.8281	6⅛	10
25/64	.3906	4⅜	7	27/32	.8438	6⅛	10
13/32	.4062	4⅜	7	55/64	.8594	6⅛	10
27/64	.4219	4⅝	7¼	7/8	.8750	6⅛	10
7/16	.4375	4⅝	7¼	57/64	.8906	6⅛	10
29/64	.4531	4¾	7½	29/32	.9062	6⅛	10
15/32	.4688	4¾	7½	59/64	.9219	6⅛	10¾
31/64	.4844	4¾	7¾	15/16	.9375	6⅛	10¾
1/2	.5000	4¾	7¾	61/64	.9531	6⅜	11
33/64	.5156	4¾	8	31/32	.9688	6⅜	11
17/32	.5312	4¾	8	63/64	.9844	6⅜	11
35/64	.5469	4⅞	8¼	1	1.0000	6⅜	11
9/16	.5625	4⅞	8¼				

CARBIDE TIPPED FOUR FLUTED TAPER SHANK DRILLS

Diameter Inches	Decimal Equivalents	Length of Flute Inches	Length Overall Inches	Shank Taper No.	Diameter Inches	Decimal Equivalents	Length of Flute Inches	Length Overall Inches	Shank Taper No.
1/2	.5000	4 3/8	8 1/4	2	1 1/32	1.031	4 7/8	9 1/2	3
17/32	.5312	4 3/8	8 1/4	2	1 1/16	1.062	4 7/8	9 1/2	3
9/16	.5625	4 3/8	8 1/4	2	1 3/32	1.093	4 7/8	10 1/2	4
19/32	.5938	4 3/8	8 1/4	2	1 1/8	1.125	4 7/8	10 1/2	4
5/8	.6250	4 3/8	8 1/4	2	1 5/32	1.1562	4 7/8	10 1/2	4
21/32	.6562	4 3/8	8 1/4	2	1 3/16	1.1875	4 7/8	10 1/2	4
11/16	.6875	4 3/8	8 1/4	2	1 7/32	1.2188	4 7/8	10 1/2	4
23/32	.7188	4 3/8	8 1/4	2	1 1/4	1.2500	4 7/8	10 1/2	4
3/4	.7500	4 3/8	8 1/4	2	1 9/32	1.2812	4 7/8	10 1/2	4
25/32	.7812	4 3/8	8 1/4	2	1 5/16	1.3175	4 7/8	10 1/2	4
13/16	.8125	4 7/8	9 1/2	3	1 11/32	1.3438	4 7/8	10 1/2	4
27/32	.8438	4 7/8	9 1/2	3	1 3/8	1.3750	4 7/8	10 1/2	4
7/8	.8750	4 7/8	9 1/2	3	1 13/32	1.4062	4 7/8	10 1/2	4
29/32	.9062	4 7/8	9 1/2	3	1 7/16	1.4375	4 7/8	10 1/2	4
15/16	.9375	4 7/8	9 1/2	3	1 15/32	1.4688	4 7/8	10 1/2	4
31/32	.9688	4 7/8	9 1/2	3	1 1/2	1.5000	4 7/8	10 1/2	4
1	1.0000	4 7/8	9 1/2	3					

REAMER SECTION

FOREWORD

Modern engineering requirements often involve the mass production of holes having smooth surfaces, accurate location, or uniform size. Quite often, holes produced by drilling alone do not entirely satisfy these requirements. To satisfy these needs, the reamer, which is a tool used for enlarging or finishing to size a previously formed hole, was developed. The reamer is one of the most important tools used in the manufacture of interchangeable parts for mass production. It provides the most economical method of securing precision fits and universal interchangeability.

Continued advances in reamer design and materials have resulted in the accurate, high-production, long-lived tools so familiar to all engaged in metal cutting. Engineering departments and laboratories of users and manufacturers alike are constantly experimenting to make further refinements of this precision cutting tool.

This section contains suggestions for proper selection and application of reamers and recommendations pointing toward economy in use. Included also, is a description of the more common standard types.

REAMER SECTION CONTENTS

NOMENCLATURE OF REAMERS

I—Definition

Reamer—A rotary cutting tool with one or more cutting elements, used for enlarging to size and contour a previously formed hole. Its principal support during the cutting action is obtained from the workpiece.

II—General Classifications

A—Classification Based on Construction.

1—*Solid Reamers*—Those made of one piece of tool material.

2—*Tipped Solid Reamers*—Those which have a body of one material with cutting edges of another material brazed or otherwise bonded in place.

3—*Inserted Blade Reamers*—Those which have replaceable mechanically retained blades. These blades may be solid or tipped and are usually adjustable.

4—*Expansion Reamers*—Those whose size may be increased by deflecting or bending segments of the reamer body.

5—*Adjustable Reamers*—Those whose size may be changed by sliding, or otherwise moving, the blades toward or away from the reamer axis.

B—Classification Based on Method of Holding, or Driving.

1—*Hand Reamers*—Those which are ordinarily used by hand. A driving square is provided at the end of the shank. The cutting end is provided with a starting taper for easy entry.

2—*Machine Reamers*—Those having shanks suitable for mounting in machines.

3—*Shell Reamers*—Machine reamers mountable on arbors specifically designed for the purpose, called "Shell Reamer Arbors."

III—Explanation of the "Hand" of Reamers

The terms "right hand" and "left hand" are used to describe both direction of rotation and direction of flute helix of reamers.

A—Hand of Rotation [or Hand of Cut].

1—*Right Hand Rotation* (*or Right Hand Cut*) is the counterclockwise rotation of a reamer revolving so as to make a cut when viewed from the cutting end.

2—*Left Hand Rotation* (*or Left Hand Cut*) is the clockwise rotation of a reamer revolving so as to make a cut when viewed from the cutting end. Special Case—Pull reamers are considered to be Right Hand cut if, when viewed from the cutting end, they must be rotated in a clockwise direction to cut.

B—Hand of Flute Helix.

Reamers with their cutting edges in planes parallel to the reamer axes are described as having "Straight Flutes."

Reamers with every other flute of opposite [right and left hand] helix are called "Alternate Helix Reamers."

Reamers with flute helix in one direction only are described as having Right Hand or Left Hand helix.

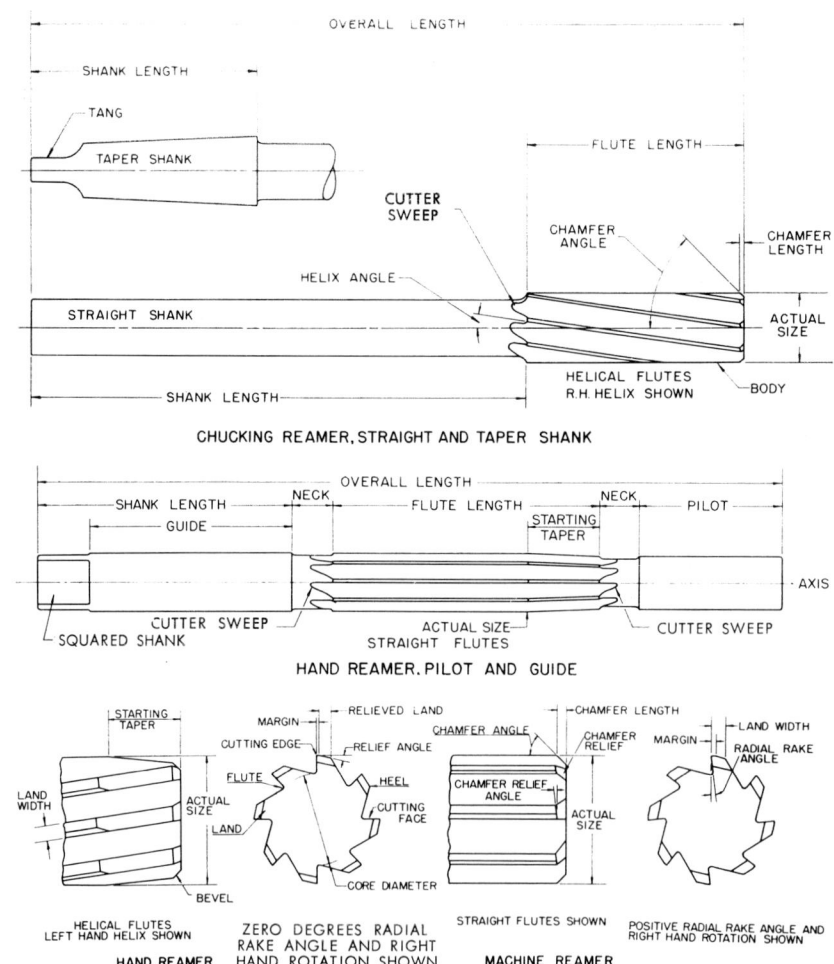

CHUCKING REAMER, STRAIGHT AND TAPER SHANK

HAND REAMER. PILOT AND GUIDE

HAND REAMER — ZERO DEGREES RADIAL RAKE ANGLE AND RIGHT HAND ROTATION SHOWN

MACHINE REAMER — POSITIVE RADIAL RAKE ANGLE AND RIGHT HAND ROTATION SHOWN

1—A Reamer has Right Hand helix when the flutes twist away from the observer in a clockwise direction when viewed from either end of the reamer.

2—A Reamer has Left Hand helix when the flutes twist away from the observer in a counterclockwise direction when viewed from either end of the reamer.

IV. NOMENCLATURE OF REAMER ELEMENTS AND OTHER TERMS RELATING TO REAMING

ACTUAL SIZE—The actual measured diameter of a reamer, usually slightly larger than the nominal size to allow for wear.

ALTERNATE—Reamer features which differ from each other in turn in a regular sequence such as cutting edges, chip breakers, chamfers, or flutes.

ANGLE OF TAPER—The included angle of taper on a taper tool or taper shank.

ANGULAR FLUTE—See FLUTES.

ARBOR HOLE—The central mounting hole in a Shell Reamer.

AXIS—The imaginary straight line which forms the longitudinal center-line of a reamer, usually established by rotating the reamer between centers.

BACK TAPER—A slight decrease in diameter, from front to back in the flute length of reamers.

BARREL—See preferred term BODY.

BELL MOUTH HOLE—A hole which is larger in diameter at the start of the hole than at some distance beyond.

BEVEL—An unrelieved angular surface of revolution. [Not to be confused with chamfer.]

BLADE—A tooth or cutting element inserted in a reamer body. It may be adjustable and/or replaceable.

BLENDING RADIUS—A relieved radius joining the chamfer and the periphery.

BODY—1—The fluted full diameter portion of a reamer, inclusive of the chamfer, starting taper and bevel.—2—The principal supporting member for a set of reamer blades, usually including the shank.

BURNISHING REAMER—A finishing reamer intended to take a light scraping cut and impart a fine finish.

CHAMBERING REAMER—A reamer [usually one of a series] for forming a shell chamber, etc.

CHAMFER—The angular cutting portion at the entering end of a reamer. [See also SECONDARY CHAMFER.]

CHAMFER ANGLE—The angle between the axis and the cutting edge of the chamfer measured in an axial plane at the cutting edge.

CHAMFER LENGTH—The length of the chamfer measured parallel to the axis at the cutting edge.

CHAMFER RELIEF ANGLE—See RELIEF.

CHAMFER RELIEF—See RELIEF.

CHIP BREAKERS—Notches or grooves in the cutting edges of some taper reamers designed to break the continuity of the chips.

CHUCKING REAMER—A type of machine reamer with relatively short straight or helical flutes on which the peripheral lands are relieved. They have a relatively small amount of back taper and have either straight or taper shanks.

CIRCULAR LAND—See preferred term MARGIN.

CLEARANCE—The space created by the relief behind the cutting edge or margin of a reamer.

CONCENTRIC—CONCENTRICITY—See preferred terms—TOTAL INDICATOR VARIATION and RELATIVE ECCENTRICITY to describe lack of concentricity between two or more reamer elements.

CORE—The central portion of a reamer below the flutes which joins the lands.

CORE DIAMETER—The diameter at a given point along the axis of the largest circle which does not project into the flutes.

CORE REAMER—A roughing reamer with deep wide flutes to provide ample room for chips in reaming cored holes.

CUTTER SWEEP—The section removed by the milling cutter or grinding wheel in entering or leaving a flute.

CUTTING EDGE—The leading edge of the land in the direction of rotation for cutting.

CUTTING FACE—The leading side of the land in the direction of rotation for cutting on which the chip impinges.

CUTTING SPEED—The peripheral lineal speed resulting from rotation, usually expressed as surface feet per minute. [sfm]

DRIFT—A flat tapered bar for forcing a taper shank out of its socket.

DRIFT SLOT—A slot through a socket at the small end of the tapered hole to receive a drift for forcing a taper shank out of the socket.

DUPLEX LEAD REAMER—A reamer with one or more flutes having a different lead than the other flutes. This produces a continuous change in flute spacing.

ECCENTRICITY [with respect to the reamer axis]—One half the Total Indicator Variation [tiv]. See RELATIVE ECCENTRICITY.

END CUTTING—A general term describing the extent to which a reamer cuts on the end. Four types are recognized—1—End Cutting on the chamfers only. —2—180° End Cutting to the bottom of the flutes [Core Diameter].—3—180° End Cutting to the center hole or a specified diameter of circle.—4—180° End Cutting to center of reamer.

EXTERNAL CENTER—The pointed end of a reamer. The included angle varies with manufacturing practice.

FEED—The axial advance in inches per revolution of the reamer with respect to the workpiece.

FLUTES—Longitudinal channels formed in the body of the reamer to provide cutting edges, permit passage of chips and allow cutting fluid to reach the cutting edges.

Angular Flute—A flute which forms a cutting face lying in a plane intersecting the reamer axis at an angle. It is unlike a helical flute in that it forms a cutting face which lies in a single plane.

Helical Flute—[Sometimes called "*spiral flute*"]. A flute which is formed in a helical path around the axis of a reamer.

Spiral Flute—1—On a taper reamer a flute of constant lead.—2—In reference to a straight reamer, see preferred term *helical flute*.

Straight Flute—A flute which forms a cutting edge lying in an axial plane.

FLUTED CHUCKING REAMER—See preferred term CHUCKING REAMER.

FLUTE LENGTH—The length of the flutes not including the cutter sweep.

FULL INDICATOR READING [fir]—See preferred term TOTAL INDICATOR VARIATION [tiv].

GAGE LINE—The axial position on a taper where the diameter is equal to the basic large end diameter of the specified taper.

GRINDING RECESS—A clearance for the edge or corner of a grinding wheel, usually necessary at a change of diameter.

GUIDE—A cylindrical portion following the flutes of a reamer to maintain alignment.

HALF-ROUND REAMER—A reamer with a transverse cross-section of approximately half a circle and having one cutting edge.

HEEL—The trailing edge of the land in the direction of rotation for cutting.

HELIX ANGLE—The angle which a helical cutting edge at a given point makes with an axial plane through the same point.

HOOK—See definition under RAKE.

INTERNAL CENTER—A 60° countersink with clearance at the bottom, in one or both ends of a tool, which establishes the tool axis.

IRREGULAR SPACING—A deliberate variation from uniform spacing of the reamer cutting edges.

LAND—The section of the reamer between adjacent flutes.

LAND WIDTH—The distance between the leading edge of the land and the heel measured at a right angle to the leading edge.

LEAD—See preferred term STARTING TAPER.

LEAD OF FLUTE—The axial advance of a helical or spiral cutting edge in one turn around the reamer axis.

LENGTH—The dimension of any reamer element measured parallel to the reamer axis.

Limits—The maximum and minimum values designated for a specific element.

Line Reamer—A reamer used to ream two or more separated holes on the same axis.

Margin—The unrelieved part of the periphery of the land adjacent to the cutting edge.

Multiple Diameter Reamer—A reamer with two or more cutting diameters.

Neck—A section of reduced diameter connecting shank to body, or connecting other portions of the reamer.

Nominal Size—The designated basic size of a reamer.

Notches—See preferred term Chip Breakers.

Oil Grooves—Longitudinal straight or helical grooves in shank, guide, or pilot for lubrication or to carry cutting fluid to the cutting edges.

Oil Holes—Holes through which a cutting fluid is fed to the cutting edges of a reamer.

Overall Length—The extreme length of the complete reamer from end to end, but not including external centers or expansion screws.

Periphery—The outside circumference of a reamer.

Pilot—A cylindrical portion preceding the entering end of the reamer body to maintain alignment.

Pull Reamer—Reamers which are designed to be pulled through long holes [such as gun barrels] while reamer or workpiece is rotated.

Radial Rake Angle—See definition under Rake.

Radial Runout—The radial variation from a true circle which lies in the diametral plane and is concentric with the reamer axis. See term Total Indicator Variation.

Rake—The angular relationship between the cutting face, or a tangent to the cutting face at a given point and a given reference plane or line.

Axial Rake—Applies to angular [not helical or spiral] cutting faces. It is the angle between a plane containing the cutting face, or tangent to the cutting face at a given point, and the reamer axis.

Helical Rake—Applies to helical and spiral cutting faces only, [not angular]. It is the angle between a plane tangent to the cutting face at a given point on the cutting edge, and the reamer axis.

Hook—A concave condition of a cutting face. The rake of a hooked cutting face must be determined at a given point.

Negative Rake—Describes a cutting face in rotation whose cutting edge lags the surface of the cutting face.

Positive Rake—Describes a cutting face in rotation whose cutting edge leads the surface of the cutting face.

Radial Rake Angle—The angle in a transverse plane between a straight cutting face and a radial line passing through the cutting edge.

Tangential Rake Angle—The angle in a transverse plane between a line tangent to a hooked cutting face at the peripheral cutting edge and a radial line passing through this point of tangency.

RECESS—See preferred term GRINDING RECESS.

RELIEF—The result of the removal of tool material behind or adjacent to the cutting edge to provide clearance and prevent rubbing [heel drag].

Axial Relief—The relief measured in the axial direction between a plane perpendicular to the axis and the relieved surface.—It can be measured by the amount of indicator drop at a given radius in a given amount of angular rotation.

Back-Off—See preferred term *Relief*.

Cam Relief—The relief from the cutting edge to the heel of the land produced by a cam action.

Chamfer Relief—The axial relief on the chamfer of the reamer.

Chamfer Relief Angle—The axial relief angle at the outer corner of the chamfer. It is measured by projection into a plane tangent to the periphery at the outer corner of the chamfer.

Eccentric Relief—A convex relieved surface behind the cutting edge.

End Relief—See preferred term *Axial Relief*.

Flat Relief—A relieved surface behind the cutting edge which is essentially flat.

Primary Relief—The relief immediately behind the cutting edge or margin. Properly called *Relief*.

Relief Angle—The angle, measured in a transverse plane, between the relieved surface and a plane tangent to the periphery at the cutting edge.

Radial Relief—Relief in a radial direction measured in the plane of rotation. It can be measured by the amount of indicator drop at a given radius in a given amount of angular rotation.

Secondary Relief—An additional relief behind the primary relief.

RELATIVE ECCENTRICITY—The distance between the axis of one portion and the axis of some other portion of a reamer.

ROSE REAMER—A type of reamer with lands which are not relieved on the periphery. It has a relatively large amount of back taper.

RUNOUT—See RADIAL RUNOUT.

SALVAGE HOLE—A central hole in the front end of a reamer of sufficient depth to provide for reconditioning.

SECONDARY CHAMFER—A slight relieved chamfer adjacent to and following the initial chamfer on a reamer.

SHANK—The portion of the reamer by which it is held and driven.

SIZE—See terms ACTUAL SIZE and NORMAL SIZE.

SLEEVE—A tapered shell designed to fit into a specified socket and to receive a taper shank smaller than the socket.

SOCKET—The tapered hole in a spindle, adaptor or sleeve, designed to receive, hold, and drive a tapered shank.

SPIRAL FLUTES—See FLUTES.

SQUARED SHANK—A cylindrical shank having a driving square on the back end.

STAGGERED FLUTES—See preferred term IRREGULAR SPACING.

STARTING RADIUS—A relieved radius at the entering end of a reamer in place of a chamfer.

STARTING TAPER—A slight relieved taper on the front end of a reamer.

STEP REAMER—A multiple diameter reamer with all lands in each step ground to the same diameter.

STRAIGHT SHANK—A cylindrical shank.

STRAIGHT FLUTE—See FLUTES.

SUBLAND REAMER—A type of multiple diameter reamer which has independent sets of lands in the same body section for each diameter.

TANG—The flatted end of a taper shank which fits a slot in the socket.

TANG DRIVE—Two opposite parallel driving flats on the extreme end of a straight shank.

TAPER PER FOOT—The difference in diameter between two points 12 inches apart measured along the axis.

TAPER SHANK—A shank made to fit a specified [conical] taper socket.

TAPER SQUARE SHANK—A taper shank whose cross section is a square.

TONGUE—See preferred term TANG.

TOTAL INDICATOR READING [tir]—See preferred term TOTAL INDICATOR VARIATION [tiv].

TOTAL INDICATOR VARIATION [tiv]—The difference between the maximum and minimum indicator readings obtained during a checking cycle.

UNDERCUT—See preferred term GRINDING RECESS.

WEB—See preferred term CORE DIAMETER.

SELECTION OF REAMERS

Factors that should influence the selection of reamers for a given job can be enumerated as:

1. Material to be reamed.
2. Diameter of hole.
3. Amount of stock to be removed.
4. Accuracy and finish desired.
5. First cost.
6. Maintenance costs.
7. Salvage value.

The material to be reamed has much to do with selecting the best type of reamer. It is evident that if the material is free cutting, reamers of fairly light construction can be used to produce holes of satisfactory quality. But, if the material is hard, tough, or stringy in texture, adequate provisions must be made to meet these conditions. The power required to drive reamers through given materials must be compensated for in their design, so that no undue deflection will take place.

Thus, Shell-Type Reamers can be used in comparatively small diameters for cutting free machining materials, but for hard or tough materials, Solid-Type Reamers are to be recommended, even for holes of fairly large size.

No definite range of sizes can be given here, because of the widely varying conditions that surround different jobs. It should be borne in mind, however, that when holes are to be finished by reaming to very close tolerances, minimum deflection is permissible in the reamer itself.

Theoretically, a Shell Reamer Arbor should fully support the reamer so that it will not deflect, but in practice this is not always the case, because of wear of the arbor or the presence of burrs, dirt or other foreign materials. On production jobs where more or less unskilled operators are used, these factors must be taken into account.

The amount of stock to be removed has a direct influence on the necessary driving force and in turn on the strength and rigidity the reamers must possess. While the driving force is not necessarily proportional to the amount of stock to be removed (as the frictional load is practically constant), the above statement is generally true.

While it is possible to produce reamed holes having a good finish, but poor accuracy, it may be said that, in general, accuracy and finish go together. Accuracy, in this case, takes into account tolerances on diameter, roundness, straightness, and absence of bell-mouth at ends of holes. To meet all of these conditions, it is necessary to use reamers with proper and adequate support for the cutting edges. Solid Reamers should be selected for holes up to sizes for which the reamer is large enough in diameter to accommodate an arbor hole and yet leave sufficient wall thickness to support fully the cutting edges against deflection under load.

It is obvious that, as diameters of reamers increase, the first cost of solid-type, or even of Shell-Type Reamers with integral cutting edges, will eventually become prohibitive. The Inserted Blade and Adjustable Type of reamer then is available. However, in changing to the latter type, the matter of maintenance costs to provide continuous rigidity and accuracy must be closely studied. Experience has shown that, unless reconditioning is carefully and competently done, the first and replacement cost advantages of these reamers are lost.

The salvage value of Solid Type Reamers will depend largely on the products that are manufactured and on the variety of sizes of holes. In some shops it is possible to regrind reamers that are worn undersize for a smaller size hole, where they can again be used with full wear life. This can only be done if the difference in sizes comes within the practical regrinding range.

The choice between a Straight Fluted and a Helical Fluted Reamer is largely a personal one. Theoretically, a helical reamer is less likely to chatter than is a straight-fluted tool, but this is not always borne out in practice, because of the multiplicity of other factors involved.

In general, right-hand helical reamers cut slightly more freely than straight-fluted or left-hand helical tools, and because of their chip clearing ability have a slight advantage in the reaming of blind holes. However, in screw machine work, if the blind holes are not too deep, left-hand helical tools work satisfactorily.

USE OF REAMERS

The conditions under which reamers are used and the results which are desired vary so widely that it is impossible to set up any guide to cover each possible combination. A few of the variables which affect reamer cutting action are:

1. Speed.
2. Feed.
3. Use of guide bushings.
4. Material.
5. Condition of machine.
6. Rigidity of set-up.
7. Rake of cutting edges.
8. Reliefs on reamer.
9. Amount of stock to be removed.
10. Finish required.
11. Tolerance of hole.

However, there are certain general principles which should be kept in mind.

SPEEDS

The most efficient speed for machine reaming is closely tied in with the type of material being reamed, the rigidity of the set-up, and the tolerance or finish required. Quite often the best speed is found to lie around two-thirds of the speed used for drilling the same material. The following table may be used for a guide:

	Speed in F.P.M.		Speed in F.P.M.
Magnesium and its alloys ..	170–270	Stainless Steel, A.I.S.I. 301,	
Aluminum and its alloys ..	130–200	304	15– 35
Plastics	70–100	Stainless Steel, A.I.S.I. 302,	
Brass and Bronze, ordinary	130–200	309, 314, 316, 325, 420 ..	15– 30
Bronze, high tensile	50– 70	Stainless Steel, A.I.S.I. 303,	
Monel Metal	25– 35	321, 347, 414, 430, 431 ..	20– 40
Cast Iron, soft	70–100	Stainless Steel, A.I.S.I. 310	15– 40
Cast Iron, hard	50– 70	Stainless Steel, A.I.S.I. 403,	
Cast Iron, chilled	20– 30	405, 406, 410	20– 50
Malleable Iron	50– 60	Stainless Steel, A.I.S.I.	
Steel, Machinery .2C to .3C	50– 70	416F, 420F	30– 60
Steel, Annealed .4C to .5C	40– 50	Stainless Steel, A.I.S.I.	
Steel, Tool, 1.2C	35– 40	430F, 440F	30– 50
Steel, Alloy	35– 40	Stainless Steel, A.I.S.I.	
Steel, Automotive Forgings	35– 40	440A, B & C, 442, 443,	
Steel, Alloy 300–400 Brinell	20– 30	446	15– 30

A lack of rigidity in the set-up may necessitate slower speeds, while occasionally a very compact, rigid operation may permit still higher speeds.

When close tolerances and fine finish are required it is usually found necessary to finish ream at considerably slower speeds.

In general, reamers do not work well when they chatter. Consequently, one primary consideration in selecting a speed is to stay low enough to eliminate chatter. Other ways of reducing chatter will be considered later, but this one rule holds: SPEEDS MUST NOT BE SO HIGH AS TO PERMIT CHATTER.

FEEDS

In reaming, feeds are usually much higher than those used for drilling. The amount of feed may vary with the material, but a good starting point would be between .0015" and .004" per flute per revolution. Too low a feed may result in glazing, excessive wear, and, occasionally, chatter. Too high a feed tends to reduce the accuracy of the hole and may also lower the quality of the finish. The basic idea is to use as high a feed as possible and still produce the required finish and accuracy.

AMOUNT OF STOCK REMOVED

Insufficient stock for reaming may result in a burnishing rather than a cutting action of the reamer. It is very difficult to generalize on this phase as it is closely tied in with type of material, feed, finish required, depth of hole, and chip capacity of the reamer. For machine reaming, .010″ on a ¼″ hole, .015″ on a ½″ hole, up to .025″ on a 1½″ hole, seem good starting points. In reaming work-hardening materials, sufficient stock must be removed so reamer cutting edges are cutting in non-work hardened or glazed material. For hand reaming, stock allowances are much smaller, partly because of the difficulty in forcing the reamer through greater stock. A common allowance is .001″ to .003″.

THE USE OF BUSHINGS

The use of bushings as guides for reamers is of great help. The ideal job would employ a fixed jig and bushing with a rigid spindle and a minimum of overhang. It is particularly important when using bushings that the spindle be accurately aligned with the bushing in order to prevent the reamer from hitting the top of the hardened bushing. In radial drill presses or in table presses equipped with sliding jigs, great care must be used. Sometimes bronze or fibre caps are placed over the bushings to act as a lead-in device, but even this will not protect the reamer unless the tool is properly centered in the bushing.

The same problem arises if work is reamed on a table press without a bushing. The use of a short entering taper on the front of the reamer, or in some cases a short pilot section, may be of help in leading the reamer into the hole. However, the practice of using a reamer as a locating device and forcing it to drag the work over into alignment is not good, and is likely to result in excessive wear or chipped edges of the reamer.

When holes are to be located at exact distances from some point or some other hole, the only sure method is to do the reaming in jigs or fixtures. In such jigs or fixtures the work is located and held securely, and the reamer is guided in bushings set in exact relation to locating points in the work.

For this type of reaming the ideal arrangement is to guide the reamer on both sides of the work, especially if the hole is comparatively long. Fig. 1 illustrates a typical set-up. A special piloted reamer is required for this purpose. Guide bushings should fit snugly to the pilots, but should not be so tight they will seize and bind. Pilots should be grooved throughout their length. These grooves serve the double purpose of permitting the cutting fluid to lubricate the pilots and to scavenge any chips that tend to wedge between the pilots and bushings.

If the holes to be reamed are short, the reamer may be guided at the entering side of the hole only. The guide bushing then may be made to fit the outside diameter of the reamer flute as illustrated in Fig. 2.

For either of these types of applications a rigid drive is satisfactory, because any misalignment of the machine spindle and the work tends to correct itself by means of the guiding bushings. By a rigid drive we mean one where the reamer shank is held directly and rigidly in the machine spindle.

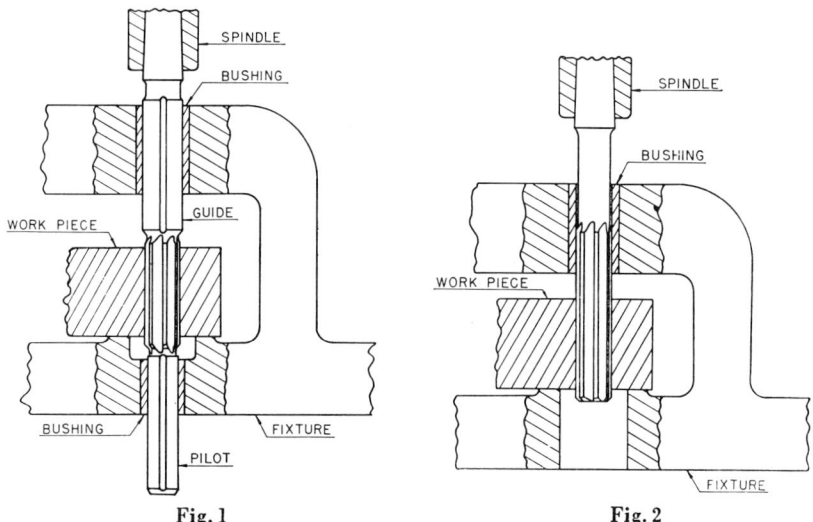

Fig. 1   Fig. 2

If, on the other hand, the reamer is to guide itself into a previously made hole, a rigid drive is no longer satisfactory, because any existing misalignment of the machine spindle with the work will result in reamed holes that are bell-mouthed, tapered, or out of round. A floating reamer driver must then be used.

Various types of floating drivers are in use, each possessing individual merits. Some types permit angular relation between the center lines of the driving spindle and the reamer. Others permit side movement of the axis of the reamer with that of the driving spindle, the two being parallel. Still others combine these two features.

The two principles that must be embodied in satisfactory floating reamer drivers are illustrated in Fig. 3. At the left are shown some of the angular positions that the reamer must be permitted to assume. At the right is illustrated what might be called parallel floating. Here the reamer axis may move away from the spindle axis while the two are in parallel.

The ideal floating driver must be able to do both of these. In other words, it must be able to adjust itself to misalignment of both types and in all positions throughout its circle of rotation. The amount of permissible angular or parallel float should be adjusted to cover the greatest amount of misalignment that is likely to occur on any given job.

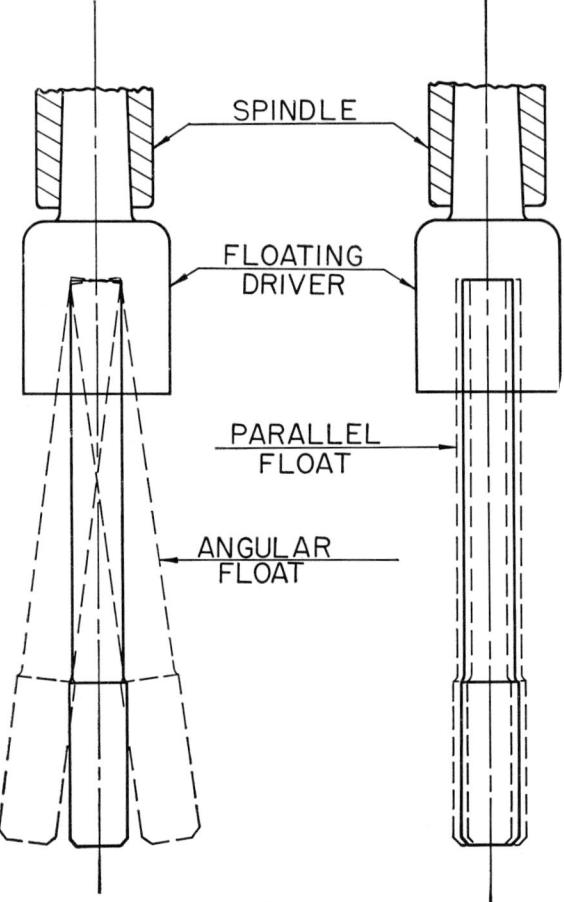

SPINDLE

FLOATING
DRIVER

PARALLEL
FLOAT

ANGULAR
FLOAT

Fig. 3

HAND REAMING

When the work piece is reasonably rigid, hand reaming may be performed by rotating the reamer by means of a double end tap wrench applied to the driving square. This type of wrench, which permits a balanced drive, should always be used in preference to a single-end wrench. The use of single-end wrenches makes it almost impossible to apply torque without disturbing the alignment of the reamer and the hole.

The reamer should be rotated slowly and *evenly*, allowing it to align itself with the hole to be reamed. Wrenches should be large enough to permit a steady torque which will help to control vibration and chatter. The feed should be steady and large compared to the feed used in machine reamers.

Feeds up to one quarter of the reamer diameter per revolution are not at all unusual.

In cases where the work piece is small enough to be handled with ease, it is often advisable to place the reamer vertically in a vise and rotate the work down over the reamer by hand. If the work is quite light it may not have enough mass to dampen vibrations. In such cases a holding device for the work may be employed which will add weight to the part to be reamed. This holding device should have two opposite handles on it, large enough in diameter to permit a steady controlled torque.

In cases where there is a large quantity of light parts to be reamed, the reamer is often mounted horizontally in a reaming machine. Most reaming machines are essentially chucks mounted on the output shaft of a motor-driven gear reducer. This drives the reamer at the necessary slow speed. The work piece is fed slowly and steadily over the reamer.

In all hand reaming done by any of the methods described, with solid, expansion, or adjustable reamers, the reamer should *never* be rotated backwards to remove it from the hole, for this results in premature dulling of the reamer. If possible, it should be passed on through the hole and removed from the far side without stopping the rotation. If this is not possible, it should be withdrawn without stopping the *forward* rotation.

THE USE OF REAMERS IN TURRET LATHES AND HAND OR AUTOMATIC SCREW MACHINES

When reaming is done in any of the machines listed above, certain problems arise because the reamer is stationary and the work revolves. This is the opposite of the condition normally encountered in drill-press reaming.

Theoretically, in this type of reaming the axis of the reamer coincides with the axis of the machine spindle, and no difficulty should be encountered. However, in actual practice, this condition seldom exists. It probably would be wise to assume that it never exists and to expect the presence of misalignment on jobs of this nature.

The misalignment is of two distinct types, illustrated in the five accompanying illustrations. Fig. 1 shows a case of true alignment.

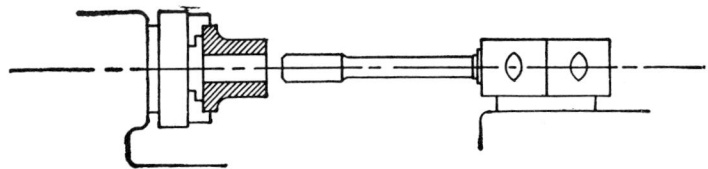

Fig. 1. *True Alignment*

In Fig. 2 is shown a case where the axis of the reamer is parallel to the axis of the spindle, but has been lowered slightly.

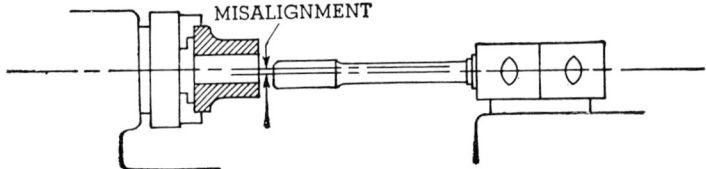

Fig. 2. *Parallel Misalignment*

In Fig. 3 the shank of the reamer is in line with the spindle, but the reamer has been clamped at a slight angle.

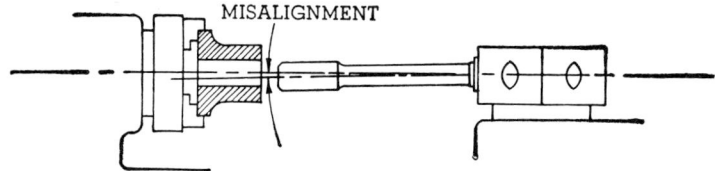

Fig. 3. *Angular Misalignment*

In actual practice both of these errors are often present at the same time in a single set-up. Such misalignment can be caused by a combination of several things:

1. Worn ways on the machine.
2. Worn or dirty holes in tool holder.
3. Worn or dirty sleeves.
4. Improper leveling of machine.
5. Improper location or adjustment of tool slides.
6. Errors in indexing mechanism.

The use of a conventional reamer held rigidly in such a machine quite often results in poor finish and oversize holes, particularly at the start of the hole. Such holes are usually called "bell-mouthed."

In Fig. 4 is shown the case of a reamer held at an angle to the spindle of the machine. As this reamer is fed into the work, the lower flute *A* will bear harder and harder against the wall of the reamed hole. This results in a scraping action by this flute which causes an oversize hole. The pressure also quite often causes a building up on this flute. When the pressure gets too high, this built-up portion breaks loose, causing gouges and tears in the reamed surface.

There are several means available to eliminate these torn, oversize holes. The first is to make corrections on the machine and tool holders. It is comparatively simple to make some of these corrections, such as cleaning tool holders, replacing worn sleeves or bushings, and relocating some types of tool holders. Others, such as correcting mis-indexing on a turret lathe or multi-spindle automatic, may require a great deal of time.

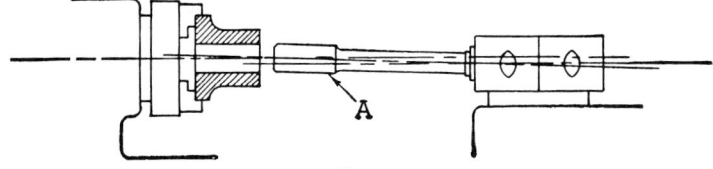

Fig. 4

The second method is to use floating holders for reamers. There are many kinds and designs of floating holders. Some correct for angular error only. Some correct for a parallel misalignment only, while some correct for both. Besides varying in function and design, these different floating holders also vary in their inherent rigidity. The problem in such holders is to allow movement in certain directions while restricting it in others. Naturally, this is not accomplished in the same manner in all designs, and the rigidity of different types of holders, whether floating or not, varies considerably.

The presence of slight errors in the machine does not necessarily mean that floating holders must be used. It must be remembered that the looseness in the machine plus the flexibility in the shank of the reamer provide some float. Also, on some types of machines, it is comparatively simple to insert blank bushings in the tool holder and rebore them from the head-stock. On other machines, adjustable rather than floating holders may be used.

This oversize condition is not always cured by the use of floating holders alone. In Fig. 5 is shown the case of a turret which has worn low and an attempt to correct this condition through the use of a simple pin drive float.

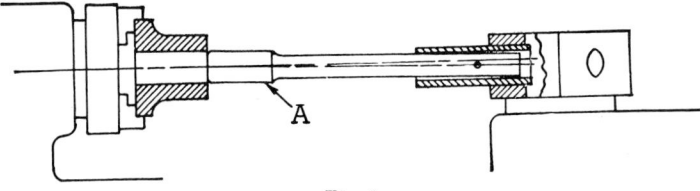

Fig. 5

Such a float allows the reamer to tilt slightly, but does not permit the shank of the reamer to move up or down or sideways. It will be noted that as the reamer feeds into the work the pressure will build up on the lower flute A as before, still causing oversize and torn holes.

In order to overcome this condition it is sometimes necessary to grind back taper on the reamer. While the use of back taper on reamers will often eliminate these rough, oversize reamed holes, it must be remembered that this back taper reduces the reamer life, particularly on close tolerance holes. This special back taper varies between .001″ and .005″ per inch of flute. Tapers such as these are far in excess of the very slight taper which has been found best for stock reamers designed to do the average reaming job. The figure of .005″, in particular, seems excessive, and in the interest of economy the back taper should be kept as low as possible without interfering with the required hole size and finish.

REAMING OF TAPERED HOLES

The reaming of tapered holes has always presented a problem in reamer design and operation because of the very nature of the work to be performed. Usually one starts out with a drilled or bored hole of uniform diameter throughout. The diameter of this hole is slightly smaller than the finished diameter of the small end of the tapered hole to be produced.

The operation of taper reaming now consists of greatly enlarging one end of this hole, with this enlargement gradually decreasing towards the other end. This means that the Taper Reamer, instead of being a finishing tool, in reality becomes a tool for heavy stock removal, and further, that this tool at the finish of the operation is engaged in the cut throughout its length.

Because of the large amount of stock to be removed and the length of the cut, taper reamers are subjected to much greater torsional strains than the ordinary straight reamer that cuts on the end only. The tendency, therefore, is to chatter, with consequent poor finish or actual destruction of the reamer. The greater the taper, the worse these conditions will become.

Taper Reamers must, therefore, be constructed as sturdily as possible. Cutting edges must be adequately backed up, and flutes must be sufficiently large, with well rounded bottoms.

For reaming tapered holes with larger tapers, better results are obtained by first rough reaming the taper, followed by a finish taper ream to merely size and smooth the hole. In general, roughing reamers are made with fewer number of flutes than the finish reamers, and are of different construction as to hand of helix and degree of spiral. Quite often it is also necessary to use chip breakers ground in the cutting edges of the roughing reamer, usually in the form of a coarse pitch square thread. These may be particularly necessary when the tapered hole is quite long and the taper relatively small. As an added precaution, the number of flutes in a finishing reamer should never be an even multiple of the number of flutes in the roughing reamer nor the same number of flutes.

For machine reaming, it has been found that a reamer having few flutes, and with cutting edges at a large spiral angle with the axis, is by far the best. Flutes may vary in number from two to five in the ordinary range of reamer sizes.

The spiral angle should be about 45° and of a hand opposite to the direction of rotation. A greater spiral angle than this does not tend to improve the quality of the cut, but does materially increase the amount of end pressure required to feed the reamer into the work.

It should be realized, however, that reamers with a high left-hand spiral and right-hand cut tend to crowd the chips ahead of themselves. For this reason this design is not usually satisfactory for blind holes. For such holes, roughing reamers are often made of flat construction with little or no spiral angle.

Occasionally when the job conditions are such that a great deal of difficulty is experienced with chatter, or the holes are definitely out of round even though not chattered, it is necessary to employ a flute construction not generally used. This particular reamer construction embodies the use of an uneven or even number of flutes coupled with irregular flute spacing of the type that no two cutting edges are diametrically opposite each other. The reamers thereby are very difficult to measure for size and taper and their use, therefore, is to be avoided unless absolutely necessary.

It should be remembered that high-spiral Taper Reamers cannot be used on hand operations because of the end pressure required, but for machine operations this is by far outweighed by smoothness of operation, better quality of work, and longer reamer life. For hand operation, Taper Reamers with straight flutes, or with about ten degrees of spiral opposite in direction to that of the cut, are recommended.

RIGIDITY

Any lack of rigidity in the holder tends to promote chatter. This chatter may exist only at the start of the hole as the reamer centers itself. In general, reamers do not work well when they chatter. Carbide Tipped Reamers, in particular, cannot stand even this momentary chatter at the start of the hole. It is quite likely to result in a chipping of the cutting corners.

Chatter may be caused by lack of rigidity in the machine, in the fixture, or in the work itself, or it may be caused by improper design of the reamer. The reamer must, therefore, be so designed that it not only will overcome chatter on its own account, but will help to counteract the lack of rigidity in the machine, fixture, or work piece.

If the hole to be reamed is fairly large in diameter, is short, and is near the outside surface of the work, the tendency to chatter can be more easily overcome, because it permits the use of a short reamer of comparatively large cross section. If, on the other hand, the hole is small in diameter, is long or inaccessible, the solution is far more difficult. The remedy usually is to make the reamer with irregular spaced flutes, or with helical flutes, or to provide pilots for support at one or both ends of the flutes. Sometimes a combination of these features will give the desired results.

The object of irregular spacing of the flutes is to break up any tendency to synchronize slippage and torsional deflection. Chatter from these causes is well known to reamer users.

Helical flutes sometimes accomplish the same object even more effectively than irregular spacing. The thing to remember here is that the reamer must, if possible, have enough helix so that two or more flutes overlap in the length of the hole being reamed. On the other hand, the helix angle should not be more than necessary, because the steeper the helix angle the more end pressure is required to feed the reamer through the work. Reamers for reaming tapered holes are an exception to the latter recommendation, because best results are obtained with a very steep spiral. The direction of spiral should be opposite to that of the cut, to prevent the reamer from pulling itself through, ahead of the feed.

Pilots are provided so that the reamer can be held in alignment, and can be supported as close to the work as possible. Sometimes the reamer is piloted in the work itself, but it is preferable to support it in the fixture by means of hardened and ground bushings. The diameters of pilots should always be kept as large as possible, to provide the greatest possible amount of rigidity to the reamer, and to prevent torsional deflection and chatter.

Chatter may be reduced in several ways, some of which are:

1. Reducing the speed.
2. Varying the feed.
3. Increasing the rigidity of the tool holder through reduction or even removal of the float.
4. Introduction of a small chamfer at the start of the hole before reaming.
5. Use of a piloted reamer.
6. Reducing the relief angle.

COOLANTS FOR REAMING

In reaming, the emphasis is usually on finish, and a coolant is chosen for this rather than for cooling. This may mean a change from the drilling coolant, but in general this list will be satisfactory:

Aluminum and its alloys: Soluble oil, kerosene and lard oil compounds, light non-viscous neutral oil, kerosene and soluble oil mixtures.

Brass: Dry, soluble oil, kerosene and lard oil compounds, light non-viscous neutral oil.

Copper: Soluble oil, winter-strained lard oil, oleic acid compounds.

Cast Iron: Dry or with a jet of compressed air for a cooling medium.

Malleable Iron: Soluble oil, non-viscous neutral oil.

Monel Metal: Soluble oil, sulfurized mineral oil.

Steel, ordinary: Soluble oil, sulfurized oil, high-EP mineral oil.

Steel, very hard and refractory: Soluble oil, sulfurized oil, turpentine.

Steel, Stainless: Soluble oil, sulfurized mineral oil.

Wrought Iron: Soluble oil, sulfurized oil, mineral-animal oil compound.

THE USE OF CARBIDE TIPPED REAMERS

In order to obtain maximum economy the user of Carbide Tipped Reamers should follow the general suggestions for the use of reamers already
specified. In addition, there are a few special considerations which should
be given to carbide tipped tools. Suggestions for regrinding of Carbide
Tipped Reamers will be found on succeeding pages.

SPEEDS AND FEEDS

The very high surface speeds which have been found necessary for turning operations with carbide tools present some difficulties if applied to reamers. In lathe work, the cutting tool may usually be made of any desired cross-
section and held in a large rigid tool-post with minimum overhang. In
reaming, however, hole size limits cross-section of the tool, and usually
considerable overhang is required. Both result in a loss of the rigidity so
necessary with very high surface speeds if chatter is to be avoided.

As already emphasized, reamers do not work well when they chatter. Carbide Tipped Reamers in particular cannot stand even a momentary chatter
at the start of a hole, as such a vibration is likely to chip the cutting edges.
Consequently, the primary consideration in selecting a speed is to STAY
LOW ENOUGH TO ELIMINATE ALL CHATTER.

In reaming with carbide tipped reamers, too low a feed may result in
glazing, excessive wear, and in some cases chatter. Feeds should always be
high enough to apply sufficient chip loads on the cutting edges. The amount
of feed may vary with the material, but a good starting point would be
between .0015″ and .003″ per flute per revolution. This may be increased
considerably on some types of work and the upper limit is usually determined by the finish required.

Insufficient stock for reaming may result in a burnishing rather than
a cutting action of the reamer. It is very difficult to specify on this
phase as it is tied in closely with type of material, feed, finish required,
depth of hole, and chip capacity of the reamer. In general, .010″ on a ¼″
hole, .015″ on ½″ holes, up to .025″ on 1½″ holes, seem good starting points.
For hand reaming, stock allowances are much smaller, partly because of the
difficulty in forcing the reamer through greater stock. A common allowance
is .001 to .003″. Carbide reamers are particularly adaptable for hand ream-

ing in materials of a highly abrasive nature and sizing holes in materials where the surface of the hole may have been slightly work-hardened from previous operations.

Carbide Reamers work best when well supported with a minimum of overhang. Any lack of rigidity in the set-up may result in a momentary vibration which may cause flaking and chipping of the carbide tips.

CAUSES OF BREAKAGE OR EXCESSIVE WEAR OF REAMERS

1. Dirt or burrs in spindle or socket in which reamer is held.
2. Misalignment of two or more parts of the set-up. This condition can cause a bell-mouthed hole plus excessive wear of reamer margins.
3. Lack of chip space in set-up or flutes of the reamer.
4. Too fast or too slow speeds.
5. Too much or too little feed.
7. Wrong type of coolant.
8. No lubricant between guide bushing and reamer.
9. Lack of lubricant.
10. Bottoming in blind holes.
11. Lack of sufficient stock to ream.
12. Too much stock to ream.
13. Entering work too fast.
14. Badly drilled holes—too rough, tapered, or bell-mouthed. Bell-mouthed holes may cause the reamer to wedge rather than cut.
15. Faulty sharpening, which may consist of: improperly ground relief on cutting edges; grinding cracks from too fast and heavy grinding; unbalanced sharpening of cutting edges, causing one or two flutes of the reamer to take the entire cutting load; saw-tooth cutting edges from too coarse a grind; and incorrect end-cutting angle.
16. Poor handling of reamer.
17. Oversize or undersize bushings.
18. Chattering of reamer.
19. Lack of rigidity in machine or work holder.
20. Improperly designed reamer for the job.

REAMER RECONDITIONING

In obtaining maximum economy from reamers, the same principles apply as in the case of most other cutting tools. One of these principles is not to allow a tool to become too dull. It is best practice to regrind the chamfer on a reamer long before it exhibits excessive wear or refuses to cut. This sharpening is usually restricted to the starting taper or chamfer. It can be done on almost any tool and cutter grinder. Care must be taken so that each flute is ground exactly even, or the tool is likely to cut oversize.

Sharpening the chamfer on a reamer by hand is not recommended because it is practically impossible to keep the cutting edges even.

The accompanying illustrations show three common types of grinds used on reamers.

On left-hand helix tools it is important that the reamer be ground so as

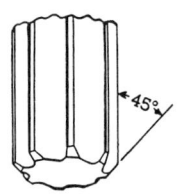

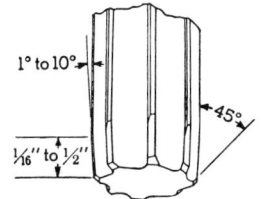

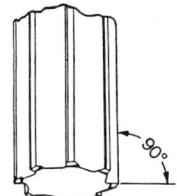

Fig. 1. *Ordinary reamer point for most jobs.*

Fig. 2. *Hand reamer grind also used on some machine reamer applications to obtain required finish or tolerance.*

Fig. 3. *Semi finish reamer grind to straighten out bent or misaligned holes. Corners must be kept sharp.*

to leave a positive effective rake angle. This can best be done by reducing the 45° angle shown in Fig. 1 to about 30°, or even less. If the reamer has a very high left-hand helix angle (20° or more), a grind similar to Fig. 2 usually is used.

In general, if the grind shown in Fig. 3 is required, a reamer having straight or right-hand helix flutes should be chosen.

Fig. 4 shows a satisfactory method of regrinding the chamfer. The use of

Fig. 4

the bushing gives a very concentric chamfer in its relation to the outside diameter of the reamer. Fig. 5 shows this same operation being done on the periphery of a straight wheel. This same operation can be performed between centers, and is shown in Fig. 6 using a cup wheel and in Fig. 7 with a straight wheel.

In order to obtain a concentric chamfer, the operator's fingers should hold the face of the flute against the locating finger with an even pressure. On

Fig. 5

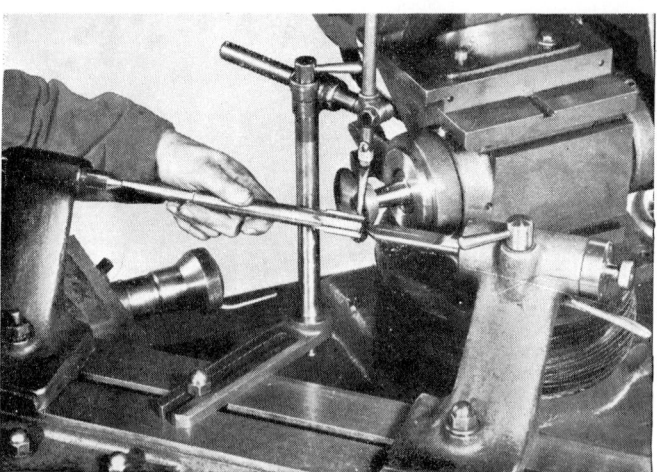

Fig. 6

Fig. 7

those set-ups, where a bushing is used, the reamer should be pressed back lightly by hand against the tail stock center. It will be found best to rotate the grinding wheel so that it cuts from the face of the flute towards the back of the land, in order to reduce any tendency of the cutting edges to flake. The answer to the question of whether a cup or straight wheel should be used will be determined to a great extent by the type of machine available. However, in general, the type of wheel to use should be the one which will allow the set-up to be as compact as possible, with a minimum amount of overhang. Relief on the chamfers of reamers varies with the job; an angle of from 7° to 12° represents normal practice.

A secondary chamfer relief angle may be necessary if the tooth width is relatively wide, such as occurs on expansion chucking reamers and carbide-tipped reamers. Secondary relief angles are not normally required on most solid-type high speed steel reamers. Secondary relief angles are approximately 50% greater than the primary relief.

MARGIN WIDTHS AND RELIEFS

If it becomes necessary to regrind the diameter of a reamer, the lands must be cleared or relieved to prevent the reamer from binding in the hole. The best relief angle and the best margin vary with the job, but the following table gives average figures.

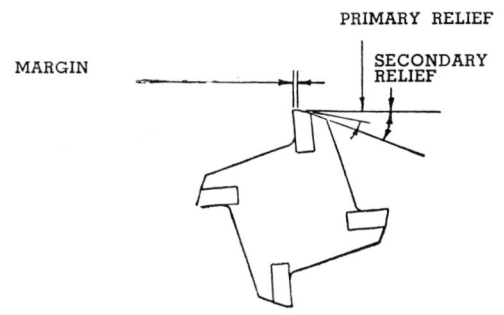

Diameter of Reamer	Margin	Primary Relief Angle
¼″	.007″	14°
½″	.009″	11°
1″	.013″	9°
1½″	.016″	7°
2″	.023″	7°

A secondary relief is usually ground on reamers as shown in the illustration above the table. This relief is only to insure the back of the land being well away from the wall of the reamed hole, in order to prevent rubbing.

REGRINDING OF CARBIDE TIPPED REAMERS

In general, the cylindrical regrinding of Carbide Reamers should be done only when absolutely necessary. Usually only the chamfer will need regrinding, and two suggested methods for doing this will be found illustrated here. Occasionally, the face of the tips will need polishing slightly to remove any rounded edges or built-up portions. This can be accomplished by the method illustrated in Fig. 8.

The use of diamond wheels is recommended for regrinding these reamers. Grades having between 200 and 300 grit perform satisfactorily; the wheel should be rotated so that it cuts from the face of the tip towards the back of the land, in order to minimize any tendency of the carbide to flake.

The relief on the chamfers of Carbide Reamers is usually less than that used on High Speed Steel Reamers, an angle of from 5° to 8° representing normal practice.

The operation shown in Figs. 9 and 10, in two different set-ups, can be done equally well by substituting a cup wheel for the straight wheel shown.

Fig. 8. *Regrinding Carbide Tipped Reamers*

SUGGESTED METHODS FOR REGRINDING CHAMFER ON CARBIDE TIPPED REAMERS

Fig. 9. *A method using the periphery of a diamond wheel and the tool mounted between centers*

Fig. 10. *A method which can be used when centers on the tool are no longer available*

PROCEDURE FOR CHECKING RELIEF ANGLES

1. Mount the reamer between centers, using a bench center equipped with a sliding indicator post.
2. Adjust the vertical indicator to bear on the junction of the margin and the primary relief as shown in Fig. 11.
3. Place another indicator below the vertical indicator on the same post to give a horizontal reading as the reamer is rotated. See Fig. 11.

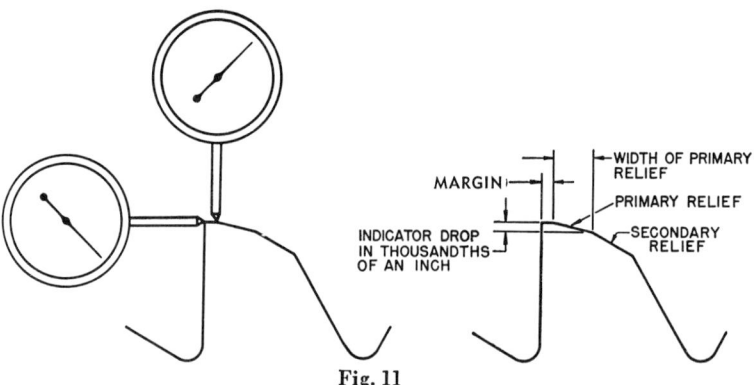

Fig. 11

4. Set up both indicators to bear on the reamer as shown, and rotate the reamer, reading the drop of the primary relief on the vertical indicator and the width of primary relief on the horizontal indicator.

5. This measured drop should be as shown in the table on page 197 for a particular degree, diameter of reamer, and margin. Should the margin become narrower or wider, the drop should be decreased or increased, respectively, to maintain a given degree of relief.

THE CARE OF REAMERS

The quality of finish and the accuracy of the holes produced, as well as the ultimate life of a reamer, depend to a large extent on the treatment it receives. No discussion of reamers, therefore, is complete without considering this important factor.

The care of reamers should extend not only to the operation itself, but it is equally important that reamers are properly cared for while in storage, and during sharpening. With this in mind, some of the precautions that must be observed if the best results are to be obtained, are enumerated below. It should be kept in mind that a reamer is a finishing tool in every sense, that it is an expensive and comparatively short-lived tool, and that it is a delicate tool as compared with tools used for roughing operations.

1. Never permit reamers to be stacked in bins or boxes without some separating layer of cardboard, wood, or the like. Individual cardboard tubes make excellent holders for reamers in storage. The edges are so hard and keen that even a light impact of the reamers against each other, or against some other hard object, will cause the edges to chip.

2. The same precautions should be observed when reamers are transported from the crib to the job and back and forth to the grinding room.

3. Provide a rack or suitable individual holder for reamers to be kept at the machine when not in use.

4. See that reamers are properly covered with oil when not in use, to prevent rusting at the cutting edges. Even a small rust spot will leave a pit or nick.

5. When resharpening, be sure to use a fine but free cutting wheel. Burning of the edge destroys the life of the reamer. A rough edge produces a rough hole, and the life of the reamer is shortened.

Indicator Drop for Various Degrees of Relief on Reamers

	Diam. of Reamer	Angle of Cutting Relief							
		5°	6°	7°	8°	9°	10°	11°	12°
1/64 Margin	1/8"	.0000	.0000	.0000	.0002	.0005	.0008	.001	.0015
	3/16"	.0000	.0004	.0006	.001	.0012	.0014	.0017	.002
	1/4"	.0004	.0007	.0009	.0012	.0015	.0018	.002	.0023
	5/16"	.0006	.0009	.0011	.0014	.0017	.002	.0022	.0025
	3/8"	.0007	.001	.0013	.0015	.0018	.0021	.0024	.0026
	7/16"	.0008	.0011	.0014	.0016	.0019	.0022	.0025	.0027
	1/2"	.0009	.0012	.0015	.0017	.002	.0022	.0025	.0028
	5/8"	.001	.0012	.0015	.0018	.0021	.0023	.0026	.0028
	3/4"	.0014	.002	.0025	.0031	.0036	.0042	.0047	.0053
	7/8"	.0017	.0022	.0028	.0033	.0038	.0044	.0049	.0055
	1"	.0018	.0023	.0029	.0034	.0039	.0045	.005	.0055
	1 1/8"	.0019	.0024	.003	.0035	.004	.0046	.0051	.0057
	1 1/4"	.002	.0025	.0031	.0036	.0041	.0047	.0052	.0058
	1 3/8"	.002	.0026	.0031	.0037	.0042	.0047	.0053	.0058
	1 1/2"	.0021	.0026	.0032	.0037	.0043	.0048	.0054	.0059
	1 5/8"	.0021	.0027	.0032	.0038	.0043	.0048	.0054	.006
	1 3/4"	.0022	.0027	.0033	.0038	.0044	.0049	.0054	.006
	1 7/8"	.0022	.0028	.0033	.0038	.0044	.0049	.0054	.006
	2"	.0022	.0028	.0033	.0039	.0044	.0049	.0055	.006
	2 1/8"	.0023	.0028	.0034	.0039	.0045	.0049	.0055	.0061
1/32 Margin	2 1/4"	.0023	.0029	.0034	.0039	.0045	.005	.0056	.0061
	2 3/8"	.0023	.0029	.0034	.004	.0045	.005	.0056	.0061
	2 1/2"	.0024	.0029	.0034	.004	.0045	.0051	.0056	.0061
	2 5/8"	.0024	.0029	.0035	.004	.0045	.0051	.0057	.0062
	2 3/4"	.0024	.0029	.0035	.004	.0045	.0051	.0057	.0062
	2 7/8"	.0024	.003	.0035	.004	.0046	.0051	.0057	.0062
	3"	.0024	.003	.0035	.004	.0046	.0051	.0057	.0062
	3 1/8"	.0024	.003	.0035	.004	.0046	.0051	.0057	.0062
	3 1/4"	.0024	.003	.0035	.0041	.0046	.0051	.0057	.0062
	3 3/8"	.0025	.003	.0035	.0041	.0046	.0051	.0057	.0062
	3 1/2"	.0025	.003	.0035	.0041	.0046	.0052	.0057	.0062
	3 5/8"	.0025	.003	.0035	.0041	.0047	.0052	.0057	.0062
	3 3/4"	.0025	.003	.0036	.0041	.0047	.0052	.0057	.0062
	3 7/8"	.0025	.003	.0036	.0041	.0047	.0052	.0057	.0063
	4"	.0025	.0031	.0036	.0041	.0047	.0052	.0057	.0063

Reamer Tolerances

Tolerance tables apply to high speed and carbon steel reamers identified under each table.

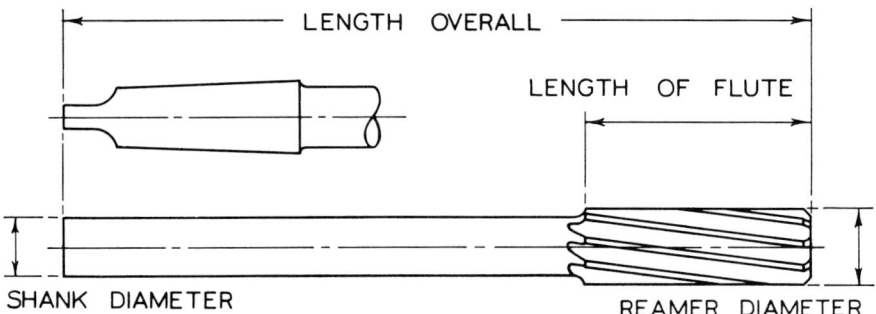

Identity of Toleranced Areas

Diameter of Reamer

Reamer Diameter Range		*Tolerance*
No. 60 to ¼″ inclusive	(.0400 to .2500)	+.0001 to +.0004
Over ¼ to 1″ inclusive	(.2500 to 1.0000)	+.0001 to +.0005
Over 1 to 3″ inclusive	(1.0000 to 3.0000)	+.0002 to +.0006

Applies to:
1—Hand reamers with straight or helical flutes & squared shanks
2—Taper shank jobbers reamers with straight flutes
3—Shell reamers with straight or helical flutes
4—Expansion chucking reamers, straight flute, straight or taper shank
5—Chucking reamers, straight or helical flute, straight or taper shank
6—Stub screw machine reamers

Length of Flute

Reamer Diameter Range		*Tolerance*
No. 60 to 1″ inclusive	(.0400 to 1.0000)	+1⁄16 to −1⁄16″
Over 1 to 2″ inclusive	(1.0000 to 2.0000)	+3⁄32 to −3⁄32″
Over 2 to 3″ inclusive	(2.0000 to 3.0000)	+1⁄8 to −1⁄8″

Applies to:
1—Hand reamers with straight or helical flutes & squared shanks
2—Taper shank jobbers reamers with straight flutes
3—Shell reamers with straight or helical flutes
4—Expansion chucking reamers, straight flute, straight or taper shank
5—Chucking reamers, straight or helical flute, straight or taper shank
6—Stub screw machine reamers
7—Expansion hand reamers, straight flute and squared shank

(Continued)

Reamer Tolerances
(Continued)

Length Overall

Reamer Diameter Range		Tolerance
No. 60 to 1″ inclusive	(.0400 to 1.0000)	$+\frac{1}{16}$ to $-\frac{1}{16}''$
Over 1 to 2″ inclusive	(1.0000 to 2.0000)	$+\frac{3}{32}$ to $-\frac{3}{32}''$
Over 2 to 3″ inclusive	(2.0000 to 3.0000)	$+\frac{1}{8}$ to $-\frac{1}{8}''$

Applies to:
1—Hand reamers with straight or helical flutes & squared shanks
2—Taper shank jobbers reamers with staight flutes
3—Shell reamers with straight or helical flutes
4—Expansion chucking reamers, straight flute or taper shank
5—Chucking reamers, straight or helical flute, straight or taper shank
6—Stub screw machine reamers
7—Expansion hand reamers, straight flute and squared shank

Shank Diameter Tolerances

Reamer Diameter Range		Tolerance
$\frac{1}{8}$ to 1″ inclusive	(.1250 to 1.0000)	—.0010 to —.0050
Over 1 to 1½″ inclusive	(1.0000 to 1.5000)	—.0015 to —.0060

*Applies to*s
1—Straight shank, solid hand reamers, straight & helical flutes
2—Straight shank, expansion reamers, straight & helical flutes

Shank Diameter Tolerances

Shank Diameter Range	Tolerance
.0390 to .4355 inclusive	+.0000 to —.0010
Over .4355 to 1.2495 inclusive	+.0000 to —.0015

*Applies to*s
1—Straight shank, solid chucking reamers, straight & helical flutes
2—Straight shank, solid rose chucking reamers, straight & helical flutes

Shank Diameter Tolerances

Type of Reamers	Shank Diameter Range	Tolerance
Expansion Chucking	.3125 to 1.7500″ incl.	—.0005 to —.0020
Stub Screw Machine	.1250 to .7500″ incl.	—.0005 to —.0020
Morse Taper	.3125 to 1.5000″ incl.	—.0005 to —.0020
Brown & Sharpe Taper	.2812 to 1.1250″ incl.	—.0005 to —.0020
Taper Pin, St. & Sp. Flute	.0781 to .6250″ incl.	—.0010 to —.0050
Taper Pin, High Sp. Flute	.0781 to .6250″ incl.	—.0005 to —.0020
Center	.1875 to .5000″ incl.	—.0005 to —.0020
Machine Countersinks	.5000	—.0005 to —.0020

(Continued)

Reamer Tolerances
(Concluded)

Shank Diameter Tolerances

Shank Diameter Range	Tolerance
.4375 to .7000 inclusive	+.0000 to —.0070
Over .7000 to 1.1250 inclusive	+.0000 to —.0090

Applies to:
1—Taper pipe reamers, straight shank, straight flutes, carbon steel

Shank Diameter Tolerances

Shank Diameter Range	Tolerance
.4375	+.0000 to —.0015
.4375 to 1.1250 inclusive	+.0000 to —.0020
Over 1.1250 to 1.8750 inclusive	+.0000 to —.0030

*Applies to*s
1—Taper pipe reamers, straight shank, spiral flutes, high speed steel

HAND REAMERS

Straight Flute, Right-Hand Cut

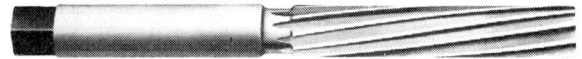

Left-Hand Helical Flute, Right-Hand Cut

Hand reamers are used for final sizing of holes. The square on the shank end is suitable for holding in a tap wrench or vise.

The cutting end is ground with a starting taper to provide easy entry.

The helical reamer is preferred where there is an interruption to the cut, such as a keyway.

Hand Reamers

Straight Flutes — Squared Shank
High Speed Steel

Standard Sizes and Dimensions

Diameter Inches	Dimensions — Inches				
	Length of Flute	Length Overall	Length of Square	Diameter of Shank	Size of Square
$\frac{1}{8}$	$1\frac{1}{2}$	3	$\frac{5}{32}$	$\frac{1}{8}$	.095
$\frac{5}{32}$	$1\frac{5}{8}$	$3\frac{1}{4}$	$\frac{7}{32}$	$\frac{5}{32}$	.115
$\frac{3}{16}$	$1\frac{3}{4}$	$3\frac{1}{2}$	$\frac{7}{32}$	$\frac{3}{16}$	.140
$\frac{7}{32}$	$1\frac{7}{8}$	$3\frac{3}{4}$	$\frac{1}{4}$	$\frac{7}{32}$	.165
$\frac{1}{4}$	2	4	$\frac{1}{4}$	$\frac{1}{4}$	.185
$\frac{9}{32}$	$2\frac{1}{8}$	$4\frac{1}{4}$	$\frac{1}{4}$	$\frac{9}{32}$	.210
$\frac{5}{16}$	$2\frac{1}{4}$	$4\frac{1}{2}$	$\frac{5}{16}$	$\frac{5}{16}$	.235
$\frac{11}{32}$	$2\frac{3}{8}$	$4\frac{3}{4}$	$\frac{5}{16}$	$\frac{11}{32}$	.255
$\frac{3}{8}$	$2\frac{1}{2}$	5	$\frac{3}{8}$	$\frac{3}{8}$	.280
$\frac{13}{32}$	$2\frac{5}{8}$	$5\frac{1}{4}$	$\frac{3}{8}$	$\frac{13}{32}$	.305
$\frac{7}{16}$	$2\frac{3}{4}$	$5\frac{1}{2}$	$\frac{7}{16}$	$\frac{7}{16}$	.330
$\frac{15}{32}$	$2\frac{7}{8}$	$5\frac{3}{4}$	$\frac{7}{16}$	$\frac{15}{32}$	.350
$\frac{1}{2}$	3	6	$\frac{1}{2}$	$\frac{1}{2}$	.375
$\frac{17}{32}$	$3\frac{1}{8}$	$6\frac{1}{4}$	$\frac{1}{2}$	$\frac{17}{32}$	.400
$\frac{9}{16}$	$3\frac{1}{4}$	$6\frac{1}{2}$	$\frac{9}{16}$	$\frac{9}{16}$	.420
$\frac{19}{32}$	$3\frac{3}{8}$	$6\frac{3}{4}$	$\frac{9}{16}$	$\frac{19}{32}$	.445
$\frac{5}{8}$	$3\frac{1}{2}$	7	$\frac{5}{8}$	$\frac{5}{8}$	.470
$\frac{21}{32}$	$3\frac{11}{16}$	$7\frac{3}{8}$	$\frac{5}{8}$	$\frac{21}{32}$	.490
$\frac{11}{16}$	$3\frac{7}{8}$	$7\frac{3}{4}$	$\frac{11}{16}$	$\frac{11}{16}$	.515
$\frac{23}{32}$	$4\frac{1}{16}$	$8\frac{1}{8}$	$\frac{11}{16}$	$\frac{23}{32}$	.540
$\frac{3}{4}$	$4\frac{3}{16}$	$8\frac{3}{8}$	$\frac{3}{4}$	$\frac{3}{4}$	.560
$\frac{7}{8}$	$4\frac{7}{8}$	$9\frac{3}{4}$	$\frac{7}{8}$	$\frac{7}{8}$	.655
1	$5\frac{7}{16}$	$10\frac{7}{8}$	1	1	.750
$1\frac{1}{8}$	$5\frac{13}{16}$	$11\frac{5}{8}$	1	$1\frac{1}{8}$	.845
$1\frac{1}{4}$	$6\frac{1}{8}$	$12\frac{1}{4}$	1	$1\frac{1}{4}$	.935
$1\frac{3}{8}$	$6\frac{5}{16}$	$12\frac{5}{8}$	1	$1\frac{3}{8}$	1.030
$1\frac{1}{2}$	$6\frac{1}{2}$	13	$1\frac{1}{8}$	$1\frac{1}{2}$	1.125

Hand Reamers
Left Hand Helical Flutes — Squared Shank
High Speed Steel

Standard Sizes and Dimensions

Diameter Inches	Dimensions — Inches				
	Length of Flute	Length Overall	Length of Square	Diameter of Shank	Size of Square
¼	2	4	¼	¼	.185
⁵⁄₁₆	2 ¼	4 ½	⁵⁄₁₆	⁵⁄₁₆	.235
³⁄₈	2 ½	5	³⁄₈	³⁄₈	.280
⁷⁄₁₆	2 ¾	5 ½	⁷⁄₁₆	⁷⁄₁₆	.330
½	3	6	½	½	.375
⁹⁄₁₆	3 ¼	6 ½	⁹⁄₁₆	⁹⁄₁₆	.420
⅝	3 ½	7	⅝	⅝	.470
¹¹⁄₁₆	3 ⅞	7 ¾	¹¹⁄₁₆	¹¹⁄₁₆	.515
¾	4 ³⁄₁₆	8 ⅜	¾	¾	.560
¹³⁄₁₆	4 ⁹⁄₁₆	9 ⅛	¹³⁄₁₆	¹³⁄₁₆	.610
⅞	4 ⅞	9 ¾	⅞	⅞	.655
¹⁵⁄₁₆	5 ⅛	10 ¼	¹⁵⁄₁₆	¹⁵⁄₁₆	.705
1	5 ⁷⁄₁₆	10 ⅞	1	1	.750
1 ⅛	5¹³⁄₁₆	11 ⅝	1	1 ⅛	.845
1 ¼	6 ⅛	12 ¼	1	1 ¼	.935
1 ⅜	6 ⁵⁄₁₆	12 ⅝	1	1 ⅜	1.030
1 ½	6 ½	13	1 ⅛	1 ½	1.125

EXPANSION HAND REAMERS

Straight Flute Expansion Hand Reamers, Right-Hand Cut

*Helical Flute Expansion Hand Reamers, Left-Hand Helix,
Right-Hand Cut*

Expansion Hand Reamers are often used where it is necessary to enlarge a hole slightly in order to secure the necessary fit. Such an operation is quite common in maintenance or assembly work. Expansion reamers are usually made with an adjusting screw by means of which the reamer can be expanded slightly to the exact size required. Such reamers are regularly made with slightly undersize pilots which guide the reamer and help to hold it in alignment with the hole to be reamed. These reamers are ground with a starting taper on the flutes which provides easy entry of the reamer into the hole.

If an intermittent cut is encountered, such as a keyway, the use of the left-hand helical-fluted tools will be found helpful.

These reamers are designed to cut a small amount over nominal size for fitting purposes, and their limits of expansion are sufficient to accomplish this. Expansion Hand Reamers often are broken in attempting to expand them an amount beyond that for which they were designed. This limit of expansion will vary, depending upon the exact design of the tool, but the following table will give a general indication:

The maximum expansion on these reamers is as follows:

$\frac{1}{4}$ to $\frac{7}{16}$ incl., 0.006 inch
$\frac{1}{2}$ to $\frac{15}{16}$ incl., 0.010 inch
1 to 1 $\frac{1}{2}$ incl., 0.012 inch

Expansion Hand Reamers

Left Hand Helical Flutes — Squared Shank

Carbon Steel

The maximum expansion on these reamers is as follows:

$\frac{1}{4}$ to $\frac{7}{16}$ incl., 0.006 inch
$\frac{1}{2}$ to $\frac{7}{8}$ incl., 0.010 inch
1   to 1 $\frac{1}{4}$ incl., 0.012 inch

Standard Sizes and Dimensions

Diameter Inches	Dimensions — Inches						
	Length Overall		Length of Flute		Length of Square	Diam. of Shank	Size of Square
	Min.	Max.	Min.	Max.			
$\frac{1}{4}$	$3\frac{1}{8}$	$4\frac{3}{8}$	$1\frac{1}{2}$	$1\frac{3}{4}$	$\frac{1}{4}$	$\frac{1}{4}$	.185
$\frac{5}{16}$	4	$4\frac{3}{8}$	$1\frac{1}{2}$	$1\frac{3}{4}$	$\frac{5}{16}$	$\frac{5}{16}$	.235
$\frac{3}{8}$	$4\frac{1}{4}$	$6\frac{1}{8}$	$1\frac{3}{4}$	2	$\frac{3}{8}$	$\frac{3}{8}$	.280
$\frac{7}{16}$	$4\frac{1}{2}$	$6\frac{1}{4}$	$1\frac{3}{4}$	2	$\frac{7}{16}$	$\frac{7}{16}$	.330
$\frac{1}{2}$	5	$6\frac{1}{2}$	$1\frac{3}{4}$	$2\frac{1}{2}$	$\frac{1}{2}$	$\frac{1}{2}$	.375
$\frac{5}{8}$	6	8	$2\frac{1}{4}$	3	$\frac{5}{8}$	$\frac{5}{8}$	.470
$\frac{3}{4}$	$6\frac{1}{2}$	$8\frac{5}{8}$	$2\frac{5}{8}$	$3\frac{1}{2}$	$\frac{3}{4}$	$\frac{3}{4}$	.560
$\frac{7}{8}$	$7\frac{1}{2}$	$9\frac{3}{8}$	$3\frac{1}{8}$	4	$\frac{7}{8}$	$\frac{7}{8}$	.655
1	$8\frac{3}{8}$	$10\frac{1}{4}$	$3\frac{1}{8}$	$4\frac{1}{2}$	1	1	.750
1 $\frac{1}{4}$	$9\frac{3}{4}$	$11\frac{3}{8}$	$4\frac{1}{4}$	5	1	1 $\frac{1}{4}$	.935

Tolerances

Element	Range	Direction	Tolerance
Length Overall	$\frac{1}{4}$ to 1 incl. $1\frac{1}{4}$	Plus or Minus Plus or Minus	$\frac{1}{16}$ $\frac{3}{32}$
Length of Flute	$\frac{1}{4}$ to 1 incl. $1\frac{1}{4}$	Plus or Minus Plus or Minus	$\frac{1}{16}$ $\frac{3}{32}$
Length of Square	$\frac{1}{4}$ to 1 incl. $1\frac{1}{4}$	Plus or Minus Plus or Minus	$\frac{1}{32}$ $\frac{1}{16}$
Diameter of Shank	$\frac{1}{4}$ to 1 incl. $1\frac{1}{4}$	Minus Minus	.0010 to .005 .0015 to .006
Size of Square	$\frac{1}{4}$ to $\frac{1}{2}$ incl. $\frac{9}{16}$ to 1 incl. $1\frac{1}{4}$	Minus Minus Minus	.004 .006 .008

All dimensions are given in inches.

Expansion Hand Reamers

Straight Flutes — Squared Shank
Carbon Steel

The maximum expansion on these reamers is as follows:
¼ to 7/16 incl., 0.006 inch
½ to 7/8 incl., 0.010 inch
1 to 1¼ incl., 0.012 inch

Standard Sizes and Dimensions

Diameter Inches	Dimensions — Inches						
	Length Overall		Length of Flute		Length of Square	Diam. of Shank	Size of Square
	Min.	Max.	Min.	Max.			
¼	3¾	4⅜	1½	1¾	¼	¼	.185
5/16	4	4⅜	1½	1⅞	5/16	5/16	.235
⅜	4¼	5⅜	1¾	2	⅜	⅜	.280
7/16	4½	5⅜	1¾	2	7/16	7/16	.330
½	5	6½	1¾	2½	½	½	.375
9/16	5⅜	6½	1⅞	2½	9/16	9/16	.420
⅝	5¾	7	2¼	3	⅝	⅝	.470
11/16	6¼	7⅝	2½	3	11/16	11/16	.515
¾	6½	8	2⅝	3½	¾	¾	.560
⅞	7½	9	3⅛	4	⅞	⅞	.655
1	8⅜	10	3⅛	4½	1	1	.750
1⅛	9	10½	3½	4¾	1	1⅛	.845
1¼	9¾	11	4¼	5	1	1¼	.935

Tolerances

Element	Range	Direction	Tolerance
Length Overall	¼ to 1 incl.	Plus or Minus	1/16
	1⅛ to 1¼ incl.	Plus or Minus	3/32
Length of Flute	¼ to 1 incl.	Plus or Minus	1/16
	1⅛ to 1¼ incl.	Plus or Minus	3/32
Length of Square	¼ to 1 incl.	Plus or Minus	1/32
	1⅛ to 1¼ incl.	Plus or Minus	1/16
Diameter of Shank	¼ to 1 incl.	Minus	.0010 to .005
	1⅛ to 1¼ incl.	Minus	.0015 to .006
Size of Square	¼ to ½ incl.	Minus	.004
	9/16 to 1 incl.	Minus	.006
	1⅛ to 1¼ incl.	Minus	.008

All dimensions are given in inches.

ADJUSTABLE HAND REAMERS

Adjustable Hand Reamer, Right-Hand Cut

Adjustable Hand Reamers are designed to fill the needs of the user who has to ream occasional holes of odd sizes. This reamer has removable, replaceable blades which fit into slots having tapered bottoms. These cutting blades are usually made of carbon steel, but quite often are made of high-speed steel. They are held in place by two locking nuts, and adjustment is made by loosening one nut while tightening the other. Because of their comparatively wide range of adjustment they find particular use in garages, repair shops, and tool rooms, as a single set of reamers will cover the complete range of hole sizes ordinarily encountered in reaming. They are usually designed with enough adjustment so that about twenty reamers are required to cover all hole sizes from about $\frac{1}{4}''$ to $3''$.

CHUCKING REAMERS

Straight Shank Chucking Reamer, Straight Flute Right-Hand Cut

Straight Shank Chucking Reamers, made of high-speed steel, are designed for use in machines such as drill presses, turret lathes, and screw machines. They are regularly pointed with a 45° chamfer and are suitable for machine reaming most of the engineering materials ordinarily encountered.

Straight shank reamers are held in collets, split bushings, or by means of a setscrew.

Chucking Reamers

Straight Flutes — Straight Shank
Fractional Sizes
High Speed Steel

Standard Sizes and Dimensions

Fractional Sizes	Decimal Equivalent Inches	Dimensions — Inches			
		Diameter of Shank		Length of Flute	Length Overall
		Maximum	Minimum		
3/64	.0469	.0455	.0445	1/2	2 1/2
1/16	.0625	.0585	.0575	1/2	2 1/2
5/64	.0781	.0720	.0710	3/4	3
3/32	.0938	.0880	.0870	3/4	3
7/64	.1094	.1030	.1020	7/8	3 1/2
1/8	.1250	.1190	.1180	7/8	3 1/2
9/64	.1406	.1350	.1340	1	4
5/32	.1562	.1510	.1500	1	4
11/64	.1719	.1645	.1635	1 1/8	4 1/2
3/16	.1875	.1805	.1795	1 1/8	4 1/2
13/64	.2031	.1945	.1935	1 1/4	5
7/32	.2188	.2075	.2065	1 1/4	5
15/64	.2344	.2265	.2255	1 1/2	6
1/4	.2500	.2405	.2395	1 1/2	6
17/64	.2656	.2485	.2475	1 1/2	6
9/32	.2812	.2485	.2475	1 1/2	6
19/64	.2969	.2792	.2782	1 1/2	6
5/16	.3125	.2792	.2782	1 1/2	6
21/64	.3281	.2792	.2782	1 1/2	6
11/32	.3438	.2792	.2782	1 1/2	6
23/64	.3594	.3105	.3095	1 3/4	7
3/8	.3750	.3105	.3095	1 3/4	7
25/64	.3906	.3105	.3095	1 3/4	7
13/32	.4062	.3105	.3095	1 3/4	7
27/64	.4219	.3730	.3720	1 3/4	7
7/16	.4375	.3730	.3720	1 3/4	7
29/64	.4531	.3730	.3720	1 3/4	7
15/32	.4688	.3730	.3720	1 3/4	7
31/64	.4844	.4355	.4345	2	8
1/2	.5000	.4355	.4345	2	8
17/32	.5312	.4355	.4345	2	8
9/16	.5625	.4355	.4345	2	8
19/32	.5938	.4355	.4345	2	8
5/8	.6250	.5620	.5605	2 1/4	9
21/32	.6562	.5620	.5605	2 1/4	9
11/16	.6875	.5620	.5605	2 1/4	9
23/32	.7188	.5620	.5605	2 1/4	9
3/4	.7500	.6245	.6230	2 1/2	9 1/2

(Concluded on following page)

Chucking Reamers

Straight Flutes — Straight Shank
Fractional Sizes
High Speed Steel
(Concluded)

Standard Sizes and Dimensions

Fractional Sizes (Cont'd)	Decimal Equivalent Inches	Dimensions — Inches		Length of Flute	Length Overall
		Diameter of Shank			
		Maximum	Minimum		
$\frac{25}{32}$	.7812	.6245	.6230	$2\frac{1}{2}$	$9\frac{1}{2}$
$\frac{13}{16}$	.8125	.6245	.6230	$2\frac{1}{2}$	$9\frac{1}{2}$
$\frac{27}{32}$	.8438	.6245	.6230	$2\frac{1}{2}$	$9\frac{1}{2}$
$\frac{7}{8}$	.8750	.7495	.7480	$2\frac{5}{8}$	10
$\frac{29}{32}$	.9062	.7495	.7480	$2\frac{5}{8}$	10
$\frac{15}{16}$	.9375	.7495	.7480	$2\frac{5}{8}$	10
$\frac{31}{32}$	.9688	.7495	.7480	$2\frac{5}{8}$	10
1	1.0000	.8745	.8730	$2\frac{3}{4}$	$10\frac{1}{2}$
$1\frac{1}{16}$	1.0625	.8745	.8730	$2\frac{3}{4}$	$10\frac{1}{2}$
$1\frac{1}{8}$	1.1250	.8745	.8730	$2\frac{7}{8}$	11
$1\frac{3}{16}$	1.1875	.9995	.9980	$2\frac{7}{8}$	11
$1\frac{1}{4}$	1.2500	.9995	.9980	3	$11\frac{1}{2}$
$1\frac{3}{8}$	1.3750	.9995	.9980	$3\frac{1}{4}$	12
$1\frac{1}{2}$	1.5000	1.2495	1.2480	$3\frac{1}{2}$	$12\frac{1}{2}$

Chucking Reamers

Straight Flutes - Straight Shank

Wire Gauge Sizes

High Speed Steel

Standard Sizes and Dimensions

Wire Gauge Sizes	Decimal Equivalent Inches	Dimensions — Inches			
		Diameter of Shank		Length of Flute	Length Overall
		Maximum	Minimum		
60	.0400	.0390	.0380	$\frac{1}{2}$	$2\frac{1}{2}$
59	.0410	.0390	.0380	$\frac{1}{2}$	$2\frac{1}{2}$
58	.0420	.0390	.0380	$\frac{1}{2}$	$2\frac{1}{2}$
57	.0430	.0390	.0380	$\frac{1}{2}$	$2\frac{1}{2}$
56	.0465	.0455	.0445	$\frac{1}{2}$	$2\frac{1}{2}$
55	.0520	.0510	.0500	$\frac{1}{2}$	$2\frac{1}{2}$
54	.0550	.0510	.0500	$\frac{1}{2}$	$2\frac{1}{2}$
53	.0595	.0585	.0575	$\frac{1}{2}$	$2\frac{1}{2}$
52	.0635	.0585	.0575	$\frac{1}{2}$	$2\frac{1}{2}$
51	.0670	.0660	.0650	$\frac{3}{4}$	3
50	.0700	.0660	.0650	$\frac{3}{4}$	3
49	.0730	.0660	.0650	$\frac{3}{4}$	3
48	.0760	.0720	.0710	$\frac{3}{4}$	3
47	.0785	.0720	.0710	$\frac{3}{4}$	3
46	.0810	.0771	.0761	$\frac{3}{4}$	3
45	.0820	.0771	.0761	$\frac{3}{4}$	3
44	.0860	.0810	.0800	$\frac{3}{4}$	3
43	.0890	.0810	.0800	$\frac{3}{4}$	3
42	.0935	.0880	.0870	$\frac{3}{4}$	3
41	.0960	.0928	.0918	$\frac{7}{8}$	$3\frac{1}{2}$
40	.0980	.0928	.0918	$\frac{7}{8}$	$3\frac{1}{2}$
39	.0995	.0928	.0918	$\frac{7}{8}$	$3\frac{1}{2}$
38	.1015	.0950	.0940	$\frac{7}{8}$	$3\frac{1}{2}$
37	.1040	.0950	.0940	$\frac{7}{8}$	$3\frac{1}{2}$
36	.1065	.1030	.1020	$\frac{7}{8}$	$3\frac{1}{2}$
35	.1100	.1030	.1020	$\frac{7}{8}$	$3\frac{1}{2}$
34	.1110	.1055	.1045	$\frac{7}{8}$	$3\frac{1}{2}$
33	.1130	.1055	.1045	$\frac{7}{8}$	$3\frac{1}{2}$
32	.1160	.1120	.1110	$\frac{7}{8}$	$3\frac{1}{2}$
31	.1200	.1120	.1110	$\frac{7}{8}$	$3\frac{1}{2}$
30	.1285	.1190	.1180	$\frac{7}{8}$	$3\frac{1}{2}$
29	.1360	.1275	.1265	1	4
28	.1405	.1350	.1340	1	4
27	.1440	.1350	.1340	1	4
26	.1470	.1430	.1420	1	4
25	.1495	.1430	.1420	1	4
24	.1520	.1460	.1450	1	4

(Concluded on following page)

Chucking Reamers

Straight Flutes — Straight Shank

Wire Gauge Sizes (Continued)

High Speed Steel

(Concluded)

Standard Sizes and Dimensions

Wire Gauge Sizes	Decimal Equivalent Inches	Dimensions — Inches			
		Diameter of Shank		Length of Flute	Length Overall
		Maximum	Minimum		
23	.1540	.1460	.1450	1	4
22	.1570	.1510	.1500	1	4
21	.1590	.1530	.1520	$1\frac{1}{8}$	$4\frac{1}{2}$
20	.1610	.1530	.1520	$1\frac{1}{8}$	$4\frac{1}{2}$
19	.1660	.1595	.1585	$1\frac{1}{8}$	$4\frac{1}{2}$
18	.1695	.1595	.1585	$1\frac{1}{8}$	$4\frac{1}{2}$
17	.1730	.1645	.1635	$1\frac{1}{8}$	$4\frac{1}{2}$
16	.1770	.1704	.1694	$1\frac{1}{8}$	$4\frac{1}{2}$
15	.1800	.1755	.1745	$1\frac{1}{8}$	$4\frac{1}{2}$
14	.1820	.1755	.1745	$1\frac{1}{8}$	$4\frac{1}{2}$
13	.1850	.1805	.1795	$1\frac{1}{8}$	$4\frac{1}{2}$
12	.1890	.1805	.1795	$1\frac{1}{8}$	$4\frac{1}{2}$
11	.1910	.1860	.1850	$1\frac{1}{4}$	5
10	.1935	.1860	.1850	$1\frac{1}{4}$	5
9	.1960	.1895	.1885	$1\frac{1}{4}$	5
8	.1990	.1895	.1885	$1\frac{1}{4}$	5
7	.2010	.1945	.1935	$1\frac{1}{4}$	5
6	.2040	.1945	.1935	$1\frac{1}{4}$	5
5	.2055	.2016	.2006	$1\frac{1}{4}$	5
4	.2090	.2016	.2006	$1\frac{1}{4}$	5
3	.2130	.2075	.2065	$1\frac{1}{4}$	5
2	.2210	.2173	.2163	$1\frac{1}{2}$	6
1	.2280	.2173	.2163	$1\frac{1}{2}$	6

Chucking Reamers

Straight Flutes — Straight Shank

Letter Sizes

High Speed Steel

Standard Sizes and Dimensions

Letter Sizes	Decimal Equivalent Inches	Dimensions — Inches			
		Diameter of Shank		Length of Flute	Length Overall
		Maximum	Minimum		
A	.2340	.2265	.2255	1½	6
B	.2380	.2329	.2319	1½	6
C	.2420	.2329	.2319	1½	6
D	.2460	.2329	.2319	1½	6
E	.2500	.2405	.2395	1½	6
F	.2570	.2485	.2475	1½	6
G	.2610	.2485	.2475	1½	6
H	.2660	.2485	.2475	1½	6
I	.2720	.2485	.2475	1½	6
J	.2770	.2485	.2475	1½	6
K	.2810	.2485	.2475	1½	6
L	.2900	.2792	.2782	1½	6
M	.2950	.2792	.2782	1½	6
N	.3020	.2792	.2782	1½	6
O	.3160	.2792	.2782	1½	6
P	.3230	.2792	.2782	1½	6
Q	.3320	.2792	.2782	1½	6
R	.3390	.2792	.2782	1½	6
S	.3480	.3105	.3095	1¾	7
T	.3580	.3105	.3095	1¾	7
U	.3680	.3105	.3095	1¾	7
V	.3770	.3105	.3095	1¾	7
W	.3860	.3105	.3095	1¾	7
X	.3970	.3105	.3095	1¾	7
Y	.4040	.3105	.3095	1¾	7
Z	.4130	.3730	.3720	1¾	7

Chucking Reamers

Straight Flutes — Straight Shank
Decimal Sizes
High Speed Steel

Standard Sizes and Dimensions

Decimal Sizes	Dimensions — Inches			
	Diameter of Shank		Length of Flute	Length Overall
	Maximum	Minimum		
.1240	.1190	.1180	⅞	3½
.1260	.1190	.1180	⅞	3½
.1865	.1805	.1795	1⅛	4½
.1885	.1805	.1795	1⅛	4½
.2490	.2405	.2395	1½	6
.2510	.2405	.2395	1½	6
.3115	.2792	.2782	1½	6
.3135	.2792	.2782	1½	6
.3740	.3105	.3095	1¾	7
.3760	.3105	.3095	1¾	7
.4365	.3730	.3720	1¾	7
.4385	.3730	.3720	1¾	7
.4990	.4355	.4345	2	8
.5010	.4355	.4345	2	8

The helical fluted chucking reamer differs from the straight-fluted chucking reamer in that it has right-hand or left-hand helical flutes. It is used in the same types of machines and has the same shank sizes, flute and over-all lengths.

Straight Shank Chucking Reamer, Right-Hand Helix, Right-Hand Cut

Because of its free reaming action, this helical-fluted reamer is of particular use in those materials ordinarily difficult to ream. It is standard in high-speed steel.

Chucking Reamers

Helical Flutes - Straight Shank

High Speed Steel

Right hand or left hand helical flutes.

Standard Sizes and Dimensions

Fractional Sizes	Decimal Equivalent Inches	Dimensions — Inches			
		Diameter of Shank		Length of Flute	Length Overall
		Maximum	Minimum		
$1/16$	.0625	.0585	.0575	$1/2$	$2\frac{1}{2}$
$5/64$	.0781	.0720	.0710	$3/4$	3
$3/32$	.0938	.0880	.0870	$3/4$	3
$7/64$	.1094	.1030	.1020	$7/8$	$3\frac{1}{2}$
$1/8$	.1250	.1190	.1180	$7/8$	$3\frac{1}{2}$
$9/64$	.1406	.1350	.1340	1	4
$5/32$	.1562	.1510	.1500	1	4
$11/64$	.1719	.1645	.1635	$1\frac{1}{8}$	$4\frac{1}{2}$
$3/16$	.1875	.1805	.1795	$1\frac{1}{8}$	$4\frac{1}{2}$
$13/64$	.2031	.1945	.1935	$1\frac{1}{4}$	5
$7/32$	.2188	.2075	.2065	$1\frac{1}{4}$	5
$15/64$	.2344	.2265	.2255	$1\frac{1}{2}$	6
$1/4$	.2500	.2405	.2395	$1\frac{1}{2}$	6
$17/64$	.2656	.2485	.2475	$1\frac{1}{2}$	6
$9/32$	.2812	.2485	.2475	$1\frac{1}{2}$	6
$19/64$	.2969	.2792	.2782	$1\frac{1}{2}$	6
$5/16$	.3125	.2792	.2782	$1\frac{1}{2}$	6
$21/64$	.3281	.2792	.2782	$1\frac{1}{2}$	6
$11/32$	.3438	.2792	.2782	$1\frac{1}{2}$	6
$23/64$	.3594	.3105	.3095	$1\frac{3}{4}$	7
$3/8$	.3750	.3105	.3095	$1\frac{3}{4}$	7
$25/64$	.3906	.3105	.3095	$1\frac{3}{4}$	7
$13/32$	.4062	.3105	.3095	$1\frac{3}{4}$	7
$27/64$	.4219	.3730	.3720	$1\frac{3}{4}$	7
$7/16$	.4375	.3730	.3720	$1\frac{3}{4}$	7
$29/64$	.4531	.3730	.3720	$1\frac{3}{4}$	7
$15/32$	.4688	.3730	.3720	$1\frac{3}{4}$	7
$31/64$	.4844	.4355	.4345	2	8
$1/2$	.5000	.4355	.4345	2	8
$17/32$	.5312	.4355	.4345	2	8
$9/16$	.5625	.4355	.4345	2	8
$19/32$	.5938	.4355	.4345	2	8
$5/8$	.6250	.5620	.5605	$2\frac{1}{4}$	9
$21/32$	.6562	.5620	.5605	$2\frac{1}{4}$	9
$11/16$	.6875	.5620	.5605	$2\frac{1}{4}$	9
$23/32$	.7188	.5620	.5605	$2\frac{1}{4}$	9
$3/4$	.7500	.6245	.6230	$2\frac{1}{2}$	$9\frac{1}{2}$

(Concluded on next page)

Chucking Reamers

Helical Flutes — Straight Shank
High Speed Steel
(Concluded)

Right hand or left hand helical flutes.

Standard Sizes and Dimensions

Fractional Sizes (Cont'd)	Decimal Equivalent Inches	Dimensions — Inches			
		Diameter of Shank		Length of Flute	Length Overall
		Maximum	Minimum		
25⁄32	.7812	.6245	.6230	2½	9½
13⁄16	.8125	.6245	.6230	2½	9½
27⁄32	.8438	.6245	.6230	2½	9½
7⁄8	.8750	.7495	.7480	2⅝	10
29⁄32	.9062	.7495	.7480	2⅝	10
15⁄16	.9375	.7495	.7480	2⅝	10
31⁄32	.9688	.7495	.7480	2⅝	10
1	1.0000	.8745	.8730	2¾	10½
1 1⁄16	1.0625	.8745	.8730	2¾	10½
1 1⁄8	1.1250	.8745	.8730	2⅞	11
1 3⁄16	1.1875	.9995	.9980	2⅞	11
1 1⁄4	1.2500	.9995	.9980	3	11½
1 5⁄16	1.3125	.9995	.9980	3	11½
1 3⁄8	1.3750	.9995	.9980	3¼	12
1 7⁄16	1.4375	1.2495	1.2480	3¼	12
1 1⁄2	1.5000	1.2495	1.2480	3½	12½

Chucking Reamers

Straight Flutes — Taper Shank
High Speed Steel

Standard Sizes and Dimensions

Diameter Inches	Dimensions — Inches		Number of Morse Taper Shank
	Length of Flute	Length Overall	
$\frac{1}{4}$	$1\frac{1}{2}$	6	1
$\frac{9}{32}$	$1\frac{1}{2}$	6	1
$\frac{5}{16}$	$1\frac{1}{2}$	6	1
$11\frac{1}{32}$	$1\frac{1}{2}$	6	1
$\frac{3}{8}$	$1\frac{3}{4}$	7	1
$13\frac{1}{32}$	$1\frac{3}{4}$	7	1
$\frac{7}{16}$	$1\frac{3}{4}$	7	1
$15\frac{1}{32}$	$1\frac{3}{4}$	7	1
$\frac{1}{2}$	2	8	1
$17\frac{1}{32}$	2	8	1
$\frac{9}{16}$	2	8	1
$19\frac{1}{32}$	2	8	1
$\frac{5}{8}$	$2\frac{1}{4}$	9	2
$21\frac{1}{32}$	$2\frac{1}{4}$	9	2
$11\frac{1}{16}$	$2\frac{1}{4}$	9	2
$23\frac{1}{32}$	$2\frac{1}{4}$	9	2
$\frac{3}{4}$	$2\frac{1}{2}$	$9\frac{1}{2}$	2
$25\frac{1}{32}$	$2\frac{1}{2}$	$9\frac{1}{2}$	2
$13\frac{1}{16}$	$2\frac{1}{2}$	$9\frac{1}{2}$	2
$27\frac{1}{32}$	$2\frac{1}{2}$	$9\frac{1}{2}$	2
$\frac{7}{8}$	$2\frac{5}{8}$	10	2
$29\frac{1}{32}$	$2\frac{5}{8}$	10	2
$15\frac{1}{16}$	$2\frac{5}{8}$	10	3
$31\frac{1}{32}$	$2\frac{5}{8}$	10	3
1	$2\frac{3}{4}$	$10\frac{1}{2}$	3
$1\frac{1}{16}$	$2\frac{3}{4}$	$10\frac{1}{2}$	3
$1\frac{1}{8}$	$2\frac{7}{8}$	11	3
$1\frac{3}{16}$	$2\frac{7}{8}$	11	3
$1\frac{1}{4}$	3	$11\frac{1}{2}$	4
$1\frac{5}{16}$	3	$11\frac{1}{2}$	4
$1\frac{3}{8}$	$3\frac{1}{4}$	12	4
$1\frac{7}{16}$	$3\frac{1}{4}$	12	4
$1\frac{1}{2}$	$3\frac{1}{2}$	$12\frac{1}{2}$	4

Chucking Reamers

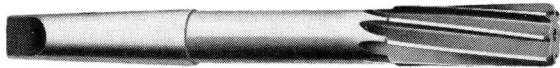

Helical Flutes - Taper Shank
High Speed Steel

Right hand or left hand helical flutes.

Standard Sizes and Dimensions

Diameter Inches	Dimensions — Inches		No. of Morse Taper Shank
	Length of Flute	Length Overall	
¼	1½	6	1
9⁄32	1½	6	1
5⁄16	1½	6	1
11⁄32	1½	6	1
3⁄8	1¾	7	1
13⁄32	1¾	7	1
7⁄16	1¾	7	1
15⁄32	1¾	7	1
½	2	8	1
17⁄32	2	8	1
9⁄16	2	8	1
19⁄32	2	8	1
5⁄8	2¼	9	2
21⁄32	2¼	9	2
11⁄16	2¼	9	2
23⁄32	2¼	9	2
¾	2½	9½	2
25⁄32	2½	9½	2
13⁄16	2½	9½	2
27⁄32	2½	9½	2
⅞	2⅝	10	2
29⁄32	2⅝	10	2
15⁄16	2⅝	10	3
31⁄32	2⅝	10	3
1	2¾	10½	3
1 1⁄16	2¾	10½	3
1 ⅛	2⅞	11	3
1 3⁄16	2⅞	11	3
1 ¼	3	11½	4
1 5⁄16	3	11½	4
1 ⅜	3¼	12	4
1 7⁄16	3¼	12	4
1 ½	3½	12½	4

(Concluded on following page)

JOBBERS' REAMERS

Taper Shank Jobbers' Reamers are designed to fill the needs of users who require a machine reamer with long flutes. They are standard in high-speed steel only. These reamers have approximately the same flute length as hand reamers, but are sharpened with a 45° chamfer and are intended for machine use.

Straight Flutes — Taper Shank
High Speed Steel

Standard Sizes and Dimensions

Diameter Inches	Dimensions — Inches		Number of Morse Taper Shank
	Length of Flute	Length Overall	
1/4	2	5 3/16	1
5/16	2 1/4	5 1/2	1
3/8	2 1/2	5 13/16	1
7/16	2 3/4	6 1/8	1
1/2	3	6 7/16	1
9/16	3 1/4	6 3/4	1
5/8	3 1/2	7 9/16	2
11/16	3 7/8	8	2
3/4	4 3/16	8 3/8	2
13/16	4 9/16	8 13/16	2
7/8	4 7/8	9 3/16	2
15/16	5 1/8	10	3
1	5 7/16	10 3/8	3
1 1/16	5 5/8	10 5/8	3
1 1/8	5 13/16	10 7/8	3
1 3/16	6	11 1/8	3
1 1/4	6 1/8	12 9/16	4
1 3/8	6 5/16	12 13/16	4
1 1/2	6 1/2	13 1/8	4

EXPANSION CHUCKING REAMERS

The expansion feature of the Expansion Chucking Reamer is designed to maintain the initial size by compensating for wear on the cutting end. When the reamer is worn, it may be expanded oversize, reground to size, and re-sharpened.

Expansion screws should never be loosened in an attempt to use the reamers for a size smaller than that for which the tools were originally circle ground.

Straight Flutes — Straight Shank
High Speed Steel

Standard Sizes and Dimensions

Diameter Inches	Dimensions — Inches		
	Diameter of Shank	Length of Flute	Length Overall
3/8	5/16	3/4	7
13/32	5/16	3/4	7
7/16	3/8	7/8	7
15/32	3/8	7/8	7
1/2	7/16	1	8
17/32	7/16	1	8
9/16	7/16	1 1/8	8
19/32	7/16	1 1/8	8
5/8	9/16	1 1/4	9
21/32	9/16	1 1/4	9
11/16	9/16	1 1/4	9
23/32	9/16	1 1/4	9
3/4	5/8	1 3/8	9 1/2
25/32	5/8	1 3/8	9 1/2
13/16	5/8	1 3/8	9 1/2
27/32	5/8	1 3/8	9 1/2
7/8	3/4	1 1/2	10
29/32	3/4	1 1/2	10
15/16	3/4	1 1/2	10
31/32	3/4	1 1/2	10

(Concluded on following page)

Expansion Chucking Reamers

Straight Flutes — Straight Shank
High Speed Steel

(Concluded)

Standard Sizes and Dimensions

Diameter Inches	Dimensions — Inches		
	Diameter of Shank	Length of Flute	Length Overall
1	$\frac{7}{8}$	$1\frac{5}{8}$	$10\frac{1}{2}$
$1\frac{1}{32}$	$\frac{7}{8}$	$1\frac{5}{8}$	$10\frac{1}{2}$
$1\frac{1}{16}$	$\frac{7}{8}$	$1\frac{5}{8}$	$10\frac{1}{2}$
$1\frac{3}{32}$	$\frac{7}{8}$	$1\frac{5}{8}$	$10\frac{1}{2}$
$1\frac{1}{8}$	$\frac{7}{8}$	$1\frac{3}{4}$	11
$1\frac{5}{32}$	$\frac{7}{8}$	$1\frac{3}{4}$	11
$1\frac{3}{16}$	1	$1\frac{3}{4}$	11
$1\frac{7}{32}$	1	$1\frac{3}{4}$	11
$1\frac{1}{4}$	1	$1\frac{7}{8}$	$11\frac{1}{2}$
$1\frac{5}{16}$	1	$1\frac{7}{8}$	$11\frac{1}{2}$
$1\frac{3}{8}$	1	2	12
$1\frac{7}{16}$	$1\frac{1}{4}$	2	12
$1\frac{1}{2}$	$1\frac{1}{4}$	$2\frac{1}{8}$	$12\frac{1}{2}$
$1\frac{9}{16}$	$1\frac{1}{4}$	$2\frac{1}{8}$	$12\frac{1}{2}$
$1\frac{5}{8}$	$1\frac{1}{4}$	$2\frac{1}{4}$	13
$1\frac{11}{16}$	$1\frac{1}{4}$	$2\frac{1}{4}$	13
$1\frac{3}{4}$	$1\frac{1}{4}$	$2\frac{3}{8}$	$13\frac{1}{2}$
$1\frac{13}{16}$	$1\frac{1}{2}$	$2\frac{3}{8}$	$13\frac{1}{2}$
$1\frac{7}{8}$	$1\frac{1}{2}$	$2\frac{1}{2}$	14
$1\frac{15}{16}$	$1\frac{1}{2}$	$2\frac{1}{2}$	14
2	$1\frac{1}{2}$	$2\frac{1}{2}$	14
$2\frac{1}{8}$	$1\frac{1}{2}$	$2\frac{3}{4}$	$14\frac{1}{2}$
$2\frac{1}{4}$	$1\frac{3}{4}$	$2\frac{3}{4}$	$14\frac{1}{2}$
$2\frac{1}{2}$	$1\frac{3}{4}$	3	15

Expansion Chucking Reamers

Straight Flutes — Taper Shank
High Speed Steel

Standard Sizes and Dimensions

Diameter Inches	Dimensions — Inches		Number of Morse Taper Shank
	Length of Flute	Length Overall	
3/8	3/4	7	1
13/32	3/4	7	1
7/16	7/8	7	1
15/32	7/8	7	1
1/2	1	8	1
17/32	1	8	1
9/16	1 1/8	8	1
19/32	1 1/8	8	1
5/8	1 1/4	9	2
21/32	1 1/4	9	2
11/16	1 1/4	9	2
23/32	1 1/4	9	2
3/4	1 3/8	9 1/2	2
25/32	1 3/8	9 1/2	2
13/16	1 3/8	9 1/2	2
27/32	1 3/8	9 1/2	2
7/8	1 1/2	10	2
29/32	1 1/2	10	2
15/16	1 1/2	10	3
31/32	1 1/2	10	3
1	1 5/8	10 1/2	3
1 1/32	1 5/8	10 1/2	3
1 1/16	1 5/8	10 1/2	3
1 3/32	1 5/8	10 1/2	3
1 1/8	1 3/4	11	3
1 5/32	1 3/4	11	3
1 3/16	1 3/4	11	3
1 7/32	1 3/4	11	3
1 1/4	1 7/8	11 1/2	4
1 5/16	1 7/8	11 1/2	4

(Concluded on following page)

Expansion Chucking Reamers

Straight Flutes — Taper Shank

High Speed Steel

(Concluded)

Standard Sizes and Dimensions

Diameter Inches	Dimensions — Inches		Number of Morse Taper Shank
	Length of Flute	Length Overall	
1 3/8	2	12	4
1 7/16	2	12	4
1 1/2	2 1/8	12 1/2	4
1 5/8	2 1/4	13	4
1 3/4	2 3/8	13 1/2	5
1 7/8	2 1/2	14	5
2	2 1/2	14	5
2 1/8	2 3/4	14 1/2	5
2 1/4	2 3/4	14 1/2	5
2 3/8	3	15	5
2 1/2	3	15	5

Tolerances

Element	Range	Direction	Tolerance
Diameter of Reamer	3/8 to 1 incl.	Plus	.0001 to .0005
	1 1/32 to 2 1/2 incl.	Plus	.0002 to .0006
Length Overall	3/8 to 1 incl.	Plus or Minus	1/16
	1 1/32 to 2 incl.	Plus or Minus	3/32
	2 1/8 to 2 1/2 incl.	Plus or Minus	1/8
Length of Flute	3/8 to 1 incl.	Plus or Minus	1/16
	1 1/32 to 2 incl.	Plus or Minus	3/32
	2 1/8 to 2 1/2 incl.	Plus or Minus	1/8

All dimensions are given in inches.

ADJUSTABLE REAMERS
With Inserted Blades

Taper Shank Adjustable Reamer

Straight Shank Adjustable Reamer

Adjustable Reamers are designed to permit the user to set them to any desired diameter within the range of the adjustment. This adjustment permits them to be set to a special diameter, and also, when the reamers are worn, allows them to be set oversize, reground, and resharpened. The high-speed steel blades may be replaced individually or in sets when necessary. For these reasons this type of reamer is particularly adapted for machine reaming on a production basis.

The method of adjustment, locking of blades in place, and overall dimensions will vary with different designs. Because of its built up construction, this tool is seldom made smaller than 1″ diameter.

SHELL REAMERS

Straight Flute Shell Reamer   *Helical Flute Shell Reamer*

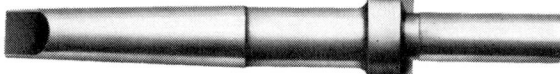

Taper Shank Arbor

Straight Shank Arbor

Shell Reamers are primarily chucking reamers and are made of high-speed steel. They are fluted almost their entire length and have a tapered hole (⅛″ taper per foot).

They are used with an arbor which is tapered to fit the hole in the reamer. This fit, together with slots in the reamer which engage lugs on the arbor, provides ample strength of drive. The large diameter of the taper on the arbor is slightly larger than the hole in the reamer to allow a space between the collar and the reamer. This space permits the reamer to be pried loose from the arbor without damaging either reamer or arbor.

Several sizes of reamers are used on one size of arbor. Only six arbors are required to fit all reamers from ¾″ diameter to 2½″, the range in which this tool is made. See page 225 for arbors.

Both expansion reamers, page 261 and adjustable reamers, page 263, are made by some manufacturers in the Shell Reamer type.

Shell Reamers

Straight or Left Hand Helical Flutes
High Speed Steel
Standard Sizes and Dimensions

Diameter Inches	Dimensions — Inches		Fitting Arbor Number
	Diameter Hole Large End	Length Overall	
3/4	3/8	2 1/4	4
7/8	1/2	2 1/2	5
1	1/2	2 1/2	5
1 1/16	5/8	2 3/4	6
1 1/8	5/8	2 3/4	6
1 3/16	5/8	2 3/4	6
1 1/4	5/8	2 3/4	6
1 5/16	3/4	3	7
1 3/8	3/4	3	7
1 7/16	3/4	3	7
1 1/2	3/4	3	7
1 9/16	3/4	3	7
1 5/8	3/4	3	7
1 11/16	1	3 1/2	8
1 3/4	1	3 1/2	8
1 13/16	1	3 1/2	8
1 7/8	1	3 1/2	8
1 15/16	1	3 1/2	8
2	1	3 1/2	8
2 1/8	1 1/4	3 3/4	9
2 1/4	1 1/4	3 3/4	9

(Concluded on following page)

Shell Reamer Arbors

Straight Shank
Standard Sizes and Dimensions

Size Number of Arbor	Fitting Size Reamer Inches	Dimensions — Inches	
		Diameter of Shank	Length Overall
4	$\frac{3}{4}$	$\frac{1}{2}$	9
5	$\frac{13}{16}$ — 1	$\frac{5}{8}$	$9\frac{1}{2}$
6	$1\frac{1}{16}$ — $1\frac{1}{4}$	$\frac{3}{4}$	10
7	$1\frac{5}{16}$ — $1\frac{5}{8}$	$\frac{7}{8}$	11
8	$1\frac{11}{16}$ — 2	$1\frac{1}{8}$	12
9	$2\frac{1}{16}$ — $2\frac{1}{2}$	$1\frac{3}{8}$	13

Taper Shank
Standard Sizes and Dimensions

Size Number of Arbor	Fitting Size Reamer Inches	Length Overall Inches	Number of Morse Taper Shank
4	$\frac{3}{4}$	9	2
5	$\frac{13}{16}$ — 1	$9\frac{1}{2}$	2
6	$1\frac{1}{16}$ — $1\frac{1}{4}$	10	3
7	$1\frac{5}{16}$ — $1\frac{5}{8}$	11	3
8	$1\frac{11}{16}$ — 2	12	4
9	$2\frac{1}{16}$ — $2\frac{1}{2}$	13	4

DRIVING SLOTS AND LUGS
SHELL REAMERS AND SHELL REAMER ARBORS

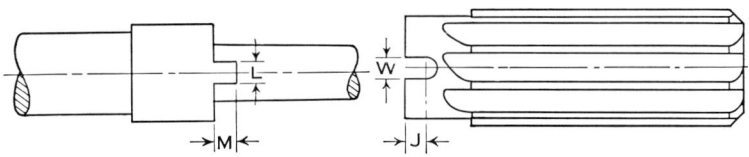

Diameter Hole in Reamer at Large End Inches	No. of Arbor	Dimensions — Inches					
		Fitting Reamer Sizes		Driving Slot		Lug on Arbor	
		From	To and Including	Width W	Depth J	Width L	Depth M
.375	4	—	$\frac{3}{4}$	$\frac{5}{32}$	$\frac{3}{16}$	$\frac{9}{64}$	$\frac{5}{32}$
.500	5	$\frac{13}{16}$	1	$\frac{3}{16}$	$\frac{1}{4}$	$\frac{11}{64}$	$\frac{7}{32}$
.625	6	$1\frac{1}{16}$	$1\frac{1}{4}$	$\frac{3}{16}$	$\frac{1}{4}$	$\frac{11}{64}$	$\frac{7}{32}$
.750	7	$1\frac{5}{16}$	$1\frac{5}{8}$	$\frac{1}{4}$	$\frac{5}{16}$	$\frac{15}{64}$	$\frac{9}{32}$
1.000	8	$1\frac{11}{16}$	2	$\frac{1}{4}$	$\frac{5}{16}$	$\frac{15}{64}$	$\frac{9}{32}$
1.250	9	$2\frac{1}{16}$	$2\frac{1}{2}$	$\frac{5}{16}$	$\frac{3}{8}$	$\frac{19}{64}$	$\frac{11}{32}$

STUB SCREW MACHINE REAMERS

High-Speed Steel

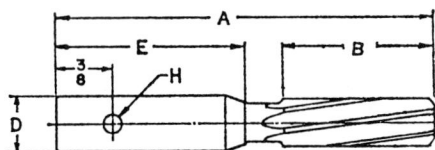

Stub Screw Machine Reamers are essentially machine reamers of short overall length, designed particularly for use in screw machine work. They are ground with a 45° chamfer on the cutting end and are standard with left-hand helix, right-hand cut.

The straight shank of these tools will fit into standard holders, the cross hole permitting them to be used in pin drive floating holders.

Stub Screw Machine Reamers are usually available either finish ground to size or semi-finished, the final grinding to size, relieving, and chamfering being done by the user.

Standard Sizes and Dimensions

Series Number	Diameter Range	Length Over-all A	Length Flute B	Shank Diameter D	Shank Length E	Pin Hole H	No. of Flutes
00	.0600 to .066 Incl.	$1\frac{3}{4}$	$\frac{1}{2}$	$\frac{1}{8}$	1	$\frac{1}{16}$	4
0	.0661 to .074 Incl.	$1\frac{3}{4}$	$\frac{1}{2}$	$\frac{1}{8}$	1	$\frac{1}{16}$	4
1	.0741 to .084 Incl.	$1\frac{3}{4}$	$\frac{1}{2}$	$\frac{1}{8}$	1	$\frac{1}{16}$	4
2	.0841 to .096 Incl.	$1\frac{3}{4}$	$\frac{1}{2}$	$\frac{1}{8}$	1	$\frac{1}{16}$	4
3	.0961 to .126 Incl.	2	$\frac{3}{4}$	$\frac{1}{8}$	1	$\frac{1}{16}$	4
4	.1261 to .158 Incl.	$2\frac{1}{4}$	1	$\frac{1}{4}$	1	$\frac{3}{32}$	4
5	.1581 to .188 Incl.	$2\frac{1}{4}$	1	$\frac{1}{4}$	1	$\frac{3}{32}$	4
6	.1881 to .219 Incl.	$2\frac{1}{4}$	1	$\frac{1}{4}$	1	$\frac{3}{32}$	6
7	.2191 to .251 Incl.	$2\frac{1}{4}$	1	$\frac{1}{4}$	1	$\frac{3}{32}$	6
8	.2511 to .282 Incl.	$2\frac{1}{4}$	1	$\frac{3}{8}$	1	$\frac{1}{8}$	6
9	.2821 to .313 Incl.	$2\frac{1}{4}$	1	$\frac{3}{8}$	1	$\frac{1}{8}$	6
10	.3131 to .344 Incl.	$2\frac{1}{2}$	$1\frac{1}{4}$	$\frac{3}{8}$	1	$\frac{1}{8}$	6
11	.3441 to .376 Incl.	$2\frac{1}{2}$	$1\frac{1}{4}$	$\frac{3}{8}$	1	$\frac{1}{8}$	6
12	.3761 to .407 Incl.	$2\frac{1}{2}$	$1\frac{1}{4}$	$\frac{1}{2}$	1	$\frac{3}{16}$	6
13	.4071 to .439 Incl.	$2\frac{1}{2}$	$1\frac{1}{4}$	$\frac{1}{2}$	1	$\frac{3}{16}$	6
14	.4391 to .470 Incl.	$2\frac{1}{2}$	$1\frac{1}{4}$	$\frac{1}{2}$	1	$\frac{3}{16}$	6
15	.4701 to .505 Incl.	$2\frac{1}{2}$	$1\frac{1}{4}$	$\frac{1}{2}$	1	$\frac{3}{16}$	6
16	.5051 to .567 Incl.	3	$1\frac{1}{2}$	$\frac{5}{8}$	$1\frac{1}{4}$	$\frac{1}{4}$	6
17	.5671 to .630 Incl.	3	$1\frac{1}{2}$	$\frac{5}{8}$	$1\frac{1}{4}$	$\frac{1}{4}$	6
18	.6301 to .692 Incl.	3	$1\frac{1}{2}$	$\frac{5}{8}$	$1\frac{1}{4}$	$\frac{1}{4}$	6
19	.6921 to .755 Incl.	3	$1\frac{1}{2}$	$\frac{3}{4}$	$1\frac{1}{4}$	$\frac{5}{16}$	8
20	.7551 to .817 Incl.	3	$1\frac{1}{2}$	$\frac{3}{4}$	$1\frac{1}{4}$	$\frac{5}{16}$	8
21	.8171 to .880 Incl.	3	$1\frac{1}{2}$	$\frac{3}{4}$	$1\frac{1}{4}$	$\frac{5}{16}$	8
22	.8801 to .942 Incl.	3	$1\frac{1}{2}$	$\frac{3}{4}$	$1\frac{1}{4}$	$\frac{5}{16}$	8
23	.9421 to 1.010 Incl.	3	$1\frac{1}{2}$	$\frac{3}{4}$	$1\frac{1}{4}$	$\frac{5}{16}$	8

TAPER PIN REAMERS

Taper Pin Reamers have a taper of ¼″ to the foot and are designed to ream holes into which standard taper pins will fit. The straight-fluted and slow left-hand spiral reamers are used for hand reaming, while the high spiral type is particularly designed for machine use. The flute construction of this last style of reamer tends to prevent chips packing in the flutes and thus reduces breakage. However, it is more difficult to feed into the work than the other two designs and consequently is not as suitable for most hand reaming applications. Best results are usually obtained if the hole to be reamed is drilled to a size a few thousandths smaller than the small diameter of the finish-reamed hole.

Taper Pin Reamers
Straight Flutes—Squared Shank
Carbon or High Speed Steel
Taper ¼ Inch per Foot

Standard Sizes and Dimensions

Size Number	Diam. Shank	Nominal Diam. Small End	Nominal Diam. Large End	Length Flutes	Length Overall	Length of Square	Diam. of Shank	Size of Square
7/0	5/64	.0497	.0666	13/16	1 13/16	5/32	5/64	.060
6/0	3/32	.0611	.0806	15/16	1 15/16	5/32	3/32	.070
5/0	7/64	.0719	.0966	1 3/16	2 3/16	5/32	7/64	.080
4/0	1/8	.0869	.1142	1 5/16	2 5/16	5/32	1/8	.095
3/0	9/64	.1029	.1302	1 5/16	2 5/16	5/32	9/64	.105
2/0	5/32	.1137	.1462	1 9/16	2 9/16	7/32	5/32	.115
0	11/64	.1287	.1638	1 11/16	2 15/16	7/32	11/64	.130
1	3/16	.1447	.1798	1 11/16	2 15/16	7/32	3/16	.140
2	13/64	.1605	.2008	1 15/16	3 3/16	1/4	13/64	.150
3	15/64	.1813	.2294	2 5/16	3 11/16	1/4	15/64	.175
4	17/64	.2071	.2604	2 9/16	4 1/16	1/4	17/64	.200
5	5/16	.2409	.2994	2 13/16	4 5/16	5/16	5/16	.235
6	23/64	.2773	.354	3 11/16	5 7/16	3/8	23/64	.270
7	13/32	.3297	.422	4 7/16	6 5/16	3/8	13/32	.305
8	7/16	.3971	.505	5 3/16	7 3/16	7/16	7/16	.330
9	9/16	.4805	.6066	6 1/16	8 5/16	9/16	9/16	.420
10	5/8	.5799	.7216	6 13/16	9 5/16	5/8	5/8	.470

Taper Pin Reamers
Left Hand Spiral Flutes
Squared Shank
High Speed Steel
Taper ¼ Inc hper Foot

Standard Sizes and Dimensions

Size Number	Diam. Shank	Nominal Diam. Small End	Nominal Diam. Large End	Length Flutes	Length Overall	Length of Square	Diam. of Shank	Size of Square
7/0	5/64	.0497	.0666	13/16	1 13/16	5/32	5/64	.060
6/0	3/32	.0611	.0806	15/16	1 15/16	5/32	3/32	.070
5/0	7/64	.0719	.0966	1 3/16	2 3/16	5/32	7/64	.080
4/0	1/8	.0869	.1142	1 5/16	2 5/16	5/32	1/8	.095
3/0	9/64	.1029	.1302	1 5/16	2 5/16	5/32	9/64	.105
2/0	5/32	.1137	.1462	1 9/16	2 9/16	7/32	5/32	.115
0	11/64	.1287	.1638	1 11/16	2 15/16	7/32	11/64	.130
1	3/16	.1447	.1798	1 11/16	2 15/16	7/32	3/16	.140
2	13/64	.1605	.2008	1 15/16	3 3/16	1/4	13/64	.150
3	15/64	.1813	.2294	2 5/16	3 11/16	1/4	15/64	.175
4	17/64	.2071	.2604	2 9/16	4 1/16	1/4	17/64	.200
5	5/16	.2409	.2994	2 13/16	4 5/16	5/16	5/16	.235
6	23/64	.2773	.354	3 11/16	5 7/16	3/8	23/64	.270
7	13/32	.3297	.422	4 7/16	6 5/16	3/8	13/32	.305
8	7/16	.3971	.505	5 3/16	7 3/16	7/16	7/16	.330
9	9/16	.4805	.6066	6 1/16	8 5/16	9/16	9/16	.420
10	5/8	.5799	.7216	6 13/16	9 5/16	5/8	5/8	.470

Taper Pin Reamers
High Spiral Flutes
High Speed Steel
Taper ¼ Inch per Foot

These reamers are standard with plain round shanks. Left hand high spiral flutes.

Standard Sizes and Dimensions

Size Number	Dimensions — Inches				
		Nominal		Length	
	Diameter Shank	Diameter Small End	Diameter Large End	Flutes	Overall
8/0	¹⁄₁₆	.0351	.0514	²⁵⁄₃₂	1 ⁵⁄₈
7/0	⁵⁄₆₄	.0497	.0666	¹³⁄₁₆	1¹³⁄₁₆
6/0	³⁄₃₂	.0611	.0806	¹⁵⁄₁₆	1¹⁵⁄₁₆
5/0	⁷⁄₆₄	.0719	.0966	1 ³⁄₁₆	2 ³⁄₁₆
4/0	⅛	.0869	.1142	1 ⁵⁄₁₆	2 ⁵⁄₁₆
3/0	⁹⁄₆₄	.1029	.1302	1 ⁵⁄₁₆	2 ⁵⁄₁₆
2/0	⁵⁄₃₂	.1137	.1462	1 ⁹⁄₁₆	2 ⁹⁄₁₆
0	¹¹⁄₆₄	.1287	.1638	1¹¹⁄₁₆	2¹⁵⁄₁₆
1	³⁄₁₆	.1447	.1798	1¹¹⁄₁₆	2¹⁵⁄₁₆
2	¹³⁄₆₄	.1605	.2008	1¹⁵⁄₁₆	3 ³⁄₁₆
3	¹⁵⁄₆₄	.1813	.2294	2 ⁵⁄₁₆	3¹¹⁄₁₆
4	¹⁷⁄₆₄	.2071	.2604	2 ⁹⁄₁₆	4 ¹⁄₁₆
5	⁵⁄₁₆	.2409	.2994	2¹³⁄₁₆	4 ⁵⁄₁₆
6	²³⁄₆₄	.2773	.354	3¹¹⁄₁₆	5 ⁷⁄₁₆
7	¹³⁄₃₂	.3297	.422	4 ⁷⁄₁₆	6 ⁵⁄₁₆
8	⁷⁄₁₆	.3971	.505	5 ³⁄₁₆	7 ³⁄₁₆
9	⁹⁄₁₆	.4805	.6066	6 ¹⁄₁₆	8 ⁵⁄₁₆
10	⅝	.5799	.7216	6¹³⁄₁₆	9 ⁵⁄₁₆

Die Makers' Reamers

Left Hand High Spiral Flutes
Taper ¾ degree included angle, or .013″ per Inch
High Speed Steel

Die Makers' Reamers are designed for machine reaming.

Standard Sizes and Dimensions

Size	Dimensions — Inches			
	Nominal Diameter		Length	
	Small End	Large End	Flutes	Overall
AAA	.055	.070	1⅛	2¼
AA	.065	.080	1⅛	2¼
A	.075	.090	1⅛	2¼
B	.085	.103	1⅜	2⅜
C	.095	.113	1⅜	2½
D	.105	.126	1⅝	2⅝
E	.115	.136	1⅝	2¾
F	.125	.148	1¾	3
G	.135	.158	1¾	3
H	.145	.169	1⅞	3¼
I	.160	.184	1⅞	3¼
J	.175	.199	1⅞	3¼
K	.190	.219	2¼	3½
L	.205	.234	2¼	3½
M	.220	.252	2½	4
N	.235	.274	3	4½
O	.250	.296	3½	5
P	.275	.327	4	5½
Q	.300	.358	4½	6
R	.335	.397	4¾	6½
S	.370	.435	5	6¾
T	.405	.473	5¼	7
U	.440	.511	5½	7¼

MORSE TAPER REAMERS

These reamers are designed for the production and maintenance of Morse Taper Sockets.

The reamers with taper shanks are usually used for machine reaming production work, and are made with straight or slow left hand spiral flutes and of high-speed-steel. The Square Shank Hand Reamers are made with straight flutes, of high speed steel and are used for final finishing or for the removal of burrs or nicks in machine spindles.

Morse Taper Finishing Reamers

Straight Flutes — Squared Shank
High Speed Steel

Standard Sizes and Dimensions

No. of Morse Taper	Nominal Diameter		Diam. Shank	Length		Length of Square	Size of Square
	Small End	Large End		Flute	Over-all		
0	.2503	.3674	5/16	2 1/4	3 3/4	5/16	.235
1	.3674	.5170	7/16	3	5	7/16	.330
2	.5696	.7444	5/8	3 1/2	6	5/8	.470
3	.7748	.9881	7/8	4 1/4	7 1/4	7/8	.655
4	1.0167	1.2893	1 1/8	5 1/4	8 1/2	1	.845
5	1.4717	1.8005	1 1/2	6 1/4	9 3/4	1 1/8	1.125

Morse Taper Finishing Reamers

Straight or Left Hand Spiral Flutes

Taper Shank

High Speed Steel

Standard Sizes and Dimensions

| No. of Morse Taper | Dimensions — Inches | | | | No. of Morse Taper Shank |
| | Diameter | | Length | | |
	Small End	Large End	Flute	Overall	
0	.2503	.3674	$2\frac{1}{4}$	$5\frac{11}{32}$	0
1	.3674	.5170	3	$6\frac{5}{16}$	1
2	.5696	.7444	$3\frac{1}{2}$	$7\frac{3}{8}$	2
3	.7748	.9881	$4\frac{1}{4}$	$8\frac{7}{8}$	3
4	1.0167	1.2893	$5\frac{1}{4}$	$10\frac{7}{8}$	4
5	1.4717	1.8005	$6\frac{1}{4}$	$13\frac{1}{8}$	5

Brown & Sharpe Taper Finishing Reamers

These reamers are designed for the production and maintenance of Brown & Sharpe Taper Sockets.

Straight or Left Hand Spiral Flutes
Squared Shank
High Speed Steel

Standard Sizes and Dimensions

No. of B & S Taper	Dimensions — Inches						
	Diameter			Length		Length of Square	Size of Square
	Small End	Large End	Shank	Flute	Over-all		
1	.1974	.3176	$\frac{9}{32}$	2 $\frac{7}{8}$	4 $\frac{3}{4}$	$\frac{1}{4}$	.218
2	.2474	.3781	$\frac{11}{32}$	3 $\frac{1}{8}$	5 $\frac{1}{8}$	$\frac{5}{16}$	.250
3	.3099	.4510	$\frac{13}{32}$	3 $\frac{3}{8}$	5 $\frac{1}{2}$	$\frac{3}{8}$	.312
4	.3474	.5017	$\frac{7}{16}$	$3\frac{11}{16}$	5 $\frac{7}{8}$	$\frac{7}{16}$	.343
5	.4474	.6145	$\frac{9}{16}$	4	6 $\frac{3}{8}$	$\frac{1}{2}$	.437
6	.4974	.6808	$\frac{5}{8}$	4 $\frac{3}{8}$	6 $\frac{7}{8}$	$\frac{5}{8}$	.468
7	.5974	.8011	$\frac{3}{4}$	4 $\frac{7}{8}$	7 $\frac{1}{2}$	$\frac{3}{4}$	.562
8	.7474	.9770	$\frac{13}{16}$	5 $\frac{1}{2}$	8 $\frac{1}{8}$	$\frac{13}{16}$	.625
9	.8974	1.1530	1	6 $\frac{1}{8}$	8 $\frac{7}{8}$	$\frac{7}{8}$	.750
10	1.0420	1.3376	1 $\frac{1}{8}$	6 $\frac{7}{8}$	9 $\frac{3}{4}$	1	.843

SPECIAL TAPER REAMERS
VARIOUS TYPES AND STYLES

Spiral-Fluted Roughing Reamer

Spiral-Fluted Finishing Reamer

High-Spiral Roughing Reamer

BRIDGE REAMERS

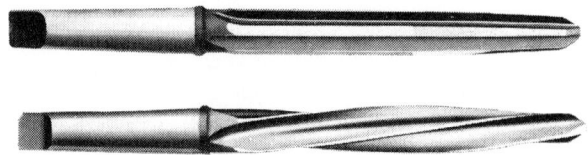

Bridge Reamers with Straight and Helical Flutes, Taper Shank

Bridge Reamers, made of high-speed steel, are especially designed for severe service and are particularly adapted for use in structural iron and steel, bridge erection, and ship construction, where extreme precision is not required. The cutting end of the flutes is tapered to permit the reamer to enter out-of-line holes often encountered in structural work. The approximate length of the tapered section is 40% of the total flute length. The normal tolerance on the full diameter is greater than that used for other types of machine reamers. Bridge Reamers are commonly used in portable electric or pneumatic equipment.

Straight Flutes

Taper Shank

High Speed Steel

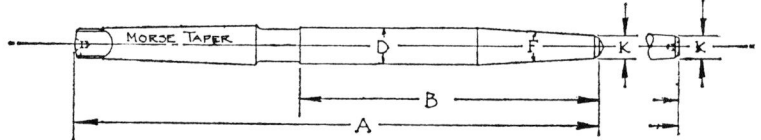

Standard Sizes and Dimensions

Nominal Diameter Inches	Dimensions — Inches				No. of Morse Taper Shank
	Approx. Diam. at Small End	Length Overall	Length of Flutes	Approx. Included Angle of Taper Flutes	
D	K	A	B	F	
$\frac{7}{16}$	$\frac{1}{4}$	$8\frac{1}{4}$	$4\frac{3}{8}$	6°	2
$\frac{9}{16}$	$\frac{11}{32}$	9	$5\frac{1}{8}$	6°	2
$\frac{11}{16}$	$\frac{25}{64}$	$11\frac{3}{4}$	$7\frac{1}{8}$	6°	3
$\frac{13}{16}$	$\frac{1}{2}$	12	$7\frac{3}{8}$	6°	3
$\frac{15}{16}$	$\frac{5}{8}$	12	$7\frac{3}{8}$	6°	3
$1\frac{1}{16}$	$\frac{3}{4}$	12	$7\frac{3}{8}$	6°	3
$1\frac{3}{16}$	$\frac{7}{8}$	12	$7\frac{3}{8}$	6°	3
$1\frac{5}{16}$	1	13	$7\frac{3}{8}$	6°	4

Bridge Reamers

Left Hand Helical Flutes
Taper Shank
High Speed Steel

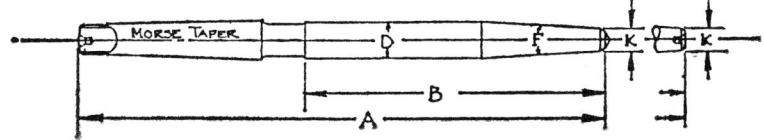

Standard Sizes and Dimensions

Nominal Diameter Inches	Dimensions — Inches				
	Approx. Diam. at Small End	Length Overall	Total Length of Flutes	Approx. Included Angle of Taper Flutes	No. of Morse Taper Shank
D	K	A	B	F	
7/16	1/4	8¼	4⅜	6°	2
1/2	9/32	9	5⅛	6°	2
9/16	11/32	9	5⅛	6°	2
5/8	3/8	10	6⅛	6°	2
11/16	25/64	11¾	7⅛	6°	3
3/4	7/16	12	7⅜	6°	3
13/16	1/2	12	7⅜	6°	3
7/8	9/16	12	7⅜	6°	3
15/16	5/8	12	7⅜	6°	3
1	11/16	12	7⅜	6°	3
1 1/16	3/4	12	7⅜	6°	3
1 1/8	13/16	12	7⅜	6°	3
1 3/16	7/8	12	7⅜	6°	3
1 1/4	15/16	13	7⅜	6°	4
1 5/16	1	13	7⅜	6°	4

Structural Reamers

Three Right Hand Helical Flutes
Taper Shank
High Speed Steel

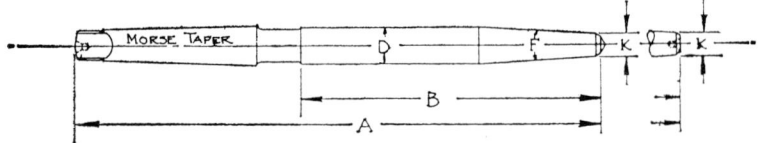

Standard Sizes and Dimensions

| Nominal Diameter Inches | Dimensions — Inches | | | | No. of Morse Taper Shank |
| | Approx. Diameter at Small End | Length Overall | Length of Flutes | Approx. Included Angle of Taper | |
D	K	A	B	F	
$1\frac{1}{16}$	$\frac{3}{8}$	$11\frac{3}{4}$	$7\frac{1}{8}$	6°	3
$1\frac{3}{16}$	$\frac{7}{16}$	12	$7\frac{3}{8}$	7°	3
$1\frac{5}{16}$	$\frac{1}{2}$	12	$7\frac{3}{8}$	7°	3
$1\frac{7}{16}$	$\frac{17}{32}$	12	$7\frac{3}{8}$	8°	3

CAR REAMERS

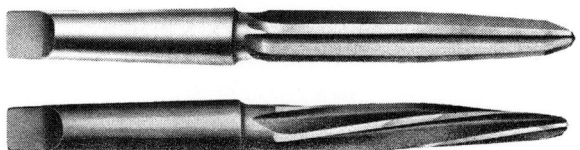

*Short Length Taper Car Reamers with Straight and
Helical Flutes, Taper Shank*

Car Reamers are of similar design to Bridge Reamers except that they have shorter flutes and overall length. The approximate length of the tapered section is 50% of the total flute length. The normal tolerance on the full diameter is greater than that used for other types of machine reamers. The shorter overall length permits their use in cramped quarters where Bridge Reamers cannot enter.

Straight Flutes
Taper Shank
High Speed Steel

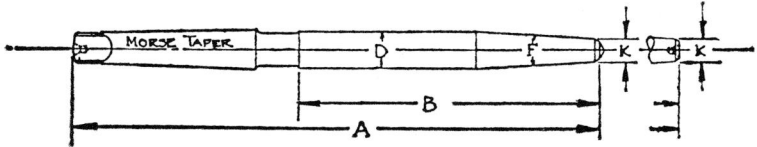

Standard Sizes and Dimensions

Nominal Diameter Inches	Dimensions — Inches					No. of Morse Taper Shank
	Approx. Diam. at Small End	Length Overall	Length of Flutes	Approx. Included Angle of Taper Flutes		
D	K	A	B	F		
$\frac{7}{16}$	$\frac{1}{4}$	$6\frac{15}{16}$	$3\frac{1}{2}$	6°		2
$\frac{9}{16}$	$\frac{9}{32}$	$7\frac{9}{16}$	4	8°		2
$\frac{11}{16}$	$\frac{3}{8}$	$8\frac{13}{16}$	$4\frac{1}{2}$	8°		3
$\frac{13}{16}$	$\frac{15}{32}$	$9\frac{1}{2}$	5	8°		3
$\frac{15}{16}$	$\frac{19}{32}$	$9\frac{1}{2}$	5	8°		3

Car Reamers

Left Hand Helical Flutes
Taper Shank
High Speed Steel

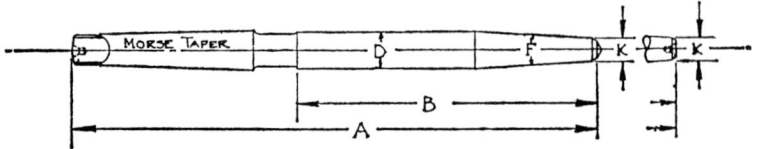

Standard Sizes and Dimensions

Nominal Diameter Inches	Dimensions — Inches				No. of Morse Taper Shank
	Approx. Diam. at Small End	Length Overall	Length of Flutes	Approx. Included Angle of Taper Flutes	
D	K	A	B	F	
$\frac{5}{16}$	$\frac{11}{64}$	$5\frac{11}{16}$	$2\frac{3}{4}$	6°	1
$\frac{3}{8}$	$\frac{15}{64}$	$5\frac{11}{16}$	$2\frac{3}{4}$	6°	1
$\frac{7}{16}$	$\frac{1}{4}$	$6\frac{15}{16}$	$3\frac{1}{2}$	6°	2
$\frac{1}{2}$	$\frac{19}{64}$	$7\frac{9}{16}$	4	6°	2
$\frac{9}{16}$	$\frac{9}{32}$	$7\frac{9}{16}$	4	8°	2
$\frac{5}{8}$	$\frac{5}{16}$	$8\frac{1}{16}$	$4\frac{1}{2}$	8°	2
$\frac{11}{16}$	$\frac{3}{8}$	$8\frac{13}{16}$	$4\frac{1}{2}$	8°	3
$\frac{3}{4}$	$\frac{13}{32}$	$9\frac{1}{2}$	5	8°	3
$\frac{13}{16}$	$\frac{15}{32}$	$9\frac{1}{2}$	5	8°	3
$\frac{15}{16}$	$\frac{19}{32}$	$9\frac{1}{2}$	5	8°	3

TAPER PIPE REAMERS
High Speed Steel

Taper Pipe Reamers are tapered ¾″ to the foot and are designed to ream holes to be tapped with American Standard taper pipe taps. They are made to fit standard pipe sizes and are of either carbon or high-speed steel.

Carbon steel reamers are made with straight flutes. High speed steel reamers are made with left-hand spiral flutes.

Standard Sizes and Dimensions

Nominal Size — Inches	Dimensions — Inches					
High Speed Steel	Diam. Large End	Diam. Small End	Length of Flutes	Diam. of Shank	Size of Square	Length Overall
⅛	.362	.316	¾	.4375	.328	2⅛
¼	.472	.406	1¹⁄₁₆	.5625	.421	2⁷⁄₁₆
⅜	.606	.540	1¹⁄₁₆	.7000	.531	2⁹⁄₁₆
½	.751	.665	1⅜	.6875	.515	3⅛
¾	.962	.876	1⅜	.9063	.679	3¼
1	1.212	1.103	1¾	1.1250	.843	3¾
1¼	1.553	1.444	1¾	1.3125	.984	4
1½	1.793	1.684	1¾	1.5000	1.125	4¼
2	2.268	2.159	1¾	1.8750	1.406	4½

BURRING REAMERS

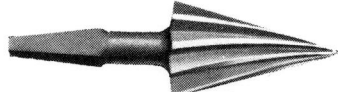

Straight Shank Burring Reamer   *Ratchet Shank Burring Reamer*

Pipe Burring Reamers are primarily intended for the removal of burrs from cut pipe or conduit. They may also be used for enlarging holes in thin sections, and countersinking or smoothing edges of holes in a variety of materials. They are usually made of carbon steel and in several sizes to accommodate the different sizes of pipe and conduit.

CENTER REAMERS AND MACHINE COUNTERSINKS

 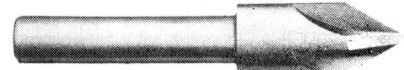

Center Reamer   *Machine Countersink*

Center reamers are designed for countersinking holes for centers, flat-head screws, and rivets.

Machine countersinks are designed for countersinking holes.

Center Reamers
(Short Countersinks)
(Fluted Type High Speed Steel)

These reamers are standard with 60°, 82°, 90°, 100° included angle.

Standard Sizes and Dimensions

Size Cut Inches	Dimensions — Inches		
	Diameter of Shank	Length of Shank	Length Overall (Approximate)
1/4	3/16	3/4	1 1/2
3/8	1/4	7/8	1 3/4
1/2	3/8	1	2
5/8	3/8	1	2 1/4
3/4	1/2	1 1/4	2 5/8

Machine Countersinks
Straight Shank High Speed Steel

These countersinks are standard with 60° or 82° included angle.

Standard Sizes and Dimensions

Size Cut Inches	Dimensions — Inches		
	Diameter of Shank	Length of Shank	Length Overall (Approximate)
1/2	1/2	2 1/4	3 7/8
5/8	1/2	2 1/4	4
3/4	1/2	2 1/4	4 1/8
7/8	1/2	2 1/4	4 1/4
1	1/2	2 1/4	4 3/8

CARBIDE-TIPPED REAMERS

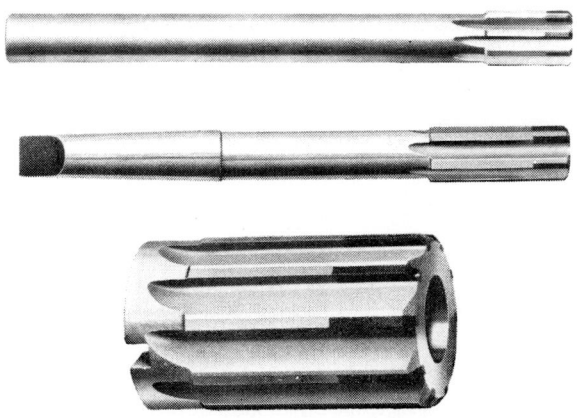

Types of Carbide-Tipped Reamers

Carbides retain a sharp edge under high temperatures, and have excellent abrasion-resistant qualities, making them well suited for use as cutting edges on reamers. They can be applied in any of the common materials, ranging from hardened steel to plastics. These tools have proven economical on long production runs. Carbide reamers are particularly advantageous when sand or scale is encountered in reaming holes in castings.

For information regarding the use and care of carbide-tipped reamers. refer to page 229.

Chucking Reamers

Carbide Tipped
Straight Flutes — Straight Shank

Standard Sizes and Dimensions

Fractional Sizes	Decimal Equivalent Inches	Dimensions — Inches			
		Diameter of Shank		Length of Flute	Length Overall
		Maximum	Minimum		
$3/16$	.1875	.1805	.1795	$1\frac{1}{8}$	$4\frac{1}{2}$
$7/32$	.2188	.2075	.2065	$1\frac{1}{4}$	5
$1/4$	.2500	.2405	.2395	$1\frac{1}{2}$	6
$9/32$	.2812	.2485	.2475	$1\frac{1}{2}$	6
$5/16$	.3125	.2792	.2782	$1\frac{1}{2}$	6
$11/32$	.3438	.2792	.2782	$1\frac{1}{2}$	6
$3/8$	.3750	.3105	.3095	$1\frac{3}{4}$	7
$13/32$	.4062	.3105	.3095	$1\frac{3}{4}$	7
$7/16$	.4375	.3730	.3720	$1\frac{3}{4}$	7
$15/32$	.4688	.3730	.3720	$1\frac{3}{4}$	7
$1/2$	.5000	.4355	.4345	2	8
$17/32$	.5312	.4355	.4345	2	8
$9/16$	.5625	.4355	.4345	2	8
$19/32$	.5938	.4355	.4345	2	8
$5/8$	.6250	.5620	.5605	$2\frac{1}{4}$	9
$21/32$	.6562	.5620	.5605	$2\frac{1}{4}$	9
$11/16$	.6875	.5620	.5605	$2\frac{1}{4}$	9
$23/32$	.7188	.5620	.5605	$2\frac{1}{4}$	9
$3/4$	.7500	.6245	.6230	$2\frac{1}{2}$	$9\frac{1}{2}$
$25/32$	.7812	.6245	.6230	$2\frac{1}{2}$	$9\frac{1}{2}$
$13/16$	.8125	.6245	.6230	$2\frac{1}{2}$	$9\frac{1}{2}$
$27/32$	.8438	.6245	.6230	$2\frac{1}{2}$	$9\frac{1}{2}$
$7/8$	.8750	.7495	.7480	$2\frac{5}{8}$	10
$29/32$	.9062	.7495	.7480	$2\frac{5}{8}$	10
$15/16$	.9375	.7495	.7480	$2\frac{5}{8}$	10
$31/32$	.9688	.7495	.7480	$2\frac{5}{8}$	10
1	1.0000	.8745	.8730	$2\frac{3}{4}$	$10\frac{1}{2}$

Expansion Chucking Reamers
Carbide Tipped
Straight Flutes — Straight Shank

The expansion feature is designed to maintain the initial size by compensating for wear at the cutting end. The expanding pins of these reamers should *never be loosened* in an attempt to use them for a size other than that designated.

Standard Sizes and Dimensions

Diameter Inches	Dimensions — Inches		
	Diameter of Shank	Length of Flute	Length Overall
$\frac{7}{16}$	$\frac{3}{8}$	$\frac{7}{8}$	7
$\frac{15}{32}$	$\frac{3}{8}$	$\frac{7}{8}$	7
$\frac{1}{2}$	$\frac{7}{16}$	1	8
$\frac{17}{32}$	$\frac{7}{16}$	1	8
$\frac{9}{16}$	$\frac{7}{16}$	$1\frac{1}{8}$	8
$\frac{19}{32}$	$\frac{7}{16}$	$1\frac{1}{8}$	8
$\frac{5}{8}$	$\frac{9}{16}$	$1\frac{1}{4}$	9
$\frac{21}{32}$	$\frac{9}{16}$	$1\frac{1}{4}$	9
$\frac{11}{16}$	$\frac{9}{16}$	$1\frac{1}{4}$	9
$\frac{23}{32}$	$\frac{9}{16}$	$1\frac{1}{4}$	9
$\frac{3}{4}$	$\frac{5}{8}$	$1\frac{3}{8}$	$9\frac{1}{2}$
$\frac{25}{32}$	$\frac{5}{8}$	$1\frac{3}{8}$	$9\frac{1}{2}$
$\frac{13}{16}$	$\frac{5}{8}$	$1\frac{3}{8}$	$9\frac{1}{2}$
$\frac{27}{32}$	$\frac{5}{8}$	$1\frac{3}{8}$	$9\frac{1}{2}$
$\frac{7}{8}$	$\frac{3}{4}$	$1\frac{1}{2}$	10
$\frac{29}{32}$	$\frac{3}{4}$	$1\frac{1}{2}$	10
$\frac{15}{16}$	$\frac{3}{4}$	$1\frac{1}{2}$	10
$\frac{31}{32}$	$\frac{3}{4}$	$1\frac{1}{2}$	10
1	$\frac{7}{8}$	$1\frac{5}{8}$	$10\frac{1}{2}$
$1\frac{1}{32}$	$\frac{7}{8}$	$1\frac{5}{8}$	$10\frac{1}{2}$
$1\frac{1}{16}$	$\frac{7}{8}$	$1\frac{5}{8}$	$10\frac{1}{2}$
$1\frac{3}{32}$	$\frac{7}{8}$	$1\frac{5}{8}$	$10\frac{1}{2}$
$1\frac{1}{8}$	$\frac{7}{8}$	$1\frac{3}{4}$	11
$1\frac{5}{32}$	$\frac{7}{8}$	$1\frac{3}{4}$	11
$1\frac{3}{16}$	1	$1\frac{3}{4}$	11
$1\frac{7}{32}$	1	$1\frac{3}{4}$	11
$1\frac{1}{4}$	1	$1\frac{7}{8}$	$11\frac{1}{2}$
$1\frac{5}{16}$	1	$1\frac{7}{8}$	$11\frac{1}{2}$
$1\frac{3}{8}$	1	2	12
$1\frac{7}{16}$	$1\frac{1}{4}$	2	12
$1\frac{1}{2}$	$1\frac{1}{4}$	$2\frac{1}{8}$	$12\frac{1}{2}$

Chucking Reamers
Carbide Tipped
Straight Flutes — Taper Shank

Standard Sizes and Dimensions

Diameter Inches	Dimensions — Inches		Number of Taper Shank
	Length of Flute	Length Overall	
¼	1½	6	1
⁹⁄₃₂	1½	6	1
⁵⁄₁₆	1½	6	1
¹¹⁄₃₂	1½	6	1
³⁄₈	1¾	7	1
¹³⁄₃₂	1¾	7	1
⁷⁄₁₆	1¾	7	1
¹⁵⁄₃₂	1¾	7	1
½	2	8	1
¹⁷⁄₃₂	2	8	1
⁹⁄₁₆	2	8	1
¹⁹⁄₃₂	2	8	1
⁵⁄₈	2¼	9	2
²¹⁄₃₂	2¼	9	2
¹¹⁄₁₆	2¼	9	2
²³⁄₃₂	2¼	9	2
¾	2½	9½	2
²⁵⁄₃₂	2½	9½	2
¹³⁄₁₆	2½	9½	2
²⁷⁄₃₂	2½	9½	2
⅞	2⅝	10	2
²⁹⁄₃₂	2⅝	10	2
¹⁵⁄₁₆	2⅝	10	3
³¹⁄₃₂	2⅝	10	3
1	2¾	10½	3
1 ¹⁄₁₆	2¾	10½	3
1 ⅛	2⅞	11	3
1 ³⁄₁₆	2⅞	11	3
1 ¼	3	11½	4
1 ⁵⁄₁₆	3	11½	4
1 ⅜	3¼	12	4
1 ⁷⁄₁₆	3¼	12	4
1 ½	3½	12½	4

Expansion Chucking Reamers
Carbide Tipped
Straight Flutes — Taper Shank

The expansion feature is designed to maintain the initial size by compensating for wear at the cutting end. The expanding pins of these reamers should *never be loosened* in an attempt to use them for a size other than that designated.

Standard Sizes and Dimensions

Diameter Inches	Decimal Equivalent	Dimensions — Inches		Number of Morse Taper Shank
		Length of Flute	Length Overall	
3/8	.3750	1	7	1
7/16	.4375	1	7	1
1/2	.5000	1	8	1
9/16	.5625	1 1/8	8	1
5/8	.6250	1 1/4	9	2
11/16	.6875	1 1/4	9	2
3/4	.7500	1 3/8	9 1/2	2
13/16	.8125	1 3/8	9 1/2	2
7/8	.8750	1 1/2	10	2
15/16	.9375	1 1/2	10	3
1	1.0000	1 5/8	10 1/2	3
1 1/16	1.0625	1 5/8	10 1/2	3
1 1/8	1.1250	1 3/4	11	3
1 3/16	1.1875	1 3/4	11	3
1 1/4	1.2500	1 7/8	11 1/2	4
1 5/16	1.3125	1 7/8	11 1/2	4
1 3/8	1.3750	2	12	4
1 7/16	1.4375	2	12	4
1 1/2	1.5000	2 1/8	12 1/2	4

Stub Screw Machine Reamers

Solid Carbide, Fluted Section

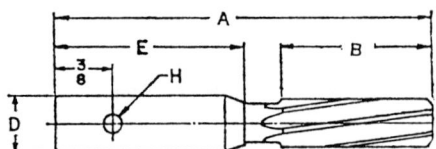

These reamers are standard with straight flutes, or with right hand or left hand helical flutes.

Standard Sizes and Dimensions

Series Number	Diameter Range	Dimensions — Inches						
		Length		Shank		Pin Hole H	No. of Flutes
		Over-all A	Flute B	Diameter D	Length E		
00	.0600 to .066 Incl.	$1\frac{3}{4}$	$\frac{1}{2}$	$\frac{1}{8}$	1	$\frac{1}{16}$	4
0	.0661 to .074 Incl.	$1\frac{3}{4}$	$\frac{1}{2}$	$\frac{1}{8}$	1	$\frac{1}{16}$	4
1	.0741 to .084 Incl.	$1\frac{3}{4}$	$\frac{1}{2}$	$\frac{1}{8}$	1	$\frac{1}{16}$	4
2	.0841 to .096 Incl.	$1\frac{3}{4}$	$\frac{1}{2}$	$\frac{1}{8}$	1	$\frac{1}{16}$	4
3	.0961 to .115 Incl.	2	$\frac{3}{4}$	$\frac{1}{4}$	1	$\frac{3}{32}$	4
3A	.1151 to .130 Incl.	$2\frac{1}{8}$	$\frac{3}{4}$	$\frac{1}{4}$	$1\frac{1}{8}$	$\frac{3}{32}$	4
4	.1301 to .158 Incl.	$2\frac{1}{4}$	$\frac{7}{8}$	$\frac{1}{4}$	$1\frac{1}{8}$	$\frac{3}{32}$	4
5	.1581 to .188 Incl.	$2\frac{1}{4}$	$\frac{7}{8}$	$\frac{1}{4}$	$1\frac{1}{8}$	$\frac{3}{32}$	4
6	.1881 to .219 Incl.	$2\frac{3}{8}$	$\frac{7}{8}$	$\frac{3}{8}$	$1\frac{1}{4}$	$\frac{1}{8}$	4
7	.2191 to .251 Incl.	$2\frac{3}{8}$	$\frac{7}{8}$	$\frac{3}{8}$	$1\frac{1}{4}$	$\frac{1}{8}$	6
8	.2511 to .282 Incl.	$2\frac{1}{2}$	1	$\frac{3}{8}$	$1\frac{1}{4}$	$\frac{1}{8}$	6
9	.2821 to .313 Incl.	$2\frac{1}{2}$	1	$\frac{3}{8}$	$1\frac{1}{4}$	$\frac{1}{8}$	6
10	.3131 to .344 Incl.	$2\frac{5}{8}$	$1\frac{1}{8}$	$\frac{1}{2}$	$1\frac{1}{4}$	$\frac{3}{16}$	6
11	.3441 to .376 Incl.	$2\frac{5}{8}$	$1\frac{1}{8}$	$\frac{1}{2}$	$1\frac{1}{4}$	$\frac{3}{16}$	6
12	.3761 to .407 Incl.	$2\frac{3}{4}$	$1\frac{1}{4}$	$\frac{1}{2}$	$1\frac{1}{4}$	$\frac{3}{16}$	6
13	.4071 to .439 Incl.	$2\frac{3}{4}$	$1\frac{1}{4}$	$\frac{1}{2}$	$1\frac{1}{4}$	$\frac{3}{16}$	6
14	.4391 to .470 Incl.	$2\frac{3}{4}$	$1\frac{1}{4}$	$\frac{5}{8}$	$1\frac{1}{4}$	$\frac{1}{4}$	6
15	.4701 to .505 Incl.	$2\frac{3}{4}$	$1\frac{1}{4}$	$\frac{5}{8}$	$1\frac{1}{4}$	$\frac{1}{4}$	6

Chucking Reamers

Solid Carbide
Straight Flutes — Straight Shank
Fractional Sizes

Shank diameters in accordance with manufacturer's practice.

Standard Sizes and Dimensions

Fractional Sizes	Decimal Equivalent	Dimensions		No. of Flutes
		Length of Flute	Length Overall	
1/64	.0156	3/16	7/8	4
1/32	.0312	1/4	1 1/8	4
3/64	.0469	3/8	1 1/2	4
1/16	.0625	3/8	1 1/2	4
5/64	.0781	1/2	1 3/4	4
3/32	.0938	1/2	2	4
7/64	.1094	5/8	2 1/4	4
1/8	.1250	5/8	2 1/4	4–6
9/64	.1406	3/4	2 1/2	4–6
5/32	.1562	3/4	2 1/2	4–6
11/64	.1719	7/8	2 3/4	4–6
3/16	.1875	7/8	2 3/4	4–6
13/64	.2031	1	3	4–6
7/32	.2188	1	3	4–6
15/64	.2344	1	3	4–6
1/4	.2500	1	3	4–6
17/64	.2656	1 1/8	3 1/4	4–6
9/32	.2812	1 1/8	3 1/4	4–6
19/64	.2969	1 1/8	3 1/4	4–6
5/16	.3215	1 1/8	3 1/4	4–6
21/64	.3281	1 1/4	3 1/2	4–6
11/32	.3438	1 1/4	3 1/2	4–6
23/64	.3594	1 1/4	3 1/2	4–6
3/8	.3750	1 1/4	3 1/2	4–6
25/64	.3906	1 1/4	3 1/2	4–6
13/32	.4062	1 1/4	3 1/2	4–6
27/64	.4219	1 1/4	3 1/2	4–6
7/16	.4375	1 1/4	3 1/2	4–6
29/64	.4531	1 1/4	3 1/2	4–6
15/32	.4687	1 1/4	3 1/2	4–6
31/64	.4844	1 1/4	3 1/2	4–6
1/2	.5000	1 1/4	3 1/2	4–6

ILLUSTRATIONS OF TYPES OF SOLID CARBIDE
AND CARBIDE TIPPED REAMERS

Solid Carbide Right Hand Helical Fluted Chucking Reamer

Solid Carbide Left Hand
Helical Fluted Chucking Reamer

Carbide Tipped High Speed
Expansion Shell Reamer

Carbide Tipped Adjustable Reamer

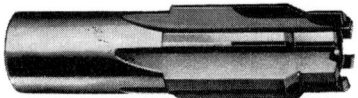

Carbide Tipped Straight Fluted Step
Reamer with Sub-land Shoulder Angles

Carbide Tipped Straight Flute
Step Reamer

GUN REAMERS

Gun reamers are specially designed oil feeding reamers that normally employ high pressure coolant to flush chips forward through the holes in parts. Aided by the coolant, they can remove larger than normal amounts of reaming stock.

The reamers are started into the holes in the work through close fitting, accurately aligned bushings. Hole alignment in the part is maintained by the bushing accuracy. The reamer tends to follow its own axis due to the geometry of its bearing pads and due to the use of relatively flat chamfer angles which minimize the tendency to follow the existing holes.

Two common styles of gun reamers are shown below.

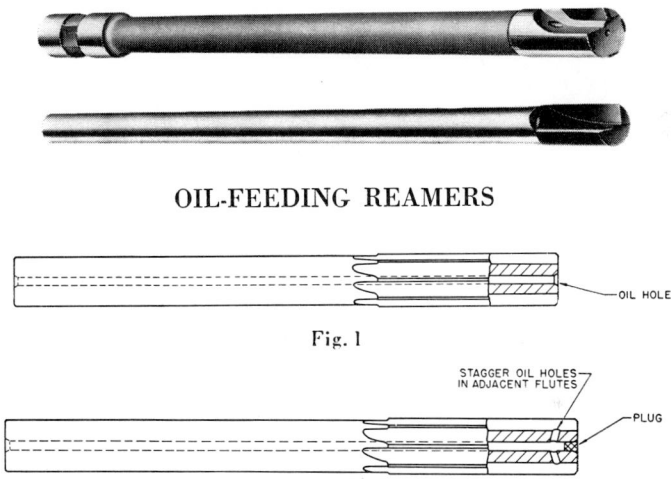

OIL-FEEDING REAMERS

Fig. 1

Fig. 2

Oil-Feeding Reamers provide a means of getting coolant to reamer cutting edges in cases where the cutting oil cannot naturally flow to them.

One type of Oil-Feeding Reamer is illustrated in Fig. 1. This type of feed is excellent in the reaming of blind holes where the oil fills up the hole ahead of the reamer and forces coolant back through the flutes carrying away heat and chips.

Fig. 2 shows an Oil Feeding Reamer with the through hole blocked off, and cross-holes placed in the flutes. This arrangement provides an ample flow of coolant to the cutting edges in cases where horizontal reaming is done to considerable depths and coolant cannot be directed to the cutting edges. In the vertical reaming of a through hole, the natural gravity feed of the coolant pvovides enough flow for a regular type reamer.

COMBINED DRILLS AND REAMERS

Quite often when holes must be drilled and reamed, the two operations can be performed with a single tool, eliminating a second handling of the part to be machined. Such a tool is known as a Combined Drill and Reamer. As the drill portion should be entirely through the work before the reamer section starts to cut, this type of tool is usually used only on shallow holes. This is most important, and in giving specifications for the tool, the drill diameter should always be longer than the depth of the hole to be drilled.

These tools are usually made with either of two types of construction.

Fig. 1

In step construction shown in Fig. 1, two drill flutes are fluted the full length of the tool. The drill section is of slightly smaller diameter than the reamer section and the drill margins extend only the length of this drill section.

The reamer section consists of one, two, or three reamer flutes milled into each land left by the drill flute milling cutter. This type of construction is subject to the same difficulties in regrinding that are encountered with step drills.

These difficulties may be overcome by using sub-land construction for Combined Drills and Reamers. (See Fig. 2.) In this type of construction the two drill flutes and cutting edges extend back through the reamer portion the full length of the tool. The reamer cutting edges are in addition to these and start back a suitable distance from the drill point. This type of construction usually makes it necessary to employ fewer reamer cutting edges than are used with step construction.

Fig. 2

In both step and subland construction the stock left for reaming should be about the same as though separate drills and reamers were used.

MULTIPLE DIAMETER REAMERS

Sometimes it is necessary to ream two or more coaxial holes of different diameters. If a separate reamer is used for each hole several operations are necessary and it is difficult to keep the holes in line. The use of a multiple diameter reamer permits the operations to be performed all at one time and insures alignment of the holes.

Quite often each reaming section is preceded by a pilot section to bear in either the drilled holes or in bushings in the fixture. It is usually found best to design the tool so that all reaming sections start to cut at the same time. Because of the distance between successive reamer sections, step construction is employed except in very rare cases.

Here are several typical multiple diameter reamers.

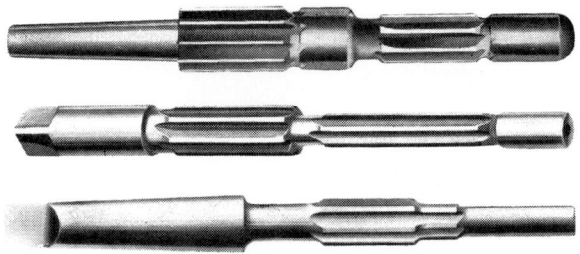

Typical Multiple Diameter Reamers

BUILT-UP LINE REAMERS

Built-Up Line reamers can be manufactured in many ways to suit a large variety of operations. The operations which can be performed by a tool of this type do not necessarily confine themselves to reaming only; Built-Up Line reamers may be designed to combine the reaming operation with boring, counterboring, or spotfacing. This type of tool, while usually designed in an adjustable style, may also be of solid construction.

Built-Up Line Reamer formed by Two Adjustable Shell Reamers on Piloted Driving Bars

VARIOUS STYLES OF SPECIAL REAMERS

Multi-Diameter Reamer

Piloted Multi-Diameter Reamer

Multi-Diameter Reamer (Straight and Helical Flute)

Piloted Expansion Line Reamer

Piloted Expansion Line Reamer with Extra Long Shank

Multi-Diameter Expansion Reamer

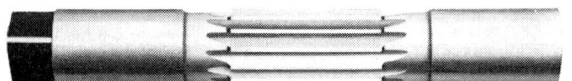

Hand Reamer with Pilot

Multi-Diameter Piloted Hand Reamer

SPECIFICATIONS FOR SPECIAL REAMERS

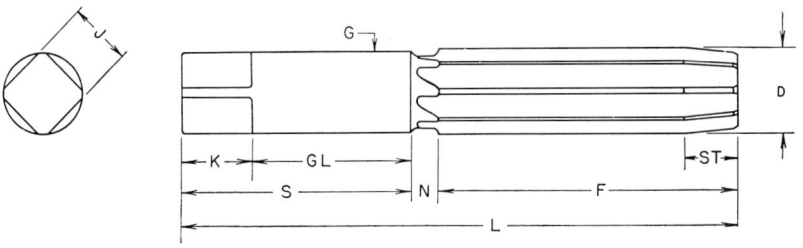

Square Shank Hand Reamer

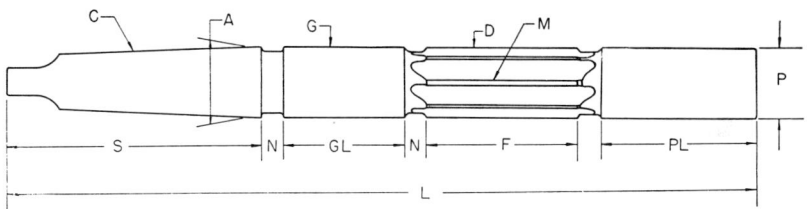

Taper Shank Reamer with Pilot and Guide

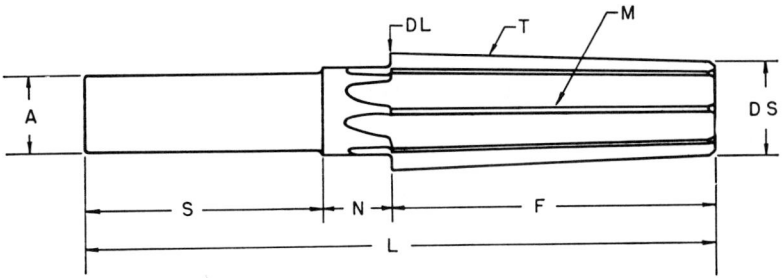

Straight Shank Taper Reamer

State kind of material being reamed. Give the dimensions called for on the sketches above.

A   Diameter of shank.

C   If regular taper shank, give TAPER NUMBER. For dimensions of Morse taper shanks, see page 762 Eng. Data.

D   Diameter of fluted section.

DL   Diameter of fluted section at large end for Taper Reamer.

DS   Diameter of fluted section at small end for Taper Reamer.

F   Length of flute.

G   Diameter of guide.

GL   Length of guide.

J   Size of square, if any.

K   Length of square, if any.

L   Length overall.

M   Kind of flute. Refer to catalog and give list number of general style of tool.

N   Length of neck.

P   Diameter of pilot.

PL   Length of pilot.

S   Length of shank.

ST   Starting taper

T   Taper per foot or included angle for taper reamers.

If taper shank is special, give diameter at small end, length of shank, diameter at large end, taper per foot, and furnish a sample or gauge if possible. If tang is special, give thickness and length.

Tolerances.—If special accuracy is required, give tolerances on the important dimensions.

SPECIFICATIONS FOR END-CUTTING REAMERS

The conventional grind for Chucking Reamers consists of a 45° chamfer and is illustrated in Fig. 1. Occasionally it is desirable to eliminate the chamfer and to grind the reamer to cut square across the end. Reamers sharpened in this manner have a cutting action much like that of an end mill. Reamers ground with a square end-cut are sometimes required to correct a bent or misaligned hole. Quite often this type of square end-cut is required to bottom a hole or to produce a square shoulder. Figs. 2 to 4 show three separate types of square end-cutting grinds for use on reamers:

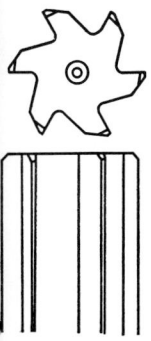 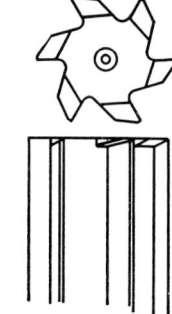 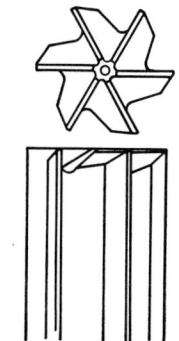 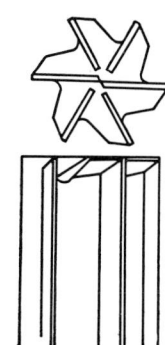

Fig. 1. *Regular 45°*   Fig. 2. *Square End-Cut*   Fig. 3. *Square End-Cut*   Fig. 4. *Square End-Cut*
Chamfer   *to Bottom of Flutes*   *to Center Hole*   *to Center*

The type of grind shown in Fig. 2 which cuts to the bottom of the flutes only, is relatively simple to apply to any reamer and should always be specified when it is unnecessary for the tool to cut to a smaller diameter than can be accommodated by the ordinary depth of flute.

When it is necessary for the reamer to cut on its end to a still smaller diameter, the tool can be ground as shown in Fig. 3. This type of grind, which permits facing down to the diameter of the center hole, is more complicated and requires additional notching on the end. This type of grind should not be specified when the grind shown in Fig. 2 can be used.

The grind shown in Fig. 4 is designed to permit facing to the extreme center of the reamer. It will be noticed that two of the flutes cut to the center. In order to grind a reamer in this manner the complete center hole must be removed and additional rather complex notching procedures must be employed. This type of point is itself rather difficult to grind and the absence of a center hole makes any reworking of the reamer a difficult job. This type of grind should not be specified when either of the grinds shown in Figs. 2 or 3 can be employed.

When specifying square end-cutting reamers, the diameter to which the reamer must cut should be given in order that the manufacturer can properly notch and sharpen the tool.

COUNTERBORE SECTION

COUNTERBORE SECTION CONTENTS

FOREWORD

Modern engineering and product designs often call for the production of holes of more than one single diameter. In some cases it is necessary to enlarge a part of a previously drilled or bored hole to accommodate a mating part, such as the head of a screw or bolt. In other cases the surface immediately around a hole must be machined to permit the accurate seating of a bolt or a mating part.

Counterbores and spotfacers of many types have been developed to fill the need for suitable tools to perform these essential operations in an efficient manner. Research and engineering constantly strive to improve on existing designs, and to develop new types of tools as the need arises.

This section attempts to acquaint the reader with the various types of tools available, the proper application of each type, and some suggestions which may lead to their economical usage.

Also included are industry standards on some types and sizes of counter-bores.

299

NOMENCLATURE OF
COUNTERBORES AND SPOTFACERS

A—COUNTERBORES

I—DEFINITION

Counterbore—A rotary, pilot guided, end cutting tool having two or more cutting lips and usually having straight or helical flutes for the passage of chips and the admission of cutting fluid.

II—GENERAL CLASSIFICATIONS

A—Classification Based on Construction.

1—*Solid Counterbores*—Those made of one piece of tool material including the pilot.

2—*Counterbores with Interchangeable Pilots*—Those having removable, mechanically held pilots.

3—*Interchangeable Counterbores*—Those having cutting portions as well as pilots removable and interchangeable, so that a series of cutters will fit the same holder, and a series of pilots will fit the same cutter.

4—*Tipped Counterbores*—Those which have a body of one material with cutting lips of another material brazed or otherwise bonded in place.

5—*Inserted Blade Counterbores*—Those which have replaceable, mechanically held blades. These blades are the cutting elements and may be either solid or tipped.

6—*Adjustable Length Counterbores*—Those which have mechanical length adjustment so as to obtain a given amount of lineal projection from the machine spindle end.

B—Classification Based on Method of Holding or Driving.

1—*Shank Type (Solid) Counterbores*—Those having either straight or taper shanks integral with counterbore cutter. They are suitable for direct mounting in machine spindles, driving sleeves or sockets.

2—*Holder Type (Interchangeable) Counterbores*—Those having intermediate holders with straight or taper shanks. The mounting of the cutters in the holders are often proprietary designs of individual makers. Interchangeable pilots are also secured in the cutters in various ways.

III—EXPLANATION OF "HAND" OF COUNTERBORES

The terms "right hand" and "left hand" are used to describe both direction of rotation and direction of flute helix.

A—Hand of Rotation [or Hand of Cut].

1—*Right Hand Rotation (of Right Hand Cut)* is the counter-clockwise rotation of a counterbore, revolving so as to make a cut, when viewed from the cutting end.

2—*Left Hand Rotation* (*or Left Hand Cut*) is the clockwise rotation of a counterbore, revolving so as to make a cut, when viewed from the cutting end.

B—Hand of Flute Helix.

1—A counterbore has *Right Hand helix* when the flutes twist away from the observer in a clockwise direction when viewed from either end.

2—A counterbore has *Left Hand helix* when the flutes twist away from the observer in a counter-clockwise direction when viewed from either end.

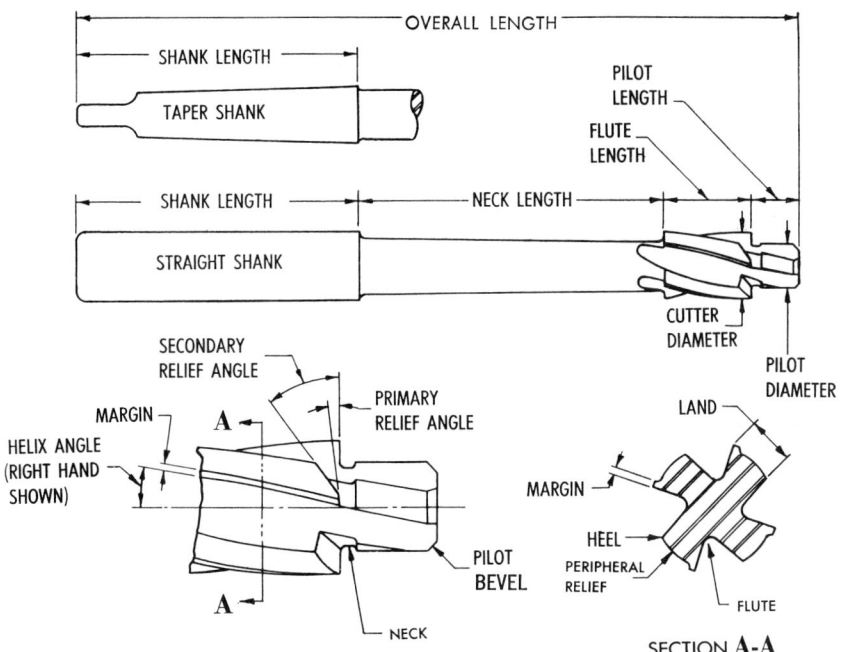

Fig. 1. *Illustrations of Terms applying to solid Counterbores.*

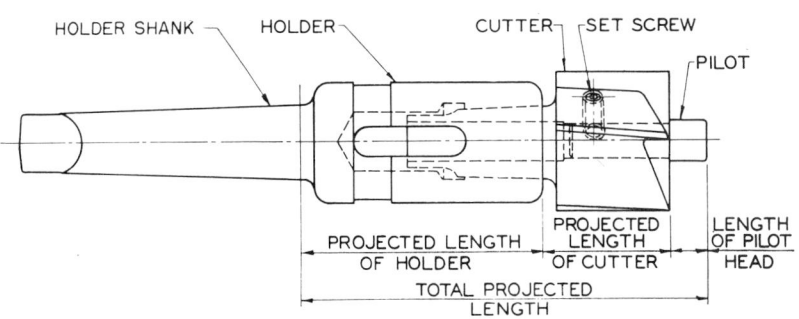

Fig. 2. *Illustrations of Terms applying to Holder Type Counterbores.*

B—SPOTFACERS

I—DEFINITION

Spotfacer—A rotary, pilot guided, end cutting tool having teeth on one or both ends. Spotfacers are mounted on either straight or taper shank pilots and are used to produce flat surfaces normal to their axis of rotation. These surfaces usually serve as seats for bolt heads, nuts or shoulders on mating members.—Basically, Spotfacers are design variations of Counterbores, often adapted to reach inaccessible areas. In open areas regular Counterbores are used for Spotfacing.

II—GENERAL CLASSIFICATIONS

A—Classification Based on Construction.

1—*Back Spotfacers*—Those having separate holders and cutters, which can be assembled in place after inserting the holder thru the hole to be spotfaced. The cutter has teeth on one end only and is pulled rather than pushed into the workpiece to make a cut.

2—*Double End Spotfacers*—Those having separate holders and cutters, which can be assembled in place after inserting the holder thru the hole to be spotfaced. The cutter has cutting teeth on both ends so that it can be both pushed and pulled to do its work. These tools are used for spotfacing two opposed internal surfaces in a workpiece.

B—Classification Based on Hand of Cut.

1—*Right Hand Cut*—See definition under *Counterbores.*
2—*Left Hand Cut*—See definition under *Counterbores.*
3—*Exception*—For Back Spotfacers the Hand of Cut is the reverse of the above definitions.

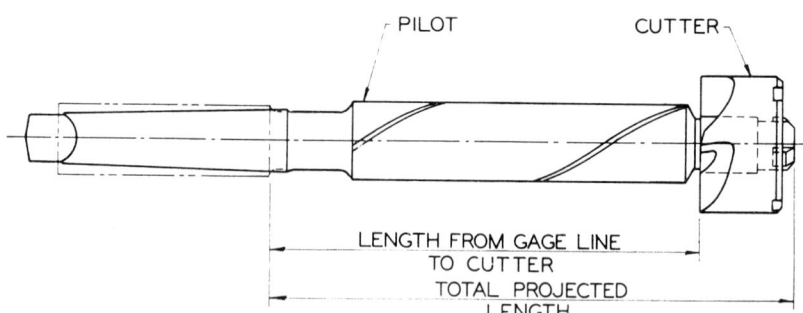

Fig. 3. *Illustrations of Terms applying to Spotfacers.*

III—VARIOUS TYPES OF COUNTERBORES AND SPOTFACERS

Fig. 4. *Shank Types of Counterbores with Integral Pilots.*

Fig. 5. *Shank Type Counterbore with Interchangeable Pilot.*

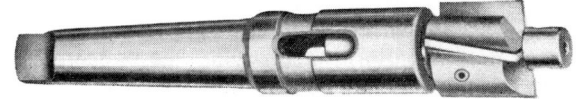

Fig. 6. *Holder Type Counterbore with Interchangeable Cutter and Interchangeable Pilot.*

Fig. 7. *Counterbores with Length Adjustment in Holder.*

Fig. 8. *Counterbore with Depth Stop on Holder.*

Fig. 9. *Multiple Diameter Counterbore.*

Fig. 10. *Inverted Spotfacer—Assembled.*

Fig. 11. *Double End Spotfacer—Assembled.*

IV—NOMENCLATURE OF COUNTERBORES AND SPOTFACERS AND OTHER TERMS RELATING TO COUNTERBORING

ACTUAL SIZE—The actual measured diameter of a counterbore, often somewhat larger than the nominal diameter.

AXIAL RAKE—The angle between a plane containing the cutting face, or a tangent to the cutting face at a given point, and the axis of an angularly fluted counterbore.

AXIAL RELIEF—The relief measured in the axial direction between a plane perpendicular to the axis and the relieved surface.

AXIS—The imaginary straight line which forms the longitudinal centerline of a counterbore.

BLADE—A tooth or cutting device inserted in a counterbore body.

BODY—The fluted full diameter portion of a solid type or interchangeable pilot type counterbore.

BODY DIAMETER CLEARANCE—That portion of the land that has been cut away so it will not rub against the walls of the hole.

CLEARANCE—The space provided behind the cutting edge or margin to eliminate undesirable contact between the counterbore and the workpiece.

CLEARANCE DIAMETER—The diameter over the cut away portion of the counterbore lands.

COUNTERBORE CUTTER—A detachable cutting portion of an interchangeable counterbore.

COUNTERBORING—A method of enlarging portions of previously formed holes.

CUTTER DIAMETER—The diameter over the margins of a counterbore measured at the cutting end.

CUTTING EDGE—The leading edge of the land in the direction of cutting.

END RELIEF—See preferred term AXIAL RELIEF.

FEED—The axial advance in inches per revolution of a counterbore with respect to the workpiece.

FLUTE LENGTH—The length of the flutes of a counterbore, not including the cutter sweep or the pilot.

FLUTED HOLDER—A fluted holding or driving member for a counterbore cutter in an interchangeable counterbore.

FLUTES—Helical or straight grooves, cut or formed in the body of a counterbore to provide cutting lips, to permit the passage of chips and to allow cutting fluid to reach the cutting lips.

GAGE LINE—The axial position on a taper where the diameter is equal to the basic large end diameter of the specified taper.

HEEL—The trailing edge of the land in the direction of rotation for cutting.

HELICAL FLUTES—[Sometimes called "SPIRAL FLUTE"]. Flutes which are formed in a helical path around the axis of a counterbore.

HELICAL RAKE—The angle between a plane tangent to the cutting face at a given point on the cutting edge, and the counterbore axis.

HELIX ANGLE—The angle which a helical cutting edge at a given point makes with an axial plane thru the same point.

HOLDER—Applied to an interchangeable type of counterbore, the extended mounting and holding member for the cutting element, or counterbore cutter.

LAND—The section of a counterbore between adjacent flutes.

LAND CLEARANCE—See preferred term BODY DIAMETER CLEARANCE.

LAND WIDTH—The distance between the leading edge and the heel of the land measured at a right angle to the leading edge.

LIP RELIEF—The axial relief on the end teeth of a counterbore.

LIP RELIEF ANGLE—The axial relief angle on the end teeth of a counterbore. It is the angle between a plane perpendicular to the axis and the relieved surface.

LIPS—The end cutting edges of a counterbore, extending from the outside diameter to the pilot diameter or the pilot hole.

MARGIN—The cylindrical portion of the land which is not cut away to provide clearance.

NECK—A section of reduced diameter connecting shank to body, or connecting other portions of a counterbore.

NEGATIVE RAKE—Describes a cutting face in rotation whose cutting edge lags the surface of the cutting face.

NOMINAL SIZE—The designated basic size of a counterbore.

OVERALL LENGTH—The extreme length of the complete counterbore from end to end, including the shank and pilot.

PILOT—A cylindrical portion, either integral or removable, at the entering end of the counterbore to maintain alignment with the hole being counterbored.

POSITIVE RAKE—Describes a cutting face in rotation whose cutting edge leads the surface of the cutting face.

PRIMARY RELIEF—The relief immediately behind the cutting edge or margin. Properly called RELIEF.

PROJECTED LENGTH—The total distance the end face of the cutting element of a counterbore extends from the end of the machine spindle.

RADIAL RAKE ANGLE—The angle in a transverse plane between a straight cutting face and a radial line passing thru the cutting edge.

RADIAL RELIEF—Relief in a radial direction measured in the plane of rotation. It can be measured by the amount of indicator drop at a given radius in a given amount of angular rotation.

RAKE—The angular relationship between the tooth face, or a tangent to the tooth face at a given point, and a given reference plane or line.

RELIEF—The result of the removal of tool material behind or adjacent to the cutting edge to provide clearance and prevent rubbing [heel drag].

RELIEF ANGLE—The angle formed between a relieved surface and a given plane tangent to a cutting edge or to a point on the cutting edge.

SHANK—The portion of a counterbore by which it is held and driven.

SIZE—See terms ACTUAL SIZE and NOMINAL SIZE.

SOCKET—The tapered hole in a spindle, sleeve or counterbore holder, designed to receive, hold, and drive a taper shank.

SPIRAL ANGLE—See preferred term—HELIX ANGLE.

SPIRAL FLUTES—See preferred term HELICAL FLUTES.

SPOTFACING—The operation of producing a usually flat surface at the terminal of, and normal to the axis of a previously made hole.

STEP COUNTERBORE—A counterbore designed to produce in one operation, two or more counterbored diameters and seats at different axial locations.

STOP COLLAR—A solid or adjustable collar on the counterbore body or holder for controlling the depth of the counterboring of a hole.

STRAIGHT FLUTE—A flute which forms a cutting edge lying in an axial plane. See FLUTES.

STRAIGHT SHANK—A cylindrical shank.

SUBLAND COUNTERBORE—A type of step counterbore which has independent sets of lands in the same body section for each diameter.

TANG—The flatted end of a taper shank which fits in a slot in the socket.

TANG DRIVE—Two opposite parallel driving flats on the extreme end of a straight shank.

TAPER SHANK—A shank made to fit a specified [conical] taper socket.

OPERATION AND PROCEDURE IN COUNTERBORING

Counterboring and spotfacing often are done on surfaces that are rough, uneven, and not at right angles to the pilot hole. This imposes additional strains on the counterbore and the pilot, frequently resulting in breakage of one or both of these. In other cases, the work surface is covered by scale or is hard because of chilling.

One remedy for this is to make sure that the work-surface conditions are as favorable as possible. When that is not practical, special attention must be given to the operation of the counterboring tools. This is particularly true at the start of the cut.

If a hand feed is used, it is important to start the cut very carefully, so that "hogging-in" does not occur, with resultant breakage. Hogging must be prevented, but sufficient feed must be used to get under the scale or sand, otherwise the cutter will dull rapidly from this highly abrasive surface. When power feeds are used under these conditions, the feed must be slow, and particularly there must be no back-lash in the feeding mechanism of the machine.

The pilot should fit snugly in the piloting hole, but must not be too tight. Tight-fitting pilots always result in scoring, seizing, and breakage; they are therefore usually made slightly undersize.

SPEEDS AND FEEDS

Due to the nature of the work, experience indicates that it is usually advisable to reduce the speeds and feeds of counterbores as compared with speeds and feeds of corresponding diameters of twist drills. This is a subject on which there is no significant volume of authentic data. Shop practice, equipment, the condition of the equipment, available power in the particular machine tool, the material being counterbored, and the width of chip, are all factors which influence the speeds and feeds used.

SPEEDS

Mild Steel	75–85 Surface Feet Per Minute
Alloy Steel	40–80 Surface Feet Per Minute
Cast Iron—Soft	120–140 Surface Feet Per Minute
Brass	150–300 Surface Feet Per Minute

FEEDS

$\frac{1}{4}$ to $\frac{3}{8}$ Dia.	.003–.005 Inch Per Rev.
$\frac{7}{16}$ to $\frac{5}{8}$ Dia.	.004–.006 Inch Per Rev.
$\frac{11}{16}$ to $\frac{7}{8}$ Dia.	.005–.007 Inch Per Rev.
$\frac{15}{16}$ to $1\frac{3}{16}$ Dia.	.006–.008 Inch Per Rev.
$1\frac{1}{4}$ to $1\frac{1}{2}$ Dia.	.007–.009 Inch Per Rev.
$1\frac{9}{16}$ to 2" Dia.	.008–.010 Inch Per Rev.

The feeds are given for average chip-widths. Wide chips call for less feed, while narrow chips permit more feed.

SHARPENING AND MAINTENANCE

Like other tools, Counterbores must be maintained in proper condition throughout their life in order to get the best results. This has particular reference to the sharpening of the end-teeth.

As these teeth are resharpened, the chip-space tends to become smaller and smaller. It is, therefore, periodically necessary to grind new chip-spaces in front of the cutting edges, thus restoring the tool to approximately its original shape. A recommended procedure for resharpening counterbore and spotfacer-cutters is illustrated in Fig. 12.

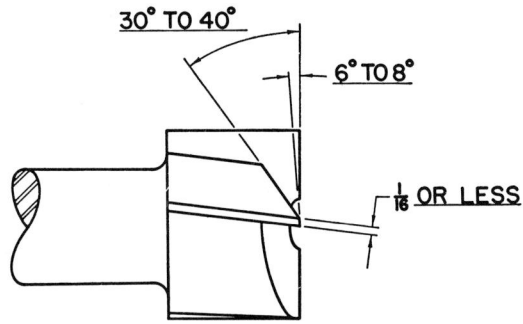

Fig. 12. *Resharpening Counterbore and Spotfacer Cutters*

It is desired that the finished tool shall have clearances as shown in Fig. 12. First there is to be a 6° to 8° relief over a narrow land of $\frac{1}{16}''$ or less. Then, behind this is a relief of 30° to 40° to provide chip-space.

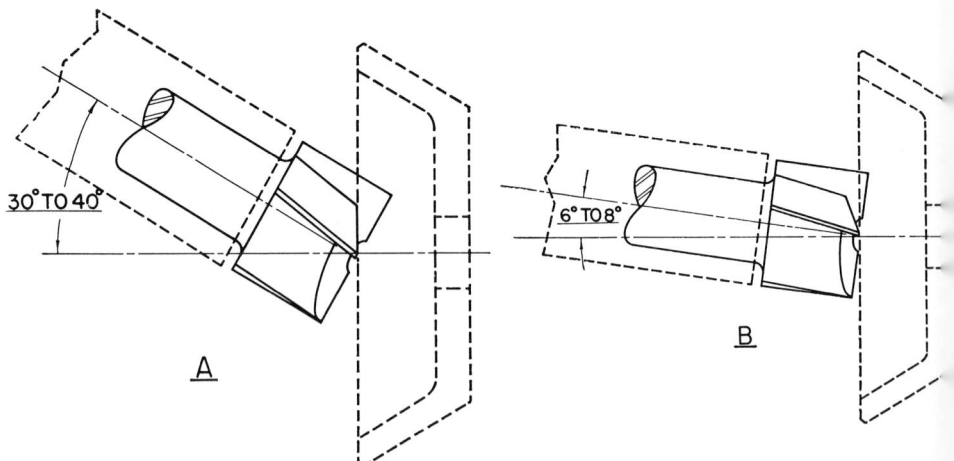

Fig. 13. *Procedure in Resharpening*

The grinding procedure is first to grind the large relief, as at *A*, Fig. 13, leaving the desired width of land. Then the machine is re-set to the proper, primary relief-angle, as at *B*. This final sharpening operation should be watched very carefully to see that all cutting edges are equally sharp and evenly located. Uneven cutting edges cause projecting teeth to be overloaded and may result in breakage. Inspection can be either optical or by an indicator. Fig. 13 shows the use of a cupped wheel in sharpening the counterbore. However, the same operation could be performed with a straight wheel.

COUNTERBORE CLINIC

Since counterboring is often a secondary operation, usually following a drill, there is sometimes a tendency to do the operation on any available equipment. This often means that the machine tool is quite worn, with excessive run-out, and with insufficient power. When trouble is encountered, the condition of the machine tool should always be carefully checked before condemning the cutter.

Counterbore cutters are usually precision-made tools on which trials and tests are made constantly in order to keep ahead of the shop requirements. New methods of manufacturing, different combinations of angles, clearances and fluting, as well as new materials of construction are investigated constantly. However, in spite of such research, they may be broken or destroyed by abuse like any other tool.

For deep-hole counterboring (usually to depths exceeding $\frac{1}{4}$ of the diameter of the counterbore), especially when there is a relatively large difference between the diameters of the counterbore and pilot, counterbore cutters with solid pilots are recommended. Most standard, interchangeable, pilot-type counterbore cutters are designed for as nearly average usage as possible. Therefore, it is advisable to consult the tool manufacturer as to the correct design when unusual manufacturing or operational conditions prevail, rather than attempting to adapt standard cutters with risk of failure.

A very common cause of failure in counterbore cutters is the lack of adequate clearance in the hole which guides the pilot of the cutter. If the pilot stops revolving for any reason, or if it stops advancing while the cutter continues to revolve and feed, the cutter may be damaged or destroyed. The pilot head is necessarily hardened to withstand wear, and this hardened head bears on the end cutting teeth (in the case of interchangeable pilot type cutters). If the pilot stops while the cutter continues to revolve under feed, it proceeds to machine the pilot head away, usually with disastrous results. The stem of the pilot may act as a wedge in such instances and actually split the cutter into several pieces.

There are several conditions which may cause the pilot to seize or stop. First among these is an undersized pilot hole resulting in an insufficient clearance between the outside diameter of the pilot and the hole. Undersize pilot holes may be caused by drilling or reaming undersize, faulty shop procedure, or jagged metal being left in the hole, due to the nature of the material. In any event, the pilot galls or binds in the undersize hole or loads with metal particles and stops entirely. If detected in time, only the cutting edges of the counterbore will be chipped or burned; otherwise total failure may result.

The prevention of this trouble lies in either increasing the size of the pilot hole, or in decreasing the size of the pilot so as to provide the necessary clearance. In grinding stock pilots to a new size, care should be taken to insure that the hardened surface is not completely removed. If there is sufficient clearance, the pilot should enter the guiding hole readily and travel its length without binding; however, too much clearance should be avoided since recent tests seem to indicate that it may cause chatter. Assuming that the pilot hole is to its nominal size, the pilot should be from .001″ to .007″ undersize, depending on its diameter. For example: .001″ undersize would be sufficient for a ⅛″ pilot, while a 3″ pilot would require .007″ clearance.

To provide maximum accuracy for the counterbore, the pilot hole should be semi-finish reamed. This will also help to remove any rough finish which might cause the pilot to gall. If final finish in the pilot hole is of importance, fluted pilots or roller pilots may be used to eliminate possible scoring.

A less common, but not unknown, failure resulting from the pilot occurs when the pilot is too long. In such instances, the pilot hits some solid surface while the cutter is still advancing. This causes the same sort of failure as found when the pilot hole is undersize.

Counterbore cutters lend themselves to another abuse which may result in failure. This is the practice of using a very small pilot in a large diameter cutter to counterbore a relatively deep hole. Unless the diameter of the pilot head is at least equivalent to the root diameter of the cutter, chips become trapped in the counterbore when the hole reaches a critical depth. This depth is approximately one quarter the diameter of the cutter. The material being counterbored is a factor in this condition, since cast iron, for example, which forms broken chips, is much less critical than one which forms continuous stringy chips. Broken chips usually escape more readily and permit the use of a much smaller pilot than could be used on AISI 1020 steel, for example.

Chip packing, caused by the use of an undersize pilot, usually results in a very poor finish in the hole being counterbored. In severe cases, sufficient force is generated to break the cutter.

It is not practical to increase the size of the hole to permit the use of a larger pilot, since the hole in which the pilot guides is determined by other considerations. In order to eliminate chip packing where the pilot diameter is less than the root diameter of the cutter, one of the following solutions may be used: On a new set-up, the counterboring operation may be designed to provide for a holder that is bushing-guided. With a counterbore cutter especially designed for the application, sufficient chip clearance is assured. On jobs already in operation, it may be possible to redesign the cutter with special deep fluting or a special form of flute which will permit the use of a detachable pilot and still provide adequate chip clearance. With standard counterbore cutters, it is frequently possible to regrind the throat sufficiently deep to permit chip flow for the particular application.

In deep-hole counterboring (that is to depths exceeding $\frac{1}{4}$ of the diameter of the counterbored holes), standard cutters may become galled on their margins, due to their widths. In extreme cases, sufficient force may be generated to cause the cutter to freeze in the work. The obvious solution to this problem is to decrease the width of the margins to $\frac{1}{64}''$ or less, depending on the diameter of the counterbore.

Used cutters sometimes fail after they have been ground, because of poor sharpening technique. As they are multi-fluted tools, it would seem obvious that free hand grinding is dangerous; nevertheless, some users still attempt it. This results in high and low spots on any one tooth, as well as tremendous variations from one tooth to another. In some instances regrinding of the rake angle is ignored completely, while in others it results in various angles between $0°$ and $15°$. These defects cause the cutter to grab and chatter and often results in breakage because one tooth is usually carrying all the load.

Another source of failure caused by faulty regrinding is the lack of chip space which occurs when the secondary clearance angle is not maintained as originally ground by the manufacturer of the cutter. Continual sharpening without periodic grinding of the secondary clearance angle reduces chip space and chip flow. In severe cases, failure of the cutter may result from the high load imposed by this condition.

Spotfacing and back-spotfacing cutters are abused when used as counterbores and are likely to give trouble because they are designed and constructed without peripheral relief. When used to counterbore to any appreciable depth, the outside diameter galls and may seize. Cutting through fillets and alongside vertical or semi-vertical surfaces produces the same results as cutting into solid metal. If such operations are necessary, a counterbore should be used, because the cutter must have side flutes and be cleared to narrow margins.

Counterbores and spotfacers are both subject to the common abuses which occur to all cutting tools. They must be accurately machine sharpened. Original grinding clearances and chip passages must be maintained, as well as clearance for the pilot. Cutters must not be hammered on the cutting edges, nor placed loosely in boxes or bins so that they bump each other. They must not be abused — to give best results.

CUTTING FLUIDS

In order to obtain efficient cooling, as large a volume of coolant as is possible should be directed into the cutting edges of the counterbore. A stream is usually directed so as to fall onto the counterbore near the surface of the work. It then flows down the flutes of the tool to the cutting edges.

Quite often when the finish in the counterbored hole is important, a cutting fluid can be selected which will materially improve this finish.

SELECTING THE CUTTING FLUID

It is suggested that cutting fluid problems be referred to a reputable manufacturer of such fluids. The following list should be used as a suggestion only:

Aluminum and its alloys:	Soluble oil; kerosene and lard oil compounds; light, non-viscous, neutral oil; kerosene and soluble oil mixtures.
Brass:	Dry; soluble oil; kerosene and lard oil compounds; light, non-viscous neutral oil.
Copper:	Soluble oil; winter-strained lard oil; oleic acid compounds.
Cast Iron:	Dry or with a jet of compressed air for a cooling medium.
Malleable Iron:	Soluble oil or non-viscous neutral oil.
Monel Metal:	Soluble oil or sulfurized mineral oil.
Steel, Ordinary:	Soluble oil; sulfurized oil; high Extreme Pressure value mineral oil.
Steel, very hard and refractory:	Soluble oil; sulfurized oil; turpentine.
Steel, Stainless:	Soluble oil or sulfurized mineral oil.
Wrought Iron:	Soluble oil; sulfurized oil; high animal-oil-content mineral-oil compound.

IMPORTANT POINTS TO CONSIDER

Intermittent cooling of cutting tools should be avoided as it may cause small checks or cracks which will result in tool failure.

The most important factor in maintaining ideal cutting fluids is to see that the fluids are kept clean at all times. If they are allowed to become contaminated and to carry large amounts of metal particles in suspension, their efficiency will be impaired. There are many commercial methods of cleaning cutting fluids from which the most suitable may be chosen. In many cases, only a regular cleaning of settling tanks and machine will be sufficient.

HORSEPOWER AND THRUST FOR COUNTERBORES

To insure adequate power in the machine tool and to insure sufficient strength and rigidity in jigs and fixtures, it is necessary to know the horsepower and thrust which counterbores require and develop.

Caution must be observed in the use of the following graphs and tables. The values obtained refer to power at the cutter and were obtained under as nearly ideal conditions as possible. With dull cutters and power loss in the machine, a safety factor of at least two must be employed. Note that the graphs were drawn for carbide-tipped counterbore cutters and that the factor employed is different for high-speed-steel counterbore cutters.

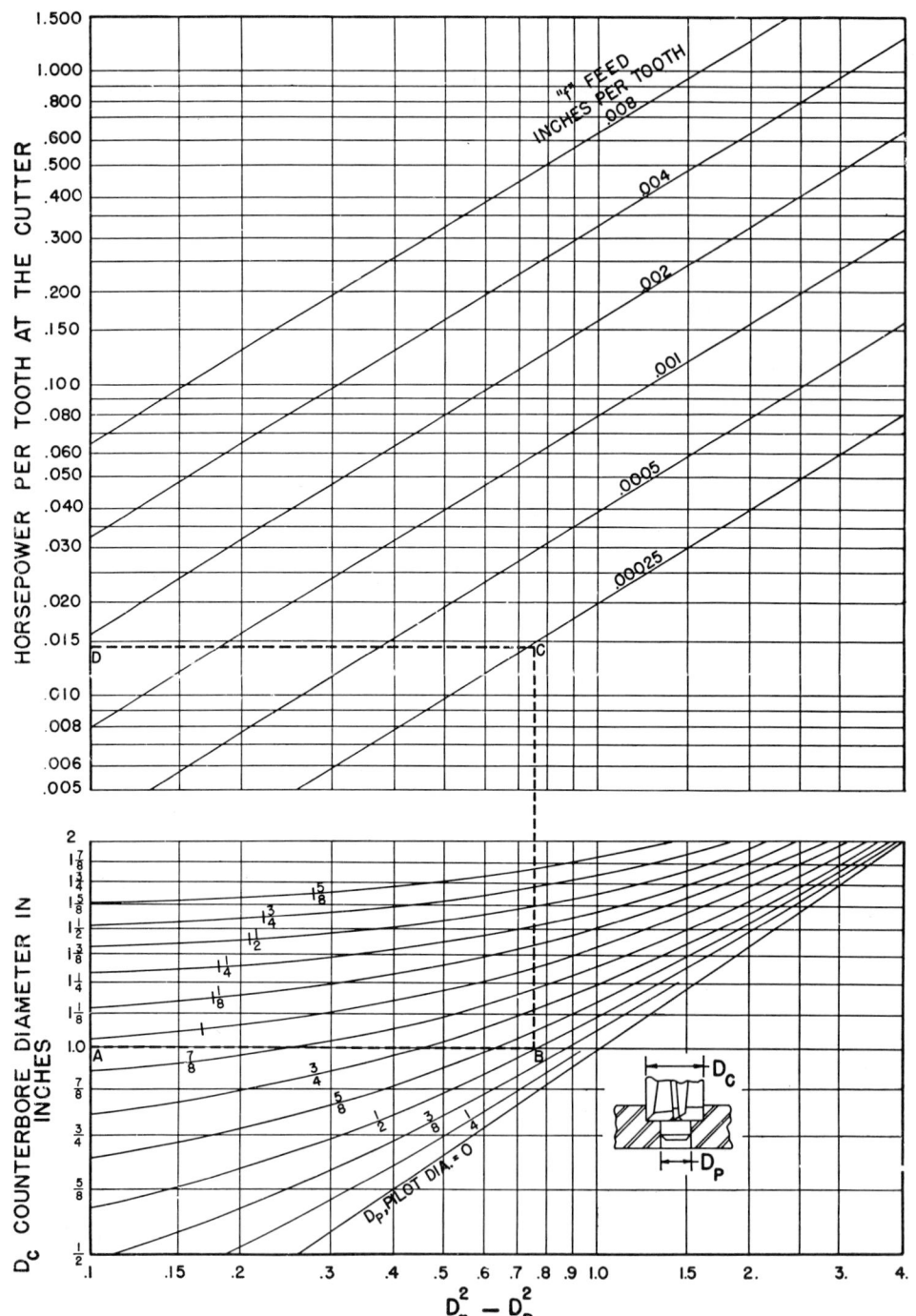

Horsepower Required Per Tooth in Counterboring AISI 1020 Steel at 100 R.P.M.,
using Carbide-Tipped Counterbores and Cutting Dry

When using carbide-tipped counterbores with no coolant for counter-boring other materials, multiply the values obtained from the graph for AISI 1020 steel by the following factors:

Material Cut	Brinell Hardness	H.P. Factor
AISI 1020, Cold Rolled	180	1.0
AISI 6150, Hot Rolled	195	1.3
AISI 6150, Heat Treated	241	2.1
Tool Steel, 1% C., Annealed	156	1.9
Cast Iron, 40,000 psi	210	.85
Monel Metal	207	1.9
Phosphorous Bronze	156	.76
Lead Brass	124	.38
Aluminum Alloy, 24ST	154	.54
Aluminum, Soft 2S	36	1.4
Magnesium, Cast H Alloy	63	.25

When using high-speed-steel counterbores, multiply the values obtained from the graph for AISI 1020 steel (cut dry with carbide-tipped counterbores) by the following factors:

Material Cut	Brinell Hardness	Cutting Fluid	H.P. Factor
AISI 1020, Cold Rolled	180	16:1 Emulsion	.85
AISI 6150, Hot Rolled	195	16:1 Emulsion	1.3
Cast Iron, 40,000 psi	210	Dry	.73

An example showing the use of the graph and conversion tables is given below:

It is required to find the horsepower necessary to counterbore hot rolled, AISI 6150 steel with a one inch carbide-tipped counterbore having a half inch pilot. The counterbore has three teeth and is run at 185 S.F.M. with .00075″ feed per revolution, which is .00025″ per tooth.

Following the heavy dotted line ABCD on the lower graph up to the .00025″ feed line on the upper graph, the horsepower per tooth is found to be .0145 at 100 R.P.M.; 185 S.F.M. = 707 R.P.M. for a 1″ diameter. The horsepower-conversion-factor for carbide-tipped counterbores in hot-rolled, AISI 6150 is 1.3.

Horsepower under given conditions $= \dfrac{707}{100} \times .0145 \times 3 \times 1.3 = 4$ H.P.

This is the horsepower required at the cutting edges of a sharp counterbore. If the counterbore becomes dull to the point where it refuses to cut, the horsepower required may double or triple. Also to be considered is the power-loss in the machine, which varies widely with the size and design.

When using carbide-tipped counterbores with no coolant for counterbor-

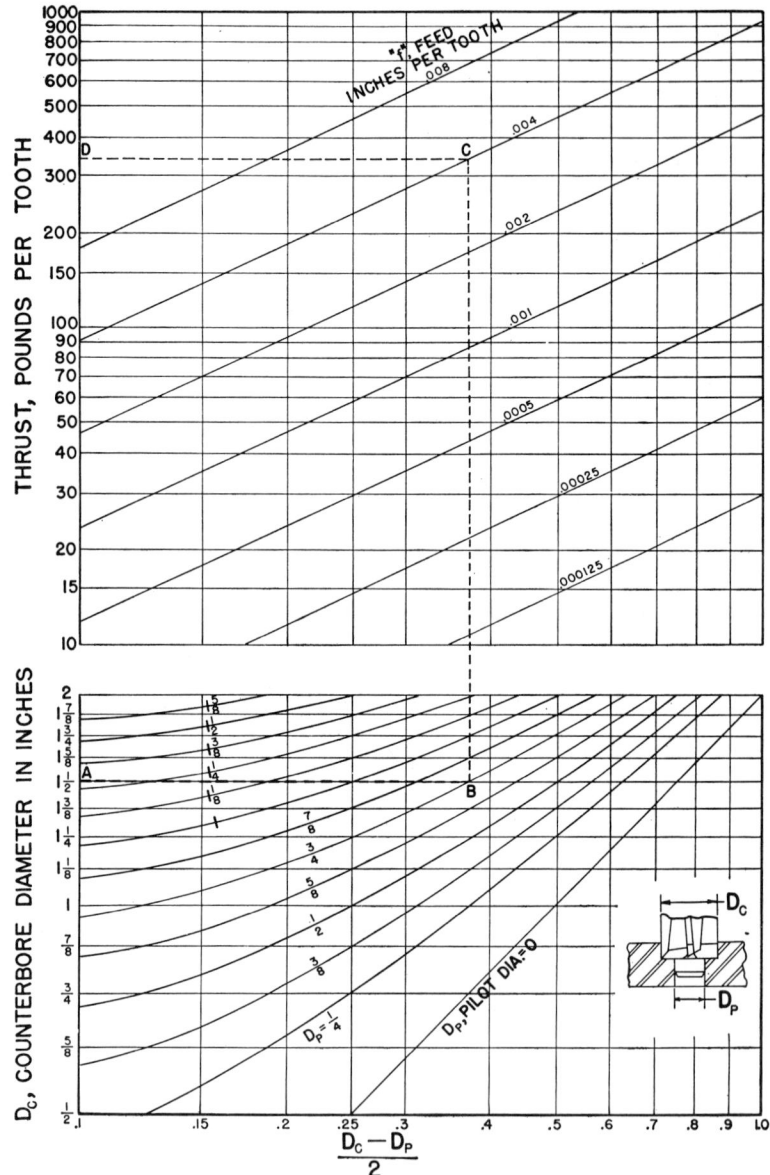

Thrust Developed in Counterboring AISI 1020 Steel Using Carbide-Tipped Counterbores, and Cutting Dry

ing other materials, multiply the values obtained from the graph for AISI 1020 steel by the following factors:

Material Cut	Brinell Hardness	Thrust Factor
AISI 1020, Cold Rolled	180	1.0
AISI 6150, Hot Rolled	195	1.8
AISI 6150, Heat Treated	241	2.1
Tool Steel, 1% C., Annealed	156	3.0
Cast Iron, 40,000 psi	210	.7
Monel Metal	207	1.4
Phosphorous Bronze	156	.82
Lead Brass	124	.27
Aluminum Alloy, 24ST	154	.82
Aluminum, Soft 2S	36	2.7
Magnesium, Cast H Alloy	63	.27

When using high-speed-steel counterbores, multiply the values obtained from the graph for AISI 1020 steel (cut dry with carbide-tipped counterbores) by the following factors:

Material Cut	Brinell Hardness	Cutting Fluid	Thrust Factor
AISI 1020, Cut Rolled	180	16:1 Emulsion	.89
AISI 6150, Hot Rolled	195	16:1 Emulsion	1.2
Cast Iron, 40,000 psi	210	Dry	.79

An example showing the use of the graph and conversion tables is given below:

It is required to find the thrust developed when counterboring AISI 1020 steel with a 1½" high-speed steel counterbore having a ¾" pilot. The counterbore has four teeth and is run at a feed of .016", or .004" per tooth.

Following the heavy dotted line ABCD on the lower graph up to the .004" feed line on the upper graph, the thrust per tooth is found to be 337 pounds. The conversion factor for high-speed steel counterbores in 1020 steel is .89.

The total thrust under the given conditions equals

$$337 \times .89 \times 4 = 1199.72 \text{ pounds.}$$

This is the thrust required to feed a counterbore with sharp cutting-edges. If the counterbore becomes dull to the point where it refuses to cut, the thrust required to feed it increases many times. Also to be considered is chip-room, which may become a problem in some cases. Thrust increases as chips clog, a condition which may be alleviated by proper sharpening and proper design.

TYPES OF COUNTERBORES AND
TYPICAL APPLICATIONS

Counterboring is the enlargement of a portion of a hole that has previously been formed. If the counterbored portion is shallow ($\frac{1}{8}''$ or less), the operation is usually known as spot-facing; for example, a flat seat for a bolt head or nut. If, on the other hand, the hole is enlarged to a considerable depth, it is designated as counterboring. Spot-facing cutters do not require peripheral relief.

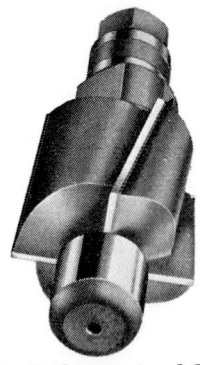 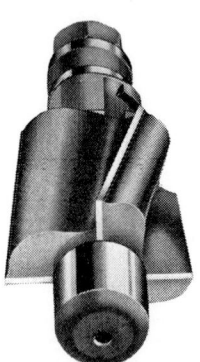 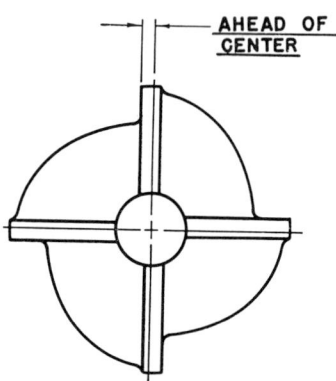

Fig. 14. *Conventional flute design provides ample chip clearance for shallow counterboring and spot facing*

Fig. 14A. *Design with deep flute and increased helix provides ample chip clearance for deep counterboring*

Fig. 15. *End teeth of counterbores are often cut ahead of center to produce a shearing action*

Helical flutes are commonly used, as this construction permits positive, effective rake-angles on the cutting edges. The end-teeth are often cut ahead of center (negative radial-rake), so that a shearing action is produced. This helps to eliminate "chatter" and "hogging" and sometimes aids in chip disposal.

In solid counterbores, Fig. 4, the body, cutter, and pilot are integral. They are generally used for cap screw and machine screw counterboring. These tools are end-cutting only and often have no peripheral relief. They have maximum rigidity and accuracy due to solid construction.

(1) Solid counterbores are made with some variations as to the number and shape of flutes.

(2) Solid counterbores are occasionally made in the "subland" and "step cutter" types as shown on pages 322–324.

(3) Solid counterbores are usually made in standard combinations, which are covered in the table on pages 327–330.

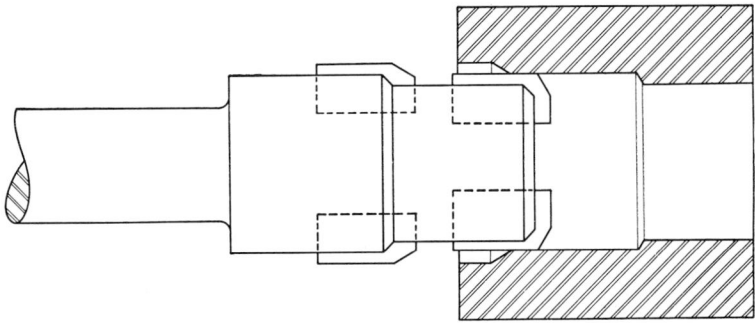

Fig. 16. *Inserted Blade Counterbore*

(4) Inserted-blade counterbore-cutters are sometimes used on large-diameter tools. In many cases, inserted-blade counterbores have proven more economical than solid counterbores due to reduced initial and replacement tool-costs. The method of holding the blade varies with the manufacturer. An illustration of this type of tool is shown in Fig. 16.

TWO-PIECE COUNTERBORES

Counterbores are often made of two-piece construction, having an integral shank and counterbore-cutter, and a pilot that is interchangeable. This type has two main advantages.

First, the alignment of the counterbore-cutter with respect to the shank is permanently true, and second, a number of pilots of different diameters can be used for a given-size counterbore, which provides flexibility in the use of the tool.

This type of counterbore is illustrated in Fig. 5.

THREE-PIECE COUNTERBORES

This type of counterbore is made in three parts: (1) the holder, (2) the counterbore-cutter, and (3) the pilot. Generally, a holder will fit a series of counterbore-cutters. In the same manner, each size counterbore-cutter will take various sizes of pilots suited to the work to be performed.

The type of holder and the method of assembly vary among manufacturers and determine whether or not the counterbore-cutter shoulders against the holder. An illustration is shown in Fig. 6.

This style of counterbore has the advantage of flexibility with regard to the number of sizes of counterbore-cutters which can be used with one holder, as well as the number of sizes of pilots used in each cutter.

CENTER-CUTTING COUNTERBORES

Because of the requirements of the particular job, it is sometimes necessary to have a flat bottom in a counterbored section rather than the conical bottom left by a drill. This may be accomplished readily by a center-cutting counterbore as shown in Fig. 17.

In counterbores which cut to center it is usual to have only two teeth reaching to the center. The other teeth are cleared at the center to provide as much chip space for the cutting teeth as is possible. Sharpening of these cutters must be performed correctly or they will not function.

Fig. 17. *Center-Cutting Counterbore*

Center-cutting counterbores are often required for other than flat bottom holes. For example, there are counterbores which form a flat bottom with a corner radius, spherical seats, and other specialized forms.

HOLDERS

A variety of holders is available for use with the three-piece counterbore, in order to augment its usefulness in production operations.

For example, holders may be made specifically to operate in a bushing. In such an application various constructions may be employed, depending on the operation, the machine tool, and the fixture used to hold the work. Such holders are usually made with helical or straight dust grooves. These grooves may serve to keep the bushing clean or to permit coolant to flow down through the bushing to the cutting edges of the tool. Frequently flutes are substituted for grooves to bring chips up through the bushing. Depending on the particular function, the grooves or flutes may be deep or shallow and have right or left hand helix. A fluted holder of this type is shown in Fig. 18.

Fig. 18. *Fluted Holder*

Other holders for use in bushings may be so machined as to leave integral-narrow-vertical-bearing margins. These margins also may be made by welding on some long-wearing material, or they may be inserts of bronze, hardened steel, or other material, to reduce friction. A holder with inserted wear strips is shown in Fig. 19.

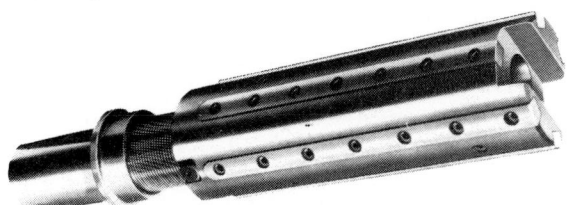

Fig. 19. *Holder with Inserted Wear Strips*

On multiple-spindle operations the shortening of counterbore cutters from wear or sharpening raises a serious problem. It is not always possible or practical to have available a series of cutters of matching length. Even though all cutters from a given machine were ground to a constant length every time they were sharpened, some provision would still have to be made for lowering the head or raising the work in order to meet the tolerances for depth of hole. To accomplish this purpose, and to prevent needless grinding of cutters for length, adjustable length holders are provided. These holders are adjusted for length by various mechanisms, all of which are based on the use of a threaded section. As the counterbore cutter decreases in length, the holder is increased in length a corresponding amount to maintain a constant overall. In this fashion each cutter is readily maintained for depth of cut by a hand operated mechanism and only sufficient grinding to restore the cutting edge is required. In Fig. 7 is shown a new counterbore and an adjustable holder. Fig. 7 illustrates how this holder is adjusted for length in order to compensate for the shortening of the cutter due to regrinding.

On certain applications it is necessary to counterbore to some depth which is related to some other surface. This usually requires a machine which is equipped with adjustable stops. Lacking such provision for machine stops, holders are provided which serve the same purpose. These are known as stop-collar (or stop-nut) holders and are shown in Figure 8. The collar fits on the counterbore holder in such a way as to be adjustable for length. In operation, the collar acts as a mechanical stop, when it comes in contact with the work or the fixture. Because it is adjustable for position, holes of constant depth may be produced irrespective of shortening because of wear or to sharpening.

A practical and often desirable variation of the counterbore stop-collar (stop-nut) holder is known as the ball bearing-thrust stop-collar holder.

This is the same as the regular stop-collar holder, except that the collar is mounted on ball thrust bearings, so that it stops revolving as soon as it comes in contact with the work or the fixture bushing. This prevents the collar from scoring or scuffing the surface on which it stops. Care must be observed to keep the stopping surface free from chips or dirt. Since the collar does not revolve, it has no opportunity to wipe away any foreign matter on the stop. Such foreign matter would cause the counterbored hole or spotfaced surface to be shallow by an amount equal to the thickness of the obstruction.

Floating holders for counterbores were developed to overcome small misalignment between the work or the bushing and the machine tool spindle and provide an excellent answer to this frequently vexing problem. Unless a floating holder is employed on large indexing set-ups, alignment must be perfect or else galling and severe wear of the bushing and holder often results. Use of these holders also insures greater concentricity with pre-existing holes than is usually possible with a non-piloted rigid holder.

Special holders may be provided which incorporate two or more of these features, in order to solve particular problems.

MULTIPLE DIAMETER COUNTERBORES

Multiple diameter counterbores are specifically designed to machine several surfaces at one pass, thereby saving machining time on long production runs. On short production runs, multiple diameter counterbores combine several individual machine set-ups into one, thereby minimizing set-up costs. The design obviously provides for maximum concentricity of all machined surfaces and long tool life, since such tools require end sharpening only. Fig. 20 shows some examples of this type of counterboring.

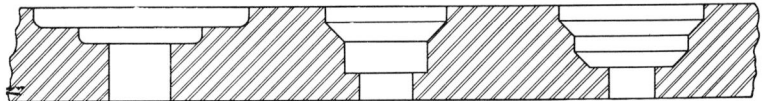

Fig. 20. *Multiple Diameter Counterboring*

MULTIPLE DIAMETER STEP-COUNTERBORES

This type of construction provides two or more diameters on the same land. These steps are separated by square, angular, or curved cutting-edges to provide the required surfaces.

The step-counterbore has many useful applications; however, where the small diameters are short, regrinding difficulties may be encountered. The cutting shoulders as well as the end-cutting teeth must be reground to maintain the correct length for each diameter. The regrinding of the shoulders produces nicks in the smaller diameter. Continued regrinding of step-counterbores results in the condition shown in Fig. 21.

Fig. 21. *Step Counterbores—(Left) New, and (Right)
Showing the Effect of Regrinding*

MULTIPLE DIAMETER SUBLAND COUNTERBORES

Subland counterbores, like step counterbores, can be made in two or more diameters. Subland, multiple diameter counterbores provide several diameters on alternate sets of blades. The life of the counterbore is greatly prolonged, because each diameter is reground independently of the others. The advantage of "subland" counterbores becomes evident when the extension of the small diameter beyond the large diameter is short. Regrinding a step-counterbore under the condition just mentioned might destroy the small diameter, thereby requiring additional, costly regrinding to establish a new

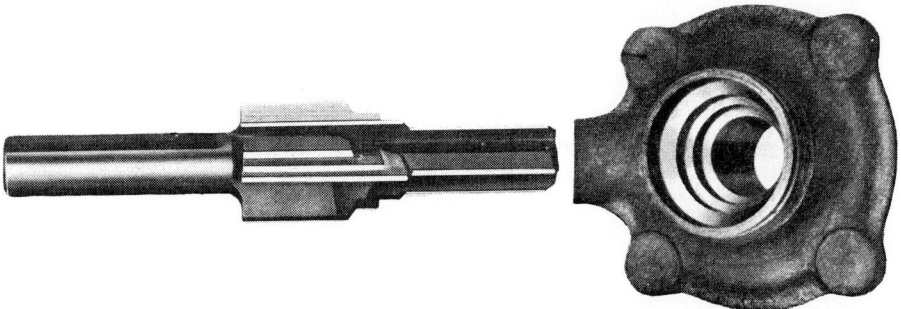

Fig. 22. *Subland Counterbore*

small diameter. A Subland counterbore, like that shown in Fig. 22 would not be so affected.

Some of these counterbores are made in complex combinations, so that many operations can be performed at one time. In Fig. 23 is shown a type of counterbore which is designed to perform six separate counterboring, facing, and chamfering operations at one time on a part.

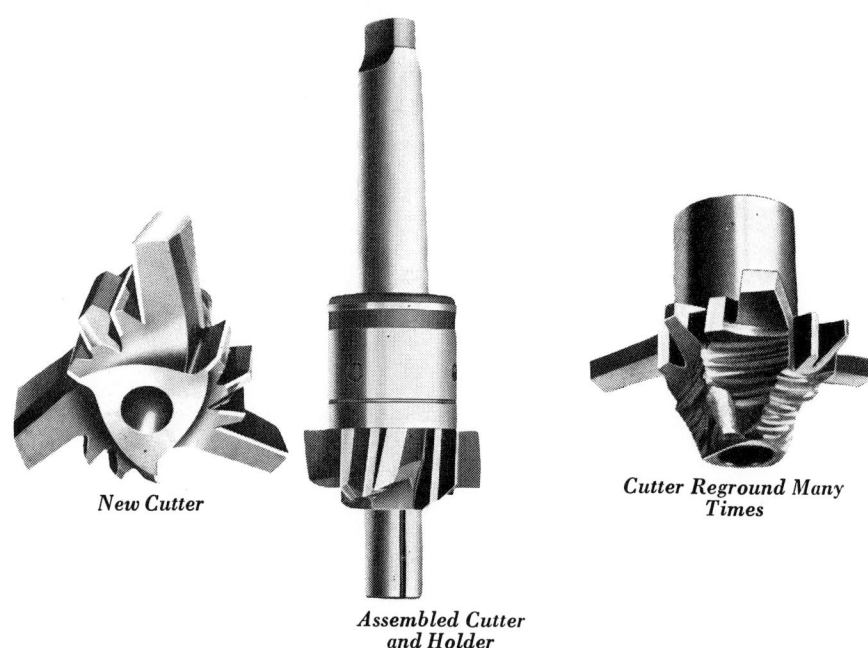

New Cutter

*Cutter Reground Many
Times*

*Assembled Cutter
and Holder*

Fig. 23. *Counterbore for Performing Six Operations at Once*

INVERTED SPOTFACERS
(See Fig. 10)

Occasionally it is necessary to spotface a hidden surface which cannot be reached with the conventional counterbore. In order to fill this need, tools known as inverted spotfacers have been developed. These tools are usually made so that the spotfacer can be quickly attached to or detached from the holder. This is necessary so that the holder can be inserted through a hole and the spotfacer attached. Inaccessible surfaces can be faced in this way as shown in Fig. 24.

There are various methods of attaching the cutter to the holder, depending upon the manufacturer's practice. Inverted spotfacers can be run right-or-left-hand, depending upon the requirements of the job. In this tool the body acts as a pilot and may have an oil-groove for lubrication. Since spotfacing is a shallow operation, the tool is end-cutting only and requires no peripheral relief.

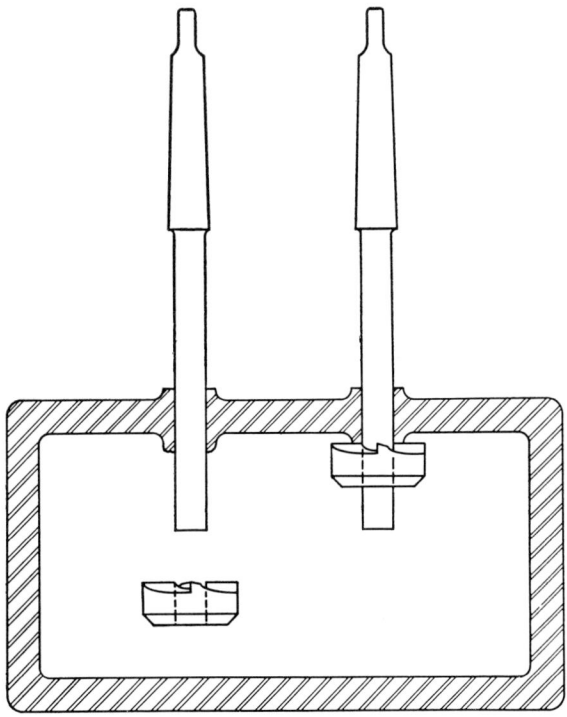

Fig. 24. *Application of Inverted Spotfacer*

DOUBLE-END SPOTFACERS
In some cases it is extremely advantageous to have a double-end spotfacer mounted on a holder, so that spotfacing and inverted spotfacing can be done without removing the cutter, and by using the same direction of rotation. (See Fig. 11.)

This type of spotfacer simplifies difficult machining jobs by finishing two internal surfaces without removing the tool or relocating the work. The job can be set up to pilot ahead of, or behind the work, or both, according to the nature of the part to be machined, Fig. 25.

CARBIDE OR HARD-ALLOY-TIPPED COUNTERBORES
Hard alloy tipped counterbores are in use in production-machining of steel, cast iron, brass, aluminum, and other materials. The efficient use of hard alloy tipped tools is dependent upon many factors, which, in addition to operating skill and experience, include such mechanical factors as rigidity in the machine and fixture, alignment of work, and adequate power to assure

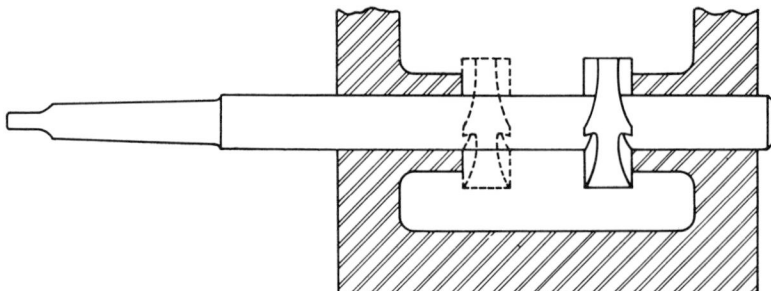

Fig. 25. *Spotfacer is moved to the left to face one surface and then to the right to face the other*

proper speeds and feeds. When hard alloy tipped tools are considered, the importance of the factors listed above cannot be over-emphasized. Many of the applications mentioned previously for high speed steel counterbores can be adapted to carbides under the proper conditions.

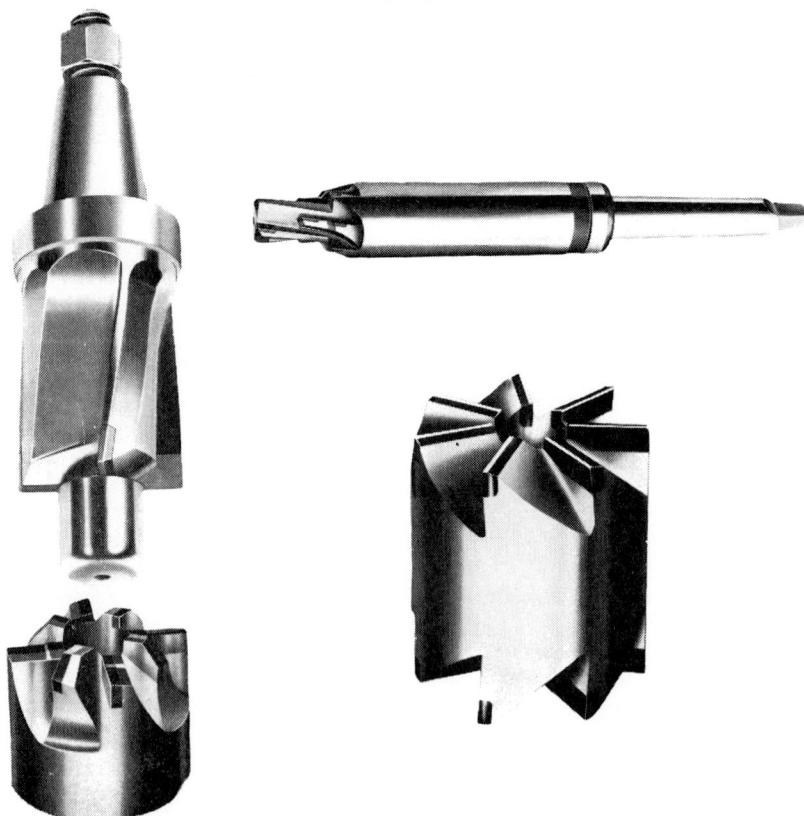

Fig. 26. *Types of Carbide or Hard-Alloy-Tipped Counterbores*

STANDARD COUNTERBORES
Solid Counterbores with Integral Pilot
Straight Shank
High Speed Steel
Fractional Screw Sizes

Straight Shank—Short Series

Standard Sizes and Dimensions

Screw Diameter	Counterbore Diameter	Pilot Diameters			Overall Length	Shank Diameter
		Nom.	+1/64	+1/32		
1/4	13/32	1/4	17/64	9/32	3½	3/8
5/16	1/2	5/16	21/64	11/32	3½	3/8
3/8	19/32	3/8	25/64	13/32	4	1/2
7/16	11/16	7/16	29/64	15/32	4	1/2
1/2	25/32	1/2	33/64	17/32	5	1/2

Straight Shank—Long Series

Standard Sizes and Dimensions

Screw Diameter	Counterbore Diameter	Pilot Diameters			Overall Length	Shank Diameter
		Nom.	+1/64	+1/32		
1/4	13/32	1/4	17/64	9/32	5½	3/8
5/16	1/2	5/16	21/64	11/32	5½	3/8
3/8	19/32	3/8	25/64	13/32	6	1/2
7/16	11/16	7/16	29/64	15/32	6	1/2
1/2	25/32	1/2	33/64	17/32	7	1/2

Tolerances:
Counterbore Dia.   +.0020 −.0020
Pilot Dia.   −.0005 −.0020
Shank Dia.   −.0005 −.0020
Overall Length $+\frac{1}{16}$ $-\frac{1}{16}$

Solid Counterbores with Integral Pilot

High Speed Steel
Machine Screw Sizes
Straight Shank

Standard Sizes and Dimensions

Screw Size	Counterbore Diameter	Pilot Diameters		Overall Length	Shank Diameter
		Nom.	+ .016		
0	.110	.060	.076	2½	7/64
1	.133	.073	.089	2½	1/8
2	.155	.086	.102	2½	5/32
3	.176	.099	.115	2½	11/64
4	.198	.112	.128	2½	5/16
5	.220	.125	.141	2½	3/16
6	.241	.138	.154	2½	7/32
8	.285	.164	.180	2½	1/4
10	.327	.190	.206	2¾	9/32
12	.372	.216	.232	2¾	5/16

Tolerances:
Counterbore Dia.   + .0020 — .0020
Pilot Dia.   — .0005 — .0020
Shank Dia.   — .0005 — .0020
Overall Length   + 1/16   — 1/16

STANDARD COUNTERBORES

Solid Counterbores

Removable Pilot, Short Length
Straight Shank, High Speed Steel

Standard Sizes and Dimensions

Diameter Inches	Length Overall	Shank Diameter	No. of Flutes	Hole Diameter
$3/16$	3	$15/64$	3	$3/32$
$7/32$	3	$15/64$	3	$3/32$
$1/4$	3 $13/16$	$15/64$	3	$3/32$
$9/32$	3 $13/16$	$17/64$	3	$3/32$
$5/16$	3 $13/16$	$19/64$	3	$3/32$
$11/32$	3 $13/16$	$5/16$	3	$3/32$
$3/8$	4 $1/16$	$5/16$	3	$5/32$
$13/32$	4 $1/16$	$3/8$	3	$5/32$
$7/16$	4 $1/16$	$3/8$	3	$5/32$
$15/32$	4 $5/16$	$7/16$	3	$3/16$
$1/2$	4 $5/16$	$7/16$	3	$3/16$
$17/32$	4 $5/16$	$1/2$	3	$3/16$
$9/16$	4 $5/16$	$1/2$	3	$3/16$
$19/32$	5 $1/8$	$1/2$	3	$3/16$
$5/8$	5 $1/8$	$1/2$	3	$3/16$
$21/32$	5 $1/8$	$1/2$	3	$3/16$
$11/16$	5 $1/8$	$1/2$	3	$3/16$
$23/32$	5 $3/8$	$1/2$	3	$1/4$
$3/4$	5 $3/8$	$1/2$	3	$1/4$
$25/32$	5 $3/8$	$5/8$	3	$1/4$
$13/16$	5 $3/8$	$5/8$	3	$1/4$
$27/32$	5 $3/8$	$3/4$	3	$1/4$
$7/8$	5 $3/8$	$3/4$	3	$1/4$
$29/32$	6 $1/8$	$3/4$	3	$1/4$
$15/16$	6 $1/8$	$3/4$	3	$1/4$
$31/32$	6 $3/8$	$3/4$	3	$5/16$
1	6 $3/8$	$3/4$	3	$5/16$
1 $1/16$	6 $3/8$	$3/4$	3	$5/16$
1 $1/8$	6 $3/8$	1	3	$5/16$
1 $3/16$	6 $3/8$	1	3	$5/16$
1 $1/4$	6 $5/8$	1	5	$3/8$
1 $3/8$	6 $5/8$	1	5	$3/8$
1 $1/2$	7 $7/8$	1 $1/4$	5	$3/8$
1 $5/8$	8 $1/8$	1 $1/4$	5	$7/16$
1 $3/4$	8 $1/8$	1 $1/4$	5	$7/16$
1 $7/8$	8 $1/8$	1 $1/2$	5	$7/16$
2	8 $3/8$	1 $1/2$	5	$1/2$

STANDARD COUNTERBORES

Solid Counterbores
Removable Pilot, Short Length
Taper Shank, High Speed Steel

Standard Sizes and Dimensions

Diameter Inches	Length Overall	Taper Shank	No. of Flutes	Hole Diameter
1/4	3 13/16	1	3	3/32
9/32	3 13/16	1	3	3/32
5/16	3 13/16	1	3	3/32
11/32	3 13/16	1	3	3/32
3/8	4 1/16	1	3	5/32
13/32	4 1/16	1	3	5/32
7/16	4 1/16	1	3	5/32
15/32	4 5/16	1	3	3/16
1/2	4 5/16	1	3	3/16
17/32	4 5/16	1	3	3/16
9/16	4 5/16	1	3	3/16
19/32	5 1/8	2	3	3/16
5/8	5 1/8	2	3	3/16
21/32	5 1/8	2	3	3/16
11/16	5 1/8	2	3	5/16
23/32	5 3/8	2	3	1/4
3/4	5 3/8	2	3	1/4
25/32	5 3/8	2	3	1/4
13/16	5 3/8	2	3	1/4
7/8	5 3/8	2	3	1/4
15/16	6 1/8	3	3	1/4
1	6 3/8	3	3	5/16
1 1/16	6 3/8	3	3	5/16
1 1/8	6 3/8	3	3	5/16
1 3/16	6 3/8	3	3	5/16
1 1/4	6 5/8	3	5	3/8
1 5/16	6 5/8	3	5	3/8
1 3/8	6 5/8	3	5	3/8
1 1/2	7 7/8	4	5	3/8
1 5/8	8 1/8	4	5	7/16
1 3/4	8 1/8	4	5	7/16
1 7/8	8 1/8	4	5	7/16
2	8 3/8	4	5	1/2
2 1/8	9 7/8	5	5	1/2
2 1/4	9 7/8	5	5	1/2
2 3/8	9 7/8	5	5	1/2
2 1/2	9 7/8	5	5	1/2
2 5/8	9 7/8	5	5	1/2
2 3/4	9 7/8	5	5	1/2
2 7/8	9 7/8	5	5	1/2
3	9 7/8	5	5	1/2

TAP AND DIE SECTION

FOREWORD

Tap and die manufacturers, recognizing the importance of standardizing screw threads and thread cutting tools, have adopted and published standards to benefit the user of these tools. The information given in this section includes some of the most frequently used standards for taps and dies developed by the Tap and Die Division of the Metal Cutting Tool Institute. Many of these are also published in the American National Standard, Taps—Cut and Ground Thread, ANSI B94.9. Nomenclature, identification, dimensions, and tolerance tables for various styles of taps and dies are given herein.

Many variables affect the size of a product thread produced by a tap. The size may depend on operating conditions, equipment, material, conditions and size of the hole before tapping, speed, alignment, tapping fluid, etc. Therefore, tap manufacturers cannot guarantee the size of product threads produced by a tap. The tap limits shown in this publication are a recommended starting point only.

Thread limits shown in the tables represent the measured size of the tap. If standard tap limits shown are not satisfactory for a particular application, efforts should be made to analyze and correct the operating conditions. Unusual operating conditions may require the user to develop specific limits suitable for these conditions.

Product thread specifications and limits for standard and special sizes and pitches may be found in the following documents:

FED-STD-H28, Federal Standard, Screw Threads for Federal Services.

ANSI B1.1, Unified Inch Screw Threads.

ANSI B1.13M, Metric Screw Threads—M Profile.

ANSI B1.18M, Metric Screw Threads for Commercial Mechanical Fasteners —Boundary Profile Defined.

ANSI B1.20.1, Pipe Threads, General Purpose (Inch).

331

ANSI B1.20.3, Dryseal Pipe Threads (Inch).

ANSI B1.21M, Metric Screw Threads—MJ Profile

MIL-S-8879, Military Specification, Screw Threads, Controlled Radius Root with Increased Minor Diameter (Inch UNJ Series).

TAP AND DIE SECTION CONTENTS

ILLUSTRATIONS OF TERMS APPLYING TO TAPS

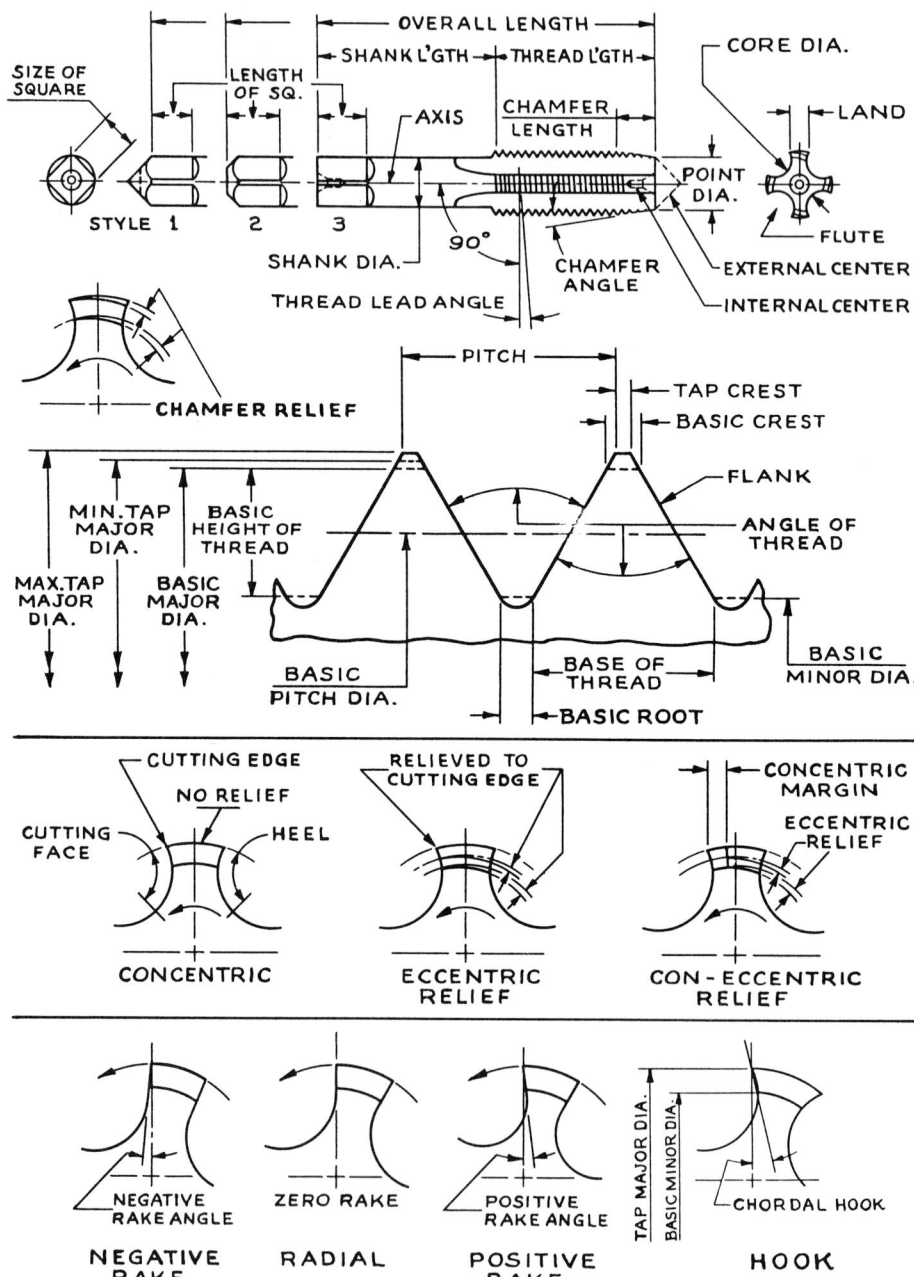

NOMENCLATURE OF
TAPS—THREAD CUTTING DIES—THREAD CHASERS

A—TAPS

I—DEFINITION

Tap—A cylindrical or conical threading tool with one or more cutting or forming elements having screw threads of a desired form on the periphery. By a combination of rotary and axial motion, the leading end produces an internal thread.

II—CLASSIFICATION BASED ON CONSTRUCTION

1—*Solid Taps*—Those made of one piece of tool material.

2—*Shell Taps*—Taps with a central hole fitting an arbor or driver. Driving means may be a cross slot or keyway on the back of the tap or keyway in the arbor hole.

3—*Sectional Taps*—Those in which a relatively short tap nib is attached to a reusable shank.

4—*Expansion Taps*—Those whose size may be changed by deflecting or bending segments of tap body.

5—*Adjustable Taps*—Those in which inserted thread cutting chasers or blades can be adjusted radially to secure a change in the tap diameter.

6—*Inserted Chaser Taps*—Those in which two or more thread cutting members are mechanically held in position in a driving body.

7—*Collapsible Taps*—Those which have inserted chasers or blades that are mechanically held within a body at the desired cutting diameter. When required thread depth is reached, the chasers or blades are retracted within the body by a tripping means so that the collapsed tap can be withdrawn from the threaded hole by axial movement.

III—CLASSIFICATION BASED ON SIZE

Taps usually are subclassified by a general term denoting the size of the tap or product major diameter. The terms are *Fractional Size*, *Machine Screw Size*, and *Metric Size*. Also combined with this is a designation for the pitch of the thread. (See Table 301, page 372.)

A—*Fractional Size* (or inch size) is the inch fractional equivalent of the product major diameter that mates with the internal thread to be tapped. The term *Hand Tap* has traditionally been applied to Fractional Size taps having the standard general purpose length. Most taps are now used in machine applications, so the term "Hand" has no significant meaning and is gradually being discontinued.

B—*Machine Screw Size* (or inch size) is a numeric designation originally established for machine screws smaller than ¼″ major diameter. The decimal equivalent of the size is specified in Table 329, page 398. Most of these taps are also used in machine applications, but like Fractional Inch Taps, some are still used by hand.

C—*Metric Size* is the diameter in millimeters of the major diameter of the product designed to conform to the metric screw thread systems. This classification includes the size range covered by machine screw and fractional sizes in the inch system.

D—The pitch designation for inch threads is the reciprocal of the thread pitch. The pitch designation for metric threads is the pitch value in millimeters. (See Table 301, Page 372.)

IV—CLASSIFICATION BASED ON FLUTING STYLE OR OTHER DESIGN FEATURES

A—*Standard Taps*—These taps are considered standard within industry and are the most common styles and types stocked by tap manufacturers.

    1. Straight Flute Taps (Commonly referred to as Hand Taps)
       These taps are used for general purpose applications.

Figure 1—Straight Fluted Tap

Figure 2—Straight Fluted Tap

    2. Spiral Point Taps
       These taps are intended for machine use. The supplementary angular flutes push the chips ahead of the tap. In their usual form they have a plug-type chamfer for through or open hole tapping. They are also made with shorter chamfer lengths for blind holes.

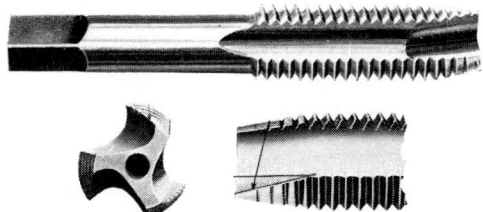

End view showing spiral point.
Figure 3—Spiral Pointed Tap

3. Spiral-Point-Only Taps

These taps are made with spiral point only. The balance of the threaded section is left unfluted. They are especially suitable for tapping thin materials.

Figure 4—Spiral-Point-Only Tap

4. Spiral Fluted Taps

These taps have helical flutes of the same hand of flute lead as the hand of cut; this configuration causes the chips to flow toward the shank of the tap. The Spiral Fluted type has a helix of $25-35°$ and is used for general purpose applications. The Fast Spiral type has a helix of $45-60°$ and is used in ductile materials.

Figure 5—Spiral Fluted Tap

Figure 6—Fast Spiral Fluted Tap

5. Thread Forming Taps

These taps are designed to develop threads in the work by forming rather than the cutting process. The tap is essentially fluteless although it may have one or more shallow grooves to facilitate lubrication or reduce hydraulic pressure in blind holes.

Figure 7—Thread Forming Tap

6. Tap Set—Taper, Plug, and Bottoming

A set of taps consists of one each of standard Taper, Plug, and Bottoming straight fluted taps of the same pitch, major diameter, and pitch diameter sizes. They are used individually or sequentially for bottoming applications.

Figure 8—Taper Tap

Figure 9—Plug Tap

Figure 10—Bottoming Tap

7. Serial Tap Set

A serial tap set is a series of special taps that are used in sequence to rough and finish a thread. The pitch diameter increases progressively with the #3 finishing tap being of the desired full size pitch diameter. They are identified by annular grooves in the shank near the square, and the number of grooves designates its serial sequence in use. Shown is a typical serial set, although various other combinations may be used.

Figure 11—Undersize Tap #1

Figure 12—Intermediate size Tap #2

Figure 13—Finishing size Tap #3

B—*Standardized Special Purpose Taps*—These taps are considered standard within the Industry, but are designed for a specific product or tapping machine.

1. Taper Pipe Taps
    Taper Pipe Taps are used for hand and machine tapping of taper pipe threads.

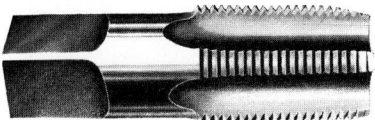

Figure 14—Taper Pipe Tap

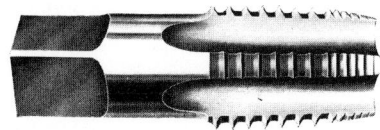

Figure 15—Interrupted Thread Taper Pipe Tap

2. Straight Pipe Taps

Straight Pipe Taps are used for hand and machine tapping of straight pipe threads.

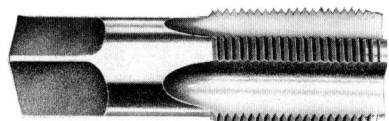

Figure 16—Straight Pipe Tap

3. Extension Taps

Extension Taps are designed with extended shanks for use where an added reach is required.

Figure 17—Extension Tap

4. Pulley Taps

Pulley Taps are similar to Extension Taps except that the shank is the same nominal diameter as the thread and the length is standardized.

Figure 18—Pulley Tap

5. Nut Taps

Nut Taps are characterized by long thread length, longer chamfers, and longer shanks. The shanks are smaller than the minor diameter of the thread.

Figure 19—Nut Tap

6. Tapper Taps

Tapper Taps are nonreversing taps designed for specific types of nut tapping machines. They may be of solid or sectional construction with straight or bent shanks smaller than the minor diameter of the parts tapped.

Figure 20—Straight Shank Tapper Tap with Plain Round Shank

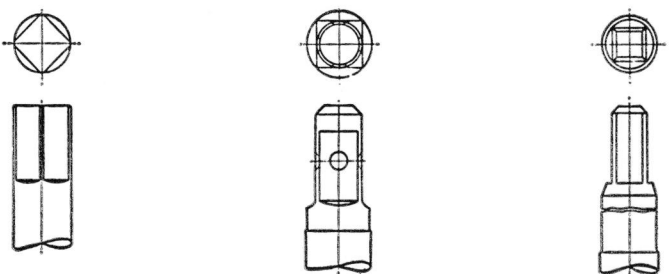

Figure 21—Alternate Shank Styles for Straight Shank Tapper Taps—
[Left]—Squared—[Center]—Acme Improved Type "C"—
[Right]—National Interchangeable Ring Lock

Figure 22—Solid Bent Shank Tapper Tap—90° Bend

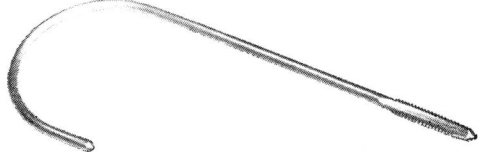

Figure 23—Solid Hook Tap—180° Bend

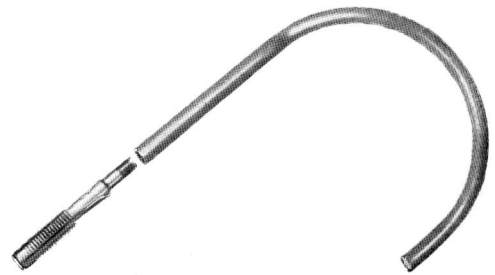

*Figure 24—Screw Type Sectional Hook Tap—180° Bend—
Showing Nib Tap and Holder*

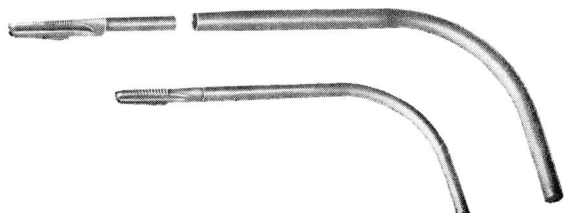

*Figure 25—Soldered Type Sectional Bent Shank Taps—90° Bend—
Showing Nib Tap and Holder*

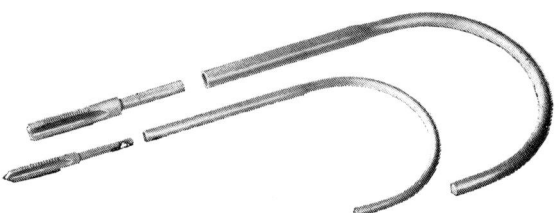

*Figure 26—Soldered Type Sectional Hook Shank Taps—
180° Bend—Showing Nib Tap and Holder*

C—*Special Taps*—These taps are examples of some of the many types of taps designed for special or unique applications. Some are stocked by the tap manufacturer, but most are usually custom designed to suit the individual customer's requirements. When designing special taps, it is usually more economical to use as many standard dimensions as possible, i.e., Overall Length, Thread Length, Shank Diameter, Size and Length of Square.

1. Shell Taps

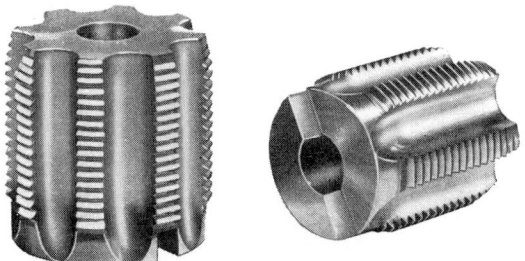

Figure 27—Shell Tap

Figure 28—Shell Tap

2. Multiple Diameter Taps

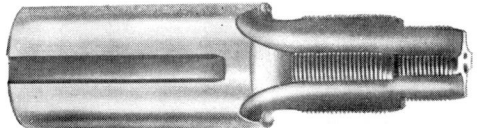

Figure 29—Multiple Diameter Tap

3. Tap with Plain Cylindrical Pilot

Figure 30—Tap with Plain Cylindrical Pilot.

4. Taper Pipe Taps—Pull Type

Pull-type Taper Pipe Taps are used where the shank is required to be on the small end of the taper thread section.

Figure 31—Taper Pipe Tap—Pull Type

5. Combined Taper Pipe Taps and Core Drills

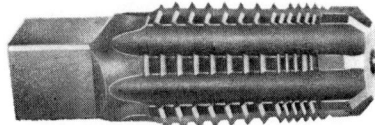

Figure 32—Combined Taper Pipe Tap and Core Drill—
Interrupted Thread

6. Combined Pipe Taps and Drills

Figure 33—Combined Pipe Tap and Drill

7. Acme Thread Taps—Set

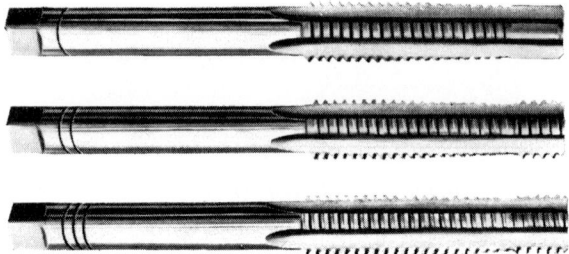

Figure 34—Set of Acme Thread Taps

8. Tandem Acme Taps

Figure 35—Tandem Acme Tap

9. Carbide-Tipped Taps

Figure 36—Carbide-Tipped Tap

10. Adjustable Inserted Blade Taps

Figure 37—Inserted Blade Tap

11. Inserted Blade Taper Pipe Taps

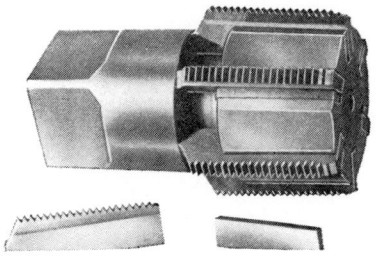

Figure 38—Inserted Blade Taper Pipe Tap

12. Collapsible Taps

Figure 39—Collapsible Tap

V—Nomenclature of Tap and Die Elements and Other Terms Relating to Screw Threads

Allowance—The prescribed difference between the design (maximum material) size and the basic size. It is numerically equal to the absolute value of the ISO term "Fundamental Deviation."

Arbor Hole—The central mounting hole in a shell tap.

Axis—The imaginary straight line that forms the longitudinal center line of the tool or threaded part.

Back Taper—See Relief, Back Taper.

Base of Thread—That which coincides with the cylindrical or conical surface from which the thread projects.

Basic Profile of Thread—The cyclical outline, in an axial plane, of the permanently established boundary between the provinces of the external and internal threads. All deviations are with respect to this boundary.

Bearing—The actual contact area of the thread form on the land of a tap, die, or chaser with the thread form on the product, exclusive of the chamfer cutting edge.

Bell-Mouth Thread—An internal thread that is larger in diameter at the start of the thread than at some distance beyond.

Blade—One of a set of flat, threaded chasers attached to, or inserted in, a tap or die body.

Body—1—The threaded full diameter portion of a solid tap, inclusive of the chamfer.—2—The principal supporting member for a set of chasers, usually including the shank.

Bore—See preferred term Arbor Hole.

Bottoming Tap—A tap having a chamfer length of 1–2 threads.

Chamfer—The tapering of the threads at the front end of each land of a chaser, tap, or die by cutting away and relieving the crest of the first few teeth to distribute the cutting action over several teeth.

Chamfer Angle—The angle formed between the chamfer and the axis of the tap or die, measured in an axial plane at the cutting edge.

Chamfer Bevel—An angular surface of revolution (which may or may not be relieved) preceding the point diameter on a tap.

Chamfer Depth—The depth of the cutting edge of the chamfer at the top of a chaser measured from the crest of the thread.

Chamfer Length—The chamfer length is measured at the cutting edge and is the axial length from the point diameter to the theoretical intersection of the tap major diameter and the chamfer angle.

Chamfer Relief Angle—The complement of the angle formed between a line tangent to the relieved surface at the cutting edge and a radial line to the same point on the cutting edge.

CHASER—One of a set of fixed or movable thread-cutting members supported by a tap or die body.

CHIP BREAKERS—Steps or notches formed in the thread crest or cutting face for the purpose of breaking the chips.

CHIP DRIVER POINT—See preferred term SPIRAL POINT.

CLASS OF THREAD—An alphanumeric designation to indicate the standard grade of tolerance and allowance specified for a thread. It is not applicable to the tools used for threading.

CLEARANCE—Any space provided to prevent undesirable contact of the tool and the workpiece.

CONCENTRIC—Having a common center.

CONCENTRICITY—See preferred terms TOTAL INDICATOR VARIATION (tiv) and RELATIVE ECCENTRICITY to describe a lack of concentricity within one or more tool elements.

CONCENTRIC MARGIN—A portion of the threaded land, adjacent to the cutting edge, that has concentric threads.

CONCENTRIC THREADS—Threads that are substantially circular for the full land width with a center coincident with the tool axis; that is, having no relief in the thread form except for that slight amount produced by back taper.

CON-ECCENTRIC THREAD—See RELIEF, CON-ECCENTRIC.

CONTROLLED ROOT DIE—A die having specified major diameter limits with or without a specified root shape.

CONTROLLED ROOT TAP—A tap having specified minor diameter limits with or without a specified root shape.

CORE—The central portion of a tap below the flutes that joins the lands.

CORE DIAMETER—The diameter of a circle that is tangent to the bottom of the flutes at a given point on the axis.

CORE TAPER—The taper in the core of a tap.

CREST—That surface of the thread that joins the flanks of the thread and is farthest from the cylinder or cone from which the thread projects.

CREST CLEARANCE—The radial distance between the root of the internal thread and the crest of the external thread of the coaxially assembled design forms of mating threads.

CUTTER SWEEP—The section removed by the milling cutter or grinding wheel in entering or leaving a flute.

CUTTING EDGE—The leading edge of the land in the direction of rotation for cutting that does the actual cutting.

CUTTING FACE—The leading side of the land in the direction of rotation for cutting on which the chip impinges.

DEPTH OF THREAD ENGAGEMENT—The radial distance, crest to crest, by which the thread forms overlap between two coaxially assembled mating threads.

DESIGN PROFILE—The maximum material profile permitted for an external or internal thread for a specified thread class or tolerance class.

DRYSEAL—A system including both external and internal threads, which is designed for use where the assembled product must withstand high fluid or gas pressures without the use of a sealing compound, or where a sealer is functionally objectionable.

ECCENTRIC—Not having a common center.

ECCENTRICITY—[With respect to tool axis.] One half of the TOTAL INDICATOR VARIATION (tiv). See also RELATIVE ECCENTRICITY.

ECCENTRIC THREAD—See RELIEF, ECCENTRIC.

END CUTTING TAP—A tap with an additional cutting edge below the chamfer.

EXPANDER—The plunger or wedge in a tap body that adjusts the lands of an expansion tap or the blades or chasers of an adjustable or collapsible tap.

EXTERNAL CENTER—The pointed end on a tap. Its included angle varies with manufacturing practice. It must not be confused with a tap chamfer or a chamfer bevel.

FACE—See CUTTING FACE.

FEMALE CENTER—See preferred term INTERNAL CENTER.

FILLET—On a thread profile the radius joining the thread flank with the thread root.

FIRST FULL THREAD—The first full thread on the cutting edge behind the chamfer. It is at this position that rake, hook, and thread elements are measured.

FISH TAIL POINT—A type of point on an end cutting tap having an internal angular relief below the chamfer.

FLANK—The part of a helical thread surface that connects the crest and the root and that is theoretically a straight line in an axial plane section.

FLANK, CLEARANCE—The flank that does not take the externally applied load in an assembly.

FLANK, FOLLOWING—The following (trailing) flank of a thread is one that is opposite to the leading flank.

FLANK, LEADING—1—The flank of a thread facing toward the chamfered end of a threading tool.—2—The flank that, when the thread is about to be assembled with a mating thread, faces the mating thread.

FLANK, LOAD—The load (pressure) flank is that which takes the externally applied axial load in an assembly. The term is used particularly in relation to buttress, square, acme, and stub acme threads.

FLANK ANGLE—The angle formed by a flank and a perpendicular to the thread axis in an axial plane.

Flatted Land—See Relief, Flatted Land.

Flute—The longitudinal channel formed in a tap to create cutting edges on the thread profile and to provide chip spaces and cutting fluid passages.

Flute, Angular—A flute lying in a plane intersecting the tool axis at an angle.

Flute, Helical—A flute with uniform axial lead and constant helix in a helical path around the axis of a cylindrical tap.

Flute, Spiral—A flute with uniform axial lead in a spiral path around the axis of a conical tap. Exception to this is its use as a proper name, which has been applied to two standard cylindrical tap series with helical flutes and which are called "Spiral Fluted Taps" and "Fast Spiral Fluted Taps."

Flute, Straight—A flute that forms a cutting edge lying in an axial plane.

Flute Angle—The angle that the face of an angular flute makes with the axial plane.

Flute Lead—The axial advance of a helical or spiral cutting edge in one turn about the axis.

Flute Lead Angle—The angle that a helical or spiral cutting edge at a given point makes with the axial plane.

Flute Length—As applied to taps, the full axial length of a flute including the cutter sweep.

Flute Taper—See preferred term Core Taper.

Form Diameter—The diameter at the point nearest the root or crest from which the flank is required to be straight.

Functional Diameter—The pitch diameter of an enveloping thread with perfect pitch, lead, and flank angles and having a specified length of engagement. This diameter includes the cumulative effect of variations in lead (pitch), flank angle, taper, straightness, and roundness. Variations at the thread crest and root are excluded. [Virtual Diameter, Effective Size, Virtual Effective Diameter, and Thread Assembly Diameter are defined the same as the preferred term Functional Diameter].

Grinding Recess—Clearance for the corner or edge of a grinding wheel at a change in tool diameter.

Guide, Nut—A cylindrical section behind the threaded body of a tap of approximately the nut minor diameter to guide the nut as it leaves the thread.

Guide, Plain—A cylindrical section, with or without oil grooves, behind the threaded body of a tap, to maintain alignment.

Guide, Threaded—A threaded section behind the cutting threads on a tap, fitting an internal thread, that acts both as a guide and as a means to advance the tap.

Gun Point—See preferred term Spiral Point.

Half Angle—See preferred term Flank Angle.

Hand of [Cut, Flutes, Threads]—A term used to denote the direction of rotation of a feature of a tool and is usually identified as Right Hand or Left Hand.

Hand of Cut—Rotation for cutting viewed from chamfered end of a tap or die is clockwise for Left Hand Cut and counterclockwise for Right Hand Cut.

Hand of Flutes—Flutes, when viewed axially, twist in a counterclockwise direction for Left Hand Flutes and in a clockwise direction for Right Hand Flutes.

Hand of Threads—A thread, when viewed axially, winds in a clockwise and receding direction for Left Hand Threads and counterclockwise and receding direction for Right Hand Threads.

Heel—The edge of the land opposite the cutting edge.

Height of Thread—The height (or depth) of thread is the distance, measured radially, between the major and minor cylinders or cones, respectively.

Helix Angle—See preferred terms Flute Lead Angle and Thread Lead Angle.

Helix Variation—The axial variation of the screw thread's actual helical path on the pitch cylinder relative to its true helix within the gaging length.

Hole, Tapped, Blind—A hole that does not pass through the workpiece and is not threaded its full depth.

Hole, Tapped, Bottoming—A blind hole that is threaded close to the bottom.

Hole, Tapped, Obstructed—A through hole that has some obstruction beyond the hole limiting the travel of the tap.

Hole, Tapped, Open—A hole that passes through the workpiece but is not threaded its full depth.

Hole, Tapped, Recessed—A blind hole with a recess larger than the tap major diameter and beyond the depth of full thread, limiting the travel of the tap.

Hole, Tapped, Stepped—A blind or open hole with a change in diameter that limits the thread depth.

Hole, Tapped, Through—A hole that passes through the workpiece and is threaded its full depth.

Hook, Chordal—A concave face having an angle of inclination specified between a chord passing through the root and crest of a thread form at the cutting face and a radial line through the crest at the cutting edge.

Hook, Tangential—A concave face having an angle of inclination specified between a line tangent to the hook surface at the cutting edge and a radial line to the same point.

HOOK ANGLE—The angle of inclination of a concave face, usually specified either as CHORDAL HOOK or TANGENTIAL HOOK.

HOOK FACE—A concave cutting face.

INCLUDED ANGLE—See preferred term THREAD ANGLE.

INTERNAL CENTER—A 60° countersink center hole with clearance at the bottom, in one or both ends of a tool, which establishes the tool axis.

INTERRUPTED THREAD TAP—A tap having an odd number of lands with alternate teeth in the thread helix removed. In some cases alternate teeth are removed only for a portion of the thread length.

LAND—1—One of the threaded sections between the flutes of a tap or die.
—2—The threaded surface of a tap chaser or die chaser.

LAND WIDTH—The chordal width of the land between the cutting edge and the heel measured normal to the cutting edge.

LEAD—The distance a helix or spiral advances axially in one complete turn.

LEAD DEVIATION—The deviation from the basic nominal lead.

LEAD, DRUNKEN—See HELIX VARIATION.

LEAD OF FLUTE—See FLUTE LEAD.

LEAD OF THREAD—The distance a screw thread advances axially in one complete turn. On a single start tap the lead and pitch are identical. On a multiple start tap the lead is the multiple of the pitch.

LENGTH OF ASSEMBLY—The axial distance over which two mating threads are designed to engage. The length includes any incomplete threads of both the external and internal member within the assembled length.

LENGTH OF THREAD—The tap length of thread includes the chamfered threads and the full threads but does not include an external center. A screw thread product length of complete thread includes a maximum of two incomplete threads at the start plus those having full form at both crest and root.

LENGTH OF THREAD ENGAGEMENT—The axial distance over which two mating threads, each having full form at both crest and root, are designed to engage.

LIMITS OF SIZE—The applicable maximum and minimum sizes.

MAJOR DIAMETER—The diameter of the major cylinder or cone, at a given position on the axis, that bounds the crests of an external thread or the roots of an internal thread.

MALE CENTER—See preferred term EXTERNAL CENTER.

MARGIN—See CONCENTRIC MARGIN.

MINOR DIAMETER—The diameter of the minor cylinder or cone, at a given position on the axis, that bounds the roots of an external thread or the crests of an internal thread.

MULTI-START THREAD—A screw thread with two or more thread starts. For this condition, pitch is equal to the thread lead divided by the number of thread starts.

Neck—A section of reduced diameter between two adjacent portions of a tool.

Non-Reversing Tap—A tap that passes completely through the part being tapped without reversal of rotation of the tap or part.

Number of Threads—See preferred term Threads per Inch.

Offset—The distance a straight cutting face is off center or the distance the measuring chord is off center on a hook cutting face.

Oil Grooves—Longitudinal straight or helical grooves in shank, guide, or pilot for lubrication or for carrying cutting fluid to the cutting edges.

Oil Holes—Holes by which a cutting fluid is fed to the cutting edges of a tool.

Overall Length—The extreme length of a complete tool from end to end, but not including adjusting screw or full external centers when required.

Percent of Thread—The ratio in percent of the actual height of thread to the value $0.75H$. This is a theoretical value based on the old American National thread profile. A 100% value cannot actually be achieved since the basic height of thread for UN and M profiles is $0.625H$ and for UNJ and MJ is $0.5625H$.

Pilot, Plain—A cylindrical portion preceding the chamfered end of the tap body to maintain alignment.

Pilot, Threaded—A threaded portion preceding the chamfered end of a tap, which facilitates starting the tap in correct relationship to a previously formed internal thread.

Pitch—The pitch of a thread having uniform spacing is the distance, measured parallel to its axis, between corresponding points on adjacent thread forms in the same axial plane and on the same side of the axis. For Unified threads the pitch in inches equals one divided by the number of threads per inch. For metric threads the pitch is equal to the specified pitch in millimeters.

Pitch Diameter—The diameter of an imaginary cylinder or cone, at a given point on the axis, of such a diameter and location of its axis that its surface would pass through the thread in such a manner as to make the thread ridge and the thread groove equal and, therefore, is located equidistant between the sharp major and minor cylinders or cones of a given thread form. On a theoretically perfect thread, these widths are equal to one half of the basic pitch (measured parallel to the axis). *Note*: When the crest of a thread is truncated beyond the pitch line, the pitch diameter, pitch cylinder, or pitch cone would be based on a theoretical extension of the thread flanks. See also terms Functional Diameter, Thread Groove

Diameter, and Thread Ridge Diameter, which are related to Pitch Diameter and are used in gaging practice.

Pitch Line—A generator of the imaginary cylinder or cone specified in the definition of Pitch Diameter.

Plug Tap—A tap with 3–5 chamfered threads.

Point Diameter—The diameter at the cutting edge of the leading end of the chamfered section.

Projection—The distance the small end of a taper thread projects through a taper thread ring gage.

Pull Tap—A tap that has its shank ahead of the chamfered teeth so that the shank passes through the hole to be tapped before cutting begins.

Qualification—The axial relation of one thread helix to another, as between a threaded pilot and the cutting threads on a tap or the relation of the thread helix on a tap to some other reference point on the tap.

Rake—The angular relationship of the straight cutting face of a tooth with respect to a radial line through the crest of the tooth at the cutting edge. Positive Rake means that the crest of the cutting face is angularly ahead of the balance of the cutting face of the tooth. Negative Rake means that the crest of the cutting face is angularly behind the balance of the cutting face of the tooth. Zero Rake means that the cutting face is directly on a radial line.

Relative Eccentricity—The distance between the geometric centerline of one portion of a tool and the geometric centerline of some other portion.

Relief—The removal of metal behind the cutting edge to provide clearance between the part being threaded and the threaded land.

Relief, Back Taper—This type of relief, usually identified only as Back Taper, is a gradual decrease in the diameter of the thread form on a tap (or a gradual increase on a die) from the chamfered end of the land toward the back, which creates a slight relief on the threads.

Relief, Center—Clearance produced on a portion of the tap land by reducing the diameter of the entire form between the cutting edge and heel.

Relief, Chamfer—The gradual decrease in land height from cutting edge to heel on the chamfered portion of the land on a tap or die to provide radial clearance for the cutting edge.

Relief, Con-Eccentric—Eccentric relief in the thread form starting back of a concentric margin.

Relief, Double Eccentric—The combination of a slight eccentric relief in the thread form starting at the cutting edge and continuing for a portion of the land width, and a greater eccentric relief for the balance of the land.

Relief, Eccentric—A relieved surface that is convex and with a uniform rate of relief. When applied to the thread form, it extends from cutting edge to heel.

Relief, Flat—A relieved surface that is essentially flat or slightly concave.

Relief, Flatted Land—Clearance produced on a portion of a tap land by truncating the thread between cutting edge and heel.

Relief, Grooved Land—Clearance produced on a tap land by forming a longitudinal groove in the center of a land.

Relief, Radial—Relief, of an undesignated type, with a decrease in diameter in a radial direction measured in the plane of rotation.

Root—That surface of the thread that joins the flanks of adjacent thread forms and is immediately adjacent to the cylinder or cone from which the thread projects.

Runout—The radial variation from a true circle that lies in a diametral plane and is concentric with the tool axis. See also Total Indicator Variation (tiv).

Screw Thread—A continuous and projecting helical ridge usually of uniform section on a cylindrical or conical surface.

Set—A set of taps consists of one each of standard taper, plug, and bottoming straight fluted taps of the same pitch and major diameter.

Set, Serial—Two or more related taps that, used in a specified sequence, progressively cut a thread of full width and height. Taps in a set frequently have a thread form modified from, or entirely different from, the basic thread form. They are identified by annular grooves on the shank near the square.

Shank—The portion of the tool by which it is held and driven.

Shank, Square—A cylindrical shank with a driving square only.

Shank, Plain Round—A cylindrical shank without a square or other driving means.

Shank, Flatted Round—A cylindrical shank with a set screw flat only.

Shaving—The excessive removal of material from the product thread profile by the tool thread flanks caused by an axial advance per revolution less than or more than the actual lead on the tool. In tapping, this results in an increase in product pitch diameter without an increase in product major diameter. In cutting an external thread with a die, shaving reduces the product pitch diameter without reducing the product minor diameter.

Size—Size is a designation of magnitude. When a value is assigned to a dimension, it is referred to as the size of that dimension. (*Note*: It is recognized that the words "dimension" and "size" are both used at times to convey the meaning of magnitude.)

Size, Actual—The measured size of a characteristic or attribute.

Size, Basic—That size from which the limits of size are derived by the application of allowances and tolerances.

Size, Functional—The size of the Functional Diameter.

Size, Nominal—The designation used for general identification.

SPINDOWN—The reduction in the minor diameter of an internal thread due to extrusion of metal below the original hole diameter.

SPIRAL FLUTE—See FLUTE, SPIRAL.

SPIRAL POINT—The angular fluting in the cutting face of the land at the chamfered end. It is formed at an angle with respect to the tap axis of opposite hand to that of rotation. Its length is usually greater than the chamfer length and its angle with respect to the tap axis is usually made great enough to direct the chips ahead of the tap. The tap may or may not have longitudinal flutes.

SQUARE—Four driving flats parallel to the axis on a tap shank forming a square or square with round corners.

TAPER PER INCH—The difference in diameter in one inch measured parallel to the axis.

TAPER SHANK—A shank made to fit a specified taper socket.

TAPER START—A tapering of the threads, with respect to the axis, which progressively reduces the diameter of the thread form for a short distance toward the entering end of the tap.

TAPER TAP—A tap having a chamfer length of 7–10 threads.

TAPER THREAD TAP—A tap for producing a tapered internal thread.

THREAD ANGLE—The angle formed by two adjacent flanks in an axial plane.

THREAD GROOVE DIAMETER—The diameter of an imaginary cylinder or cone, the surface of which would pass through the thread profiles at such points as to make the width of the thread groove (measured parallel to the axis) equal to one half of the basic pitch. It is the diameter yielded by measuring over or under cylinders (wires) or spheres (balls) inserted in the thread groove on opposite sides of the axis and computing the thread groove diameter as thus defined. [SIMPLE EFFECTIVE DIAMETER and SIMPLE PITCH DIAMETER are defined the same as the preferred term THREAD GROOVE DIAMETER.]

THREAD LEAD ANGLE—On a straight thread it is the angle made by the helix of the thread at the pitch diameter with the axial plane. On a taper thread it is the angle made by the conical spiral at the pitch line with the axial plane at a given axial position.

THREADS PER INCH [TPI]—The reciprocal of the pitch in inches.

THREAD RIDGE DIAMETER—The diameter of an imaginary cylinder or cone, the surface of which would pass through the thread profiles at such points as to make the width of the thread ridge (measured parallel to the axis) equal to one half of the basic pitch.

TOLERANCE—The total amount of variation permitted for the size of a dimension. It is the difference between the maximum limit of size and the minimum limit of size. Also see TOLERANCE LIMIT.

TOLERANCE CLASS (METRIC)—The combination of a tolerance position with a tolerance grade. It specifies the allowance (fundamental deviation), pitch diameter tolerance (flanks diametral displacement), and the crest diameter tolerance.

TOLERANCE GRADE (METRIC)—A numerical symbol that designates the tolerance of crest diameters and pitch diameters applied to the design profile.

TOLERANCE POSITION (METRIC)—A letter symbol that designates the position of the tolerance zone in relation to the basic size. This position provides the allowance (fundamental deviation).

TOLERANCE ZONE—The zone between the maximum and minimum limits of size.

TOPPING (DIE OR TAP)—See CONTROLLED ROOT (DIE OR TAP).

TOTAL INDICATOR VARIATION [tiv]—The difference between maximum and minimum indicator readings obtained during a checking cycle.

TRUNCATION, CREST—The crest truncation of a thread is the radial distance between the sharp crest and the cylinder or cone that bounds the crest.

TRUNCATION, ROOT—The root truncation of a thread is the radial distance between the sharp root and the cylinder or cone that bounds the root.

UNDERCUT—See preferred term GRINDING RECESS.

WEB—See preferred term CORE.

B—THREAD CUTTING DIES

I—DEFINITION

Threading Die—A multiedged, internally threaded cutting tool used for cutting external screw threads.

II—GENERAL CLASSIFICATIONS

A—Classification based on construction. Threading dies are often described by using terms that refer to their construction characteristics.

1—*Round Adjustable Dies*—Round dies are made with either of two types of adjustment, open adjusting or screw adjusting.

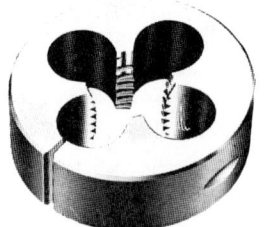 

Figure 40—Round Adjustable Die

2—*Solid Square Dies*—Square dies without adjustment features.

Figure 41—Solid Square Die

3—*Hexagon Rethreading Dies*—These are of hexagonal shape without adjustment features.

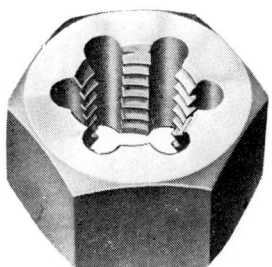

Figure 42—Hexagon Rethreading Die

4—*Two Piece Adjustable Dies*—These are made in halves for adjustability and in several styles.

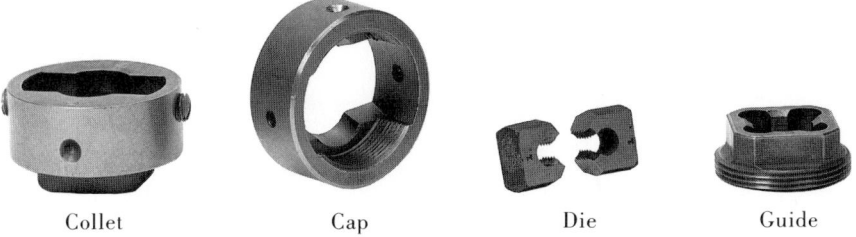

Collet   Cap   Die   Guide

Figure 43—Collet Assembly for Two Piece Adjustable Die

5—*Spring Dies, Cap Adjusting*—These are adjustable thread-cutting dies of spring tempered construction, the lands of which are tapered externally at the chamfered end to facilitate adjustment by means of an internally tapered cap.

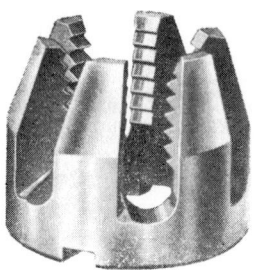

Figure 44—Spring Die

B—Classification based on purpose or use. Threading dies are sometimes described by terms that refer to their use or the purpose for which they are made.

1—*Bolt Dies*—Any of the solid square, round, or two-piece dies primarily for threading bolts or screws, either by hand or by machine.

2—*Pipe Dies*—Made expressly for producing pipe threads and may be of any type of construction for either hand or machine use.

3—*Rethreading Dies*—Used in repair work for restoring damaged threads. They are generally of hexagonal shape, for hand use with open end, socket, or box type wrenches.

4—*Machine Dies*—Intended for use in threading machines, hand or automatic screw machines, these are generally of the adjustable type. Some types of round, square, or two-piece dies are used in pipe-threading machines.

C—Classification based on method of holding or mounting.

1—*Hand Dies*—Any of the round, two-piece, square, or hexagonal types, which are primarily intended for hand use.

2—*Machine Dies*—These include all types of spring, round, square, or two-piece dies, expressly intended for machine use.

III—Illustrations of Terms Applying to Thread Cutting Dies

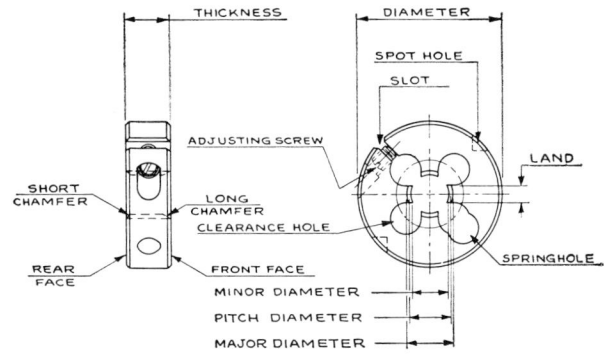

Figure 45—Round Screw Adjusting Die

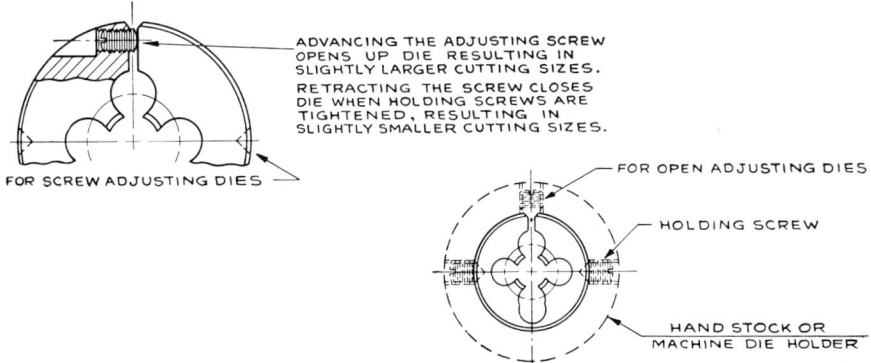

Figure 46—Adjustment of Round Dies

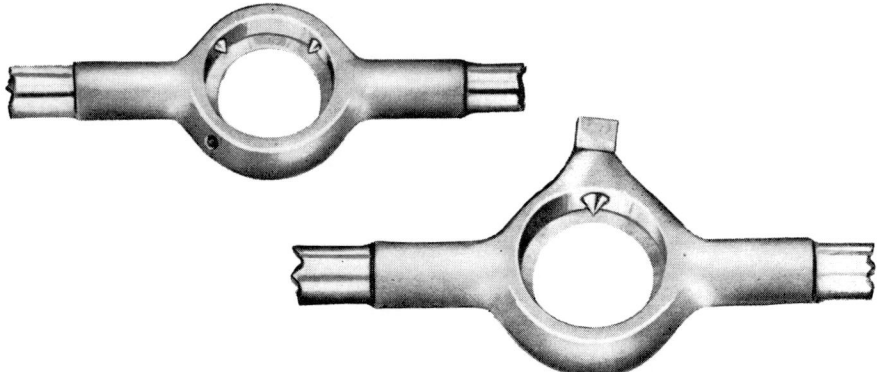

Figure 47—Hand Stocks for Round Dies

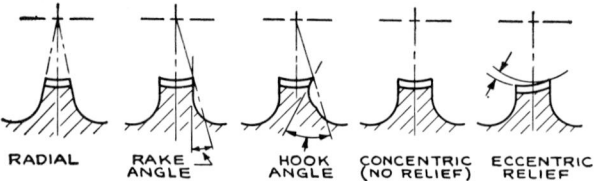

Figure 48—Rake and Hook Angles and Thread Relief

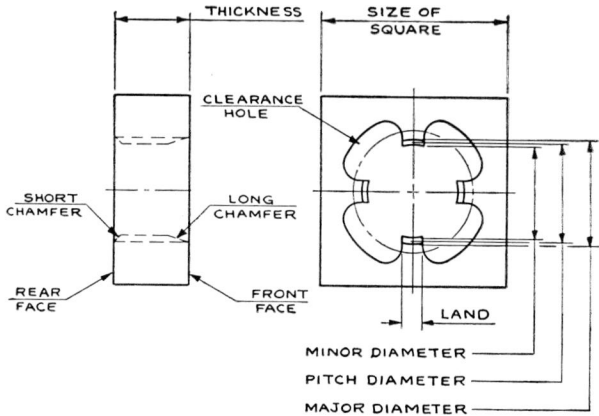

NON-ADJUSTABLE TYPE TO FIT SPECIFIC TYPES OF
THREADING MACHINES, AS WELL AS FOR HAND USE.

Figure 49—Solid Square Die

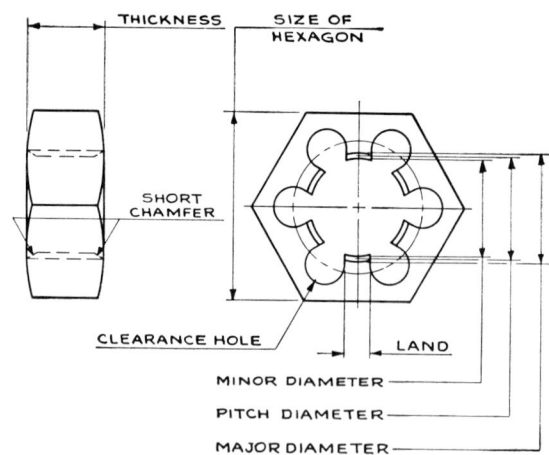

NON-ADJUSTABLE TYPE, FOR HAND USE WITH OPEN END,
SOCKET, OR BOX TYPE WRENCHES.

Figure 50—Hexagon Rethreading Die

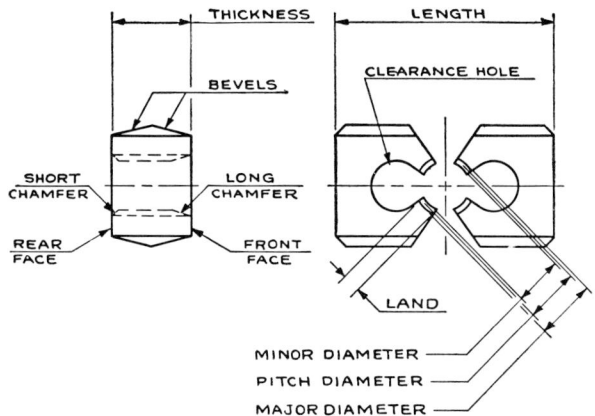

Figure 51—Two-Piece Adjustable Die

CAP ADJUSTING TYPE SHOWN

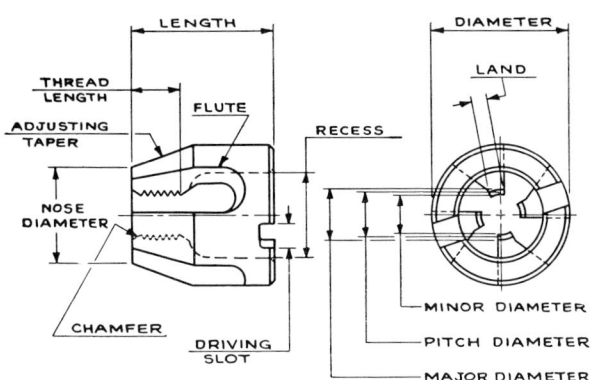

Figure 52—Spring Die

Figure 53—Spring Die Holder

IV—NOMENCLATURE AND DEFINITIONS

ADJUSTMENT—The closing or opening of the die lands to obtain the desired screw thread size.

CHAMFER—The entrance cone ground on the leading edge [front or rear face] of the die lands to distribute the cutting action over several teeth.

CLEARANCE HOLES—Holes through the die between the lands to provide chip space.

SPOT HOLES—Indentations on the periphery of a die for adjusting, holding, and driving.

SPRING HOLE—The clearance hole [or holes] opposite the adjusting slot in round adjustable dies. Die body is spring tempered at these points to provide flexibility for adjustment.

For terms not defined above see General Nomenclature.

C—TAP AND DIE SETS (SCREW PLATES)

Tap and die sets are assortments of taps and dies, which are usually packaged in containers with recesses for each individual tool. In addition to the taps and dies, they contain appropriate tap wrenches and die stocks for operating the complete assortment. Tap and die sets are furnished in numerous combinations of cutting sizes to accommodate the user's individual requirements, and are available as listed in manufacturers' catalogs.

D-1—DIE HEAD THREAD CHASERS—BLADE TYPE

I—DEFINITION

Blade Type Die Head Chasers—Made in sets, these are the thread-cutting blades that fit directly into the chaser slots of the Die Head body without the use of chaser holders. [Figures 54 and 55.]

II—GENERAL CLASSIFICATIONS

A—Classification based on construction.

1—*Milled or Ground Chasers*—Chasers that have the threaded land formed flat with a positive relief angle.

2—*Tapped Chasers*—Chasers that have the threaded land formed concave as developed by a tap or diehob.

3—*Regular Chasers*—Chasers that are practically flush with the face of the Die Head.

4—*Projection Chasers*—Chasers that extend beyond the face of the Die Head; intended for close to shoulder threading.

B—Classification based on hand of rotation or cut and hand of thread lead angle.

1—Terms *"Right Hand"* and *"Left Hand"* are used to describe both rotation or cut and hand of thread lead angle.

[a] The hand of rotation or cut must be in the same direction as the hand of the thread lead angle.

III—NOMENCLATURE AND DEFINITIONS OF MISCELLANEOUS TERMS

NOTE—For illustrations of these terms see Figures 56, 57, and 58.

BACK SIDE—The side opposite the cutting face (front side).

BOTTOM—The surface opposite the top.

FRONT SIDE—The side on which the cutting face is formed.

HEIGHT—Distance from the top to the bottom of the chaser.

LENGTH—Greatest dimension from the crest of the thread to the outer end of the chaser measured parallel to the top.

OUTER END—The surface of the chaser opposite the threaded end.

RELIEF ANGLE [MILLED OR GROUND CHASERS]—The straight relief angle at which the land is formed in relation to a plane perpendicular to the sides of the chaser.

TAPER CHASERS—Chasers that produce a tapered thread.

THICKNESS—Distance between the front side and the back side of the chaser.

TOP—The surface joining the front side and back side at the chamfered end of the chaser.

TOP RELIEF ANGLE—Relief angle ground on the top or projection of a chaser forming an additional cutting edge and adjoining the chamfer.

(See following figures.)

IV—ILLUSTRATIONS OF TYPES AND TERMS

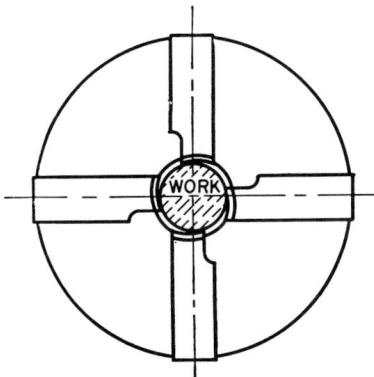

Figure 54—Tapped Chasers

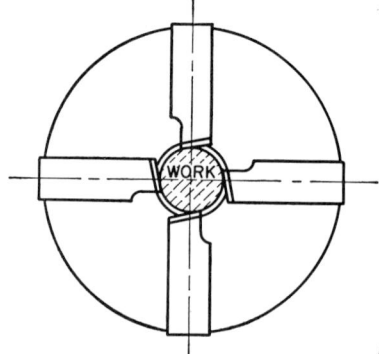

Figure 55—Milled or Ground Chasers

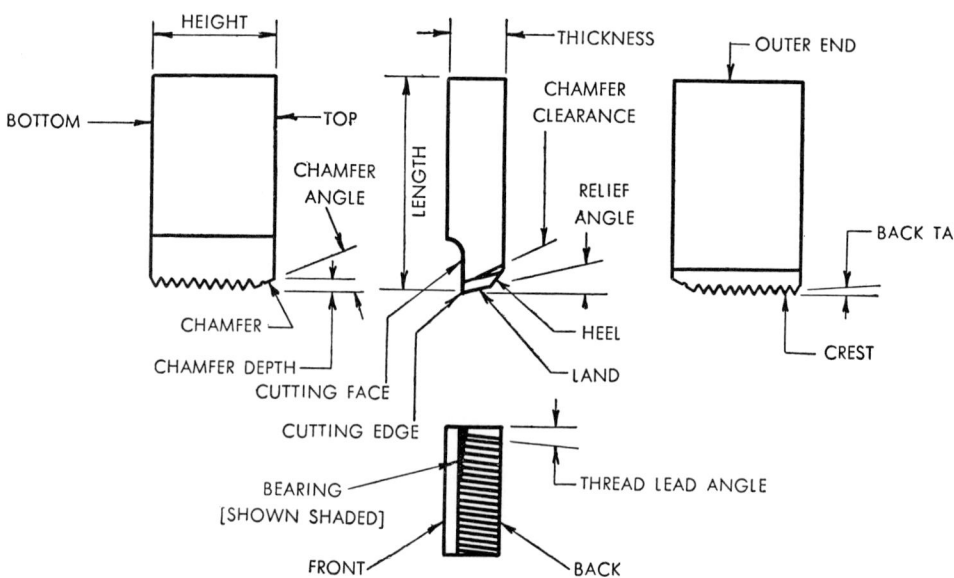

Figure 56—Die Head Chasers, Blade Type, Milled or Ground (RH Shown)

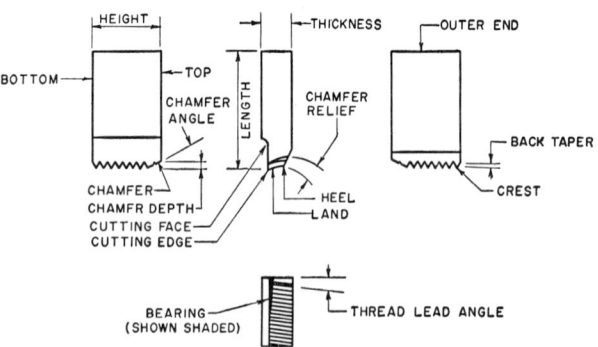

Figure 57—Die Head Chasers, Blade Type, Tapped (RH Shown)

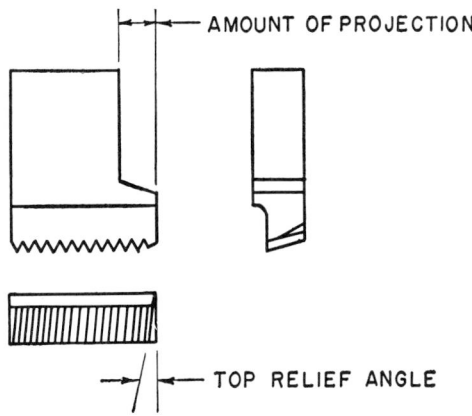

Figure 58—Die Head Chasers, Blade Type, Milled, Ground, or
Tapped, With Projection Milled Shown

D-2—DIE HEAD THREAD CHASERS—TANGENT TYPE

I—DEFINITION

Tangent Type Die Head Chasers—Made in sets, these are thread-cutting blades threaded lengthwise on one side and mounted in holders so that the threaded side at its inner end contacts the work tangentially forming the cutting edge. The holders in turn fit into the slots of the Die Head. [Figure 59.]

II—GENERAL CLASSIFICATIONS

A—Classification based on construction.

1—*Helix Chasers*—Chasers threaded to the thread lead angle of the thread to be cut.

2—*Nonhelix Chasers*—Chasers threaded parallel to the edge of the chaser; the thread lead angle being incorporated in the chaser holders.

B—Classification based on hand of rotation or cut and hand of thread lead angle.

1—Terms *"Right Hand"* and *"Left Hand"* are used to describe both rotation or cut and hand of thread lead angle.

[a] The hand of rotation or cut must be in the same direction as the hand of the thread lead angle.

III—NOMENCLATURE AND DEFINITIONS OF MISCELLANEOUS TERMS

NOTE—For illustrations of these terms see Figure 60.

BOTTOM—The surface opposite the top.

HEIGHT—Distance from the top to the bottom of the chaser.

Length—Overall dimension measured from the top.

Outer End—The surface of the chaser opposite the end that does the cutting.

Taper Chasers—Chasers that produce a tapered thread.

Thickness—Distance between the threaded side and opposite side of the chaser.

Top—The plain surface adjoining that on which the chamfer is ground.

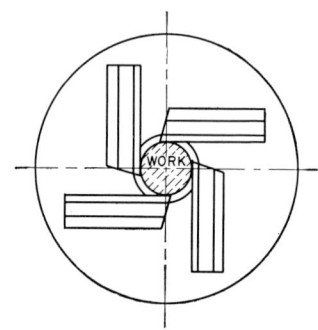

Figure 59—Tangent Chasers

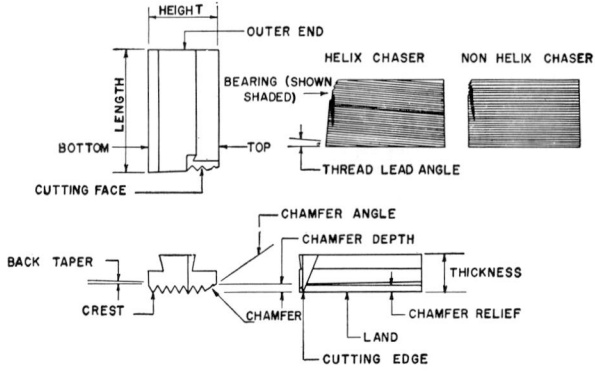

Figure 60—Die Head Chasers, Tangent Type (RH Shown)

D-3—DIE HEAD CHASERS—CIRCULAR TYPE

I—Definition

Circular Type Die Head Chasers—Made in sets these are made with annular thread forms and a single flute, which provides the cutting edge. They are mounted on holders which in turn fit into the slots of the Die Head. [Figure 61.]

II—General Classifications

A—Classification based on construction.

1—*Nonhelix Chasers*—Chasers that are threaded parallel to the top of the chaser; the thread lead angle being incorporated in the chaser holders.

B—Classification based on hand of thread lead angle.

1—Terms *"Right Hand"* and *"Left Hand"* are used to describe both rotation or cut and hand of thread lead angle.

[a] The hand of rotation or cut must be in the same direction as the hand of the thread lead angle.

III—Nomenclature and Definitions of Miscellaneous Terms

Note—For illustrations of these terms see Figure 62.

Bearing—The surface along the cutting edge indicating where the chaser contacts the work.

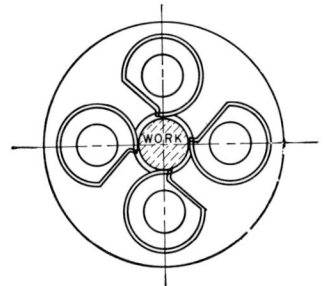

Figure 61—Circular Chasers

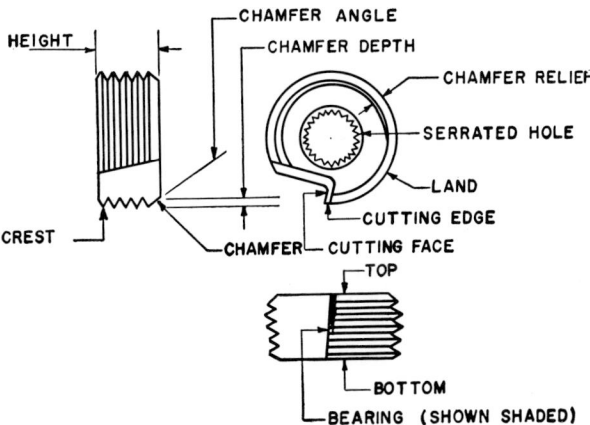

Figure 62—Die Head Chasers: Circular Type (RH shown)

Bottom—The surface opposite the top.

Height—Distance from the top to the bottom of the chaser.

Top—The plain surface adjoining that on which the chamfer is ground.

D-4—TAP CHASERS—BLADE TYPE

I—Definition

Blade Type Tap Chasers—Made in sets, these are the thread-cutting blades that fit directly into the chaser slots of the tap body without the use of chaser holders. [Figure 63.]

II—General Classifications

A—Classification based on construction.

1—*Regular Chasers*—Chasers that do not extend beyond the front end of the tap body.

2—*Projection Chasers*—Chasers that extend beyond the front end of the tap body; intended for shoulder or bottoming tapping.

B—Classification based on hand of rotation or cut and hand of thread lead angle.

1—Terms *"Right Hand"* and *"Left Hand"* are used to describe both rotation or cut and hand of thread lead angle.

[a] The hand of rotation or cut must be in the same direction as the hand of the thread lead angle.

III—Nomenclature and Definitions of Miscellaneous Terms

Note—For illustrations of these terms see Figures 64 and 65.

Back Side—The side opposite the cutting face (front side).

Bottom—The surface opposite the top.

Front Side—The side on which the cutting face is formed.

Height—Distance from the top to the bottom of the chaser.

Inner End—The surface of the chaser opposite the threaded end.

Length—Greatest dimension from the crest of the thread to the inner end of the chaser measured parallel to the top.

Taper Chasers—Chasers that produce a tapered thread.

Thickness—Distance between the front side and back side of the chaser.

Top—The surface joining the front side and back side at the chamfered end of the chaser.

Top Relief Angle—Relief angle ground on the top or projection of a chaser forming an additional cutting edge adjoining the chamfer.

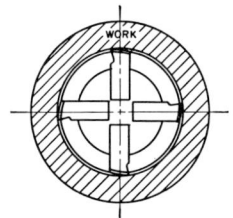

Figure 63—Tap Chasers

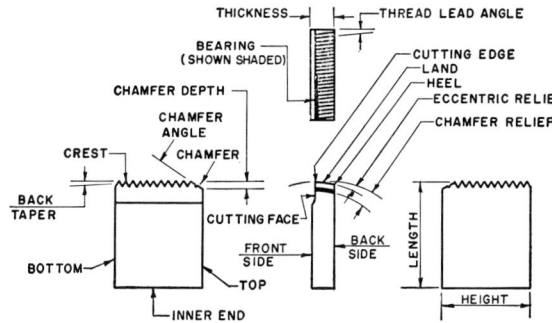

Figure 64—Regular Tap Chasers, Blade Type (RH Shown)

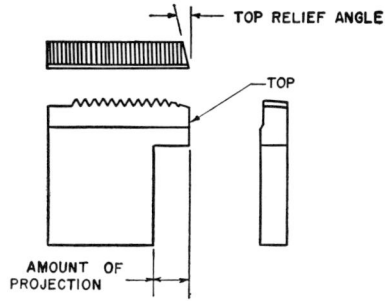

Figure 65—Projection Tap Chaser, Blade Type

D-5—DIE HEAD THREAD CHASERS—INSERT TYPE

I—Definition

Insert-type die chasers, made in sets, are thread cutting blades (inserts) threaded on one surface. The inserts can be mounted directly into a solid adjustable die head or into chaser carriers (holders) that fit an opening die head. The inserts are held in place by means of a screw threaded in the die body or into the carrier. Carriers fit into slots in the opening die head and are actuated by a cam action.

II—CONSTRUCTION CLASSIFICATION

Insert chasers have a cut or ground threadform with a positive relief angle. Chasers extend beyond the face of the die body or carrier to allow for close-to-shoulder threading.

Right-hand and left-hand describe rotation or cut and the hand of thread lead angle. Rotation must be in the same direction as hand of the thread lead angle.

III—NOMENCLATURE

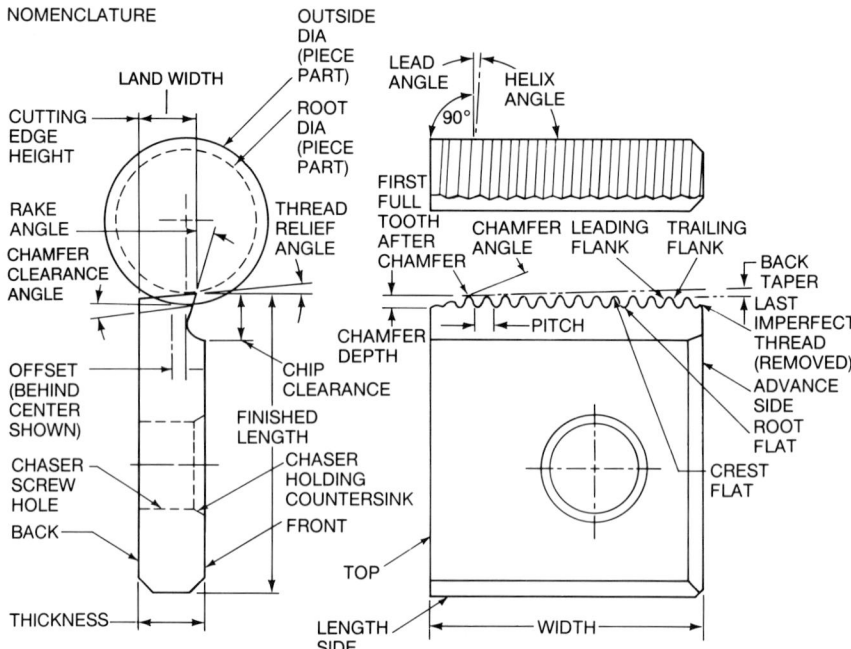

CARRIER (HOLDER)—Made in sets, the chaser carrier is a one-piece unit designed to hold the insert chasers securely, which are actuated by cams or cam surfaces. The insert chaser is held in place in the carrier by means of a screw through the insert into the carrier. The hole in the chaser is offset from that in the carrier, which serves to pull the chaser firmly into correct position in the carrier. The carrier, which will accommodate a range of standard and special insert chasers, fits in the slots of the die head.

The design of the carriers allows the chasers to be held in the correct thread cutting position. When worn, the insert chasers can be easily removed and resharpened or replaced at a relatively low cost compared to solid, one-piece-style chasers.

INDEX OF TABLES

TABLE 301

Standard System of Marking

1. Ground Thread Taps, Inch Screw Threads

    Ground thread taps specified in the inch system are marked with the nominal size, number of threads per inch, the proper symbol to identify the thread form, "HS" for high speed steel, "G" for ground thread, and designators for tap pitch diameter and special features, such as for Left Hand and Multi-start threads. Tap pitch diameter designators, system of limits, and special features are specified in Table 301A. Thread symbol designators are specified in Table 301B for both the Unified and British screw thread series.

2. Ground Thread Taps, Metric Screw Threads—M Profile

    Ground thread taps made with Metric Screw Threads—M Profile are marked with an upper case "M," followed by the nominal size in millimeters and the pitch in millimeters separated by the sign "×." Also included in the marking is "HS" for high speed steel, "G" for ground thread, and designators for tap pitch diameter, and special features, such as for Left Hand and Multi-start threads. Tap pitch diameter designators, system of limits, and special features are specified in Table 301A.

3. Cut Thread Taps and Dies, Inch Screw Threads

    Cut thread taps and dies specified in the inch system are marked with the nominal size, number of threads per inch, and the proper symbol to identify thread form. High speed steel taps are marked "HS," while carbon steel taps need not be marked with the steel designation. Thread symbol designators are specified in Table 301B for both the Unified and British screw thread sizes.

Note: In general, metal cutting tools must be marked with the country of origin. Reference should be made to Part 134, Customs Regulations (19 CFR Part 134) to determine the specific requirements as to the permanent marking of imported tools.

TABLE 301A

**Standard System of Tap Thread Limits
and Identification, Ground Thread**

1. UNIFIED INCH SCREW THREADS

A. H or L Limits

When the maximum tap pitch diameter limit is over basic pitch diameter by an even multiple of .0005 or the minimum tap pitch diameter limit is under basic pitch diameter by an even multiple of .0005, the taps are marked with the letters H or L, respectively, followed by a limit number. The limit numbers are determined as follows:

$$\text{H limit number} = \frac{\text{Amount max. tap PD limit is over basic PD}}{.0005}$$

$$\text{L limit number} = \frac{\text{Amount min. tap PD limit is under basic PD}}{.0005}$$

For tap PD tolerances, see Table 331, Column D. For standard taps, PD limits for various H limit numbers are shown in Tables 327 and 329.

When Unified Inch Screw Thread taps are ordered without a pitch diameter or an H or L limit number, the tap pitch diameter is normally determined from Table 331, and is marked with the appropriate H limit number.

Examples of tap marking with H or L limit numbers:

$3/8-16$ NC HS G H1
Max. tap PD = Basic PD + .0005
Min. tap PD = Max. tap PD − .0005

$1\frac{1}{4}-7$ NC HS G H4
Max. tap PD = Basic PD + .0020
Min. tap PD = Max. tap PD − .0010

$2-16$ NEF HS G H8
Max. tap PD = Basic PD + .0040
Min. tap PD = Max. tap PD − .0015

$3/8-16$ NC HS G L2
Min. tap PD = Basic PD − .0010
Max. tap PD = Min. tap PD + .0005

TABLE 301A

Standard System of Tap Thread Limits
and Identification, Ground Thread

1. UNIFIED INCH SCREW THREADS (con't.)

B. OVERSIZE OR UNDERSIZE

When the maximum tap PD over basic PD or the minimum tap PD under basic PD is not an even multiple of .0005, the tap PD is normally designated as an amount oversize or undersize. The amount oversize is added to the basic PD to establish the <u>MINIMUM</u> tap PD. The amount undersize is subtracted from the basic PD to establish the <u>MINIMUM</u> tap PD. The PD tolerance from Table 331, column D, is added to the minimum tap PD to establish the maximum tap PD in both cases.

Example: $\boxed{\text{7/16}-14 \text{ NC plus } .0017 \text{ HS G}}$
Min. tap PD = Basic PD + .0017
Max. tap PD = Min. tap PD + .0005

Whenever possible in the case of oversize or other special tap PD requirements, the maximum and minimum tap PD requirements should be specified on the order.

C. SPECIAL TAP PITCH DIAMETER

Taps not made to H or L limit numbers, nor to Table 331, nor to the formula for oversize or undersize taps, may be marked with the letter S enclosed by a circle or some other special identifier.

Example: $\boxed{\text{3/8}-16 \text{ NC HS G } Ⓢ}$

D. LEFT HAND TAPS

Taps with left hand threads are marked "LEFT HAND" or "LH."

Example: $\boxed{\text{3/8}-16 \text{ NC LH HS G H3}}$

E. MULTI-START THREADS

Taps with multiple-start threads are marked with the lead designated as a fraction and also "Double," "Triple," etc. The Unified Screw Thread form symbol is always designated as "NS" for multiple start threads.

Example: $\boxed{\text{3/8}-16 \text{ NS Double 1/8 Lead HS G H5}}$

TABLE 301A

Standard System of Tap Thread Limits and Identification, Ground Thread

2. METRIC SCREW THREADS—M PROFILE

All calculations for metric taps are done using millimeter values. When inch values are needed, they are translated from the three-place millimeter tap diameters only after calculations are complete.

A. "D" OR "DU" LIMITS

When the maximum tap pitch diameter limit is over basic pitch diameter by an even multiple of .013 mm (.000512 inch reference), or the minimum tap pitch diameter limit is under basic pitch diameter by an even multiple of .013 mm, the taps are marked with the letter "D" or "DU," respectively, followed by a limit number. The limit number is determined as follows:

$$\text{D limit number} = \frac{\text{Amount max. tap PD limit is over basic PD in mm}}{.013 \text{ mm}}$$

$$\text{DU limit number} = \frac{\text{Amount min. tap PD limit is under basic PD in mm}}{.013 \text{ m}}$$

For D limit increments which are based on .013 mm, see Table 341, Column Y. For tap PD tolerances, see Table 341, Column Z. For standard taps, PD limits for various D limit numbers are shown on Table 337.

When metric screw thread—M profile ground thread taps are ordered without a pitch diameter, or a D or DU limit number, the tap pitch diameter will normally be determined from Table 341, and the tap will be marked with the appropriate D limit number.

Examples of tap marking with D or DU limit numbers:

| M1.6 × 0.35 HS G D3 |

Max. tap PD = Basic PD + .039 mm
Min. tap PD = Max. tap PD − .015 mm

TABLE 301A

Standard System of Tap Thread Limits and Identification, Ground Thread

2. METRIC SCREW THREADS—M PROFILE (CON'T.)

A. "D" OR "DU" LIMITS (CON'T.)

M12 × 1.75 HS G D6

Max. tap PD = Basic PD + .078 mm
Min. tap PD = Max. tap PD − .031 mm

M39 × 4 HS G D10

Max. tap PD = Basic PD + .130 mm
Min. tap PD = Max. tap PD − .052 mm

M6 × 1 HS G DU4

Min. tap PD = Basic PD − .052 mm
Max. tap PD = Min. tap PD + .025 mm

B. METRIC OVERSIZE OR UNDERSIZE TAPS, TAPS WITH SPECIAL PITCH DIAMETER, AND LEFT HAND taps follow the marking system given above for inch taps.

Examples:

M12 × 1.75 plus .044 HS G

M10 × 1.5 HS G ○ itr   S

M10 × 1.5 LH HS G D6

C. MULTI-START THREADS

Metric taps with multiple start threads are marked with the lead designated in millimeters preceded by the letter "L," the pitch in millimeters preceded by the letter "P," and the words (2 starts), (3 starts), etc.

Examples:

M16 × L4-P2 (2 starts) HS G D8

M14 × L6-P2 (3 starts) HS G D7

Table 301B
Thread Series Designations

Standard Tap Marking	Product Thread Designation	Thread Series	References American National Standards (ANSI/ASME)
ACME-C	ACME-C	Acme threads, centralizing	B1.5
ACME-G	ACME-G	Acme threads, general purpose (see also "STUB ACME")	B1.5
AMO	AMO	American Standard microscope objective threads	B1.11
NPT	ANPT	Aeronautical National Form taper pipe threads	(MIL-P-7105B)
BUTT	BUTT	Buttress Threads, pull type	B1.9
PUSH-BUTT	PUSH-BUTT	Buttress Threads, push type	B1.9
F-PTF	F-PTF	Dryseal fine taper pipe thread series	B1.20.3 (Appendix C)
M	M	Metric Screw Threads—M Profile, with basic ISO 68 profile.	B1.13M, B1.18M
M	MJ	Metric Screw Threads—MJ Profile, with rounded root of radius 0.15011P to 0.18042P	B1.21M
NC	NC5 IF	Class 5 interference fit–Internal ferrous material	B1.12
NC	NC5 INF	Class 5 interference fit–Internal nonferros material	B1.12
NGO (RH or LH)	NGO	National gas outlet threads	ANSI/CGA V-1 (B57.1)
NGS	NGS	National gas straight threads	ANSI/CGA V-1 (B57.1)
NGT	NGT	National gas taper threads	ANSI/CGA V-1 (B57.1)
NH	NH	American Standard hose coupling threads of full form.	B2.4
NPS	NPSC	American Standard straight pipe threads in pipe couplings	B1.20.1
NPSF	NPSF	Dryseal American Standard fuel internal straight pipe threads	B1.20.3

(Continued on following page)

Table 301B
Thread Series Designations (continued)

Standard Tap Marking	Product Thread Designation	Thread Series	References American National Standards (ANSI/ASME)
NPSH	NPSH	American Standard straight hose coupling threads for joining to American Standard taper pipe threads	B2.4
NPSI	NPSI	Dryseal American Standard intermediate internal straight pipe threads	B1.20.3
NPSL	NPSL	American Standard straight pipe threads for loose-fitting mechanical joints with locknuts	B1.20.1
NPS	NPSM	American Standard straight pipe threads for free-fitting mechanical joints for fixtures	B1.20.1
NPT	NPT	American Standard taper pipe threads for general use	B1.20.1
NPTF	NPTF	Dryseal American Standard taper pipe threads	B1.20.3
NPTR	NPTR	American Standard taper pipe threads for railing joints	B1.20.1
PTF SHORT	PTF-SAE SHORT	Dryseal SAE short taper pipe threads	B1.20.3
PTF SPL SHORT	PTF-SPL SHORT	Dryseal special short taper pipe threads	B1.20.3 (Appendix C)
PTF SPL X-SHORT	PTF-SPL EXTRA SHORT	Dryseal special extra short taper pipe threads (see also "SPL-PTF")	B1.20.3 (Appendix C)
M	S	ISO miniature screw threads 0.25 to 1.4 mm. inc.	
SGT	SGT	Special gas taper threads	ANSI/CGA V-1 (B57.1)
SPL-PTF	SPL-PTF	Dryseal special taper pipe threads	B1.20.3 (Appendix C)
STUB ACME	STUB ACME	Stub Acme threads	B1.8
STUB ACME M1	STUB ACME M1	Stub Acme Modified Form 1	B1.8
STUB ACME M2	STUB ACME M2	Stub Acme Modified Form 2	B1.8
N	UN	Unified Inch Screw Thread, constant-pitch series	B1.1
NC	UNC	Unified Inch Screw Thread, coarse-pitch series	B1.1

(Continued on following page)

UNF	NF	Unified Inch Screw Thread, fine-pitch series	B1.1
UNEF	NEF	Unified Inch Screw Thread, extra-fine-pitch series	B1.1
UNJ	N	Unified Inch Screw Thread, constant-pitch series, with rounded root of radius 0.15011P to 0.18042P (Ext. thd. only.)	B1.15 (Draft), MIL-S-8879A
UNJC	NC	Unified Inch Screw Thread, coarse pitch series, with rounded root of radius 0.15011P to 0.18042P (Ext. thd. only.)	B1.15 (Draft), MIL-S-8879A
UNJF	NF	Unified Inch Screw Thread, fine pitch series, with rounded root of radius 0.15011P to 0.18042P (Ext. thd. only.)	B1.15 (Draft), MIL-S-8879A
UNJEF	NEF	Unified Inch Screw Thread, extra-fine pitch series, with rounded root of radius 0.15011P to 0.18042P (Ext. thd. only)	B1.15 (Draft), MIL-S-8879A
UNR	N	Unified Inch Screw Thread, constant-pitch series, with rounded root of radius not less than 0.108P (Ext. thd. only)	B1.1
UNRC	NC	Unified Inch Screw Thread, coarse thread series, with rounded root of radius not less than 0.108P (Ext. thd. only)	B1.1
UNRF	NF	Unified Inch Screw Thread, fine pitch series, with rounded root of radius not less than 0.108P (Ext. thd. only)	B1.1
UNREF	NEF	Unified Inch Screw Thread, extra-fine pitch series, with rounded root of radius not less than 0.108P (Ext. thd. only)	B1.1
UNM	UNM	Unified miniature thread series, .030 to 1.40 mm	B1.10
UNS	NS	Unified Inch Screw Thread, special diameter pitch, or length of engagement	B1.1
BA	BA	British Association	BS93
BSC	BSC	British cycle	BS811
BSF	BSF	British Whitworth Fine	BS84

(Continued on following page)

Table 301B
Thread Series Designations (continued)

Standard Tap Marking	Product Thread Designation	Thread Series	References American National Standards (ANSI/ASME)
BSW	BSW	British Whitworth Coarse	BS84
OBSOLETE	BSPP	British Straight Pipe	Obsolete
OBSOLETE	BSPT	British Taper Pipe	Obsolete
WHIT	WHIT	British Whitworth Special	BS84
		British Pipe Threads	
Standard Tap Marking*			
G¼	G	British Internal Straight Pipe for mechanical joints	BS2779
Rc¼	Rc	British Internal Taper Pipe for pressure tight joints	BS21
Rp¼	Rp	British Internal Straight Pipe for pressure tight joints	BS21
Standard Die Marking*			
G¼A	G × A	British External Straight Pipe for mechanical joints	BS2779
G¼B	G × B	British External Straight Pipe for mechanical joints	BS2779
R¼	R	British External Taper Pipe for pressure tight joints	BS21

* Example to show length of size designation

Table 302
TAPS
General Dimensions

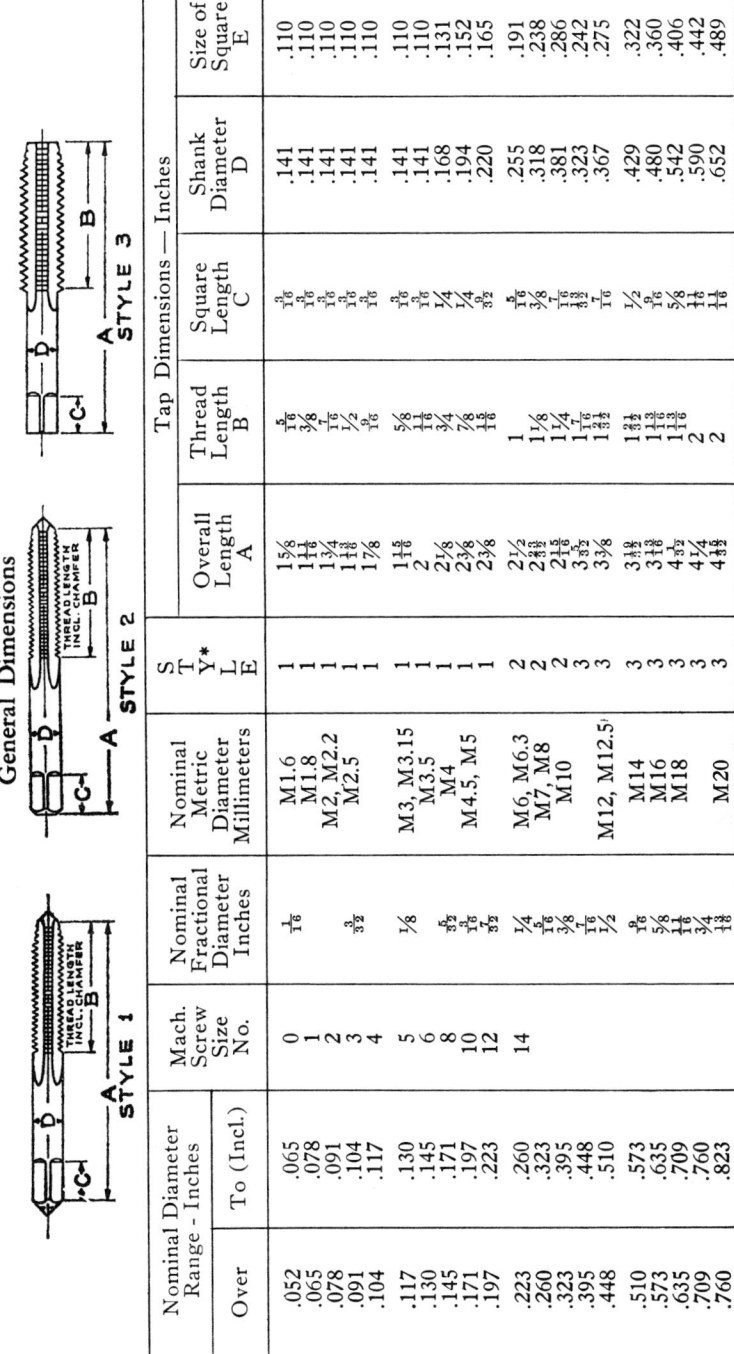

STYLE 1   STYLE 2   STYLE 3

Nominal Diameter Range - Inches Over	To (Incl.)	Mach. Screw Size No.	Nominal Fractional Diameter Inches	Nominal Metric Diameter Millimeters	STYLE	Overall Length A	Thread Length B	Square Length C	Shank Diameter D	Size of Square E
.052	.065	0	1/16	M1.6	1	1 5/8	5/16	3/16	.141	.110
.065	.078	1		M1.8	1	1 11/16	3/8	3/16	.141	.110
.078	.091	2		M2, M2.2	1	1 3/4	7/16	3/16	.141	.110
.091	.104	3	3/32	M2.5	1	1 13/16	1/2	3/16	.141	.110
.104	.117	4			1	1 7/8	9/16	3/16	.141	.110
.117	.130	5	1/8	M3, M3.15	1	1 15/16	5/8	3/16	.141	.110
.130	.145	6		M3.5	1	2	11/16	3/16	.141	.110
.145	.171	8	5/32	M4	1	2 1/8	3/4	1/4	.168	.131
.171	.197	10	3/16	M4.5, M5	1	2 3/8	7/8	1/4	.194	.152
.197	.223	12	7/32		1	2 3/8	15/16	9/32	.220	.165
.223	.260	14	1/4	M6, M6.3	2	2 1/2	1	5/16	.255	.191
.260	.323		5/16	M7, M8	2	2 23/32	1 1/8	3/8	.318	.238
.323	.395		3/8	M10	2	3 5/16	1 1/4	7/16	.381	.286
.395	.448		7/16		3	3 3/32	1 7/16	13/32	.323	.242
.448	.510		1/2	M12, M12.5	3	3 3/8	1 21/32	7/16	.367	.275
.510	.573		9/16	M14	3	3 1/2	1 21/32	1/2	.429	.322
.573	.635		5/8	M16	3	3 11/16	1 11/16	9/16	.480	.360
.635	.709		11/16	M18	3	4 3/8	1 11/16	5/8	.542	.406
.709	.760		3/4		3	4 1/4	2	11/16	.590	.442
.760	.823		13/16	M20	3	4 1/2	2	11/16	.652	.489

Tap Dimensions — Inches

Table 302 (Continued)
TAPS
General Dimensions

Nominal Diameter Range - Inches		Mach. Screw Size No.	Nominal Fractional Diameter Inches	Nominal Metric Diameter Millimeters	STYLE*	Tap Dimensions — Inches				
Over	To (Incl.)					Overall Length A	Thread Length B	Square Length C	Shank Diameter D	Size of Square E
.823	.885		7/8	M22	3	4 11/16	2 23/32	3/4	.697	.523
.885	.948		15/16	M24	3	4 23/32	2 23/32	3/4	.760	.570
.948	1.010		1	M25	3	5 1/8	2 1/2	13/16	.800	.600
1.010	1.073		1 1/16	M27	3	5 1/8	2 1/2	7/8	.896	.672
1.073	1.135		1 1/8		3	5 1/16	2 9/16	7/8	.896	.672
1.135	1.198		1 3/16	M30	3	5 1/16	2 9/16	1	1.021	.766
1.198	1.260		1 1/4		3	5 3/4	2 9/16	1	1.021	.766
1.260	1.323		1 5/16	M33	3	5 3/4	2 9/16	1 1/16	1.108	.831
1.323	1.385		1 3/8		3	6 1/16	3	1 1/16	1.108	.831
1.385	1.448		1 7/16	M36	3	6 1/16	3	1 1/8	1.233	.925
1.448	1.510		1 1/2		3	6 3/8	3	1 1/8	1.233	.925
1.510	1.635		1 5/8	M39	3	6 11/16	3 3/16	1 1/8	1.305	.979
1.635	1.760		1 3/4	M42	3	7	3 3/16	1 1/4	1.430	1.072
1.760	1.885		1 7/8		3	7 15/16	3 9/16	1 1/4	1.519	1.139
1.885	2.010		2	M48	3	7 5/8	3 9/16	1 3/8	1.644	1.233
2.010	2.135		2 1/8		3	8	3 9/16	1 3/8	1.769	1.327
2.135	2.260		2 1/4	M56	3	8 1/4	3 9/16	1 7/16	1.894	1.420
2.260	2.385		2 3/8		3	8 1/2	4	1 7/16	2.019	1.514
2.385	2.510		2 1/2	M64	3	8 3/4	4	1 1/2	2.100	1.575
2.510	2.635		2 5/8		3	8 3/4	4	1 1/2	2.225	1.669
2.635	2.760		2 3/4		3	9 1/4	4	1 9/16	2.350	1.762
2.760	2.885		2 7/8	M72	3	9 1/4	4	1 9/16	2.475	1.856
2.885	3.010		3		3	9 3/4	4 9/16	1 5/8	2.543	1.907
3.010	3.135		3 1/8	M80	3	9 3/4	4 9/16	1 5/8	2.668	2.001
3.135	3.260		3 1/4		3	10	4 9/16	1 3/4	2.793	2.095

Nominal Diameter Range - Inches		Mach. Screw Size No.	Nominal Fractional Diameter Inches	Nominal Metric Diameter Millimeters	STYLE	Tap Dimensions — Inches				
Over	To (Incl.)					Overall Length A	Thread Length B	Square Length C	Shank Diameter D	Size of Square E
3.260	3.385		3⅜	M90	3	10	4 9/16	1¾	2.883	2.162
3.385	3.510		3½		3	10¼	4 11/16	2	3.008	2.256
3.510	3.635		3⅝		3	10¼	4 11/16	2	3.133	2.350
3.635	3.760		3¾		3	10½	5 1/16	2⅛	3.217	2.413
3.760	3.885		3⅞		3	10½	5 1/16	2⅛	3.342	2.506
3.885	4.010		4	M100	3	10¾	5 1/16	2¼	3.467	2.600

*Styles shown are for ground thread taps. See notes for cut thread taps.

*STYLE — See notes for cut thread tap centers.

SPECIAL TAPS

Unless otherwise specified:
Special taps over 1.010" to 1.510" diameter inclusive, having 14 or more threads per inch or 1.75 millimeter pitch and finer, and sizes over 1.510" diameter with 10 or more threads per inch or 2.5 millimeter pitch and finer, will be made to general dimensions shown in Table 303.

Special cut and ground thread taps for Unified and American National form threads will be made to limits shown in Tables 330 and 331.

NOTES

Cut thread taps, sizes .395" and smaller, have optional style center on thread and shank ends.

Ground thread taps, sizes .395" and smaller, have external center on thread end (may be removed on bottoming taps). Sizes .223" and smaller, have external center on shank end; sizes .224" thru .395" have truncated partial cone centers on shank end (length of cone approximately ¼ of diameter of shank). Sizes over .395" have internal center in thread and shank ends.

For standard thread limits and tolerances for Unified and American National form threads see Tables 325 and 327.
For eccentricity tolerances of tap elements see table 317.

Table 302 (Concluded)
Tolerances

Element	Nominal Diameter Range — Inches		Direction	Tolerance — Inches	
	Over	To (Incl.)		Cut Thread	Ground Thread
Length Overall — A	.052	1.010	Plus or Minus	$\frac{1}{32}$	$\frac{1}{32}$
	1.010	4.010	Plus or Minus	$\frac{1}{16}$	$\frac{1}{16}$
Length of Thread — B	.052	.223	Plus or Minus	$\frac{3}{64}$	$\frac{3}{64}$
	.223	.510	Plus or Minus	$\frac{1}{16}$	$\frac{1}{16}$
	.510	1.510	Plus or Minus	$\frac{3}{32}$	$\frac{3}{32}$
	1.510	4.010	Plus or Minus	$\frac{1}{8}$	$\frac{1}{8}$
Length of Square — C	.052	1.010	Plus or Minus	$\frac{1}{32}$	$\frac{1}{32}$
	1.010	4.010	Plus or Minus	$\frac{1}{16}$	$\frac{1}{16}$
Diameter of Shank — D	.052	.223	Minus	.004	.0015
	.223	.635	Minus	.005	.0015
	.635	1.010	Minus	.005	.002
	1.010	1.510	Minus	.007	.002
	1.510	2.010	Minus	.007	.003
	2.010	4.010	Minus	.009	.003
Size of Square — E	.052	.510	Minus	.004	.004
	.510	1.010	Minus	.006	.006
	1.010	2.010	Minus	.008	.008
	2.010	4.010	Minus	.010	.010

Table 303
Special Fine Pitch Taps
Short Series

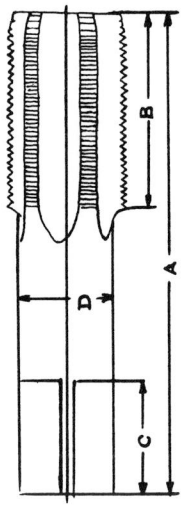

Unless otherwise specified, special taps 1.010″ to 1.510″ diameter inclusive, have 14 or more threads per inch or 1.75 millimeter pitch and finer, and sizes over 1.510″ diameter with 10 or more threads per inch, or 2.5 millimeter pitch and finer, will be made to the general dimensions shown below:

General Dimensions

Nominal Diameter Range — Inches		Nominal Fractional Diameter Inches	Nominal Metric Diameter Millimeters	Tap Dimensions — Inches				
Over	To (Incl.)			Overall Length A	Thread Length B	Square Length C	Shank Diameter D	Size of Square E
1.010	1.073	1 1/16	M27	4	1 1/2	7/8	.896	.672
1.073	1.135	1 1/8		4	1 1/2	7/8	.896	.672
1.135	1.198	1 3/16	M30	4	1 1/2	1	1.021	.766
1.198	1.260	1 1/4		4	1 1/2	1	1.021	.766
1.260	1.323	1 5/16	M33	4	1 1/2	1	1.108	.831
1.323	1.385	1 3/8		4	1 1/2	1	1.108	.831
1.385	1.448	1 7/16	M36	4	1 1/2	1	1.233	.925
1.448	1.510	1 1/2		5	1 1/2	1	1.233	.925
1.510	1.635	1 5/8	M39	5	2	1 1/8	1.305	.979
1.635	1.760	1 3/4	M42	5	2	1 1/4	1.430	1.072
1.760	1.885	1 7/8		5	2	1 1/4	1.519	1.139
1.885	2.010	2	M48	5	2	1 3/8	1.644	1.233
2.010	2.135	2 1/8		5 1/4	2	1 3/8	1.769	1.327
2.135	2.260	2 1/4	M56	5 1/4	2	1 7/16	1.894	1.420
2.260	2.385	2 3/8		5 1/4	2	1 7/16	2.019	1.514

Table 303 (Concluded)

2.385	2.510	2½	M64	5¼	2	1½	2.100	1.575
2.510	2.635	2⅝		5½	2	1½	2.100	1.575
2.635	2.760	2¾		5½	2	1½	2.100	1.575
2.760	2.885	2⅞	M72	5½	2	1½	2.100	1.575
2.885	3.010	3		5½	2	1½	2.100	1.575
3.010	3.135	3⅛		5¾	2	1½	2.100	1.575
3.135	3.260	3¼	M80	5¾	2	1½	2.100	1.575
3.260	3.385	3⅜		5¾	2	1½	2.100	1.575
3.385	3.510	3½		5¾	2	1½	2.100	1.575
3.510	3.635	3⅝	M90	6	2	1¾	2.100	1.575
3.635	3.760	3¾		6	2	1¾	2.100	1.575
3.760	3.885	3⅞		6	2	1¾	2.100	1.575
3.885	4.010	4	M100	6	2	1¾	2.100	1.575

NOTES

For tolerances see Table 302.

For standard thread limits and tolerances for Unified and American National form thread see Tables 327A, 330 and 331.

For eccentricity tolerances of tap elements see Table 317.

TABLE 303A

SPECIAL EXTENSION TAPS

Machine Screw, Fractional, Metric and Pipe

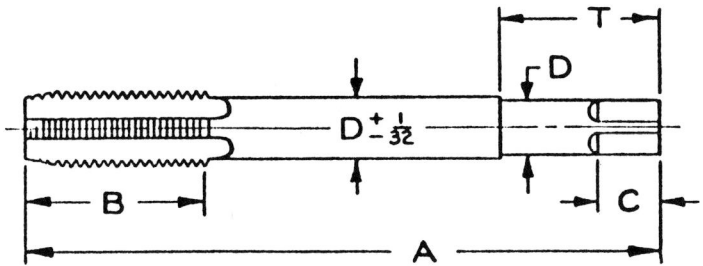

Unless otherwise specified, special extension taps will be furnished with dimensions and tolerences as shown for machine screw, fractional and metric taps in Table 302 and 303 and Pipe Taps in Table 311.

Exceptions: (1) Types of centers optional with manufacturer, (2) Tolerances on shank diameter-D ("T" length), as tabulated on sheet 2.

Shank eccentricity tolerances (Table 317) applies only to "T" length shown in the following Table.

Nominal Diameter Range, Inches		Mach. Screw Size No.	Nominal Fractional Diameter Inches	Nominal Metric Diameter Millimeters	Nominal Pipe Size	"T"
Over	To (Incl.)					
.052	.104	0-3	$\frac{1}{16}$-$\frac{3}{32}$	M1.6-M2.5		$\frac{7}{8}$
.104	.117	4				1
.117	.145	5-6	$\frac{1}{8}$	M3-M3.5		$1\frac{1}{8}$
.145	.171	8	$\frac{5}{32}$	M4		$1\frac{1}{4}$
.171	.223	10-12	$\frac{3}{16}$-$\frac{7}{32}$	M4.5-M5	$\frac{1}{16}$-$\frac{1}{8}$-$\frac{1}{4}$	$1\frac{3}{8}$
.223	.260	14	$\frac{1}{4}$	M6-M6.3		$1\frac{1}{2}$
.260	.323		$\frac{5}{16}$	M7-M8		$1\frac{9}{16}$
.323	.395		$\frac{3}{8}$	M10		$1\frac{5}{8}$
.395	.510		$\frac{7}{16}$-$\frac{1}{2}$	M12-M12.5	$\frac{3}{8}$-$\frac{1}{2}$	$1\frac{11}{16}$
.510	.573		$\frac{9}{16}$	M14	$\frac{3}{4}$	$1\frac{7}{8}$
.573	.635		$\frac{5}{8}$	M16	1	2
.635	.709		$\frac{11}{16}$	M18		$2\frac{1}{8}$
.709	.760		$\frac{3}{4}$		$1\frac{1}{4}$	$2\frac{1}{4}$
.760	.823		$\frac{13}{16}$	M20	$1\frac{1}{2}$	$2\frac{3}{8}$
.823	.885		$\frac{7}{8}$	M22		$2\frac{1}{2}$
.885	1.010		$\frac{15}{16}$-1	M24-M25		$2\frac{5}{8}$
1.010	1.135		$1\frac{1}{16}$-$1\frac{1}{8}$	M27	2	$2\frac{3}{4}$
1.135	1.260		$1\frac{3}{16}$-$1\frac{1}{4}$	M30	$2\frac{1}{2}$	$2\frac{7}{8}$
1.260	1.510		$1\frac{5}{16}$-$1\frac{1}{2}$	M33-M36		3
1.510	1.760		$1\frac{5}{8}$-$1\frac{3}{4}$	M39-M42	3	$3\frac{1}{8}$
1.760	2.010		$1\frac{7}{8}$-2	M48		$3\frac{1}{4}$
2.010	2.260		$2\frac{1}{8}$-$2\frac{1}{4}$	M56		$3\frac{3}{8}$
2.260	2.510		$2\frac{3}{8}$-$2\frac{1}{2}$			$3\frac{1}{2}$
2.510	2.760		$2\frac{5}{8}$-$2\frac{3}{4}$	M64	$3\frac{1}{2}$	$3\frac{5}{8}$
2.760	3.010		$2\frac{7}{8}$-3	M72		$3\frac{3}{4}$

(Concluded on following page)

Table 303A (Concluded)

3.010	3.260		3⅛-3¼	M80		3⅞
3.260	3.510		3⅜-3½		4	4
3.510	3.760		3⅝-3¾	M90		4⅛
3.760	4.010		3⅞-4	M100		4¼

Tolerances

For Shank Diameter — D (T-Length)

Nominal Diameter Range		Tolerance Direction	Tolerance	
Over	To (Incl.)		Cut Thread	Ground Thread
Machine Screw, Fractional, Metric Taps				
.052	.223	Minus	.004	.003
.223	.635	Minus	.005	.003
.635	1.010	Minus	.005	.004
1.010	1.510	Minus	.007	.004
1.510	2.010	Minus	.007	.006
2.010	4.010	Minus	.009	.006
Pipe Taps				
¹⁄₁₆	⅛	Minus	.007	.003
⅛	½	Minus	.007	.004
½	1	Minus	.009	.004
1	4	Minus	.009	.006

TABLE 311
Pipe Taps
Straight and Taper

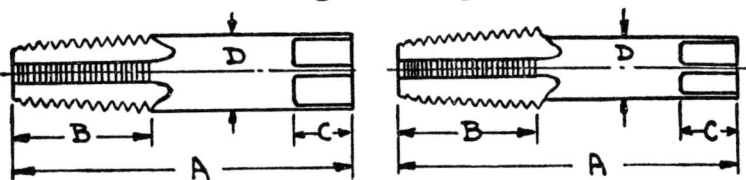

General Dimensions

Nominal Size Inches	Dimensions — Inches				
	Length Overall A	Length of Thread B	Length of Square C	Diameter of Shank D	Size of Square E
$\frac{1}{16}$	$2\frac{1}{8}$	$\frac{11}{16}$	$\frac{3}{8}$	.3125	.234
$\frac{1}{8}$	$2\frac{1}{8}$	$\frac{3}{4}$	$\frac{3}{8}$	.3125	.234
$\frac{1}{8}$	$2\frac{1}{8}$	$\frac{3}{4}$	$\frac{3}{8}$	.4375	.328
$\frac{1}{4}$	$2\frac{7}{16}$	$1\frac{1}{16}$	$\frac{7}{16}$	.5625	.421
$\frac{3}{8}$	$2\frac{9}{16}$	$1\frac{1}{16}$	$\frac{1}{2}$	.7000	.531
$\frac{1}{2}$	$3\frac{1}{8}$	$1\frac{3}{8}$	$\frac{5}{8}$	.6875	.515
$\frac{3}{4}$	$3\frac{1}{4}$	$1\frac{3}{8}$	$\frac{11}{16}$	.9063	.679
1	$3\frac{3}{4}$	$1\frac{3}{4}$	$\frac{13}{16}$	1.1250	.843
$1\frac{1}{4}$	4	$1\frac{3}{4}$	$\frac{15}{16}$	1.3125	.984
$1\frac{1}{2}$	$4\frac{1}{4}$	$1\frac{3}{4}$	1	1.5000	1.125
2	$4\frac{1}{2}$	$1\frac{3}{4}$	$1\frac{1}{8}$	1.8750	1.406
$2\frac{1}{2}$	$5\frac{1}{2}$	$2\frac{9}{16}$	$1\frac{1}{4}$	2.2500	1.687
3	6	$2\frac{5}{8}$	$1\frac{3}{8}$	2.6250	1.968
$3\frac{1}{2}$	$6\frac{1}{2}$	$2\frac{11}{16}$	$1\frac{1}{2}$	2.8125	2.108
4	$6\frac{3}{4}$	$2\frac{3}{4}$	$1\frac{5}{8}$	3.0000	2.250

Tolerances

Element	Range	Direction	Tolerance	
			Cut Thread	Ground Thread
Length Overall—A	$\frac{1}{16}''$ to $\frac{3}{4}''$ incl.	Plus or Minus	$\frac{1}{32}''$	$\frac{1}{32}''$
	$1''$ to $4''$ incl.	Plus or Minus	$\frac{1}{16}''$	$\frac{1}{16}''$
Length of Thread—B	$\frac{1}{16}''$ to $\frac{3}{4}''$ incl.	Plus or Minus	$\frac{1}{16}''$	$\frac{1}{16}''$
	$1''$ to $1\frac{1}{4}''$ incl.	Plus or Minus	$\frac{3}{32}''$	$\frac{3}{32}''$
	$1\frac{1}{2}''$ to $4''$ incl.	Plus or Minus	$\frac{1}{8}''$	$\frac{1}{8}''$
Length of Square—C	$\frac{1}{16}''$ to $\frac{3}{4}''$ incl.	Plus or Minus	$\frac{1}{32}''$	$\frac{1}{32}''$
	$1''$ to $4''$ incl.	Plus or Minus	$\frac{1}{16}''$	$\frac{1}{16}''$
Diameter of Shank—D	$\frac{1}{16}''$ to $\frac{1}{8}''$ incl.	Minus	.0070''	.0015''
	$\frac{1}{4}''$ to $\frac{1}{2}''$ incl.	Minus	.0070''	.0020''
	$\frac{3}{4}''$ to $1''$ incl.	Minus	.0090''	.0020''
	$1\frac{1}{4}''$ to $4''$ incl.	Minus	.0090''	.0030''
Size of Square—E	$\frac{1}{16}''$ to $\frac{1}{8}''$ incl.	Minus	.0040''	.0040''
	$\frac{1}{4}''$ to $\frac{3}{4}''$ incl.	Minus	.0060''	.0060''
	$1''$ to $4''$ incl.	Minus	.0080''	.0080''

Note
For thread limits and tolerances see Tables 334, 335, 335A and 338.
For Eccentricity Tolerances of Tap Elements — See Table 317.

TABLE 331 Special Taps
Ground Thread — Unified and American National Form

The following tables and formulae are used in determining the limits and tolerances for ground thread taps having special diameter or special pitch or both and having a thread lead angle not in excess of 5°, unless otherwise specified.

Lead Tolerance

A maximum lead deviation of plus or minus .0005″ within any two threads not farther apart than 1″ is permitted.

Angle Tolerance

Threads per inch	Deviation in Half Angle
4 to 5½ incl.	20′ Plus or Minus
6 to 9 incl.	25′ Plus or Minus
10 to 80 incl.	30′ Plus or Minus

Formulae

Max. Major Dia. = Basic plus A   Max. Pitch Dia. = Min. plus D

Min. Major Dia. = Max. minus B   Min. Pitch Dia = Basic plus C

In the above formulae: —

A = Constant to Add = 0.130P for all Pitches

B = Major Diameter Tolerance   = 0.087P For 48 Through 80 TPI

= 0.076P For 36 Through 47 TPI

= 0.065P For 4 Through 35 TPI

C = Amount over basic for minimum pitch diameter

D = Pitch diameter tolerance

Note: When the tap major diameter must be determined from a specified tap pitch diameter, the maximum major diameter equals the minimum specified pitch diameter minus Constant C, plus 0.64952P, plus Constant A.

Threads per Inch	A	B	C To 5/8″ incl.	C Over 5/8″ To 2½″ incl.	C Over 2½″	D To 1″ incl.	D Over 1″ To 1½″ incl.	D Over 1½″ To 2½″ incl.	D Over 2½″
80	.0016	.0011	.0005	.0010	.0015	.0005	.0010	.0010	.0015
72	.0018	.0012	.0005	.0010	.0015	.0005	.0010	.0010	.0015
64	.0020	.0014	.0005	.0010	.0015	.0005	.0010	.0010	.0015
56	.0023	.0016	.0005	.0010	.0015	.0005	.0010	.0010	.0015
48	.0027	.0018	.0005	.0010	.0015	.0005	.0010	.0010	.0015
44	.0030	.0017	.0005	.0010	.0015	.0005	.0010	.0010	.0015
40	.0032	.0019	.0005	.0010	.0015	.0005	.0010	.0010	.0015
36	.0036	.0021	.0005	.0010	.0015	.0005	.0010	.0010	.0015
32	.0041	.0020	.0010	.0010	.0015	.0005	.0010	.0010	.0015
28	.0046	.0023	.0010	.0010	.0015	.0005	.0010	.0010	.0015
24	.0054	.0027	.0010	.0010	.0015	.0005	.0010	.0015	.0015
20	.0065	.0032	.0010	.0010	.0015	.0005	.0010	.0015	.0015
18	.0072	.0036	.0010	.0010	.0015	.0005	.0010	.0015	.0015
16	.0081	.0041	.0010	.0010	.0015	.0005	.0010	.0015	.0020
14	.0093	.0046	.0010	.0015	.0015	.0005	.0010	.0015	.0020
13	.0100	.0050	.0010	.0015	.0015	.0005	.0010	.0015	.0020
12	.0108	.0054	.0010	.0015	.0015	.0005	.0010	.0015	.0020
11	.0118	.0059	.0010	.0015	.0020	.0005	.0010	.0015	.0020
10	.0130	.0065	. . . .	.0015	.0020	.0005	.0010	.0015	.0020
9	.0144	.0072	. . . .	.0015	.0020	.0005	.0010	.0015	.0020
8	.0162	.0081	. . . .	.0015	.0020	.0005	.0010	.0015	.0020
7	.0186	.0093	. . . .	.0015	.0020	.0010	.0010	.0020	.0025
6	.0217	.0108	. . . .	.0015	.0020	.0010	.0010	.0020	.0025
5½	.0236	.0118	. . . .	.0015	.0020	.0010	.0015	.0020	.0025
5	.0260	.0130	. . . .	.0015	.0020	.0010	.0015	.0020	.0025
4½	.0289	.0144	. . . .	.0015	.0020	.0010	.0015	.0020	.0025
4	.0325	.0162	. . . .	.0015	.0020	.0010	.0015	.0020	.0025

For intermediate pitches use value for next coarser pitch for C & D, but use formulas for A & B

TABLE 359
TAPS

STANDARD CHAMFERS

The chamfer length is measured at the cutting edge and is the axial length from the point diameter to the theoretical intersection of the tap major diameter and the chamfer angle. Wherever chamfer length is specified in terms of number of threads, this length is measured in number of pitches as shown. The point diameter is approximately equal to the thread minor diameter. Standard types are illustrated below where:

a = minimum length of chamfer
b = maximum length of chamfer
p = pitch

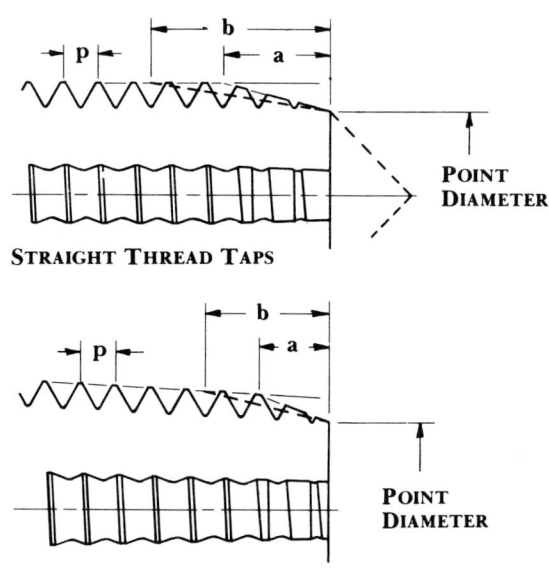

STRAIGHT THREAD TAPS

TAPER PIPE TAPS

		CHAMFER LENGTH	
TYPE OF TAP		a	b
Straight Thread Taps:	BOTTOMING	1p	2p
	PLUG	3p	5p
	TAPER	7p	10p
TAPER PIPE TAPS		2p	3½p

TABLE 359-A

Thread Forming Taps
Standard Entry Taper Lengths

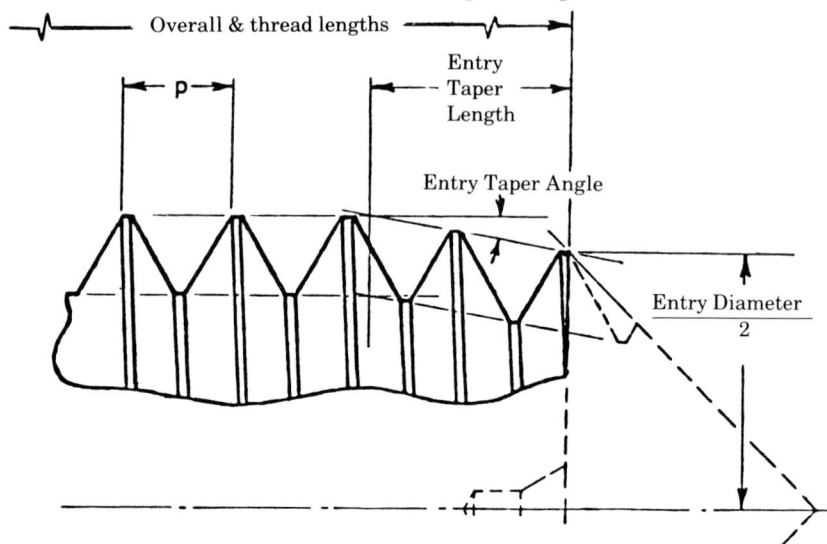

Entry taper length is measured on the full diameter of the thread form-ing lobes and is the axial distance from the entry diameter position to the theoretical intersection of tap major diameter and entry taper angle.

Beveled end threads provided on taps having internal center or incom-plete threads retained when external center is removed (not shown), optional with manufacturer.

Whenever entry taper length is specfied in terms of number of threads, this length is measured in number of pitches (p).

Bottoming length = 1 to 2½ pitches.
Plug length = 3 to 5 pitches.

Entry diameter, measured at the thread crest nearest the front of the tap, is an appropriate amount smaller than the diameter of the hole drilled for tapping.

TABLE 317

Eccentricity Tolerances of Tap Elements
When Tested on Dead Centers

Applicable to Tables 302, 303, 303A, 304, 310, 311, 322 and 345

Element	Range Sizes are Inclusive			Cut Thread		Ground Thread	
	Inch & Mch.Screw	Pipe	Metric	Eccentricity	t.i.v.*	Eccentricity	t.i.v.*
Square (at central point)	#0-1/2"	1/16-1/8"	M1.6—M12	.0030	.0060	.0030	.0060
	over 1/2" thru 4"	1/4-4"	Over M12 thru M100	.0040	.0080	.0040	.0080
Shank	#0-5/16"	1/16"	M1.6—M8	.0030	.0060	.0005	.0010
	Over 5/16" thru 4"	1/8-4"	Over M8 thru M100	.0040	.0080	.0008	.0016
Major Diameter	#0-5/16"	1/16"	M1.6—M8	.0025	.0050	.0005	.0010
	Over 5/16" thru 4"	1/8-4"	Over M8 thru M100	.0040	.0080	.0008	.0016
Pitch Diameter (at first full thrd.)	#0-1/2"	1/16"	M1.6—M8	.0025	.0050	.0005	.0010
	Over 1/2" thru 4"	1/8-4"	Over M8 thru M100	.0040	.0080	.0008	.0016
Chamfer**	#0-1/2"	1/16-1/8"	M1.6—M12	.0020	.0040	.0010	.0020
	Over 1/2" thru 4"	1/4-4"	Over M12 thru M100	.0030	.0060	.0015	.0030

*t.i.v. =Total indicator variation. Figures are given for both eccentricity and total indicator variation to avoid misunderstanding.

**Chamfer should preferably be inspected by light projection to avoid errors due to indicator contact points dropping into the thread grooves.

Table 327
Fractional Size Taps—Ground Thread
Unified and American National Form - Thread Limits

Size	Threads Per Inch			Major Diameter			Basic Pitch Diam.	Pitch Diameter Limits											
								H1 Limit		H2 Limit		H3 Limit		H4 Limit		H5 Limit		H6 Limit	
	NC UNC	NF UNF	NS UNS	Basic	Minimum	Maximum		Minimum	Maximum	Minimum	Maximum	Minimum	Maximum	Minimum	Maximum	Minimum	Maximum	Minimum	Maximum
1/4	20	....	....	.2500	.2533	.2565	.2175	.2175	.2180	.2180	.2185	.2185	.2190	....	....	.2195	.2200	....	....
1/4	....	28	....	.2500	.2523	.2546	.2268	.2268	.2273	.2273	.2278	.2278	.2283	.2283	.2288	....	....	....	....
5/16	18	....	....	.3125	.3161	.3197	.2764	.2764	.2769	.2769	.2774	.2774	.2779	....	....	.2784	.2789	....	....
5/16	....	24	....	.3125	.3152	.3179	.2854	.2854	.2859	.2859	.2864	.2864	.2869	.2869	.2874	....	....	....	....
3/8	16	....	....	.3750	.3790	.3831	.3344	.3344	.3349	.3349	.3354	.3354	.3359	....	....	.3364	.3369	....	....
3/8	....	24	....	.3750	.3777	.3804	.3479	.3479	.3484	.3484	.3489	.3489	.3494	.3494	.3499	....	....	....	....
7/16	14	....	....	.4375	.4422	.4468	.3911	....	....	.3916	.3921	.3921	.3926	....	....	.3931	.3936	....	....
7/16	....	20	....	.4375	.4408	.4440	.4050	....	....	....	....	.4060	.4065	....	....	.4070	.4075	....	....
1/2	13	....	....	.5000	.5050	.5100	.4500	.4500	.4505	.4505	.4510	.4510	.4515	....	....	.4520	.4525	....	....
1/2	....	20	....	.5000	.5033	.5065	.4675	.4675	.4680	.4680	.4685	.4685	.4690	....	....	.4695	.4700	....	....
9/16	12	....	....	.5625	.5679	.5733	.5084	....	....	....	....	.5094	.5099	....	....	.5104	.5109	....	....
9/16	....	18	....	.5625	.5661	.5697	.5264	....	....	.5269	.5274	.5274	.5279	....	....	.5284	.5289	....	....
5/8	11	....	....	.6250	.6309	.6368	.5660	....	....	.5665	.5670	.5670	.5675	....	....	.5680	.5685	....	....
5/8	....	18	....	.6250	.6286	.6322	.5889	....	....	.5894	.5899	.5899	.5904	....	....	.5909	.5914	....	....
11/16	....	....	11	.6875	.6934	.6993	.6285	....	....	....	....	.6295	.6300	....	....	....	....	....	....
11/16	....	....	16	.6875	.6915	.6956	.6469	....	....	....	....	.6479	.6484	....	....	....	....	....	....
3/4	10	....	....	.7500	.7540	.7581	.6850	....	....	.6855	.6860	.6860	.6865	....	....	.6870	.6875	....	....
3/4	....	16	....	.7500	.7565	.7630	.7094	.7094	.7099	.7099	.7104	.7104	.7109	....	....	.7114	.7119	....	....
7/8	9	....	....	.8750	.8822	.8894	.8028	....	....	....	....	....	....	.8043	.8048	....	....	.8053	.8058
7/8	....	14	....	.8750	.8797	.8843	.8286	....	....	.8291	.8296	....	....	.8301	.8306	....	....	.8311	.8316
1	8	....	....	1.0000	1.0081	1.0162	.9188	....	....	....	....	....	....	.9203	.9208	....	....	.9213	.9218
1	....	12	....	1.0000	1.0054	1.0108	.9459	....	....	....	....	....	....	.9474	.9479	....	....	....	....
1	....	....	14	1.0000	1.0047	1.0093	.9536	....	....	....	....	....	....	.9551	.9556	....	....	....	....

Size	Threads Per Inch			Major Diameter			Pitch Diameter Limits		
	NC UNC	NF UNF	NS	Basic	Minimum	Maximum	Basic Pitch Diam.	H4 Limit Minimum	H4 Limit Maximum
1⅛	7	:	...	1.1250	1.1343	1.1436	1.0322	1.0332	1.0342
1⅛	:	12	...	1.1250	1.1304	1.1358	1.0709	1.0719	1.0729
1¼	7	:	...	1.2500	1.2593	1.2686	1.1572	1.1582	1.1592
1¼	:	12	...	1.2500	1.2554	1.2608	1.1959	1.1969	1.1979
1⅜	6	:	...	1.3750	1.3859	1.3967	1.2667	1.2677	1.2687
1⅜	:	12	...	1.3750	1.3804	1.3858	1.3209	1.3219	1.3229
1½	6	:	...	1.5000	1.5109	1.5217	1.3917	1.3927	1.3937
1½	:	12	...	1.5000	1.5054	1.5108	1.4459	1.4469	1.4479

Lead Tolerance

A maximum lead deviation of plus or minus .0005″ within any two threads not farther apart than 1″ is permitted.

Angle Tolerance

Threads Per Inch	Deviation in Half Angle
6 to 9 incl.	25′ plus or minus
10 to 28 incl.	30′ plus or minus

Formulae (Approximate)

Maximum Major Diameter = Basic plus A
Minimum Major Diameter = Maximum minus B
For values of A and B see Table 331

Pitch Diameter Limits for Taps Thru 1″ Diameter:

H1 = Basic to basic plus .0005″
H2 = Basic plus .0005″ to basic plus .001″
H3 = Basic plus .001″ to basic plus .0015″
H4 = Basic plus .0015″ to basic plus .002″
H5 = Basic plus .002″ to basic plus .0025″
H6 = Basic plus .0025″ to basic plus .003″

Pitch Diameter Limits for Taps over 1″ Diameter Thru 1½″ Diameter:

H4 = Basic plus .001″ to basic plus .002″

TABLE—327A
Fractional Size Taps

Ground Thread—Unified and American Form
Sizes 1⅝ to 4 Inclusive
Thread Limits Computed from Table 331

Size	Threads Per Inch				Major Diameter			Pitch Diameter Limits			
	NC UNC	8N 8UN	12N 12UN	16N 16UN	Basic	Mini-mum	Maxi-mum	Basic	Mini-mum	Maxi-mum	Limit No.
1⅝	..	8	..	..	1.6250	1.6331	1.6412	1.5438	1.5453	1.5468	H6
1⅝	..	..	12	..	1.6250	1.6304	1.6358	1.5709	1.5724	1.5739	H6
1⅝	..	..	..	16	1.6250	1.6290	1.6331	1.5844	1.5854	1.5869	H5
1¾	5	..	..	..	1.7500	1.7630	1.7760	1.6201	1.6216	1.6236	H7
1¾	..	8	..	..	1.7500	1.7581	1.7662	1.6688	1.6703	1.6718	H6
1¾	..	..	12	..	1.7500	1.7554	1.7608	1.6959	1.6974	1.6989	H6
1¾	..	..	..	16	1.7500	1.7540	1.7581	1.7094	1.7104	1.7119	H5
1⅞	..	8	..	..	1.8750	1.8831	1.8912	1.7938	1.7953	1.7968	H6
1⅞	..	..	12	..	1.8750	1.8804	1.8858	1.8209	1.8224	1.8239	H6
1⅞	..	..	..	16	1.8750	1.8790	1.8831	1.8344	1.8354	1.8369	H5
2	4½	..	..	..	2.0000	2.0145	2.0289	1.8557	1.8572	1.8592	H7
2	..	8	..	..	2.0000	2.0081	2.0162	1.9188	1.9203	1.9218	H6
2	..	..	12	..	2.0000	2.0054	2.0108	1.9459	1.9474	1.9489	H6
2	..	..	..	16	2.0000	2.0040	2.0081	1.9594	1.9604	1.9619	H5
2¼	4½	..	..	..	2.2500	2.2645	2.2789	2.1057	2.1072	2.1092	H7
2¼	..	8	..	..	2.2500	2.2581	2.2662	2.1688	2.1703	2.1718	H6
2¼	..	..	12	..	2.2500	2.2554	2.2608	2.1959	2.1974	2.1989	H6
2¼	..	..	..	16	2.2500	2.2540	2.2581	2.2094	2.2104	2.2119	H5
2½	4	..	..	..	2.5000	2.5163	2.5325	2.3376	2.3391	2.3411	H7
2½	..	8	..	..	2.5000	2.5081	2.5162	2.4188	2.4203	2.4218	H6
2½	..	..	12	..	2.5000	2.5054	2.5108	2.4459	2.4474	2.4489	H6
2½	..	..	..	16	2.5000	2.5040	2.5081	2.4594	2.4604	2.4619	H5
2¾	4	..	..	..	2.7500	2.7663	2.7825	2.5876	2.5896	2.5921	H9
2¾	..	8	..	..	2.7500	2.7581	2.7662	2.6688	2.6708	2.6728	H8
2¾	..	..	12	..	2.7500	2.7554	2.7608	2.6959	2.6974	2.6994	H7
2¾	..	..	..	16	2.7500	2.7540	2.7581	2.7094	2.7109	2.7129	H7
3	4	..	..	..	3.0000	3.0163	3.0325	2.8376	2.8396	2.8421	H9
3	..	8	..	..	3.0000	3.0081	3.0162	2.9188	2.9208	2.9228	H8
3	..	..	12	..	3.0000	3.0054	3.0108	2.9459	2.9474	2.9494	H7
3	..	..	..	16	3.0000	3.0040	3.0081	2.9594	2.9609	2.9629	H7
3¼	4	..	..	..	3.2500	3.2663	3.2825	3.0876	3.0896	3.0921	H9
3¼	..	8	..	..	3.2500	3.2581	3.2662	3.1688	3.1705	3.1728	H8
3¼	..	..	12	..	3.2500	3.2554	3.2608	3.1959	3.1974	3.1994	H7
3¼	..	..	..	16	3.2500	3.2540	3.2581	3.2094	3.2109	3.2129	H7

TABLE—327A
Fractional Size Taps

Ground Thread—Unified and American National Form
Sizes 1⅝ to 4 Inclusive
Thread Limits Computed from Table 331
(Concluded)

Size	Threads Per Inch				Major Diameter			Pitch Diameter Limits			
	NC UNC	8N 8UN	12N 12UN	16N 16UN	Basic	Mini-mum	Maxi-mum	Basic	Mini-mum	Maxi-mum	Limit No.
3½	4	..	..	..	3.5000	3.5163	3.5325	3.3376	3.3396	3.3421	H9
3½	..	8	..	..	3.5000	3.5081	3.5162	3.4188	3.4208	3.4228	H8
3½	..	..	12	..	3.5000	3.5054	3.5108	3.4459	3.4474	3.4494	H7
3½	..	..	..	16	3.5000	3.5040	3.5081	3.4594	3.4609	3.4629	H7
3¾	4	..	..	..	3.7500	3.7663	3.7825	3.5876	3.5896	3.5921	H9
3¾	..	8	..	..	3.7500	3.7581	3.7662	3.6688	3.6708	3.6728	H8
3¾	..	..	12	..	3.7500	3.7554	3.7608	3.6959	3.6974	3.6994	H7
3¾	..	..	..	16	3.7500	3.7540	3.7581	3.7094	3.7109	3.7129	H7
4	4	..	..	..	4.0000	4.0163	4.0325	3.8376	3.8396	3.8421	H9
4	..	8	..	..	4.0000	4.0081	4.0162	3.9188	3.9208	3.9228	H8
4	..	..	12	..	4.0000	4.0054	4.0108	3.9459	3.9474	3.9494	H7
4	..	..	..	16	4.0000	4.0040	4.0081	3.9594	3.9609	3.9629	H7

Lead Tolerance

A maximum lead deviation of plus or minus .0005″ within any two threads not farther apart than 1″ is permitted.

Angle Tolerance

Thread Per Inch	Deviation in Half Angle
4 to 5½ incl.	20′ plus or minus
6 to 9 incl.	25′ plus or minus
10 to 28 incl.	30′ plus or minus

FORMULAE

Maximum Major Diameter = Basic plus A
Minimum Major Diameter = Maximum minus B
Maximum Pitch Diameter = Minimum plus D
Minimum Pitch Diameter = Basic plus C
For Values of A, B, C, and D see Table 331

TABLE 329—Machine Screw Taps—Ground Thread—Unified and American National Form—Thread Limits

Size	NC UNC	NF UNF	NS UNS	Major Basic	Major Minimum	Major Maximum	Basic Pitch Diam.	H1 Minimum	H1 Maximum	H2 Minimum	H2 Maximum	H3 Minimum	H3 Maximum	H7 Minimum	H7 Maximum
0		80		.0600	.0605	.0616	.0519	.0519	.0524	.0524	.0529				
1	64			.0730	.0736	.0750	.0629	.0629	.0634	.0634	.0639				
1		72		.0730	.0736	.0748	.0640	.0640	.0645	.0645	.0650				
2	56			.0860	.0866	.0883	.0744	.0744	.0749	.0749	.0754				
2		64		.0860	.0866	.0880	.0759			.0764	.0769				
3	48			.0990	.0999	.1017	.0855	.0855	.0860	.0860	.0865				
3		56		.0990	.0997	.1013	.0874	.0874	.0879	.0879	.0884				
4			36	.1120	.1135	.1156	.0940			.0945	.0950				
4	40			.1120	.1133	.1152	.0958	.0958	.0963	.0963	.0968				
4		48		.1120	.1129	.1147	.0985	.0985	.0990	.0990	.0995				
5	40			.1250	.1263	.1282	.1088	.1088	.1093	.1093	.1098				
5		44		.1250	.1263	.1280	.1102			.1107	.1112				
6	32			.1380	.1401	.1421	.1177	.1177	.1182	.1182	.1187	.1187	.1192	.1207	.1212
6		40		.1380	.1393	.1412	.1218	.1218	.1223	.1223	.1228				
8	32			.1640	.1661	.1681	.1437	.1437	.1442	.1442	.1447	.1447	.1452	.1467	.1472
8		36		.1640	.1655	.1676	.1460			.1465	.1470				
10	24			.1900	.1927	.1954	.1629	.1629	.1634	.1634	.1639	.1639	.1644	.1659	.1664
10		32		.1900	.1921	.1941	.1697	.1697	.1702	.1702	.1707	.1707	.1712	.1727	.1732
12	24			.2160	.2187	.2214	.1889					.1899	.1904		
12		28		.2160	.2183	.2206	.1928					.1938	.1943		

* Major Diameter for H7 limit taps is .002″ larger than values shown in columns 6 & 7.

(Concluded on following page)

Table 329 (Concluded)

Machine Screw Taps

Ground Thread—Unified and American National Form Thread Limits

(Concluded)

Lead Tolerance

A maximum lead deviation of plus or minus .0005″ within any two threads not farther apart than 1″ is permitted.

Angle Tolerance

20 to 80 threads per inch incl. = 30′ plus or minus in ½ angle.

Formulae

Maximum Major Diameter = Basic plus A
Minimum Major Diameter = Maximum minus B

In the above formulae:—

A = Constant to Add = 0.130P for all Pitches

B = Major Diameter Tolerance = 0.087P For 48 Through 80 TPI
= 0.076P For 36 Through 47 TPI
= 0.065P For 4 Through 35 TPI

For values of A and B see Table 331.

Pitch Diameter Limits

H1 Limit = Basic to basic plus .0005″
H2 Limit = Basic plus .0005″ to basic plus .001″
H3 Limit = Basic plus .001″ to basic plus .0015″
H7 Limit = Basic plus .003″ to basic plus .0035″

Table 337
Standard Metric Taps
Ground Thread Tap Limits
(Millimeters)

Nominal Dia. mm	Pitch mm	Major Diameter (mm)			Pitch Diameter Limits (mm)										
		Basic	Min.	Max.	Basic	D3 Min.	D3 Max.	D4 Min.	D4 Max.	D5 Min.	D5 Max.	D6 Min.	D6 Max.
1.6	0.35	1.600	1.628	1.653	1.373	1.397	1.412						
2	0.4	2.000	2.032	2.057	1.740	1.764	1.779						
2.5	0.45	2.500	2.536	2.561	2.208	2.232	2.247						
3	0.5	3.000	3.040	3.065	2.675	2.699	2.714						
3.5	0.6	3.500	3.548	3.573	3.110	—	—	3.142	3.162				
4	0.7	4.000	4.056	4.097	3.545	—	—	3.577	3.597				
4.5	0.75	4.500	4.560	4.601	4.013	—	—	4.045	4.065				
5	0.8	5.000	5.064	5.105	4.480	—	—	4.512	4.532				
6	1	6.000	6.080	6.121	5.350	—	—	—	—	5.390	5.415		
7	1	7.000	7.080	7.121	6.350	—	—	—	—	6.390	6.415		
8	1.25	8.000	8.100	8.164	7.188	—	—	—	—	7.222	7.253		
10	1.5	10.000	10.120	10.184	9.026	—	—	—	—	—	—	9.073	9.104
12	1.75	12.000	12.140	12.204	10.863	—	—	—	—	—	—	10.910	10.941

(Continued on following page)

Nominal Dia. mm	Pitch mm	Major Diameter (mm)			Pitch Diameter Limits (mm)						
		Basic	Min.	Max.	Basic	D7 Min.	D7 Max.	D8 Min.	D8 Max.	D9 Min.	D9 Max.
14	2	14.000	14.160	14.224	12.701	12.751	12.792				
16	2	16.000	16.160	16.224	14.701	14.751	14.792				
20	2.5	20.000	20.200	20.264	18.376	18.426	18.467				
24	3	24.000	24.240	24.340	22.051	—	—	22.114	22.155		
30	3.5	30.000	30.280	30.380	27.727	—	—	—	—	27.792	27.844
36	4	36.000	36.320	36.420	33.402	—	—	—	—	33.467	33.519

Basic Pitch Diameter is the same as Min. Pitch Diameter of Internal Thread Class 6H—Table 19, ANSI B1.13M-1979

ANGLE TOLERANCE

Pitch (mm)	Deviation in Half Angle
Over 0.25 to 2.5 Inc.	30' Plus or Minus
Over 2.5 to 4 Inc.	25' Plus or Minus
Over 4 to 6 Inc.	20' Plus or Minus

A maximum lead deviation of ± 0.013 mm within any two thread not farther apart than 25 mm is permitted.

(Continued on following page)

Table 337
Standard Metric Taps
Ground Thread Tap Limits
(Inches)

Nominal Dia. mm	Pitch mm	Major Diameter (Inches)			Pitch Diameter Limits (Inches)								
		Basic	Min.	Max.	Basic	D3		D4		D5		D6	
						Min.	Max.	Min.	Max.	Min.	Max.	Min.	Max.
1.6	0.35	.06299	.06409	.06508	.05406	.05500	.05559						
2	0.4	.07874	.08000	.08098	.06850	.06945	.07004						
2.5	0.45	.09843	.09984	.10083	.08693	.08787	.08846						
3	0.5	.11811	.11969	.12067	.10531	.10626	.10685						
3.5	0.6	.13780	.13969	.14067	.12244	—	—	.12370	.12449				
4	0.7	.15748	.15969	.16130	.13957	—	—	.14083	.14161				
4.5	0.75	.17717	.17953	.18114	.15799	—	—	.15925	.16004				
5	0.8	.19685	.19937	.20098	.17638	—	—	.17764	.17843				
6	1	.23622	.23937	.24098	.21063	—	—	—	—	.21220	.21319		
7	1	.27559	.27874	.28035	.25000	—	—	—	—	.25157	.25256		
8	1.25	.31496	.31890	.32142	.28299	—	—	—	—	.28433	.28555		
10	1.5	.39370	.39843	.40094	.35535	—	—	—	—	—	—	.35720	.35843
12	1.75	.47244	.47795	.48047	.42768	—	—	—	—	—	—	.42953	.43075

(Concluded on following page)

Nominal Dia. mm	Pitch mm	Major Diameter (Inches)			Pitch Diameter Limits (Inches)						
		Basic	Min.	Max.	Basic	D7		D8		D9	
						Min.	Max.	Min.	Max.	Min.	Max.
14	2	.55118	.55748	.56000	.50004	.50201	.50362				
16	2	.62992	.63622	.63874	.57878	.58075	.58236				
20	2.5	.78740	.79528	.79780	.72346	.72543	.72705				
24	3	.94488	.95433	.95827	.86815	—	—	.87063	.87224		
30	3.5	1.18110	1.19213	1.19606	1.09161	—	—	—	—	1.09417	1.09622
36	4	1.41732	1.42992	1.43386	1.31504	—	—	—	—	1.31760	1.31965

Basic Pitch Diameter is the same as Min. Pitch Diameter of Internal Thread Class 6H—Table 21, ANSI B1.13M-1979

ANGLE TOLERANCE

Pitch (mm)	Deviation in Half Angle
Over 0.25 to 2.5 Inc.	30' Plus or Minus
Over 2.5 to 4 Inc.	25' Plus or Minus
Over 4 to 6 Inc.	20' Plus or Minus

A maximum lead deviation of ± .0005" within any two thread not farther apart than 1" is permitted.

TABLE 338
Taper Pipe Tap, Thread Limits,
Cut & Ground Thread

Cut Thread Taps for NPT · Ground Thread Taps for NPTF

American National Standard Taper Pipe Thread Form (NPT) · Aeronautical National Form Taper Pipe Thread (ANPT) · Ground Thread Taps for NPT and ANPT

Dryseal American National Standard Taper Pipe Thread Form (NPTF)

Nominal Size Inches	Threads per Inch	Tap Thread Limits					Reference Dimensions	
		*Projection Inches	Projection Tolerance + or −	Taper per Foot Limits			L₁ Length	Tap Drill Size** NPT, ANPT, NPTF
				Cut & Ground Minimum	Cut Maximum	Ground Maximum		
¹⁄₁₆	27	.312	¹⁄₁₆	²³⁄₃₂	²⁷⁄₃₂	²⁵⁄₃₂	.160	C
¹⁄₈	27	.312	¹⁄₁₆	²³⁄₃₂	²⁷⁄₃₂	²⁵⁄₃₂	.1615	Q
¹⁄₄	18	.459	¹⁄₁₆	²³⁄₃₂	²⁷⁄₃₂	²⁵⁄₃₂	.2278	⁷⁄₁₆
³⁄₈	18	.454	¹⁄₁₆	²³⁄₃₂	²⁷⁄₃₂	²⁵⁄₃₂	.240	⁹⁄₁₆
¹⁄₂	14	.579	¹⁄₁₆	²³⁄₃₂	¹³⁄₁₆	²⁵⁄₃₂	.320	⁴⁵⁄₆₄
³⁄₄	14	.565	¹⁄₁₆	²³⁄₃₂	¹³⁄₁₆	²⁵⁄₃₂	.339	²⁹⁄₃₂
1	11½	.678	³⁄₃₂	²³⁄₃₂	¹³⁄₁₆	²⁵⁄₃₂	.400	1⁹⁄₆₄
1¼	11½	.686	³⁄₃₂	²³⁄₃₂	¹³⁄₁₆	²⁵⁄₃₂	.420	1³¹⁄₆₄
1½	11½	.699	³⁄₃₂	²³⁄₃₂	¹³⁄₁₆	²⁵⁄₃₂	.420	1²³⁄₃₂
2	11½	.667	³⁄₃₂	²³⁄₃₂	¹³⁄₁₆	²⁵⁄₃₂	.436	2³⁄₁₆
2½	8	.925	³⁄₃₂	⁴⁷⁄₆₄	⁵¹⁄₆₄	²⁵⁄₃₂	.682	2³⁹⁄₆₄
3	8	.925	³⁄₃₂	⁴⁷⁄₆₄	⁵¹⁄₆₄	²⁵⁄₃₂	.766	3¹⁵⁄₆₄
3½	8	.938	¹⁄₈	⁴⁷⁄₆₄	⁵¹⁄₆₄	²⁵⁄₃₂	.821	· · · ·
4	8	.950	¹⁄₈	⁴⁷⁄₆₄	⁵¹⁄₆₄	²⁵⁄₃₂	.844	· · · ·

*Distance small end of tap projects through L₁ Taper Thread Ring Gage.

**Recommended sizes given permit direct tapping without reaming the hole, but only give a full thread for approx. L₁ distance.

Lead Tolerance

Cut Thread = A maximum lead deviation of plus or minus .003" in one inch of thread is permitted.

Ground Thread = A maximum lead deviation of plus or minus .0005" within any two threads not farther apart than 1" is permitted.

TABLE 338 (Concluded)

Taper Pipe Taps

American National Standard Taper Pipe Thread Form (NPT)
Aeronautical National Form Taper Pipe Thread (ANPT)
Dryseal American National Standard Taper Pipe Thread Form (NPTF)

Angle Tolerance

Threads per Inch	Tolerance		
	Half Angle		Full Angle
	Cut Thread	Ground Thread	Cut Thread
8	40' Plus or Minus	25' Plus or Minus	60'
11½ to 27 inclusive	45' Plus or Minus	30' Plus or Minus	68'

Widths of Flats at Tap Crests and Roots

Threads Per Inch	Tap Flat Width at	Column 1 NPT-Cut & Grd. Thd. ANPT — Grd. Thd.		Column II NPTF — Grd. Thd.	
		Minimum	Maximum	Minimum	Maximum
27	Major Dia.	.0014	.0041	.0040	.0055
	Minor Dia.	—	.0041	—	.0040
18	Major Dia.	.0021	.0057	.0050	.0065
	Minor Dia.	—	.0057	—	.0050
14	Major Dia.	.0027	.0064	.0050	.0065
	Minor Dia.	—	.0064	—	.0050
11½	Major Dia.	.0033	.0073	.0060	.0083
	Minor Dia.	—	.0073	—	.0060
8	Major Dia.	.0048	.0090	.0080	.0103
	Minor Dia.	—	.0090	—	.0080

Minimum minor diameter flats are not specified.
May be as sharp as practicable.

Cut thread taps made to Column 1 are marked NPT but are not recommended for ANPT application. Ground thread taps made to Column 1 are marked NPT and may be used for NPT and ANPT applications. Ground thread taps made to Column II are marked NPTF and used for Dryseal application.

TABLE 341

FORMULAE AND TOLERANCES FOR GROUND THREAD TAP LIMITS FOR METRIC THREADS

General

The following tables and formulae are used in determining the limits and tolerances for ground thread metric taps unless otherwise specified. They apply only to metric threads having a 60° form with a P/8 flat at the major diameter of the basic thread form. They apply to both standard and special metric taps.

All calculations for metric taps are done using millimeter values as shown. When inch values are needed, they are translated from the 3 place millimeter tap diameters, only after calculations are complete.

Lead Tolerance

A maximum lead deviation of plus or minus .013 mm within any two threads not farther apart than 25 mm is permitted.

Angle Tolerance

Pitch (mm)	Deviation in Half Angles
over 0.25 to 2.5 Incl.	30' Plus or Minus
over 2.5 to 4 Incl.	25' Plus or Minus
over 4 to 6 Incl.	20' Plus or Minus

Formulae

Min. Major Dia. = Basic plus W

Max. Major Dia. = Min. plus X

Max. Pitch Dia. = Basic plus Y

Min. Pitch Dia. = Max. minus Z

W = Constant to add to Basic Major Diameter (W = .080P)

X = Major Diameter Tolerance

Y = Amount over Basic for Maximum Pitch Diameter

Z = Pitch Diameter Tolerance

Note: When the tap major diameter must be determined from a specified tap pitch diameter, the minimum major diameter equals the maximum specified tap pitch diameter minus Constant Y plus 0.64952P, plus Constant W.

(Concluded on Following Page)

Table 341 (Concluded)
(Values for W, X, Y, and Z in millimeters)

P Pitch mm	W (.08P)	X	Y M1.6 To M6.3 Inc.	Y Over M6.3 To M25 Inc.	Y Over M25 To M90 Inc.	Y Over M90	Z M1.6 To M6.3 Inc.	Z Over M6.3 To M25 Inc.	Z Over M25 To M90 Inc.	Z Over M90
0.3	.024	.025	.039	.039	.052	.052	.015	.015	.020	.020
0.35	.028	.025	.039	.039	.052	.052	.015	.015	.020	.020
0.4	.032	.025	.039	.052	.052	.052	.015	.015	.020	.025
0.45	.036	.025	.039	.052	.052	.052	.015	.020	.020	.025
0.5	.040	.025	.039	.052	.052	.065	.015	.020	.025	.025
0.6	.048	.025	.052	.052	.065	.065	.020	.020	.025	.025
0.7	.056	.041	.052	.052	.065	.065	.020	.020	.025	.025
0.75	.060	.041	.052	.065	.065	.078	.020	.025	.025	.031
0.8	.064	.041	.052	.065	.065	.078	.020	.025	.025	.031
0.9	.072	.041	.052	.065	.065	.078	.020	.025	.025	.031
1	.080	.041	.065	.065	.078	.078	.025	.025	.031	.031
1.25	.100	.064	.065	.065	.078	.091	.025	.031	.031	.041
1.5	.120	.064	.065	.078	.078	.091	.025	.031	.031	.041
1.75	.140	.064	—	.078	.091	.104	—	.031	.041	.041
2	.160	.064	—	.091	.091	.104	—	.041	.041	.041
2.5	.200	.064	—	.091	.104	.117	—	.041	.041	.052
3	.240	.100	—	.104	.104	.130	—	.041	.052	.052
3.5	.280	.100	—	.104	.117	.130	—	.041	.052	.052
4	.320	.100	—	.104	.117	.143	—	.052	.052	.064
4.5	.360	.100	—	—	.130	.143	—	.052	.052	.064
5	.400	.100	—	—	.130	.156	—	—	.064	.064
5.5	.440	.100	—	—	.143	.156	—	—	.064	.064
6	.480	.100	—	—	.143	.156	—	—	.064	.064

For intermediate pitches use value for next coarser pitch.

MINOR DIAMETER OF TAPPED HOLES

Taps are different from other cutting tools since few operating variables are possible. The feed per revolution is fixed by the lead or pitch of the tap. The rate of metal removed per tooth is thus governed by effective chamfer length, number of flutes, revolutions per minute, and minor diameter of the product. In addition, application of cutting fluid is difficult since chips cannot always be removed from the cutting zone and the cross-sectional area of the tap is often small compared to the load imposed. Since freedom of choice as to relief, rake, and shear angles is frequently limited, and the tap must usually stop and reverse in the cut, it should be realized that every reasonable precaution should be taken to favor the tap.

There are four general groups of tapped holes.

First and largest—Holes into which a threaded part is inserted for fastening purposes only and left either for the life of the part or until repairs are needed.

Second—Holes which are used to adjust parts of a machine where the screw or bolt may be tightened or loosened many times.

Third—Holes in which a screw is used to move a slide or nut and hold it at a desired location.

Fourth—Holes for studs.

The first group includes the greater percentage of holes with Unified Form of Thread. It is the group for which reasonable recommendations on minor diameter can be made. Some holes in the second group can be included but other applications may require individual attention.

In Unified Inch Screw Threads ANSI B1.1, Metric Screw Threads ANSI B1.13M, and Screw Threads for Federal Services, Handbook H-28, the maximum minor diameter runs from about 53% engagement with a basic thread plug on a No. 0-80, 61% Engagement on an M1.6 × 0.35 to about 66% engagement on ½" (M12) diameter and larger.

Tests have shown that tapping torque increases when the minor diameter decreases. Two principal reasons for this should be emphasized.

First, the increase in material removed is shown by Figure 66—where a 10-24 basic screw thread is shown by the heavy outline and the Class 2B maximum tapped hole size is shown by the dot and dash lines.

As the thread height increases, the width of chip and amount of material removed increases rapidly. The sketch shows that on a basic thread form a 50% thread height represents the removal of only 31.3% of the basic area, while an increase to 75% thread height increases the area to be removed to 60.9%, or practically double the first amount. With a tapped hole increased to the maximum pitch diameter, the figures increase to approximately 40% and 72% of the area.

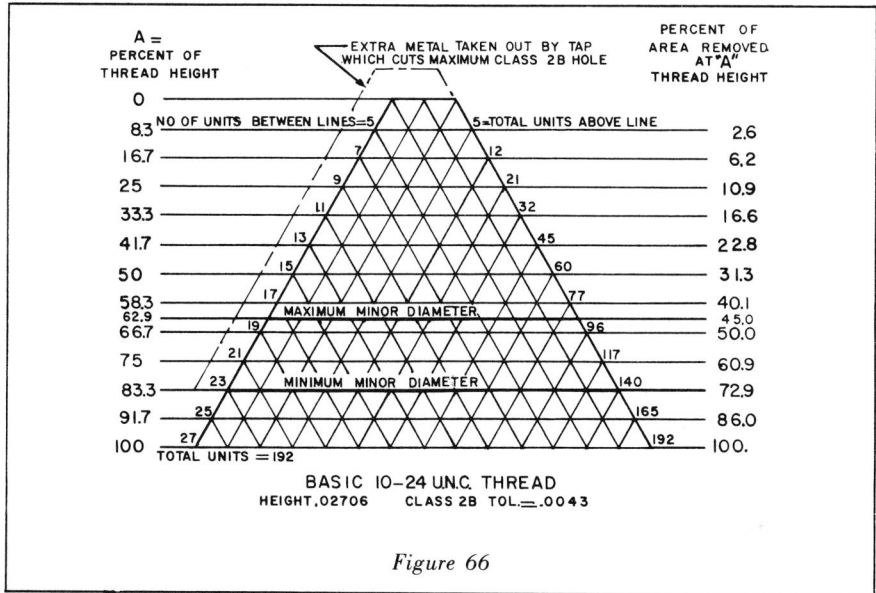

Figure 66

The second cause of increase in torque (except in the case of spiral point or spiral flute taps) is when a tap enters a drilled hole and starts cutting chips that will usually remain in the flutes as the tap advances. If stringy chips result, they roll over and over between the minor diameter of the hole and the bottom of the flutes, causing considerable friction. As the drilled hole becomes smaller, the amount of chips to be removed becomes so great that the friction generated may require as much power as does actual cutting.

This friction increase is apparent where the power used in cutting cast iron with its fine crumbling chips, increases in proportion to the metal removed, but the curling chips from No. 1020 AISI steel clog the flutes of the tap, resulting in an increase in torque out of proportion to the metal removed and detrimental to the tap.

In addition to causing friction, these curling chips score and tear the thread resulting in rough and oversize holes, and leads to work rejection. This is especially true of the Unified Coarse Thread series in sizes ½" and smaller and the Metric Thread Sizes M12 and smaller.

While minor diameter limits, shown in B1.1, B1.13M, and H-28 result in a satisfactory threaded hole, there are cases where a larger hole will save time and tool cost. Tests show that in general the bolt, or external thread, breaks at about 55% thread engagement, and there is very little increase in the strength of the nut when the thread height is increased.

It is, therefore, an advantage to tap users to keep the minor diameter as large as possible. If tapping difficulties continue, the Fine Thread Series should be considered, since the volume of chips is much smaller and the strength of the internal thread is practically the same.

It must be remembered that usually the part to be tapped is the most valuable and that the balance of strength should be in the tapped hole.

The following table shows both the theoretical percentage of thread[1] represented by the drill size and the percentage that would normally be obtained in drilling based on test data.

[1] Percent of thread is the ratio of the actual height of thread to the value .64952P.

Formula For Obtaining Tap Drill Sizes For Cutting Taps

$$\begin{array}{l}\text{Drilled Hole} \\ \text{Size (inches)}\end{array} = \begin{array}{l}\text{Basic Major Dia.} \\ \text{of thread (Inch)}\end{array} - .0130 \times \frac{\text{Percentage of Full Thread}}{\text{No. of Threads per Inch}}$$

$$\begin{array}{l}\text{Drilled Hole} \\ \text{Size (mm)}\end{array} = \begin{array}{l}\text{Basic Major Dia.} \\ \text{of thread (mm)}\end{array} - \frac{\text{Percentage of Full Thread} \times \text{mm Pitch}}{76.98}$$

Formula For Obtaining Tap Drill Sizes For Thread Forming Taps

$$\begin{array}{l}\text{Drilled Hole} \\ \text{Size (inches)}\end{array} = \begin{array}{l}\text{Basic Major Dia.} \\ \text{of thread (Inch)}\end{array} - .0068 \times \frac{\text{Percentage of Full Thread}}{\text{No. of Threads per Inch}}$$

$$\begin{array}{l}\text{Drilled Hole} \\ \text{Size (mm)}\end{array} = \begin{array}{l}\text{Basic Major Dia.} \\ \text{of thread (mm)}\end{array} - \frac{\text{Percentage of Full Thread} \times \text{mm Pitch}}{147.06}$$

Formula For Obtaining Percentage Of Full Thread For Other Drill Sizes

$$\frac{\text{Percentage of}}{\text{Full Thread}} = \frac{\text{No. of Threads}}{\text{per inch}} \times \left(\frac{\text{Basic Major Dia.}}{\text{of Thread (inch)}} \frac{\text{Drill Hole}}{\text{Size (inch)}} \right)$$

$$\frac{\text{Percentage of}}{\text{Full Thread}} = \frac{76.98}{\text{Pitch (mm)}} \times \left(\frac{\text{Basic Major Dia.}}{\text{(mm)}} - \frac{\text{Drill Hole Size}}{\text{(mm)}} \right)$$

Note: Drill size should be smaller than hole size by the probable amount the drill will cut oversize. See tables on following pages.

Tap Drill Sizes[1] For Unified Inch Screw Threads

Tap Size	Tap Drill Size	Decimal Equiv. of Tap Drill (Inches)	Theoretical Percent of Thread (%)	Probable Mean Oversize (Inches)	Probable Hole Size (Inches)	Probable Percent of Thread* (%)
0-80	56	.0465	83	.0015	.0480	74
	3/64	.0469	81	.0015	.0484	71
	1.20mm	.0472	79	.0015	.0487	69
1-64	1.25mm	.0492	67	.0015	.0507	57
	54	.0550	89	.0015	.0565	81
	1.45mm	.0571	78	.0015	.0586	71
1-72	53	.0595	67	.0015	.0610	59
	1.5mm	.0591	77	.0015	.0606	68
	53	.0595	75	.0015	.0610	67
2-56	1.55mm	.0610	67	.0015	.0625	68
	51	.0670	82	.0017	.0687	74
	1.75mm	.0689	73	.0017	.0706	66
2-64	50	.0700	69	.0017	.0717	62
	1.80mm	.0709	65	.0017	.0726	58
	50	.0700	79	.0017	.0717	70
3-48	1.80mm	.0709	74	.0017	.0726	66
	49	.0730	64	.0017	.0747	56
	48	.0760	85	.0019	.0779	78
3-56	5/64	.0781	77	.0019	.0800	70
	47	.0785	76	.0019	.0804	69
	2.00mm	.0787	75	.0019	.0806	68
4-40	46	.0810	67	.0019	.0829	60
	45	.0820	63	.0019	.0839	56
	46	.0810	78	.0019	.0829	69
8-32	3.40mm	.1339	74	.0029	.1368	67
	29	.1360	69	.0029	.1389	62
8-36	29	.1360	78	.0029	.1389	70
10-24	3.5mm	.1378	72	.0029	.1407	65
	27	.1440	85	.0032	.1472	79
	3.70mm	.1457	82	.0032	.1489	76
10-32	26	.1470	79	.0032	.1502	74
	25	.1495	75	.0032	.1527	69
	24	.1520	70	.0032	.1552	64
12-24	5/32	.1563	83	.0032	.1595	75
	22	.1570	81	.0032	.1602	73
	21	.1590	76	.0032	.1622	68
12-28	11/64	.1719	82	.0035	.1754	75
	17	.1730	79	.0035	.1765	73
	16	.1770	72	.0035	.1805	66
1/4-20	16	.1770	84	.0035	.1805	77
	15	.1800	78	.0035	.1835	70
	4.60mm	.1811	75	.0035	.1846	67
	14	.1820	73	.0035	.1855	66
1/4-28	9	.1960	83	.0038	.1998	77
	8	.1990	79	.0038	.2028	73
	7	.2010	75	.0038	.2048	70
	13/64	.2031	72	.0038	.2069	66
	5.40mm	.2126	81	.0038	.2164	72

Machine Screw and Fractional Taps — Recommended Drill Sizes

Tap Size	Drill¹	Decimal Equiv.	% of Thread	Probable Oversize	Probable Hole Size	Probable % of Full Thread*
4-40	45	.0820	73	.0019	.0839	65
	2.10mm	.0827	70	.0019	.0846	62
	2.15mm	.0846	62	.0019	.0865	54
	44	.0860	80	.0020	.0880	74
	2.20mm	.0866	78	.0020	.0886	72
	43	.0890	71	.0020	.0910	65
4-48	2.30mm	.0906	66	.0020	.0926	60
	2.35mm	.0925	72	.0020	.0926	72
	42	.0935	68	.0020	.0955	61
5-40	3/32	.0938	68	.0020	.0958	60
	2.40mm	.0945	65	.0020	.0965	57
	40	.0980	83	.0023	.1003	76
	39	.0995	79	.0023	.1018	71
	38	.1015	72	.0023	.1038	65
	2.60mm	.1024	70	.0023	.1047	63
5-44	38	.1015	79	.0023	.1038	72
	2.60mm	.1024	77	.0023	.1047	69
	37	.1040	71	.0023	.1063	63
6-32	37	.1040	84	.0023	.1063	78
	36	.1065	78	.0023	.1088	72
	7/64	.1094	70	.0026	.1120	64
6-40	35	.1100	69	.0026	.1126	63
	34	.1100	67	.0026	.1136	60
	34	.1110	83	.0026	.1136	75
	33	.1130	77	.0026	.1156	69
	2.90mm	.1142	73	.0026	.1168	65
	32	.1160	68	.0026	.1186	60

Tap Size	Drill¹	Decimal Equiv.	% of Thread	Probable Oversize	Probable Hole Size	Probable % of Full Thread*
5/16-18	3	.2130	80	.0038	.2168	72
	F	.2570	77	.0038	.2608	72
	6.60mm	.2598	73	.0038	.2636	68
5/16-24	G	.2610	71	.0041	.2651	66
	H	.2660	86	.0041	.2701	78
	6.80mm	.2677	83	.0041	.2718	75
3/8-16	I	.2720	75	.0041	.2761	67
	7.80mm	.3071	84	.0044	.3115	78
	7.90mm	.3110	79	.0044	.3154	73
3/8-24	5/16	.3125	77	.0044	.3169	72
	O	.3160	73	.0044	.3204	68
	21/64	.3281	87	.0044	.3325	79
	8.40mm	.3307	82	.0044	.3351	74
	Q	.3320	79	.0044	.3364	71
	8.50mm	.3346	75	.0044	.3390	67
7/16-14	T	.3580	86	.0046	.3626	81
	23/64	.3594	84	.0046	.3640	79
	9.20mm	.3622	81	.0046	.3668	76
7/16-20	9.30mm	.3661	77	.0046	.3707	72
	U	.3680	75	.0046	.3726	70
	9.40mm	.3701	73	.0046	.3747	68
1/2-13	W	.3860	79	.0046	.3906	72
	25/64	.3906	72	.0046	.3952	65
	10.50mm	.4134	87	.0047	.4181	82
1/2-20	27/64	.4219	78	.0047	.4266	73
	29/64	.4531	72	.0047	.4578	65

¹Recommended sizes for cutting taps. For forming taps see formula on pg. 410. Hole sizes shown may not suit UNJ and MJ hole requirements.

*Probable percent of full thread produced in tapped hole using standard drill sizes.

Tap Drill Sizes[1] For Unified Inch Screw Threads

*Probable percent of full thread produced in tapped hole using standard drill sizes.

Tap Size	Tap Drill Size	Decimal Equiv. of Tap Drill (Inches)	Theoretical Percent of Thread %	Probable Mean Oversize (Inches)	Probable Hole Size (Inches)	Probable Percent of Thread* %
9/16-12	15/32	.4688	87	.0048	.4736	82
9/16-18	31/64	.4844	72	.0048	.4892	68
	1/2	.500	87	.0048	.5048	80
5/8-11	17/32	.5313	79	.0049	.5362	75
5/8-18	9/16	.5625	87	.0049	.5674	80
3/4-10	41/64	.6406	84	.0050	.6456	80
3/4-16	21/32	.6563	72	.0050	.6613	68
	11/16	.6875	77	.0050	.6925	71
	17.50mm	.6890	75	.0050	.6940	69
7/8-9	49/64	.7656	76	.0052	.7708	72
7/8-14	51/64	.7969	84	.0052	.8021	79
1"-8	55/64	.8594	87	.0059	.8653	83
1"-12	7/8	.875	77	.0059	.8809	73
	29/32	.9063	87	.0059	.9122	81
	59/64	.9219	72	.0060	.9279	67
1"-14	59/64	.9219	84	.0060	.9279	78
1 1/8-7	31/32	.9688	84	.0062	.9750	81
	63/64	.9844	76	.0067	.9911	72
1 1/8-12	1 1/32	1.0313	87	.0071	1.0384	80

Tap Size	Tap Drill Size	Decimal Equiv. of Tap Drill (Inches)	Theoretical Percent of Thread %	Probable Mean Oversize (Inches)	Probable Hole Size (Inches)	Probable Percent of Thread* %
1 1/4-7	1 3/32	1.0938	84	REAMING RECOMMENDED		
1 1/4-12	1 7/64	1.1094	76			
	1 11/64	1.1719	72			
1 3/8-6	1 13/64	1.2031	79			
1 3/8-12	1 7/32	1.2188	72			
	1 19/64	1.2969	72			
1 1/2-6	1 21/64	1.3281	79			
1 1/2-12	1 11/32	1.3438	72			
	1 27/64	1.4219	72			

[1]Recommended sizes for cutting taps. For forming taps see formula on pg. 410. Hole sizes shown may not suit UNJ and MJ hole requirements.

Tap Drill Sizes[1] for Metric M-Profile Screw Threads

*Probable percent of full thread produced in tapped hole using standard drill sizes.

Metric Tap Size	Tap Drill Size	Decimal Equiv. of Tap Drill (Inches)	Theoretical Percent of Thread %	Probable Mean Oversize (Inches)	Probable Hole Size (Inches)	Probable Percent of Thread* %
M1.6×0.35	1.20mm	.0472	88	.0014	.0486	80
	1.25mm	.0492	77	.0014	.0506	69
M2×0.4	1/16	.0625	79	.0015	.0640	72
	1.60mm	.0630	77	.0017	.0647	69
	52	.0635	74	.0017	.0652	66
M2.5×0.45	2.05mm	.0807	77	.0019	.0826	69
	46	.0810	76	.0019	.0829	67
	45	.0820	71	.0019	.0839	63
M3×0.5	40	.0980	79	.0023	.1003	70
	2.5mm	.0984	77	.0023	.1007	68
	39	.0995	73	.0023	.1018	64
M3.5×0.6	33	.1130	81	.0026	.1156	72
	2.9mm	.1142	77	.0026	.1168	68
	32	.1160	71	.0026	.1186	63
M4×0.7	3.2mm	.1260	88	.0029	.1289	80
	30	.1285	81	.0029	.1314	73
	3.3mm	.1299	77	.0029	.1328	69
M4.5×0.75	3.7mm	.1457	82	.0032	.1489	74
	26	.1470	79	.0032	.1502	70
	25	.1495	72	.0032	.1527	64
M5×0.8	4.2mm	.1654	77	.0032	.1686	69
	19	.1660	75	.0032	.1692	68
M6×1	10	.1935	84	.0038	.1973	76
	9	.1960	79	.0038	.1998	71
	5mm	.1969	77	.0038	.2007	70
	8	.1990	73	.0038	.2028	65

Metric Tap Size	Tap Drill Size	Decimal Equiv. of Tap Drill (Inches)	Theoretical Percent of Thread %	Probable Mean Oversize (Inches)	Probable Hole Size (Inches)	Probable Percent of Thread* %
M7×1	A	.2340	81	.0038	.2378	74
	6mm	.2362	77	.0038	.2400	70
	B	.2380	74	.0038	.2418	66
M8×1.25	6.7mm	.2638	80	.0041	.2679	74
	17/64	.2656	77	.0041	.2697	71
	H	.2660	77	.0041	.2701	70
	6.8mm	.2677	74	.0041	.2718	68
M10×1.5	8.4mm	.3307	82	.0044	.3351	76
	Q	.3320	80	.0044	.3364	75
	8.5mm	.3346	77	.0044	.3390	71
M12×1.75	10.20mm	.4016	79	.0047	.4063	75
	Y	.4040	76	.0047	.4087	71
	13/32	.4062	74	.0047	.4109	69
M14×2	15/32	.4688	81	.0048	.4736	76
	12mm	.4724	77	.0048	.4772	72
M16×2	35/64	.5469	81	.0049	.5518	76
	14mm	.5512	77	.0049	.5561	72
M20×2.5	11/16	.6875	78	.0050	.6925	74
M24×3	17.5mm	.6890	77	.0052	.6942	73
	13/16	.8125	86	.0052	.8177	82
	21mm	.8268	76	.0054	.8322	73
M30×3.5	53/64	.8281	76	.0054	.8335	73
	1 1/32	1.0312	83	.0071	1.0383	80
	26.5mm	1.0433	77	.0071	1.0504	73
M36×4	1 3/64	1.0469	75	.0072	1.0541	70
	1 17/64	1.2656	74	REAMING RECOMMENDED		

[1]Recommended sizes for cutting taps. For forming taps see formula on pg. 410. Hole sizes shown may not suit UNJ and MJ hole requirements.

THE "OVER WIRE" METHOD OF THREAD MEASUREMENT

The most universally accepted method for measuring pitch diameter is over thread measuring wires. This is referred to as "Measurement Over Wires" (M_w). The wire is placed in the thread groove and the measurement taken over it. Usually, three wires are used as shown in Figure 67.

The measurement over wire is determined by adding a constant to the pitch diameter or, where the measurement over wire is known, subtracting the constant to determine the pitch diameter.

For 60° threads, when the helix angle is less than 5°, the formula for the constant (C) is as follows:

$$C = 3W - .86603P$$

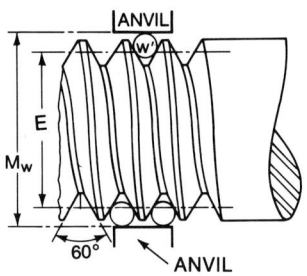

Figure 67

$$E = M_w + \frac{\cot a}{2n} - w(1 + \csc a)$$

in which

   E   = pitch diameter
   M_w = measurement over wires
   a   = one-half included angle of thread
   n   = number of threads per inch
   w   = diameter of wires
   λ' = angle between axis of wire and plane perpendicular to axis of thread

For Unified and Metric M Threads, the "best size" wire and the corresponding constants are shown in the following tables. For further information, refer to ANSI B1.2 and ANSI B1.16M.

Measuring Wires—Standard
"Best Size" for 60° Inch Threads

Threads per Inch	Best Size Wire (w) (Inch)	Constant (c) (Inch)
80	.00722	.01083
72	.00802	.01203
64	.00902	.01353
56	.01031	.01547
48	.01203	.01805
44	.01312	.01968
40	.01443	.02164
36	.01604	.02406
32	.01804	.02706
28	.02062	.03093
24	.02406	.03610
20	.02887	.04331
18	.03208	.04813
16	.03608	.05411
14	.04124	.06186
13	.04441	.06661
12	.04811	.07216
11½	.05020	.07529
11	.05249	.07874
10	.05774	.08662
9	.06415	.09622
8	.07217	.10826
7	.08248	.12372
6	.09623	.14435

See page 418 for metric threads.

Measuring Wires—Standard
"Best Size" for 60° Metric Threads

Pitch (mm)	Best Size Wire (w) (mm)	Constant (c) (mm)
0.25	.1443	.2165
0.3	.1732	.2598
0.35	.2021	.3031
0.4	.2309	.3464
0.45	.2598	.3897
0.5	.2887	.4330
0.6	.3464	.5196
0.7	.4041	.6062
0.75	.4330	.6495
0.8	.4619	.6928
0.9	.5196	.7794
1.0	.5774	.8660
1.25	.7217	1.0825
1.5	.8660	1.2990
1.75	1.0104	1.5155
2.0	1.1547	1.7321
2.5	1.4434	2.1651
3.0	1.7321	2.5981
3.5	2.0207	3.0311
4.0	2.3094	3.4641
4.5	2.5981	3.8971
5.0	2.8868	4.3301
5.5	3.1754	4.7631
6.0	3.4641	5.1962

See page 417 for inch threads.

APPLICATION AND MAINTENANCE OF TAPS

SELECTION OF STANDARD TAPS

Standard stock taps are available in various styles to meet the majority of tapping requirements. These styles include both cutting and forming styles of taps. The cutting taps include straight fluted, spiral fluted, and spiral pointed. They produce the thread by cutting away workpiece material in the form of chips, leaving the threads. The straight fluted taps contain most of the chips that are produced within the flutes. The spiral fluted taps tend to pull the chips up and out of the threaded hole, and the spiral pointed taps tend to push the chips forward through the bottom of the hole. Forming taps produce threads by plastic deformation of the workpiece material. Because they do not cut chips, flutes are not required for the containment or disposal of the chips. Some manufacturers provide one or more grooves to allow equalization of hydraulic pressures, particularly when tapping blind holes that contain tapping fluid.

The choice of the style of tap for a particular tapping job will require consideration of some or all of the following factors:

1. The material to be tapped.
2. The material hardness.
3. Whether through or blind hole.
4. The depth or length of thread being tapped.
5. The percentage of thread height to be produced.
6. The type of tapping fluid.

Constants for Calculating Screw Thread Pitch Diameter

(60° Thread Only)

Basic Pitch Diameter equals Basic Major Diameter minus Basic Thread Height.

Unified Inch Screw Threads

Threads per inch	P Pitch (in inches)	Basic Thread Height (inches) .64952P	Threads per inch	P Pitch (in inches)	Basic Thread Height (inches) .64952P
2	.500000	.324760	12	.083333	.054127
2¼	.444444	.288676	13	.076923	.049963
2⅜	.421053	.273482	14	.071429	.046394
2½	.400000	.259808	16	.062500	.040595
2⅝	.380952	.247436	18	.055556	.036084
2¾	.363636	.236189	20	.050000	.032476
2⅞	.347826	.225920	24	.041667	.027063
3	.333333	.216506	27	.037037	.024056
3¼	.307692	.199852	28	.035714	.023197
3½	.285714	.185577	30	.033333	.021651
4	.250000	.162380	32	.031250	.020297
4½	.222222	.144338	36	.027778	.018042
5	.200000	.129904	40	.025000	.016238
5½	.181818	.118094	44	.022727	.014762
6	.166667	.108253	48	.020833	.013532
7	.142857	.092788	50	.020000	.012990
8	.125000	.081190	56	.017857	.011599
9	.111111	.072169	60	.016667	.010825
10	.100000	.064952	64	.015625	.010149
11	.090909	.059047	72	.013889	.009021
11½	.086957	.056480	80	.012500	.008119

Constants for Calculating Screw Thread Pitch Diameter

(60° Thread Only)

Basic Pitch Diameter equals Basic Major Diameter minus

Basic Thread Height.

Metric Screw Threads—M Profile

P Pitch (mm)	Basic Thread Height (mm) .64952P	P Pitch (mm)	Basic Thread Height (mm) .64952P
0.25	.16238	1.5	.97428
0.3	.19486	1.75	1.13666
0.35	.22733	2	1.29904
0.4	.25981	2.5	1.62380
0.45	.29228	3	1.94856
0.5	.32476	3.5	2.27332
0.6	.38971	4	2.59808
0.7	.45466	4.5	2.92284
0.75	.48714	5	3.24760
0.8	.51962	5.5	3.57236
1	.64952	6	3.89712
1.25	.81190		

Note: Metric Screw Thread diameters are calculated using millimeters and converted to inches when necessary only as a final operation.

ORDERING SPECIAL TAPS

When ordering special taps, it is advisable to furnish as much information as possible so that the tap manufacturer can supply the correct tap with the best design for that application. Whenever possible, part prints and detailed specifications about the job should be furnished. The following form may prove helpful; fill in as many details as possible.

1. DETAILS RELATING TO THE PRODUCT THREADS

Size _____ _____ (Include nominal size, threads per inch or pitch, and thread form)

For multiple threads: pitch _____ lead _____

Class of thread _____ _____

Fill in the next three lines only if special internal thread product limits are needed

Major diameter (max.) _____ (min.) _____

Pitch diameter (max.) _____ (min.) _____

Minor diameter (max.) _____ (min.) _____

2. DETAILS RELATING TO PART DIMENSIONS

Material to be tapped _____. Hardness

_____

Tap drill size or hole size _____
Fill in one condition illustrated below and/or submit part print

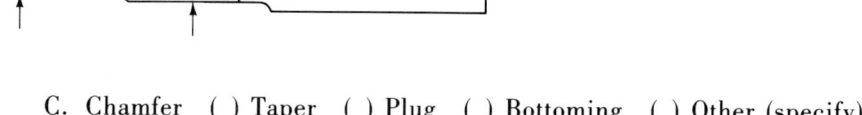

Which end will tap enter?

3. DETAILS RELATING TO TAP

If tap manufacturer is to make a recommendation, check here ( ). Otherwise, specify details below

   A. Style of Tap

   ( ) Straight flute (hand)
   ( ) Spiral point
   ( ) Spiral flute   ( ) Other (specify details)
   ( ) Forming

   B. Dimensions of Tap   ( ) Industry standards   ( ) Other (specify below)

Number of flutes _____

   C. Chamfer ( ) Taper ( ) Plug ( ) Bottoming ( ) Other (specify)

GENERAL OPERATING RECOMMENDATIONS

   A. *Hand tapping.* When using taps by hand, align the tap with the properly drilled hole and avoid reversing the tap before reaching the end of the cut. When reversed, a cutting tap may cause part chips to become wedged in the tap teeth and cause tap breakage. Always use plenty of the recommended tapping fluid, directed toward the cutting area. If tapping blind holes, fill the hole with tapping fluid.

B. *Machine tapping.* The type of machine available, its capacity and mechanical condition, and its adaptability have an influence on successful tapping operations. As an example, the choice between selecting a carbon steel, high-speed steel, or carbide tap is somewhat dependent on the machine's speed capabilities. Some of the points to consider when establishing the optimum operating conditions are the following:

1. The weight, sensitivity, and capacity of the machine.
2. The speed range available.
3. The available horsepower or torque.
4. The condition and alignment of the spindle.
5. The condition and accuracy of the work table and part fixture.
6. The feed mechanism (hand, cam, gears, lead screw, or numerically controlled).

The type of machine to select is obviously the one with the features best suited for each particular application, and one that will be the most cost effective and will produce the quality required.

TAP HOLDING DEVICES

Tap holders are necessary to provide the tap driving means between the tapping machine spindle and the tap. They can be simple tap chucks, split-sleeve drivers or more complicated tension-compression, torque control, floating drivers. There are also multispindle tapping devices; leadscrew tapping devices; reversing and nonreversing devices; quick release type of drivers; and drivers with various combinations of these features.

No matter how simple or complicated the driver, the taps require positive driving means, which are usually holders that drive by the tap square, while gripping the tap shank to provide alignment.

When consistent tapping to a specific depth is required, such as with blind holes having only a few threads allowed between the minimum thread depth requirement and the maximum depth of drilled hole allowed, it is best to choose the holding devices that locate from the back surface of the tap length rather than from the inboard end of the tap square, unless individual adjustment for changing length with each change of the tap is built into the setup or holder. As an example, a $\frac{3}{8}$-inch tap has an overall length tolerance of $\pm \frac{1}{32}$ inch, the depth variation from tap to tap at the extreme opposite ends of this tolerance could, therefore, be $\frac{1}{16}$ inch, which is equal to one pitch on a $\frac{3}{8}$-16 inch tap. The length of square on a $\frac{3}{8}$-inch tap also has a length tolerance of $\pm \frac{1}{32}$ inch. If the holder locates on the inboard end of the square, both tolerances, square length and overall length, can accumulate to make a depth variation from tap to tap at the extreme opposite ends of these tolerances equal ⅛ inch, which is two pitches in length for a ⅜-16 tap.

RECOMMENDED TAPPING SPEEDS

Tapping speeds not only affect productivity, but they can influence the tap life and the part quality. The feed of a tap cannot be varied independently as can be done with other cutting tools. The tap feed is fixed by the number of threads per inch or the lead of the tap threads.

Excessive speed can cause early tap failure due to the high temperature resulting. Also, failure can be accelerated due to the welding of small particles of the workpiece material to the tap surfaces. Factors to be considered when determining the best tapping speed are the following:

A. Type of material to be tapped.
B. Hardness of material to be tapped.
C. Length or depth of hole.
D. Whether through or blind hole.
E. Length of tap chamfer.
F. Lead of the thread.
G. Tapping fluid used.
H. Condition and type of machine.
I. Percentage of full thread to be cut.

The tabulated recommended tapping speeds should be modified for the following:

- Speeds may have to be lowered as the length of hole increases, owing to the accumulation of chips that increase friction and interfere with lubrication.
- Speeds can be increased when using spiral point, spiral flute, or forming taps.
- Speeds should be decreased when using bottoming taps.
- Speeds can be increased as the percentage of full thread is decreased.
- Speeds should be decreased for coarse thread taps.
- Taper pipe taps should be run at between one-half and three-quarters the speed of a comparable tap of the same diameter and pitch.
- The quantity and quality of tapping fluids may affect the tapping speed significantly.

The recommended peripheral speeds shown in the following table are only starting points.

GENERAL RECOMMENDATIONS FOR TAPPING SPEED

Material	Recommended Tapping Speed (fpm)
Aluminum	50–150
Aluminum bronze	20–80
Brass	50–200
Bronze	50–100
Cast iron	50–100
Copper	30–60
Fiber	80–90
Magnesium	75–200
Malleable Iron	35–60
Monel metal	20–40
Plastic—thermoplastic	50–100
thermosetting	50–100
Steel—free machining	40–80
Stainless steel	5–35
Titanium alloys	10–40
Zinc die casting	60–150

Speeds for Metric Taps—Conversion Chart: Feet per minute (fpm) to Revolutions per minute (rpm)

Metric size (mm)	Decimal Size[a]	Feet per minute																	
		5'	10'	15'	20'	25'	30'	35'	40'	50'	60'	70'	80'	90'	100'	125'	150'	175'	200'
1.6	.06295	303	605	910	1210	1520	1820	2120	2430	3030	3640	4240	4850	5450	6050	7600	9100	10600	12100
2	.07874	242	485	730	970	1210	1460	1700	1940	2430	2910	3400	3880	4370	4850	6050	7300	8500	9700
2.5	.09843	194	388	580	775	970	1160	1360	1550	1940	2330	2720	3100	3496	3880	4856	5800	6800	7750
3	.11611	162	323	485	645	810	970	1130	1290	1620	1940	2260	2590	2910	3230	4046	4850	5650	6450
3.5	.13780	138	277	416	555	695	830	970	1110	1390	1660	1940	2220	2490	2770	3460	4160	4850	5540
4	.15748	121	242	364	485	605	730	850	970	1210	1450	1700	1940	2180	2430	3030	3640	4240	4850
4.5	.17717	108	216	323	431	540	645	755	860	1080	1290	1510	1720	1940	2160	2690	3230	3770	4310
5	.19685	97	194	291	388	485	580	680	775	970	1160	1360	1550	1750	1940	2430	2910	3400	3880
6	.23622	80	162	242	323	404	485	565	645	810	970	1130	1290	1450	1620	2020	2430	2830	3230
7	.27559	69	138	208	277	346	416	485	553	695	830	970	1110	1250	1390	1730	2080	2430	2770
8	.31496	60	121	182	242	303	364	424	485	605	730	850	970	1090	1210	1520	1820	2120	2430
10	.39370	48	97	146	194	242	291	340	388	485	580	680	775	875	970	1210	1400	1700	1940
12	.47244	40	80	121	162	202	242	263	323	404	485	565	645	730	810	1010	1210	1410	1620
14	.55116	35	69	104	139	173	208	242	277	346	416	485	555	625	695	865	1040	1210	1390
16	.62992	30	61	91	121	152	182	212	242	303	364	424	485	545	605	760	910	1060	1210
20	.78740	24	48	73	97	121	146	170	194	242	291	340	386	436	485	605	730	850	970
24	.94488	20	40	61	81	101	121	141	162	202	242	283	323	364	404	505	605	705	810
30	1.18110	16	32	48	65	81	97	113	129	162	194	226	259	291	323	464	485	565	645
36	1.41732	13	27	40	54	67	81	94	108	135	162	189	216	242	270	337	404	472	540

[a] Decimal Size Based on Nominal O.D.

Speeds for Inch Taps—Conversion Chart: Feet per minute (fpm) to Revolutions per minute (rpm)

Number Size	Fractional Size	Decimal Size[a]	Feet per minute																	
			5'	10'	15'	20'	25'	30'	35'	40'	50'	60'	70'	80'	90'	100'	125'	150'	175'	200'
0		.060	318	635	955	1275	1590	1910	2230	2550	3180	3820	4460	5100	5750	6350	7950	9550	11100	12700
1		.073	262	525	785	1050	1310	1570	1830	2090	2620	3140	3550	4190	4710	5250	6550	7850	9150	10500
2		.086	222	444	665	890	1110	1330	1550	1780	2220	2660	3110	3550	4000	4440	5550	6650	7750	8900
3		.099	193	386	580	770	965	1160	1350	1540	1930	2310	2700	3090	3470	3860	4820	5800	6800	7700
4		.112	170	341	510	680	855	1020	1190	1360	1700	2050	2390	2730	3070	3410	4260	5110	5950	6800
5	⅛	.125	153	306	458	610	765	915	1070	1220	1530	1830	2140	2440	2750	3060	3820	4580	5350	6100
6		.138	138	277	415	555	690	830	965	1110	1380	1660	1940	2210	2490	2770	3460	4150	4840	5550
8		.164	116	233	350	466	580	700	815	930	1160	1400	1630	1860	2100	2330	2910	3490	4080	4660
10		.190	100	201	302	402	500	605	705	805	1000	1210	1410	1610	1810	2010	2510	3020	3520	4020
12		.216	88	177	265	354	442	530	620	710	885	1060	1240	1410	1590	1770	2210	2650	3090	3540
14		.242	79	158	237	316	394	474	550	630	790	945	1100	1260	1420	1580	1970	2370	2760	3160
	¼	.250	76	153	229	306	382	458	535	610	765	915	1070	1220	1370	1530	1910	2290	2670	3060
	⁵⁄₁₆	.312	61	122	183	244	306	367	428	489	610	735	855	980	1100	1220	1530	1830	2140	2440
	⅜	.375	51	102	153	204	255	306	356	407	510	610	715	815	915	1020	1270	1530	1780	2040
	⁷⁄₁₆	.437	43	87	131	175	218	262	306	349	436	525	610	700	785	875	1090	1310	1530	1750
	½	.500	38	76	114	153	191	229	267	306	382	458	535	610	690	765	955	1150	1340	1530
	⅝	.625	31	61	92	122	153	183	214	244	306	367	428	489	550	611	765	917	1070	1220
	¾	.750	25	51	76	102	127	153	178	204	255	306	356	407	458	510	635	765	890	1020
	⅞	.875	22	44	65	87	109	131	153	175	218	262	306	349	393	436	545	655	765	875
	1	1.00	19	38	57	76	95	114	134	153	191	229	267	306	344	382	477	575	670	765

[a] Decimal Size Based on Nominal O.D.

Suggested Tapping Fluids[a]

Aluminum	Water soluble or oil/chemical especially for aluminum
Ampco metals	Soluble oil
Armco iron	Soluble oil or light base oil
Bakelite	Dry, air jet
Brass	Soluble oil, light duty oil
Bronze	
Aluminum	Soluble oil
Leaded	Soluble or light base oil
Silicon	Soluble oil
Phosphor	Light base oil
Cast iron	
Gray	Dry, air jet or soluble oil
Ductile	Soluble oil or chemical-type coolant
Malleable	Soluble oil or chemical-type coolant
Pearlitic	Soluble oil or chemical-type coolant
Copper	Light base oil
Beryllium copper	Light base oil
Magnesium alloys	Oil specially recommended for magnesium
Plastics	
Thermoplastic	Dry, air jet or water soluble oil
Thermosetting	Dry, air jet
Steel	
Low carbon	Sulfur base oil
High carbon	Sulfo- or chlorinated oil
Cast	Sulfo- or chlorinated oil
Chromium	Sulfo- or chlorinated oil
Cobalt	Sulfo- or Chlorinated oil
Free cutting	Sulfur base oil
Hardened	Sulfo- or chlorinated oil
Manganese	Sulfo- or chlorinated oil
Molybdenum	Sulfo- or chlorinated oil
Nickel	Sulfo- or chlorinated oil
Tool	Sulfo- or chlorinated oil
Vanadium	Sulfo- or chlorinated oil
Stainless steel	Sulfo- or chlorinated oil
Sintered metal	Soluble or light base oil
High-temperature alloys	Sulfo- or Chlorinated oil
Titanium	Sulfo- or chlorinated oil
Zinc	Soluble oil

[a] Note: Forming taps require tapping fluids having higher lubricity characteristics than those fluids used for cutting taps.

TAPPING FLUIDS

Longer tap life, greater production, better size control, smoother and more accurate threads, less resharpening, and more efficient control of chips are some of the positive results traceable to the application of proper tapping fluids. Unfortunately, many tap users do not appreciate the important role that proper fluids play in tapping operations. There is no single fluid that can be applied efficiently and economically for all applications. The user of taps should take advantage of the services and recommendations offered by most suppliers of tapping fluids.

In addition to selecting the proper fluid, some thought should be given to the method of its application. It should be applied at the cutting area under pressure. In horizontal tapping, where the tap is stationary and the part revolves, it is desirable to use more than one stream of fluid—at least one on the top and one on the bottom. Whether the method of tapping is vertical, horizontal, or inclined, the important thing is to have the tapping fluid reach the cutting or forming area of the tap at all times. If the fluid is automatically applied only on the forward stroke of the tap, it should be timed so that the fluid reaches the hole before the tap starts to cut or form.

Filters should be used to keep the fluid clean. If filters are not used, care must be taken to replace the fluid when it is contaminated. Also, the tank and piping should be thoroughly cleaned at regular intervals.

GENERAL RECOMMENDATIONS FOR CUTTING FACE ANGLES

Stock cutting taps are provided with a flute and cutting face design that experience has proven to be best for the majority of tapping requirements in both metallic and nonmetallic materials. It is advisable to try stock taps first. It may be necessary to provide special taps with a cutting face angle designed for a specific application. The things to consider when selecting the best cutting face angle are the following:

1. The material to be tapped.
2. The material hardness.
3. The style of tap.
4. The diameter and pitch of the tap.

Generally, the softer materials require a larger cutting face angle and the harder materials require a smaller cutting face angle.

Following is a chart that gives a starting point to use when it is necessary to establish a cutting face angle. The recommendations shown do not represent a tolerance. They are a range to consider when establishing a nominal cutting face angle, to which the tap manufacturers tolerance is applied.

General Recommendations for Cutting Face Angle

Material	Cutting Face Angle (deg)
Aluminum	10–15
Aluminum bronze	0–5
Brass	0–5
Bronze	5–10
Cast iron	0–5
Copper	15–20
Fiber	5–10
Magnesium	10–15
Malleable iron	5–10
Monel metal	7–12
Plastic	
thermoplastic	5–10
thermosetting	0–5
Steel—free machining	5–10
Stainless steel	10–15
Titanium alloys	7–12
Zinc die casting	10–15

TAPPING THE DIFFICULT TO MACHINE MATERIALS

Some of the more difficult to machine materials include the high-temperature alloys, the high tensile steels, and titanium alloys. Following are some recommendations that may be helpful in attempting to tap these difficult to machine materials.

1. *High-Temperature Alloys.*

The high temperature alloys include nickle-base, cobalt-base, and iron-base alloys used primarily in the aerospace industries and in other applications exposed to extremely high temperatures. Some are work hardening materials and have a high resistance to cutting; they usually require a special tap design. Work hardening and frictional drag need to be minimized by using narrow lands or thread relief (relief to cutting edge, extra back taper, or both). Fine pitch threads and the lowest possible percentage of thread height are recommended. Very slow tapping speeds (5–10 fpm) are recommended. Bottoming holes should be avoided, and the depth of the tapped hole should be kept to a minimum. Surface-treated taps generally help, and sulfoor chlorinated tapping fluids are recommended.

2. *Titanium Alloys.*

The unusual thermal, mechanical, and chemical properties of titanium and its alloys cause difficulties in tapping. These properties are low specific heat, high plasticity of heated chips, pronounced tendency to gall or weld, and high abrasiveness. Taps for titanium alloys should have minimum contact surfaces with the workpiece, which can be accomplished with narrow lands, increased back taper, eccentric or coneccentric relief, or interrupted threads. Surface treatments often assist in minimizing galling and abrasive wear. A good flow of sulfo- or chlorinated oil is recommended. Tapping speeds should be relatively slow (10−40 fpm). Blind or bottoming holes should be avoided. The depth of tapped holes and the percentage of thread should be kept to a minimum.

3. *High Tensile Steel.*

The high tensile steels include alloys that need to be tapped after hardening to the range of 40−50 R_c (sometimes harder). Long chamfer and 0° or negative cutting face angles have helped to minimize problems. Shallow through holes, fine pitch threads, and the largest acceptable tap drill size have been among the most successful applications. Bottoming holes should be avoided. Very slow tapping speeds are necessary (less than 10 fpm). Sulfo- or chlorinated oils are recommended. Surface treatments that improve abrasive resistance are generally helpful in prolonging tap life.

TAP SURFACE TREATMENTS

Tap manufacturers have developed heat treating techniques for high speed steels that produce high speed ground thread taps with great resistance to heat and abrasion without the need for a surface treatment. There are tap applications, however, where the addition of so-called "surface treatments" can add appreciably to the life of a tap. These surface treatments may be roughly classified as nitrided, oxided, chrome plated, titanium nitrided, and vapor blasted.

NITRIDED. This treatment develops a "case" on the tap surface, the hardness of which is greater than the basic hardness of the tap. It is applied to the tap when increased resistance to abrasion is desired. It has been used successfully in a wide variety of materials from plastics to high alloy steels of high hardness.

Careful thought should be given a tapping problem before specifying taps having a nitrided surface treatment. It is usually best to give the tap manufacturer full details of the application and let him advise if nitrided taps can be used advantageously.

OXIDED. There are several types of oxide coatings, most of which produce a black oxide surface. In general, they do not increase the surface hardness of a tap. They do add some degree of lubricity that, together with their oil-absorbing powers, aids materially in decreasing "galling" or the picking up of metal on the tap thread flanks. When applied at elevated temperatures, beneficial results are obtained through the relieving of possible grinding stresses in the tap. Recommended for taps used primarily in tapping steel.

CHROME PLATED. The application of a very thin plate of hard chromium to the thread surfaces of a high speed ground thread tap decreases friction between these thread surfaces and the material being tapped. This results in better "chip flow" and increased tap life.

Chrome plated taps have been used in tapping a wide variety of materials including mild steel, tool steel, plastics, nonferrous materials, and hard rubber.

TITANIUM NITRIDE. This refractory surface treatment is relatively new. It reduces friction and chip welding, and acts as a thermal insulator between the chip and the tool. The areas of application where it appears to have the most potential are in ferrous materials below Rc 40 in hardness and in nonferrous metals.

VAPOR BLAST OR LIQUID HONED. Taps that have been given this surface treatment have a gray, satin finish similar to a dull chrome plate. This surface is produced by exposing the taps to a blast of very fine abrasive, in the form of a "slurry." Through careful control it is possible to remove minute surface scratches and burrs resulting in a finish on the thread flanks and cutting edges similar to that produced by honing.

Taps that have been vapor blasted or liquid honed can be used to tap any material for which the use of an untreated high speed ground thread tap has been satisfactory.

RESHARPENING AND GRINDING TAPS

Taps, like all other cutting tools, become dull through constant use. When dull they are likely to chip, break, produce rough and poor threads, or cut oversize. Furthermore, taps that are dull cut much harder and require more power to drive. They also frequently slow down the tapping machine. This condition can be easily remedied by resharpening. As a rule, the chamfered portion or point is the only part of the tap that requires sharpening. In many instances, this portion is reground by hand either on the periphery or side of the grinding wheel. This method, however, has the disadvantage of producing an uneven grind and usually results in the teeth on one or two of the lands carrying all the burden, thereby placing an excessive strain on the tap with resultant greater power consumption and undue tap breakage. Also, with the uneven grind, the tap tends to cut oversize.

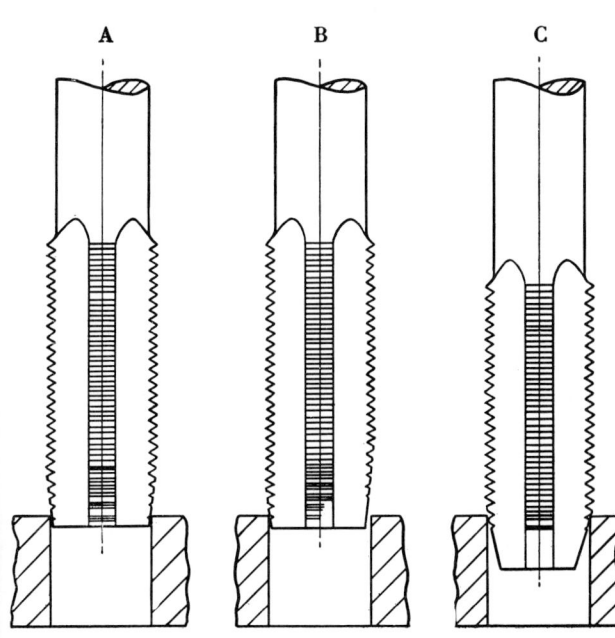

A

Chamfer correctly ground. All lands of chamfered portion in contact. Point diameter large enough to permit only one thread to enter, thus getting full cutting benefit of practically entire chamfer.

B

Chamfer incorrectly ground. Lands of chamfered portions are uneven as to height, thus forcing entire cutting burden on lesser number of lands. Result—poor threads, increased power consumption, and high tap breakage.

C

Chamfer correctly ground as to height of lands, but point diameter too small, forcing entire cutting burden on small portion of chamfered section. Result —greater power consumption. and shorter tap life due to dulling of cutting edges.

The above sketches show the results of both correct and incorrect chamfer grinding.

The ideal way is to resharpen taps on a regular tap grinding machine. There are several such machines available, which reproduce the original grind accurately. This results in easier cutting, more accurate tapping, and longer tap life.

Where a tap-grinding machine is not available, and the tap must be resharpened by hand, great care should be exercised to reproduce as nearly as possible the original grind of the tap manufacturer. To get best results, a new tap should be used for comparison, and the operator should note

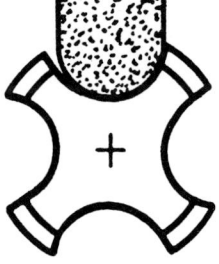

Grinding Flutes in Taps

carefully the number of threads chamfered, the angle, and depth of eccentric relief back of the cutting edge. A soft, 80-grain grinding wheel should be used for this purpose.

For average cutting conditions the recommended rate of relief for taper and plug style chamfers is $4-6°$ and slightly higher for a bottoming-style chamfer. Taps designed for special applications and materials may require special chamfer lengths and relief; consequently, it is good practice to measure and follow the design of the chamfer as supplied by the manufacturer.

Occasionally, large diameter taps require grinding in the flutes to touch up the cutting edge as well as regrinding of the chamfer. Also, the hook of a standard tap may require altering by regrinding the cutting face when used for a special application. A simple method for this form of grinding is to mount the tap between centers on a universal grinder and pass the tap back and forth under the grinding wheel. To prevent the tap from turning, the back face of the land can be held against the blade of a universal tooth rest. A hard 60-grain grinding wheel should be used. The wheel should be of the saucer or dish type when the cutting face to be ground is radial or has a straight rake; when the cutting face is hooked, a straight grinding wheel should be used and its periphery formed to suit the type of hook necessary. Forming or dressing the grinding wheel can be accomplished more accurately with a diamond, taking no deeper cuts than .001″ or .002″ per pass of the diamond. Sketches on p. 433 show how to touch up (regrind) the cutting faces of taps.

If a universal grinder is not available, the cutting face may be ground by hand, although in grinding by hand special care should be exercised to make sure the tap passes under the grinding wheel, as nearly as possible, at right angles to the axis of the grinding-wheel spindle. The original outline of the flute in the tap can be used as a guide for shaping the wheel.

When considering resharpening standard taps, it must be remembered that a resharpened tap seldom has more than 60% of the life potential of a new tap, since resharpening does not correct chipped edges, material that welds to the flanks of tap threads, or size loss due to back taper as the tap is cut back. This should be weighed in relation to the cost of a new tap compared to costs of collecting, categorizing, and sharpening used taps, plus protecting and redistributing them.

OFFSET FOR GRINDING TAP FLUTES

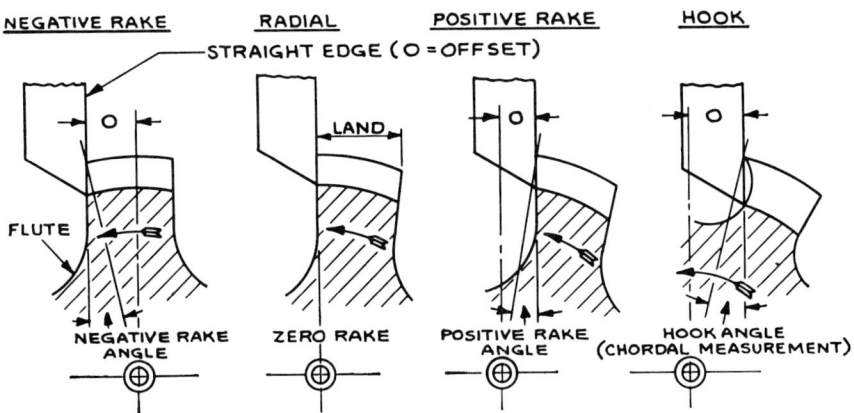

SIZE OF TAP	ANGLE OF RAKE OR CHORDAL HOOK															
	1°	2°	3°	4°	5°	6°	7°	8°	9°	10°	12°	14°	16°	18°	20°
	OFFSET FROM CENTERLINE OF TAP														
#6	.001	.002	.004	.005	.006	.007	.009	.010	.011	.012	.015	.018	.020	.023	.026
#8	.001	.003	.004	.006	.007	.009	.010	.012	.013	.015	.018	.021	.024	.027	.030
#10	.002	.003	.005	.007	.008	.010	.012	.014	.015	.017	.020	.024	.028	.031	.035
#12	.002	.004	.006	.008	.010	.012	.014	.015	.017	.020	.023	.028	.032	.036	.040
¼	.002	.004	.007	.009	.011	.013	.016	.018	.020	.022	.027	.032	.037	.041	.046
⁵⁄₁₆	.003	.006	.008	.011	.014	.017	.020	.022	.025	.028	.034	.040	.046	.052	.058
⅜	.003	.007	.010	.013	.017	.020	.023	.027	.030	.033	.040	.047	.055	.062	.069
⁷⁄₁₆	.004	.008	.012	.016	.019	.023	.027	.031	.035	.039	.047	.055	.064	.070	.081
½	.004	.009	.013	.018	.022	.027	.031	.035	.040	.045	.054	.063	.073	.082	.092
⁹⁄₁₆	.005	.010	.015	.020	.025	.030	.035	.040	.045	.050	.061	.071	.082	.093	.104
⅝	.006	.011	.017	.022	.028	.033	.039	.044	.050	.056	.067	.079	.091	.103	.115
¾	.007	.013	.020	.027	.033	.040	.047	.053	.060	.067	.081	.095	.109	.123	.138
⅞	.008	.015	.023	.031	.039	.046	.054	.062	.070	.078	.094	.110	.127	.144	.161
1	.009	.018	.026	.035	.044	.053	.062	.071	.080	.089	.107	.126	.144	.164	.184
1⅛	.010	.020	.030	.040	.050	.060	.070	.080	.090	.100	.121	.142	.163	.185	.207
1¼	.011	.022	.033	.044	.055	.066	.077	.089	.100	.111	.134	.157	.181	.205	.230
1⅜	.012	.024	.036	.049	.061	.073	.085	.098	.110	.122	.147	.173	.199	.225	.252
1½	.013	.026	.040	.053	.066	.080	.093	.106	.120	.133	.161	.189	.217	.246	.276
⅛ Pipe	.003	.007	.010	.013	.017	.020	.023	.027	.030	.033	.040	.047	.054	.061	.069
¼ "	.005	.009	.014	.019	.023	.028	.033	.038	.042	.047	.057	.067	.077	.087	.097
⅜ "	.006	.012	.018	.024	.029	.035	.041	.047	.053	.059	.071	.084	.096	.109	.122
½ "	.007	.015	.022	.029	.037	.044	.051	.059	.066	.074	.089	.104	.120	.136	.152
¾ "	.009	.018	.027	.037	.046	.055	.064	.074	.083	.092	.111	.130	.150	.170	.190
1 "	.011	.023	.034	.046	.057	.069	.080	.092	.103	.115	.139	.163	.188	.212	.238
1¼ "	.015	.029	.043	.058	.072	.087	.101	.116	.131	.146	.176	.206	.237	.269	.302
1½ "	.017	.033	.050	.066	.083	.100	.116	.133	.150	.167	.202	.236	.272	.308	.345

TROUBLE SOURCES IN TAPPING

The various steps incidental to good tapping have already been considered, and, as a guide for locating the source of tapping troubles which may be encountered, it is suggested that the points listed below be checked.

Type of Machine

> Are spindle, fixture, and work in alignment?
> Was tap started correctly?
> Is drive uneven because of slipping belts?
> Is machine powered properly?
> Are tap and drilled hole in alignment?
> Is there undue wear on sliding parts?

Tap Holding Device

> Is worn or wrong type of holder being used?
> Is holder in alignment with drilled hole?

Type of Holes to be Tapped

> If a blind hole, is there sufficient untapped space at the bottom for the accumulation of chips?
> Does condition call for a two- or three-fluted tap?

Class of Fit Required

> If the tap produces an oversize hole, has the proper tap been selected for the class of fit desired?
> If proper tap is being used, is there any play in the work or tap holding spindles?
> Do the work and tap line up accurately?
> Is the tap dull?

Tapping Different Materials

> Has the tap the proper cutting face for the particular material being tapped?
> Is the tap of the proper design or type?

Proper Hole Sizes before Tapping

> Is the drilled hole of the proper size?
> Is the drilled hole perfectly round?
> Is the axis of the hole parallel to the axis of the tap?

Tapping Fluid

Has the proper tapping fluid been employed?

Does the tapping fluid flood the tap sufficiently while engaged in the hole?

> Is there sufficient force behind the tapping fluid to wash away the chips?
> If applied with a brush, has the tapping fluid a sufficiently heavy body to adhere to the tap? (A light lubricant will be thrown off the revolving tap before it enters the hole.)

TYPES OF STANDARD SCREW THREADS

There are numerous standard and special screw thread forms that are used for various applications. These must be properly identified and used in order to ensure interchangeability of threaded parts. The following are the identifying features of some of the more common forms and their source standard, as of the publication date of this document.

GENERAL PURPOSE SCREW THREADS

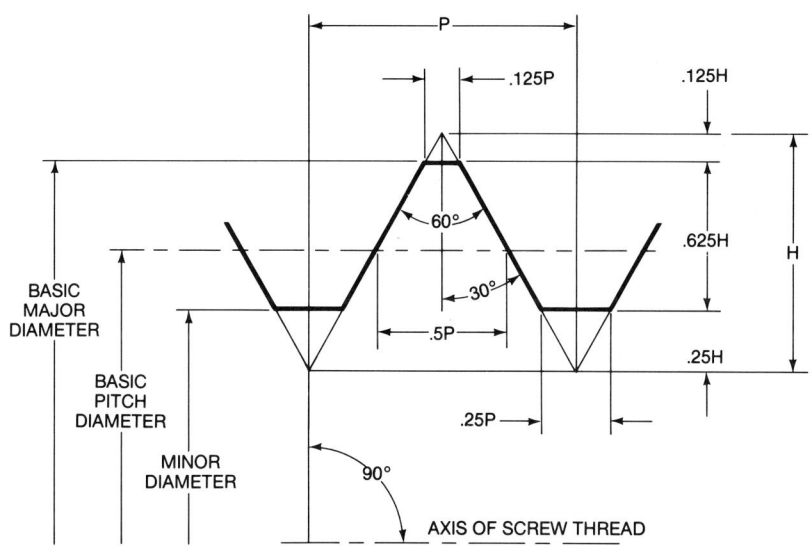

Fig. 68—Basic Profile for Unified Inch Screw Threads (ANSI NSI B1.1).
Basic Profile for ISO General Purpose Metric Screw Threads (ISO 68).
Basic Profile for Metric Screw Threads—M Profile (ANSI B1.13M).

The Basic Profile as detailed in Fig. 68 is identical for Unified Inch, ISO Metric, and M Profile Metric Screw Threads. However, the Design Profile for maximum material condition for internal and external threads, the system of limits, and the gaging practice used for each may differ. The appropriate document should be consulted when applying these thread forms.

AMERICAN SYMMETRICAL SCREW THREAD

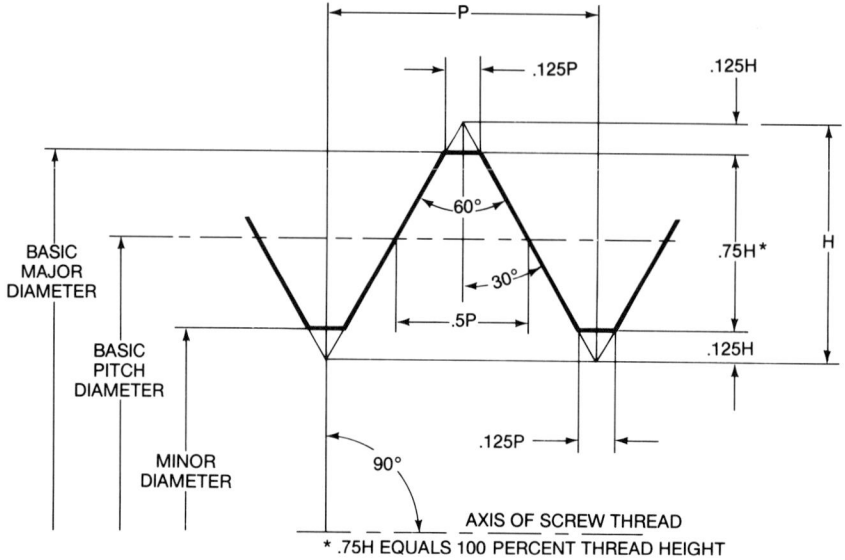

Figure 69—Basic Profile for American Symmetrical Screw Threads

This is not a Unified or Metric thread standard, but is the form used for calculating percentage of thread height. The long established practice in the United States of considering 100 percent thread height equal to 0.75H is based on the American Symmetrical Thread having the Basic Profile shown in Fig. 69.

UNIFIED INCH SCREW THREADS

Unified Inch Screw Thread Standard (ANSI B1.1) is an integrated system of threads for fastening purposes in mechanisms and structures. Its outstanding characteristic is general interchangeability of threads achieved through the standardization of thread form, diameter-pitch combinations, and limits of size. The Unified Standard originated in an Accord signed at Washington, D.C., by representatives of Standardizing Bodies of Canada, The United Kingdom, and the United States on November 18, 1948. Unified Standard threads supersede American National standard threads.

INTERCHANGEABILITY. Unified and the superseded American National threads have substantially the same thread form and are mechanically interchangeable. The principal differences relate to application of allowances, variation of tolerances with size, differences in amounts of pitch diameter tolerances for external and internal threads, and differences in thread designations.

DESIGNATIONS. Unified thread sizes (specific combinations of diameter and pitch) are identified by the letters "UN" in the thread symbol. In the Unified standards, pitch diameter tolerances for external threads differ from those for internal threads; the letter "A" used in the thread symbol denotes an external thread and the letter "B" an internal thread. Where the letters "U", "A", or "B" do not appear in thread designations, threads conform to the superseded American National standard threads.

Limits and tolerances for Unified screws (external threads) are designated by the letter "A" resulting in Class 1A, Class 2A, and Class 3A screws, while the nut (internal threads) limits and tolerances are designated by "B" resulting in Class 1B, Class 2B, and Class 3B.

TOLERANCES. In the Unified series, the pitch diameter tolerance of the tapped hole is always ⅓ times the tolerance of the screw for the corresponding class. In the American National Standard, pitch diameter tolerances on both nut and screw were equal: the nut above and screw below basic.

SCREW THREAD CLASSES. Thread classes are distinguished from each other by the amount of tolerance and allowance.

Class 1A and Class 1B. The combination of Class 1A for external threads and Class 1B for internal threads is intended to cover the manufacture of threaded parts where quick and easy assembly is necessary or desired and an allowance is provided to permit ready assembly.

Class 2A and Class 2B. The combination of Class 2A for external threads and Class 2B for internal threads designed for screws, bolts, and nuts is also suitable for a variety of other applications. A similar allowance is provided which minimizes galling and seizure encountered in assembly and use. It also accommodates to a limited extent, plating, finishes, or coatings.

Class 3A and Class 3B. The combination of Class 3A for external threads and Class 3B for internal threads is provided for those applications where closeness of fit and accuracy of lead and angle of thread are important. These threads are obtained consistently only by use of high quality production equipment supported by a very efficient system of gaging and inspection. No allowance is provided.

METRIC SCREW THREADS—M PROFILE

Metric Screw Threads—M Profile Screw Thread Standard (ANSI B1.13M) is a system of metric screw threads for general fastening purposes in mechanisms and structures. It is in basic agreement with ISO screw thread standards. ANSI B1.13M was approved as an American National Standard on February 12, 1979.

Similar threads are shown in ANSI B1.18M except that these threads are specified with Boundary Profile tolerances instead of Pitch Diameter sizes as shown in ANSI B1.13M. The Boundary Profile tolerances are derived from class 6H fits for the internal threads and would use the same tap "D" limits as shown here for class 6H threads to ANSI B1.13M.

INTERCHANGEABILITY. Metric Screw Threads—M Profile have the same thread form and are mechanically interchangeable with 60° ISO metric screw threads.

DESIGNATIONS. Metric Screw Threads—M Profile are designated by the letter "M" followed by the nominal size in millimeters, and the pitch in millimeters, separated by the sign "×."

TOLERANCE GRADE. The tolerance grade number is used to designate the amount of product tolerance. The system provides for a series of tolerance grade for each of the four screw thread parameters. a. Minor diameter of internal thread (grades 4, 5, 6, 7, and 8). b. Major diameter of external threads (grades 4, 6, and 8). c. Pitch diameter of internal threads (4, 5, 6, 7 and 8). d. Pitch diameter of external threads (3, 4, 5, 6, 7, 8 and 9). The underlined tolerance grades are preferred and used for normal lengths of engagement.

TOLERANCE POSITION. Letters are used to designate the "position" of the product tolerance relative to basic diameters. An upper case letter is used for internal threads, and a lower case letter for external threads. Internal threads can have position G or H, and external threads e, g, or h. The underlined letters are preferred. Tolerance classes 6H/6g are intended for metric applications where inch classes 2A/2B have been used. See ANSI B1.13M for complete details.

Examples:
M12 × 1.25-6H (Class 6 internal threads—tolerance position H—no allowance)

M4 × 0.7-6g (Class 6 external threads—tolerance position g—small allowance)

CONTROLLED RADIUS SCREW THREADS

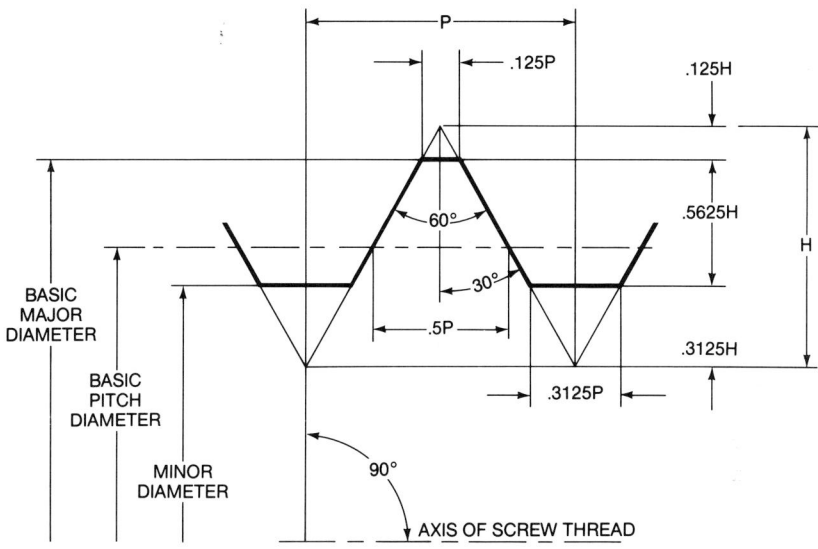

Fig. 70 Basic Profile of UNJ Thread Form.
Basic Profile of MJ Thread Form.

UNIFIED INCH SCREW THREAD—UNJ THREAD FORM

The major characteristic of the UNJ inch series of thread is the 0.15P to 0.18P controlled radius at the root of the external axis thread. The minor diameters of the external and internal threads increased above the ANSI B1.1 UN thread form to accommodate the external thread maximum root radius. The Basic Profile is shown in Fig. 70.

This thread is intended for aerospace inch threaded parts and for high stressed applications requiring high fatigue strength or no allowance application.

The thread specifications are detailed in MIL-S-8879 and in proposed American National Standard ANSI B1.15.

METRIC SCREW THREADS—MJ PROFILE

The major characteristic of the MJ metric series of thread is the 0.15P to 0.18P controlled radius at the root of the external thread. The minor diameters of the external and internal threads are increased above the ANSI B1.13M. The Basic Profile is shown in Fig. 70.

The application and design parallel those of the UNJ thread form.

The thread specifications are detailed in American National Standard ANSI B1.21M.

AMERICAN NATIONAL STANDARD TAPER PIPE THREAD

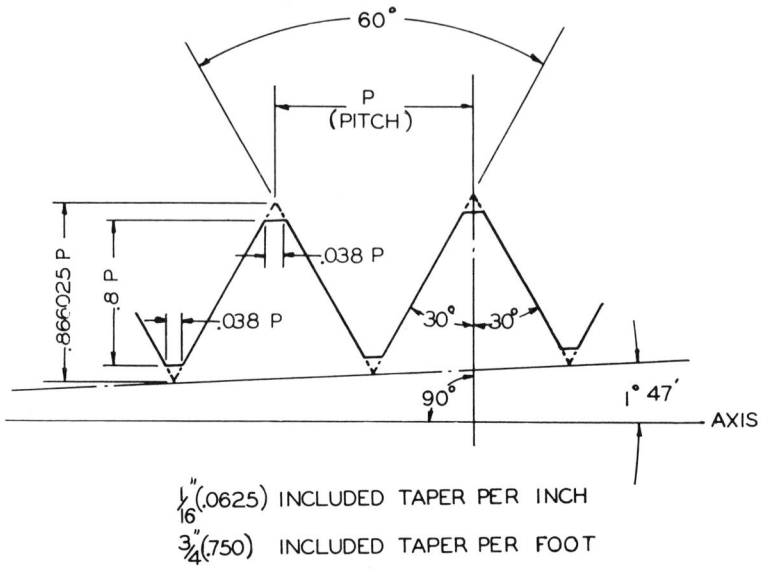

Fig. 71.
Basic Thread Form, NPT (American National Standard Taper Pipe Threads, General Purpose)

The American National Standard Taper Pipe Thread (NPT) is commonly used for the normal type of pipe joint employing tapered external and internal threads. For complete specifications, refer to ANSI B1.20.1 or Screw Thread Standards For Federal Services (Handbook H28).

This thread is designed for use where the assembled product is not required to withstand high fluid or gas pressures and when the use of a sealing compound to provide a leak-proof joint is not functionally objectionable.

The National Pipe Thread product truncation limits for both the external and internal thread are identical, so that hand-tight engagement can result in thread flank contact only or crest and root contact only. Wrench tightening normally produces an assembly with crest and root clearances, although at extremes of the truncation limits entire profile contact of mating threads may result. Excessive wrench tightening of National Taper Pipe Threads in an attempt to secure a leak-proof joint without the use of a sealing compound is inadvisable in that the result may be malformed or stripped threads.

NPTF (DRYSEAL AMERICAN NATIONAL STANDARD TAPER PIPE THREAD); PTF–SAE–SHORT (DRYSEAL SAE SHORT TAPER PIPE THREAD)

The Dryseal American National Standard Taper Pipe Threads have the same basic dimensions as the American National Standard Taper Pipe Threads except for the crest and root flats. The external threads are tapered, while the internal threads may be tapered or straight. They are used for pressure tight joints that must be assembled without the use of lubricant or sealer. The joints are designed to assemble wrench tight with metal-to-metal contact of the entire thread profile to eliminate spiral leakage.

Dryseal American National Standard Taper Pipe Threads are designated NPTF; in some cases, such threads are used with a reduced thread length and are designated as PTF–SAE Short, PTF–Special Short, or PTF–Special Extra Short, depending on the amount of reduction of the thread length.

For complete specifications refer to latest SAE Handbook, ANSI B1.20.3, or H28 Handbook.

NPTF Application

The Dryseal American National Standard Pipe Thread for fuel connection has both external and internal application. It is designed for use where the assembled product must withstand high fluid or gas pressures without the use of a sealing compound or where a sealer is functionally objectionable.

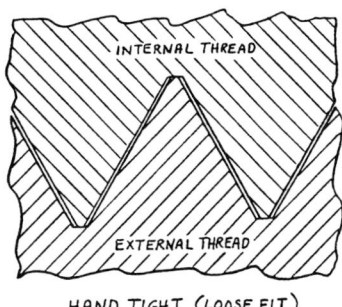

HAND TIGHT (LOOSE FIT)

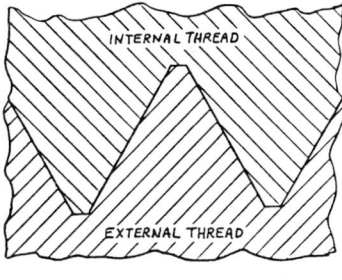

WRENCH TIGHT (SEAL FIT)

The Dryseal pipe thread joints (which are to be made up without sealing compound) consist of external taper and internal taper or straight pipe threads having the same general form and dimensions as NPT threads, but differing in the amount of truncation at the crest and root. The amount of truncation is designed to produce complete mating profile or interference at crests and roots of external and internal threads when the joint is made up hand tight. When made up wrench tight, the crest and roots are crushed sufficiently to bring the flanks of external and internal threads in contact. It is this feature that eliminates the need of a sealer. In many applications, however, a lubricant may be necessary for wrench makeup without damage to the threads.

PTF–SAE–Short Application

The Dryseal SAE Short Taper Pipe Thread has both external and internal application. Its thread form and general characteristics are the same as the Standard NPTF—shortening of the length of engagement is effected by eliminating 1 pitch (thread) from the small end of the external pipe thread. The internal thread is shortened by eliminating 1 pitch (thread) at the large end of the coupling. It is intended that the SAE short external-taper pipe thread be assembled with Dryseal SAE Intermediate Internal Straight Pipe Threads; it will, however, assemble with NPTF (Dryseal) Internal Taper Threads. The SAE Short Internal Taper Pipe Thread is intended for assembly with NPTF (Dryseal) External Taper Pipe Thread. The depth of drilled hole is shortened and the internal thread must be tapped with SAE Short Taps. The term "short" applies to the projection of the tap through the standard L-1 Ring Thread Gage.

ANPT (AERONAUTICAL NATIONAL FORM TAPER PIPE THREAD MIL-P-7105)

The Aeronautical National Form Taper Pipe Thread (ANPT) is similar to the American National Standard Taper Pipe Thread (NPT). The difference is in the gaging methods specified, which provide a rigid control of the thread elements of the product. For complete specifications refer to Military Specification, Pipe Threads, Taper, Aeronautical National Form, MIL-P-7105.

ANPT Application

The Aeronautical National Form Taper Pipe Thread has both external and internal application. It is intended for use where the assembled product, made up using a sealing compound, is satisfactory for pressure-tight joints or where a sealer is not functionally objectionable.

The use of this thread system has generally been replaced with the Dryseal NPTF form, which can, in many applications, be used with a lubricant–sealer for better assurance of sealing.

AMERICAN NATIONAL STANDARD STRAIGHT PIPE THREAD

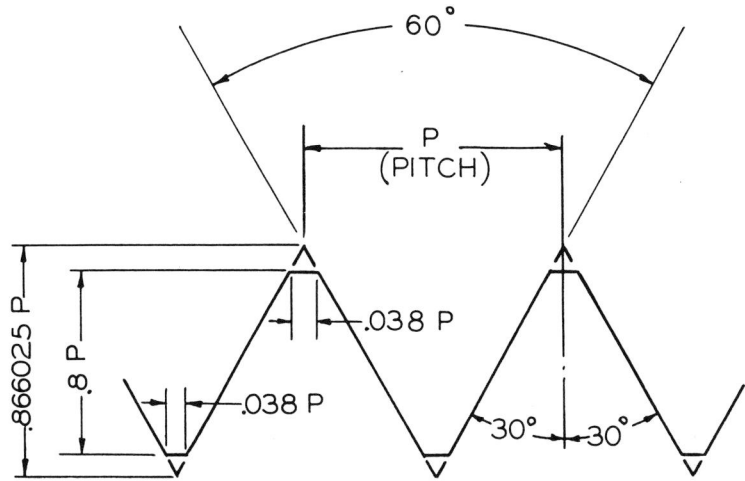

Fig. 72. Basic Thread Form

Several types of American National Standard Straight Pipe Threads are commonly used for specific types of service. They have the same thread form and basic dimensions as American National Standard Taper Pipe Threads and are designated as (NPS) with an additional letter added to indicate the application.

American National Standard Straight Pipe Coupling Threads (NPSC) are used for low-pressure pipe couplings that assemble with American National Standard Taper Pipe (NPT) external threads.

American National Standard Straight Pipe Threads for Free-Fitting Mechanical Joints (NPSM) are used for assembling pipe and fittings that are not required to withstand fluid pressures. Such joints consist of straight internal and external threads.

American National Standard Straight Pipe Threads for Locknut Connections (NPSL) are used where loose-fitting connections are required. They were developed for the seal joints commonly used in tank nipple connections. The internal thread will assemble freely with the largest full thread that can be cut on the outside diameter of a standard pipe.

For complete specifications refer to ANSI 1.20.1 or Screw Thread Standards For Federal Services, Handbook H28.

DRYSEAL AMERICAN NATIONAL STRAIGHT PIPE THREADS

Dryseal American National Standard Straight Pipe Threads are designated NPSF or NPSI and have a thread form the same as that of the NPTF thread. The NPSI thread is larger than the NPSF thread by one turn of the NPTF thread gage.

NPSF Internal Threads

Dryseal American National Standard Fuel Internal Straight Pipe Thread is a straight internal thread intended for assembly with NPTF external-taper pipe. It covers a range of sizes $\frac{1}{16}''$ to $1''$ inclusive. The product truncation at the crest and root is held to a tolerance that will provide for crest and root interference when assembled with NPTF External Taper Pipe.

NPSI Internal Threads

Dryseal American National Standard Intermediate Internal Straight Pipe Threads are intended for assembly with NPTF SAE Short External Taper Pipe. The specification covers a range of sizes $\frac{1}{16}''$ to $1''$ inclusive. The product truncation at the crest and root is held to a tolerance that will provide for crest and root interference when assembled with PTF SAE Short External Taper Pipe.

AMERICAN NATIONAL STANDARD STRAIGHT PIPE THREADS FOR HOSE COUPLINGS

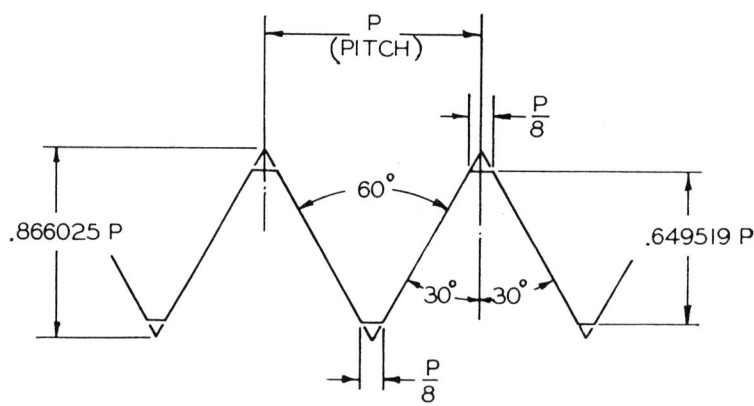

Fig. 73. Basic Thread Form

American National Standard Straight Pipe Threads for Hose-Couplings (NPSH) and American National Fire Hose Coupling Thread (NH) have the same basic thread form as the Unified Thread. For complete specifications refer to ANSI B2.4 or Screw Thread Standards For Federal Services, Handbook H28.

Hose-Coupling Threads apply to threaded parts of hose couplings, valves, nozzles, and other fittings used in direct connection with hose intended for fire protection, or for domestic, industrial, and general service in nominal sizes ½", ⅝", ¾", 1", 1¼", 1½", and 2".

The American National Fire-Hose Coupling Thread (NH) is recommended for use on all couplings and hydrant connections for fire protection systems, and other purposes where hose couplings and connections are required, in nominal sizes 2½", 3", 3½" and 4½".

For complete specifications refer to Screw Thread Standards For Federal Services, Handbook H28.

AMERICAN STANDARD GAS CYLINDER VALVE THREADS

Several screw threads are used in connection with valves and cylinders for compressed gases. Standards have been established for the Valve outlet threads and for the Valve to cylinder connection.

Valve outlet threads of various types are used to prevent accidental connection to equipment or gases that might create a hazard.

Straight threads for Valve outlets are designated NGO (National Gas Outlet) and have the American National thread form. Tapered threads for such outlets are designated NGT (National Gas Taper) and are similar to American Standard Pipe Threads except in length.

Threads for the Valve to cylinder connection are NGT threads, SGT (Special Gas Taper) threads, or NGS (National Gas Straight) threads. SGT threads are similar to NGT except that the taper is ⅛" per inch (1½" per foot) on diameter. NGS threads are also specified as NGIS (National Gas Inlet Straight) threads and are similar to American Standard Straight pipe threads for mechanical joints (NPSM).

For complete specifications refer to ANSI B57.1.

AMERICAN PETROLEUM INSTITUTE THREADS

American Petroleum Institute Threads (API) are of many types and are used in connection with oil well drilling equipment.

For complete specifications refer to the API Standards published by the American Petroleum Institute.

BRITISH STANDARD WHITWORTH THREAD

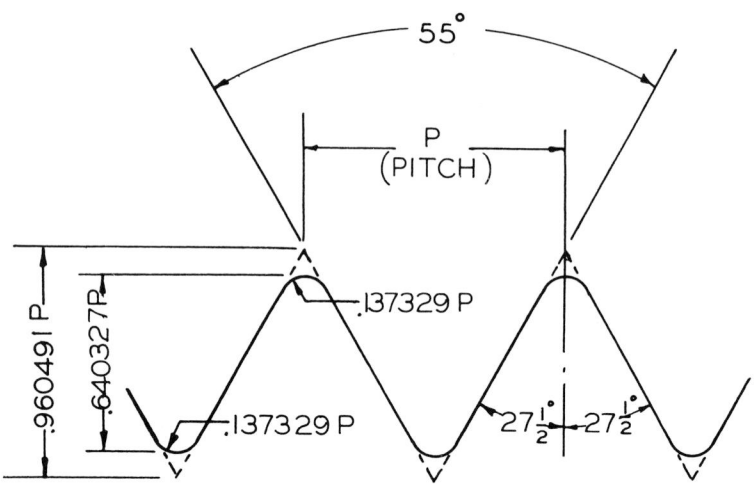

Fig. 74. Basic Thread Form

The British Standard Whitworth Thread Form originated in Great Britain.

The British Standard Whitworth Thread (BSW) is a coarse thread series and its application is similar to the Unified and American National Coarse Thread Series.

The British Standard Fine Thread (BSF) is used in applications similar to the Unified and American National Fine Thread Series.

For complete specifications refer to British Standards Institute publication BS84.

AMERICAN TRUNCATED WHITWORTH THREAD

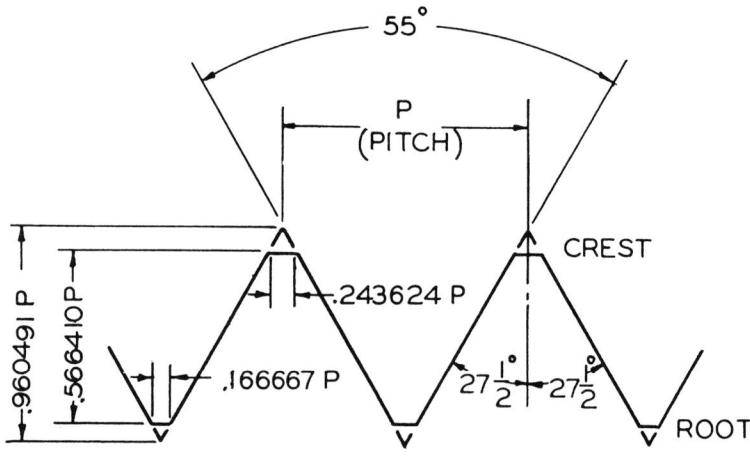

Fig. 75. Basic Thread Form

The American Truncated Whitworth Thread Form differs from that of the British Standard Whitworth Thread Form only in that it has a flat at both crest and root. They are interchangeable.

BRITISH STANDARD PIPE THREADS

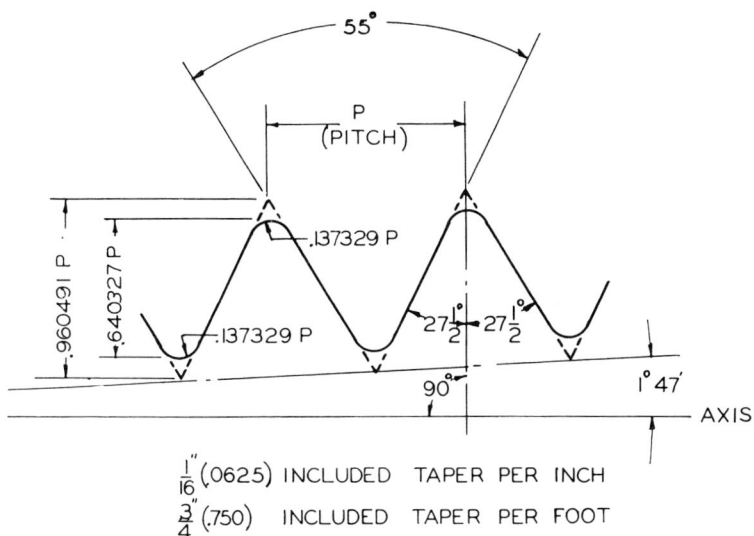

Fig. 76. Basic Thread Form

British Standard Pipe Threads have the Whitworth form of thread and are tapered ¹⁄₁₆″ per inch (¾″ per foot) on the diameter. They are used in Great Britain and other countries for the same purpose as the American Standard Taper Pipe Thread. For complete specifications refer to British Standards Institute publication BS21 and ISO 7.

British Standard Parallel Pipe Threads (BSPP) also have the Whitworth thread form but are parallel to the axis and are used in Great Britain and other countries for general use. For complete specifications, refer to publications BS21, BS2779, and ISO 7.

AMERICAN STANDARD ACME THREAD

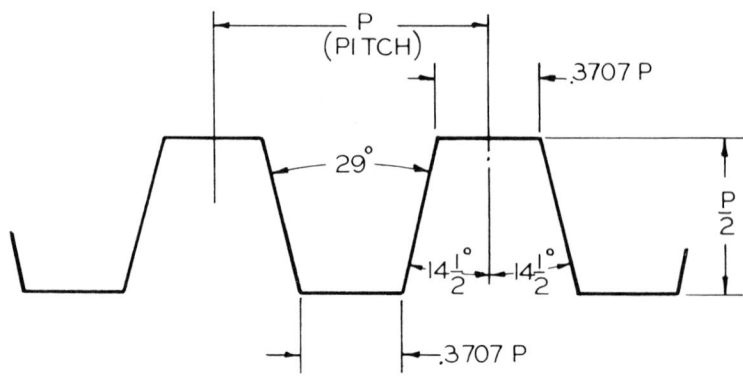

Fig. 77. Basic Thread Form

American Standard Acme Threads are used chiefly to produce traversing motions on machines, tools, etc.

American Standard Acme General Purpose Threads have clearance on all diameters to allow free movement and may be used in assemblies where axial alignment of both threaded members is maintained by their mounting or bearings. Clearance is provided at the thread flanks by making the pitch diameter of the external thread smaller than basic by a specified allowance. Major and minor diameter clearance is obtained by increasing the thread height at the root of both external and internal threads.

American Standard Acme Centralizing Threads have clearance at the pitch and minor diameters, but a very limited clearance at the major diameter so that alignment of the screw and nut is controlled. Alignment by this means prevents wedging of the thread flanks when positive alignment is not provided by the screw and nut mountings.

For complete specifications refer to ANSI B1.5 or Screw Thread Standards For Federal Services, Handbook H28.

STUB ACME THREAD

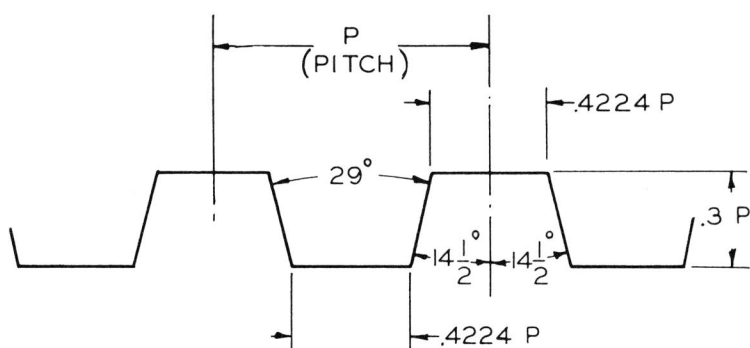

Fig. 78. Basic Thread Form

Stub Acme Threads are used for purposes similar to Acme General Purpose Threads where a coarse-pitch thread of shallow height is required due to mechanical or metallurgical considerations.

Clearances and allowances similar to those for Acme General Purpose Threads are provided.

For complete specifications, refer to ANSI B1.8 or Screw Thread Standards For Federal Services, Handbook H28.

60° STUB THREAD

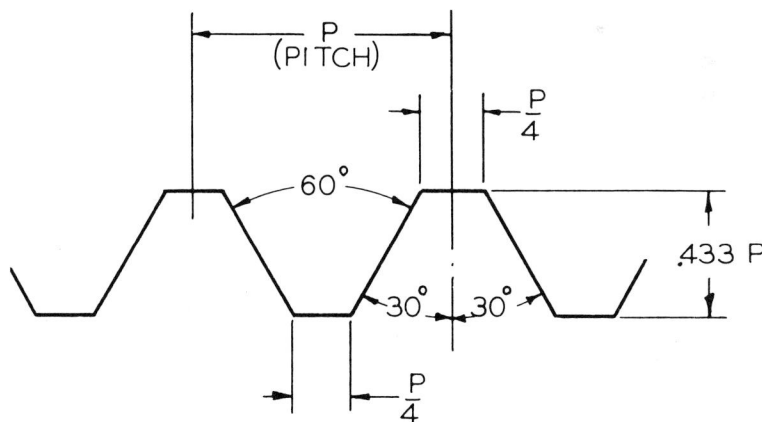

Fig. 79. Basic Thread Form

60° Stub Threads are often substituted for Acme Threads where the axial loads do not require the smaller flank angle. They are also easier to produce. This form of thread is also used where a shallow thread height is desired such as on "Fitting-Up" bolts and nuts.

Clearances similar to those used on Acme Threads are provided at the major and minor diameters.

For complete specifications refer to Screw Thread Standards For Federal Services, Handbook H28 and Bolt, Nut, and Rivet Standards of the Industrial Fasteners Institute.

MODIFIED SQUARE THREAD

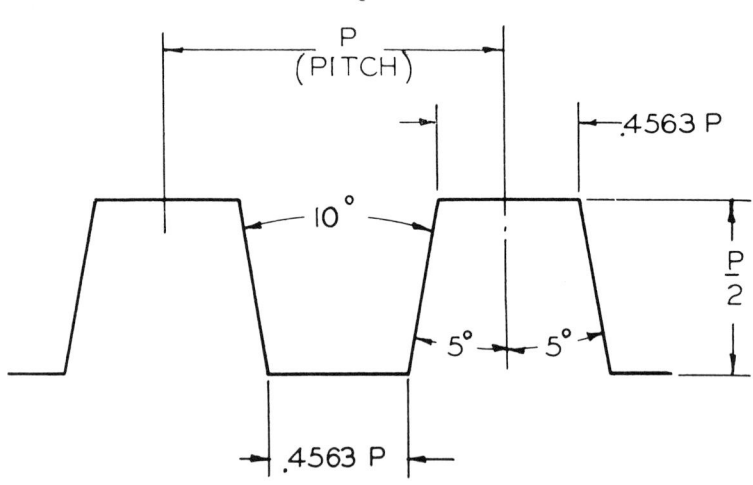

Fig. 80. Basic Thread Form

Modified Square Threads are equivalent to Square Threads so far as practical considerations are concerned, and are easier to produce. However, they are more difficult to produce than Acme Threads and should not be used where Acme Threads can serve the same purpose.

For complete specifications refer to Screw Thread Standards For Federal Services, Handbook H28.

AMERICAN STANDARD BUTTRESS THREAD

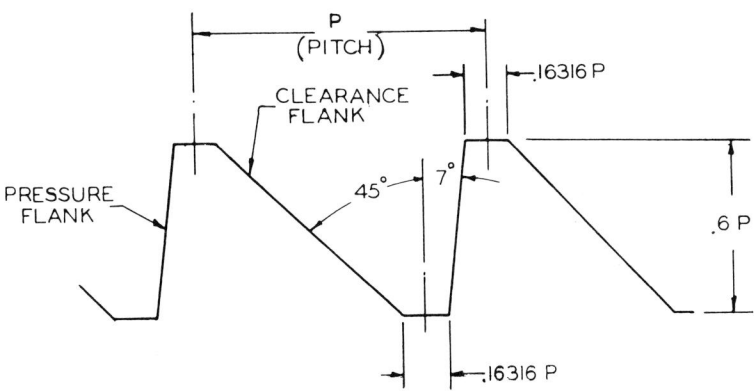

Fig. 81. Basic Thread Form

The Buttress Thread has certain advantages in application involving a high axial load in one direction only. As the pressure flank of the thread is made nearly perpendicular to the thread axis, the radial component of the axial thrust is reduced to a minimum.

For complete specifications refer to ANSI B1.9 and Screw Thread Standards for Federal Services, Handbook H28.

MILLING CUTTER AND END MILL SECTION
(Milling Cutters and End Mills)

FOREWORD

Information in this section, compiled by engineering specialists, is presented as a guide for milling cutter users.

Selection of cutters best suited for a particular job, has a bearing on efficiency, life and production.

Work materials, horsepower, machine and fixture conditions, accuracy demands, desired production and tool-sharpening methods, require certain milling cutter characteristics.

It is important to keep cutters in prime cutting condition.

Operating conditions best for cutter performance and life, methods of correct sharpening to retain original tool accuracy throughout the cutter life, are discussed herein.

Proper design of milling cuttters is not discussed as this is best served by the cutter manufacturers.

MILLING CUTTER and END MILL SECTION CONTENTS

(continued)

MILLING CUTTER and END MILL SECTION CONTENTS
(concluded)

NOMENCLATURE OF MILLING CUTTERS

I—DEFINITION

Milling Cutter—A rotary cutting tool provided with one or more cutting elements called teeth, which intermittently engage the workpiece and remove material by relative movement of the workpiece and cutter.

II—GENERAL CLASSIFICATIONS

A—Classification Based on Construction.

1—*Solid Cutters*—Those made of one piece of tool material such as High Speed Steel.

2—*Tipped Solid Cutters*—Have a body of one material with tips or cutting edges of another material brazed or otherwise bonded in place.

3—*Inserted Blade Cutters*—Have replaceable mechanically retained blades, either solid or tipped, and usually adjustable.

4—*Indexable Insert Cutters*—Have replaceable, mechanically retained inserts, which are sometimes adjustable.

B—Classification Based on Type of Relief on Cutting Edges.

1—*Profile Sharpened Cutters*—Those on which the relief is obtained, and which are resharpened, by grinding a narrow land back of the cutting edges. Profile sharpened cutters may produce flat, curved or irregular surfaces.

2—*Form Relieved Cutters*—Those which are so relieved that, by grinding only the faces of the teeth, the original form is maintained throughout the life of the cutters. Form relieved cutters may produce flat, curved or irregular surfaces. The relieved surfaces may be ground or unground.

C—Classification Based on Method of Mounting.

1—*Arbor Type Cutters*—Those which have a hole for mounting on an arbor and usually have a keyway to receive a driving key. Sometimes called SHELL TYPE.

2—*Shank Type Cutters*—Have a straight or tapered shank to fit into the machine tool spindle or adapter.

3—*Face Type Cutters*—Those mounted directly on and driven from the machine spindle nose.

III—EXPLANATION OF THE "HAND" OF MILLING CUTTERS

The terms "right hand" and "left hand" are used to describe hand of rotation, hand of cutter, and hand of flute helix.

A—Hand of Rotation (or Hand of Cut).

1—*Right Hand Rotation (or Right Hand Cut)*—The counter-clockwise rotation of a cutter revolving so as to make a cut when viewed from a position in front of a horizontal milling machine and facing the spindle.

2—*Left Hand Rotation (or Left Hand Cut)*—The clockwise rotation of a cutter revolving so as to make a cut when viewed from a position in front of a horizontal milling machine and facing the spindle.

B—Hand of Cutter.

Some types of cutters require special consideration when referring to their hand. These are principally cutters with unsymmetrical forms, face type cutters or cutters with threaded holes.

1—*Symmetrical Cutters*—May be reversed on the arbor in the same axial position and rotated in the cutting direction without altering the contour produced on the workpiece, and may be considered as either right or left hand.

2—*Unsymmetrical Cutters*—Reverse the contour produced on the workpiece when reversed on the arbor in the same axial position and rotated in the cutting direction.

[a] *Single Angle Milling Cutters (Arbor Type)*

1—When the larger diameter side faces the viewer, a single angle milling cutter is Right Hand if it must be rotated counter-clockwise to make a cut. When the larger diameter side faces the viewer, a single angle milling cutter is Left Hand if it must be rotated clockwise to make a cut.

2—*Special Case—Threaded Holes*—The hand of rotation of single angle milling cutters with threaded holes is established as shown in Figure 1 and is not necessarily the same as the hand of the cutter.

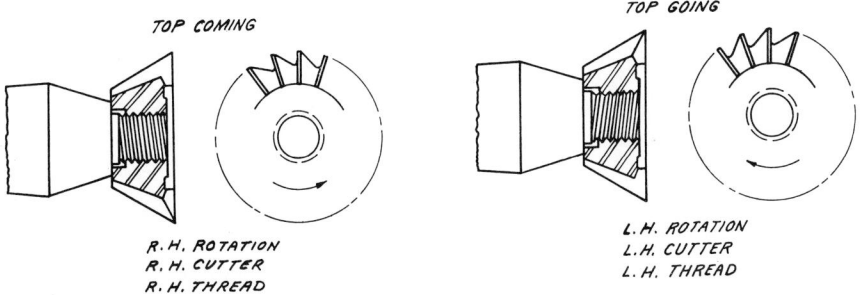

Figure 1—Hand of Rotation of Single Angle Milling Cutters with Threaded Holes.

[b] *Single Corner Rounding Cutters (Arbor Type)*—When the small diameter side faces the viewer, the cutter is right hand if rotated counterclockwise to make a cut, and is left hand if rotated clockwise to make a cut.

[c] *Other Unsymmetrical Cutters (Arbor Type)*—The hand of other unsymmetrical milling cutters, including those with threaded holes, should be clearly specified in relation to the desired form. The recommended way of establishing the hand of rotation is to note "Top Coming" or "Top Going" on a profile view of the desired cutter.

C—Hand of Flute Helix.

1—*Straight Flutes*—Milling cutters with their cutting edges in planes parallel to the cutter axis.

2—Milling cutters with flute helix in one direction only are described as having Right Hand or Left Hand helix:

[a] *Right Hand Helix*—When the flutes twist away from the observer in a clockwise direction when viewed from either end of the cutter.

[b] *Left Hand Helix*—When the flutes twist away from the observer in a counter-clockwise direction when viewed from either end of the cutter.

3—*Staggered Tooth Cutters*—Milling cutters with every other flute of opposite [right and left hand] helix.

IV—Various Types of Milling Cutters

Plain Milling Cutters [Figures 2–9]

Plain Milling Cutters are cutters of cylindrical shape, having teeth on the peripheral surface only. They may be solid, indexable, inserted blade or tipped. They are usually profile sharpened, but may be form-relieved.

Figure 2—Light Duty Plain Milling Cutters of narrow width usually have straight teeth. The wider cutters have teeth with a helix angle usually less than 25 degrees. Both are relatively fine tooth cutters.

Figure 3—Heavy-Duty Plain Milling Cutters have coarse teeth and helix angles ranging between 25 and 45 degrees.

Figure 4—High Helix Plain Milling Cutters have coarse teeth and helix angles greater than 45°, but not over 52°.

Figure 5——High Helix Plain Milling Cutter with side teeth added for producing a shoulder with the side of the cutter.

Figure 6—Shank Type Helical Plain Milling Cutter with outer bearing. These are relatively small diameter plain milling cutters with a straight or tapered shank on one end and an outboard bearing portion on the opposite end.

Figure 7——Plain Milling Cutter (Slab Type) with chip breakers.

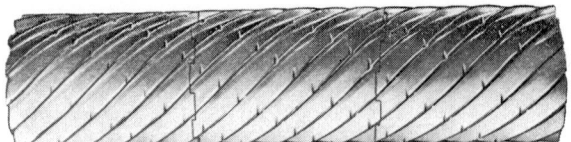

Figure 8—Plain Milling Cutter (Slab Type) with interlocking sections and chip breakers.

Figure 9—Helical Tipped Plain Milling Cutter.

Side Milling Cutters [Figures 10–19]

Side Milling Cutters have peripheral and side teeth and may be solid, tipped, inserted blade, or indexable design with indexable inserts. They are usually profile sharpened, but may be form-relieved.

Figure 10—Side Milling Cutters have side teeth on both sides. The peripheral teeth are usually straight.

Figure 11—Staggered Tooth Side Milling Cutters have peripheral teeth of alternate right and left hand helix and alternate side teeth.

Figure 12—Half Side Milling Cutters have side teeth on one side only. The peripheral teeth are usually helical. Made both right hand and left hand cut. (Right hand shown)

Figure 13—Interlocking Side Milling Cutters are made in two or more sections. Mating sections are similar to half side mills or staggered tooth side mills with uniform or alternate helical teeth so designed that the paths of cutters overlap. See Figure 15 for variations of design.

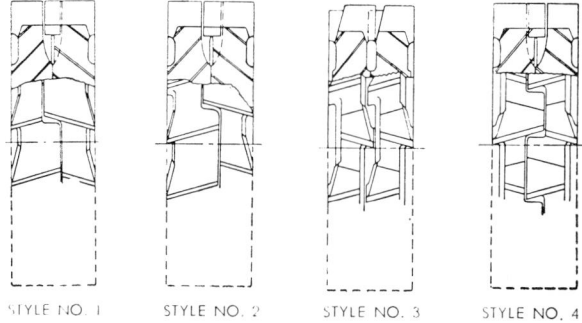

STYLE NO. 1   STYLE NO. 2   STYLE NO. 3   STYLE NO. 4

Figure 14—Variations in design of Interlocking Side Milling Cutters.—Herringbone Design, Style 1, with outside corners in line.—Herringbone Design, Style 2, with outside corners not axially aligned.—Staggered Tooth Design, Style 3, wherein each section has staggered teeth, and the sections are interchangeable.—Staggered Tooth Design, Style 4, wherein each section has staggered teeth with outside corners in line. These sections are not interchangeable.

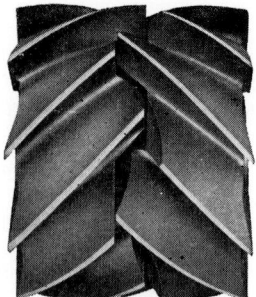

*Figure 15—High Helix Inter-
locking Side Milling Cutter with
deep interlock for wide range
of width adjustment.*

*Figure 16—Tipped Side Mill-
ing Cutter.*

*Figure 17—Staggered Tooth
Side Milling Cutter, with coarse
teeth and relatively high rake.*

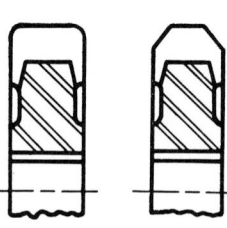 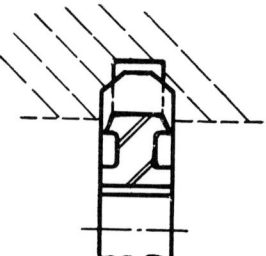

[Left] —*Figure 18 —Side Milling Cutter with radius or corner angle.*
[Right]—*Figure 19 —Side Milling Cutter (Sub-land Type),
with corner angles on alternate teeth.*

Shell Mills [Figures 20–21]

Shell Mills are facing cutters cylindrical in shape with teeth on the periphery and teeth on the end. They are designed for mounting on stub arbors and to mill flat surfaces normal to their axis of rotation or a shoulder parallel to their axis. They are either solid, tipped, inserted blade, or indexable insert design.

Figure 20—Solid Shell Mill.

Figure 21—Inserted Blade Shell Mill.

Saws [Figures 22–26]

Metal Slitting Saws are similar to plain or side milling cutters but are relatively thin. They are usually profile sharpened and may be solid or tipped.

Figure 22—Plain Metal Slitting Saws have peripheral teeth only, with sides concaved.

Figure 23—Side Tooth Metal Slitting Saws have both peripheral and side teeth.

Figure 24—Staggered Tooth Metal Slitting Saws have peripheral teeth of alternate right and left hand helix and alternate side teeth.

*Figure 25——Plain Metal Slitting Saws with alternate
tooth form.*

*Figure 26—Screw Slotting Cutters are saws having fine pitch peripheral teeth only
They are resharpened by grinding the entire tooth.*

Angle Milling Cutters [Figures 27–30]

Angle Milling Cutters have peripheral teeth, the cutting edges of which
lie in conical surfaces. They may also have side teeth and can be solid,
tipped or of inserted blade construction. They are usually profile sharpened, but may be form relieved.

Figure 27—Single Angle Milling Cutter (Arbor Type).

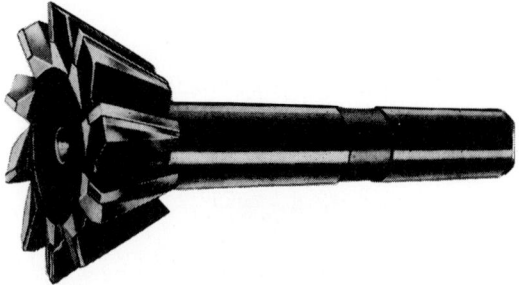

Figure 28—Single Angle Milling Cutter (Shank Type, Tipped).
(Also called "Dovetail Cutter".)

Figure 29—Double Angle Milling Cutter, solid.

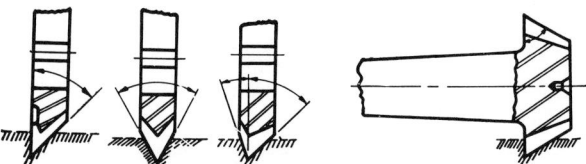

Figure 30—Method of specifying Cutter Angles on Angle Milling Cutters. Note that unsymmetrical angles are designated from a plane perpendicular to the axis.

End Mills [Figures 31–39]

End Mills have straight or helical teeth on the periphery, and usually have end teeth. They may have straight or tapered shanks. Straight Shank End Mills may be single or double end. They may be solid, inserted blade, indexable insert, or tipped construction and may have square, chamfer, radius or ball ends.

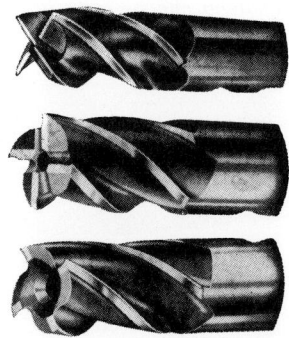

Figure 31—General Purpose End Mills, with 4 or more teeth. They may be made with cup type end, end teeth cut to center holes or ccounterbore, or cut to center.

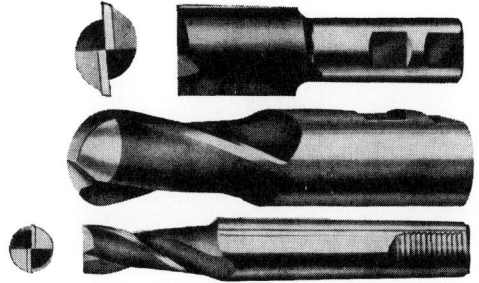

Figure 32—Two Flute End Mills. End teeth are designed to cut to center.

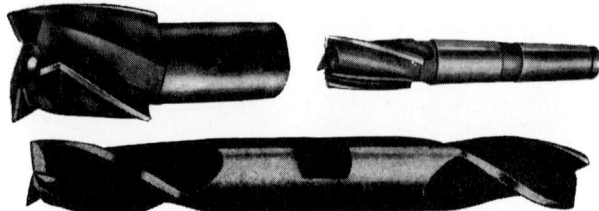

Figure 33—Three Flute End Mills. End teeth may or may not be designed to cut to center.

Figure 34—High Helix End Mills are designed primarily for cutting aluminum and other light metals.

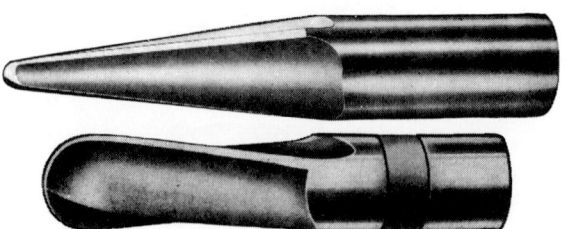

Figure 35—Die Sinking Cutters are straight or tapered end mills with square, ball or radius end teeth, so-called because they are principally used for milling, or "sinking" die or mold cavities.

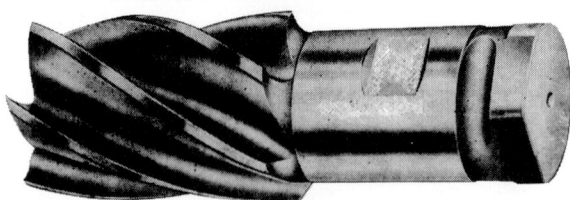

Figure 36—Combination Shank used for large diameter Heavy Duty End Mills.

*Figure 37—Router Bits are shank type cutters
designed for routing.*

*Figure 38—Tracer Milling Cutters are end mills
designed for tracer milling.*

Roughing End Mills are designed to remove a large volume of material quickly, at reduced horsepower and to produce broken chips. The periphery of the mill varies from a truncated thread form to full wave form.

*Figure 39—Hollow Mill, so-called because, being hollow, the workpiece passes into it
after being turned to a diameter.*

T-Slot Milling Cutters

T-Slot Milling Cutters are profile sharpened side milling cutters provided with straight or taper shanks. The cutter portion is usually staggered-tooth. They may be solid or tipped.

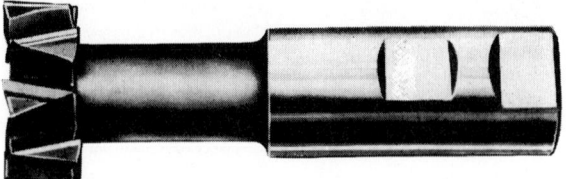

Figure 40—Straight Shank T-Slot Milling Cutter.

Woodruff Keyseat Milling Cutters

Woodruff Keyseat Cutters are profile sharpened and normally used for milling Woodruff keyseats. They may be solid or tipped.

Figure 41—Shank-Type Woodruff Keyseat Cutters are relatively small diameter plain milling cutters provided with straight shanks.

Figure 42—Arbor-Type Woodruff Keyseat Cutters are usually staggered-tooth side milling cutters.

Form Relieved Milling Cutters [Figures 43–54]

A—Form Relieved Milling Cutters are cam relieved cutters and are sharpened by grinding the faces of the teeth. They may be solid, tipped or of inserted blade construction.

Figure 44—Concave Milling Cutters produce convex surfaces of circular contour
equal to a half circle or less.

Figure 43—Convex Milling Cutters produce
concave surfaces of circular contour equal
to half a circle or less.

Figure 45 —Form Re-
lieved Cutter, Interlocking
Concave type, with side
relief and chip breakers.

Figure 46—Corner Rounding Milling Cutters produce convex surfaces of circular contour equal to a quarter circle or less.

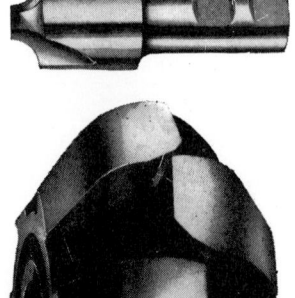

Figure 47—Shank Type Corner Rounding Cutter.

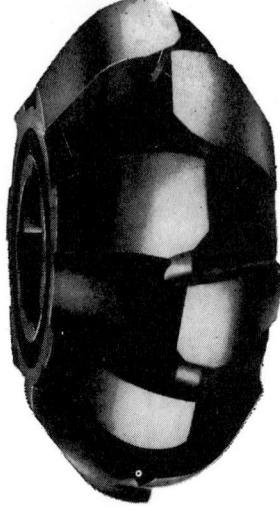

Figure 48—Form Relieved Cutter, Staggered Tooth with positive alternate axial rake and positive radial rake for heavy cuts.

B—Gear Milling Cutters are form relieved cutters to cut the spaces between gear teeth. They may be range type or single purpose type.

Figures 49-50—Two types of roughing gear milling cutters. These are used for roughing out spaces between gear teeth. Various styles are made, differentiated chiefly by the type of chip breakers.

Figure 51—Finishing Gear Cutters finish the spaces between gear teeth. They may be single gear milling cutters, for milling one space at a time or gangs for milling two or more spaces simultaneously. They are also made in gangs consisting of roughers and finishers.

C.—Sprocket Milling Cutters.

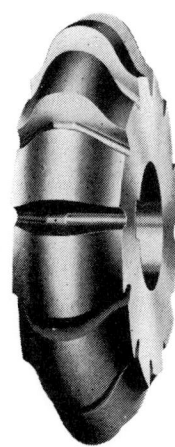

Figure 52—Single Roller Chain Sprocket Milling Cutter, for milling the spaces between sprocket teeth one space at a time.

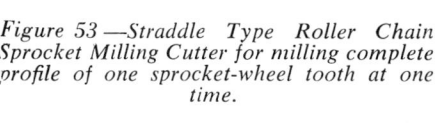

Figure 53—Straddle Type Roller Chain Sprocket Milling Cutter for milling complete profile of one sprocket-wheel tooth at one time.

D--Spline Milling Cutters [Arbor Type].

Figure 54—Single Spline milling cutters cut the space between two adjacent splines.

Thread Milling Cutters [Figures 55–62]

Thread Milling Cutters cut internal or external threads of a specific form. Single cutters may be profile sharpened or form relieved. One piece multiple cutters are form relieved. They may be solid or tipped.

Figure 55—Single.Thread Milling Cutter, profile sharpened type, generally used for milling fine pitch worms or screw threads.

Figure 56—Single Thread Milling Cutter, profile sharpened type, with alternate teeth cut away for chip space. Generally used for coarse pitch worms or screw threads.

Figure 57—Single Thread Milling Cutter, form relieved type.

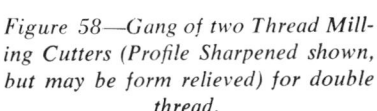

Figure 58—Gang of two Thread Milling Cutters (Profile Sharpened shown, but may be form relieved) for double thread.

Multiple Thread Milling Cutters cut the entire thread length in one revolution of the workpiece. These cutters have no lead and are regularly made for all standard thread types. They are used for both internal and external thread milling, and can be designed for straight or tapered threads. They have form relieved teeth and are arbor or shank type. They are sometimes erroneously called "hobs" or "thread hobs."

Figure 59—Arbor Type Multiple Thread Milling Cutter. Used where space permits mounting on an arbor.

Figure 60—Shank Type Multiple Thread Milling Cutter. Used where space does not permit the use of an arbor type cutter.

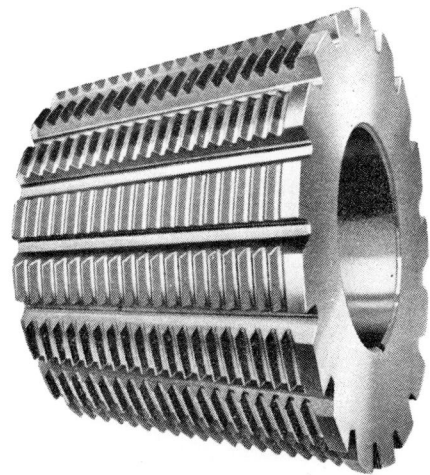

*Figure 61—Tapered Multiple Thread Milling Cutter. Used for milling tapered threads.
Cutter shown has plain section for removing a section of the thread from the workpiece.*

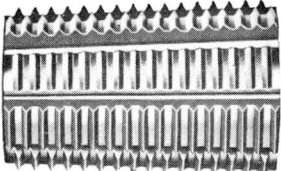

*Figure 62—Alternate Tooth
Type Multiple Thread Milling Cutter. The cutting points
are spaced alternately on adjacent cutter teeth thus avoid-
ing wedging and tearing the
work.*

Form Milling Cutter Gangs [Figures 63–66]

Frequently, complex cutters rather than being made in one piece, are
'ganged," or grouped together. Possible combinations are infinite; they
may be form relieved or profile sharpened type, tipped or inserted blade.
A few examples are shown below:

*Figure 63—Gang of Profile
Sharpened Cutters, for milling a simple form, consisting
of a plane surface, an angular
surface, and two vertical
sides.*

*Figure 64—Gang of Profile
Sharpened Cutters for milling
a complex form.*

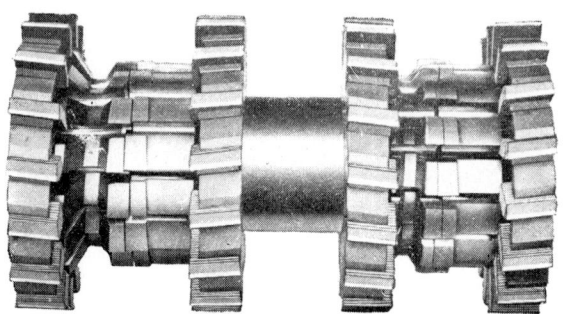

*Figure 65—Gang of Form Relieved Cutters combined with
Profile Sharpened Inserted Blade Cutters.*

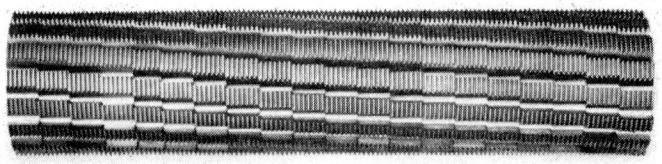

*Figure 66—Gang of Alternate Tooth Type Form Relieved Cutters. Alternate tooth
design produces sharp points on workpiece.*

Inserted Blade and Indexable Insert Cutters [Figures 67–73]

Inserted Blade and Indexable Insert cutters are milling cutters whose design incorporates a mechanical means of clamping a cutting blade to a steel body. The types of cutters available are similar to those designed for solid high speed steel or brazed carbide tip cutters. Exceptions to this are those having size or geometry restraints resulting from the mechanical features or size of the clamping device and blade geometry.

Inserted Blade Milling Cutters

Inserted Blade milling cutters are those cutters of a design having a blade clamped within a slot in a cutter body by means of screws and wedges or by means of the taper configuration of the blade. The blades are usually mounted in a fixed orientation and are profile sharpened in the cutter body in that position throughout the life of the blade. Various design features, however, may permit the blade to be advanced radially and/or axially prior to sharpening so that the original diameter or thickness may be maintained.

Figure 67—
Typical Wedge and Screw Construction

Figure 68—
Tapered Blade Construction

The blades are usually flat or tapered with cutting rakes obtained by the blade position within the body. Some designs may incorporate a helical wedging surface and/or cutting face to produce a helical cutting edge. Serrations, either radial or axial, are often formed in the back of the blade to provide a positive locking feature or a means of incremental adjustment. The blade material is usually either high speed steel or carbide brazed to a pocket in a steel blade.

The specific blade, wedges, and body slot geometry are usually unique with each cutter manufacturer so that replacement blades are seldom interchangeable between various suppliers of this type of cutter.

The object of the inserted blade construction is economy, for both the initial and subsequent costs of maintenance. By using a lower cost material for the body than that which is required at the cutting edge, this feature becomes extremely advantageous when cutters of large physical size are required.

The blades can be quickly replaced or easily adjusted to compensate for wear. This design permits using different blade materials such as high speed steels, premium high speed steels, or the various grades of sintered carbide as job conditions change or dictate. Another feature is that any one or number of blades, when damaged, can be repaired or replaced, usually without disturbing the "good" blades.

The most commonly used inserted blade cutters are face and shell mills designed to be mounted directly to the machine spindle or mounted on a stub arbor. When shell type mills are restricted in size, the arbor (shank) is integrated with the tool and termed an end mill. Other types also used are the various styles of side mills, plain mills, angle cutters and interlocking combinations of these. Cutters with complex forms are occasionally used, but usually the restrictions imposed by the blade assembly limits their use.

Figure 69—Inserted Blade (Solid Blade) Face Milling Cutter.

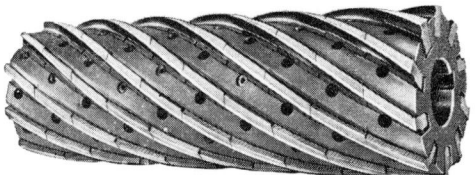

Figure 70——Inserted Blade Plain Milling Cutter.

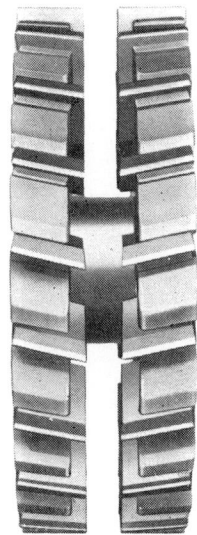

Figure 71—Inserted Blade Half Side Milling Cutters. A pair (Right and Left Hand) as used for straddle milling.

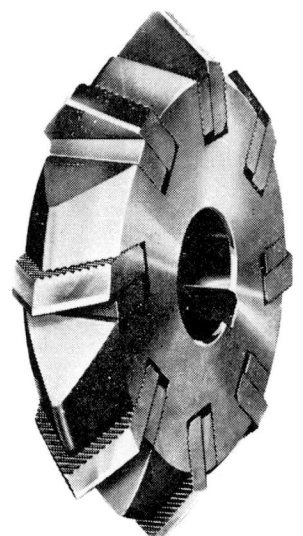

Figure 72—Double Angle Milling Cutter with inserted blades

Indexable Insert Milling Cutters

Indexable insert milling cutters are similar to inserted blade milling cutters with respect to both the range of cutter styles and the principle of clamping a cutting blade to a cutter body as described in the "Inserted Blade Milling Cutter" section.

Figure 73—Typical Indexable Insert Face Milling Cutter.

The differences in blade construction and design, however, are considerable. The blade material is usually solid carbide as opposed to high speed steel or carbide brazed to some type of steel. Cutting relief angles are either preformed into the blade or the blade may be mounted in the cutting body in such a position as to provide both relief and rake. Grooves or other configurations are sometimes also preformed into the face of the blade to control the chip formation or to provide cutting rake. In addition, the most identifying characteristic of this type of blade is that the relief, rake, and corner configuration are repeated in two or more positions about the periphery of the blade. This permits indexing the blades to additional cutting positions—hence the term "indexable insert." On some designs using plain square, triangular, or round blades the indexable feature is the result of positioning with the body, which produces the desired relief and rake for each index.

The resulting design features of the blade for this style of inserted blade cutter produce a cutting edge that, when it becomes dull with use, can be indexed to present a new cutting edge without the need for resharpening.

When all combinations of cutting edges for a specific blade are used, a new blade is usually selected to replace the worn one. Used blades are seldom reconditioned—hence the term "throw-away" as an adjective frequently used to describe these blades. This type of cutter, when properly applied, requires little labor and down time to recondition; therefore, it is very economical to use.

The physical size and design constraints required to make the blade indexable somewhat restricts the application of this type to facing, slotting, or side cutting applications. It is possible, however, to position blade segments about a formed body or to use shaped blades to approximate contoured forms.

Various proprietary means of clamping the blades to the body have been developed for this style of cutter. Some used a wedge placed within a body slot, some use an external clamp, and some use a locking type screw that is inserted through a central hole in the blade. Because of the variety of blade designs and clamping configurations, interchangeability of indexable insert blades may not always be possible between various manufacturers' milling cutter body designs.

V—Nomenclature of Milling Cutter Elements and Other Terms Relating to Milling

Abrasion—The gradual wearing away of the cutting lips by contact with the work material.

Adapter—A device for fitting a tool to, and driving it from, the spindle nose of a machine tool.

Alternate—Cutter features which differ from each other in turn in a regular sequence such as cutting edges, chip breakers, chamfers, or teeth.

Angular Flute—A space between two cutter teeth which forms a cutting edge lying in a plane intersecting the tool axis at an angle. It is unlike a helical flute in that it forms a cutting edge which lies in a single plane. See Figure 74.

Approach Distance—The linear distance in the direction of feed between the point of initial cutter contact and the point of full cutter contact.

Arbor—A device to mount in or on the spindle of a machine tool, and which is designed to carry and drive an arbor type cutting tool.

Arbor Collar—A hollow cylinder which fits an arbor, and which is used to position a cutter.

Arbor Support—A brace or carrier to support the outer end or an intermediate point of an arbor.

Axial Rake—See definition under Rake.

Axial Relief—See definition under Relief.

Axial Runout—The total variation in an axial direction of a cutter element from a true plane of rotation.

Axis—The line about which a cutter rotates.

Back-Off—See preferred term: Relief.

Bearing Collar—An arbor collar which fits over an arbor and in an arbor support bearing of a milling machine.

Bi-Axial Rake—See definition under Rake.

Bi-Negative Rake—See definition under Rake.

Blade—A tooth or cutting device inserted in a holder or cutter body. It may be adjustable and replaceable.

Blade Setting Angle—See preferred term Cone Angle.

Body—The carrier or head for holding blades or inserts; or that part of a solid or brazed tip cutter exclusive of the teeth or shank.

Bolt Circle—The circle on which the bolt holes for mounting a face mill to a machine tool spindle [or adapter] are located.

Built-Up Edge—An adhering deposit of work material on the tooth face adjacent to the cutting edge.

Cam—See definition under Relief.

Cam Relief—See definition under Relief.

Centering Plug—A plug fitting both spindle and cutter to insure concentricity of cutter mounting.

Chamfer—1—A beveled surface to eliminate an otherwise sharp corner —2—A relieved angular cutting edge at a tooth corner.

Chamfer Angle—The angle between a beveled surface and the axis of the cutter.

Chamfer Length—The extent of the chamfer measured parallel to the axis of the cutter.

Chatter—1—An irregularity in cutting action—2—The undesirable finish on the workpiece caused by irregularity in the cutting action.

Chips—The pieces of material removed from the workpiece by the action of milling cutter teeth.

Chip Breakers—Notches or grooves in the cutting edges designed to reduce the size of the chips.

Chipping—The breakdown of a cutting edge by loss of fragments broken away during the cutting action.

Chip Thickness—The thickness of a chip removed by a cutting edge. It is not to be confused with feed per tooth. See Figure 76.

Circular Land—See preferred term Cylindrical Land, under Land.

Circular Pitch—In reference to Milling Cutters—The circular distance between adjacent teeth measured at the periphery of the cutter.

Circular Pitch Variation—The amount of departure from true or specified circular pitch.

CLEARANCE—The additional space provided behind the relieved land of a cutter tooth to eliminate undesirable contact between the cutter and workpiece.

CLEARANCE SURFACES—Angular or curved surfaces behind the relieved land. See Figure 75.

CLIMB MILLING—Rotation of a cutter in the same direction as the feed at the point of contact. Climb milling produces the thickest part of a wedge shaped chip first. See Figure 76.

CLOGGING—The undesirable packing of chips in the flutes of a cutting tool.

COLLAR—See ARBOR COLLAR and BEARING COLLAR.

CONCAVE—1—Departure from a plane where the central portion is below or interior to the outer edges, or—2—Departure from parallelism on a cutter where the diameter is less at the middle than at the ends, or—3—Specifically a cutter designed to produce a convex surface of circular contour. See Figure 44.

CONCAVITY—See definition under RELIEF.

CONE ANGLE—The nominal angle with the cutter axis, along which the blades are moved for adjustment.

CONTOUR MILLING—Milling of a specified shape, essentially with the periphery of a cutter, without axial movement of the cutter relative to the workpiece.

CONVENTIONAL MILLING—Rotation of a cutter in the opposite direction to the feed at the point of contact. Conventional milling produces the thinnest part of a wedge shaped chip first. See Figure 76.

CONVEX—1—Departure from a plane where the central portion is above or exterior to the outer edges, or—2—Departure from parallelism on a cutter where the diameter is greater at the middle than at the ends, or—3—Specifically a cutter designed to produce a concave surface of circular contour. See Figure 43.

CORE DIAMETER—The diameter of a circle which is tangent to the bottoms of the flutes.

CORNER ANGLE—On face milling cutters, the angle between an angular cutting edge of a cutter tooth and the axis of the cutter, measured by rotation into an axial plane. See Figure 78.

COUNTERBORE—As applied to a milling cutter—An enlargement of the cutter bore at one or both ends to provide space for a nut, screw or bolts, or to provide clearance for a shoulder on arbor or spindle. A recess to facilitate manufacturing and sharpening where a cutter has end teeth.

CRATER—A depression in a tooth face eroded by chip contact.

CUTTER SWEEP—The sections removed by the milling cutter or grinding wheel in entering or leaving the flute. See Figure 79.

CUTTING EDGE—The leading edge of the cutter tooth.

CUTTING EDGE ANGLE—The angle which a cutting edge makes with an axial plane at any given point. A constant lead will produce a constant cutting edge angle on a cylindrical cutter, and a varying cutting edge angle on a tapered cutter. A varying lead can be used to produce a constant cutting edge angle on a tapered cutter.

CUTTING SPEED—The peripheral lineal speed resulting from rotation, usually expressed as surface feet per minute [sfm].

CYLINDRICAL LAND—See definition under LAND.

DEPTH OF CUT—The perpendicular distance between the original and the final surfaces of the workpiece being milled, or the thickness of the layer being removed.

DEPTH OF FORM—On a cutter the difference between the radial distances to the highest and lowest points of the cutting edges. See Figure 77.

DISH—See preferred term CONCAVITY under RELIEF.

DOWN MILLING—See preferred term CLIMB MILLING.

DRAW BAR—A rod which holds the cutter, arbor or adapter in the spindle.

DRIVE SLOT—A slot in the side or end of a cutter, arbor or adapter to receive a driving key.

DUPLICATING—See preferred term TRACER MILLING.

ECCENTRIC RELIEF—See definition under RELIEF.

EFFECTIVE DIAMETER—The maximum width of the flat surface which a face mill will produce at one pass. It is the face diameter of a face mill measured between the inner ends of the chamfer or corner angle. See Figure 78.

EFFECTIVE RAKE—See definition under RAKE.

END RELIEF—See preferred term AXIAL RELIEF under RELIEF.

ENTRANCE ANGLE—The angle formed between a centerline on the cutter which is perpendicular to the direction of feed and a radial line through a point on the cutting edge where the tooth first contacts the workpiece. See Figure 80.

FACE—1—The axial cutting length of a plain mill or similar cutter. See Figure 81—2—The cutting side of a face mill. See Figure 78.

FACE CUTTING EDGE—That edge of the tooth on a face mill [or similar cutter] which travels in a plane perpendicular to the axis. It is the edge which sweeps the milled surface in the normal operation of a face milling cutter. See Figure 78.

FACE CUTTING EDGE ANGLE—The angle of concavity between the face cutting edge and the face plane of a face mill. It serves as relief to prevent the face cutting edges from rubbing in the cut. See Figure 78.

FACE OR END MILLING—Milling of a surface which is perpendicular to the cutter axis.

FACE RELIEF—See preferred term AXIAL RELIEF under RELIEF.

FACE RIDGE—The intersection of the main tooth face and the K land. See Figure 80.

FEED—The rate of change of position of the milling cutter as a whole relative to the work while cutting. Usually expressed in inches per minute [ipm].

FEED PER REVOLUTION [fpr]—The feed in inches per revolution of the cutter.

FEED PER TOOTH [fpt]—The feed in inches per revolution divided by the number of active teeth in the cutter. See Figure 76.

FILLET—The bottom surface of the flute. See Figure 75.

FINGER—See preferred term TOOTH REST.

FLAT RELIEF—See definition under RELIEF.

FLUTE—The chip space between the back of one tooth and the face of the following tooth. See Figure 75.

FLUTE LENGTH—As applied to End Mills, the effective axial length of cutting edge. [Not including the cutter sweep.] See Figure 79.

FLUTE SPACING—See preferred term CIRCULAR PITCH.

FLY-CUTTING—A milling operation performed with a rotating single tooth cutter.

FORM CUTTER—Any cutter, profile sharpened or cam relieved, shaped to produce a specified form on the work.

FORM TOOL—As related to milling cutters, a tool used to shape a cutter blank or to produce the form on a cam relieved cutter.

FRONT TAPER—The converse of BACK TAPER defined under RELIEF.

GASH—1—See preferred term FLUTE—2—A term applied to the secondary cuts on a tool to provide chip space at corners and ends. See Figure 79.

GULLET—See preferred term FILLET.

HEEL—1—The back edge of the relieved land. See Figure 79—2—The inner end of a face cutting edge. See Figure 78.

HEEL DRAG—An interference between the heel and the work.

HELICAL—A term describing a cutting edge or flute which progresses uniformly around a cylindrical surface in an axial direction.

HELICAL RAKE—See definition under RAKE.

HELIX ANGLE—The cutting edge angle which a helical cutting edge makes with a plane containing the axis of a cylindrical cutter. See Figure 81.

HOOK—See definition under RAKE.

IN-CUT OR IN-MILLING—See preferred term CLIMB MILLING.

INDEXABLE INSERT—An insert having two or more cutting edges about the periphery of the blade that can be indexed to present each edge in proper cutting position without additional grinding.

INSERT—A cutting blade, having one or more preformed cutting edges, which can be inserted into a body in proper cutting position without additional grinding.

INTERLOCKING—Mating cutter sections on which side projections on one section mesh with those on an adjacent section to provide cutting edge overlap. See Figure 82.

INTERMITTENT—See preferred term ALTERNATE.

"K" LAND—A relatively narrow land on the face of a tooth from the cutting edge inward which is at a lesser rake angle than the main face of the tooth. It is the surface of the tooth on which the chip is intended to impinge so as to reduce contact between the chip and the whole tooth face. See Figure 80.

KEEPER SLOT—A cross slot in a tool shank or adapter to receive a tapered retaining key. See Figure 79.

KELLERING—See preferred term TRACER MILLING.

KEY—A mechanical member which transmits torque from the driving element to the tool.

KEYSEAT—The pocket, usually in the driving element, in which the key is retained.

KEYWAY—The pocket or slot, usually in the driven element, to provide a driving surface for the key.

LAND—The narrow surface of a profile sharpened cutter tooth immediately behind the cutting edge. See Figure 83.

Cylindrical—A narrow portion of the peripheral land, adjacent to the cutting edge, having no radial relief.

Relieved—The portion of the land adjacent to the cutting edge, which provides a relief.

Unrelieved—A narrow portion of the land adjacent to the cutting edge, having no relief.

LEAD—1—The axial advance of a helical cutting edge in one turn around the axis—2—The relieved angular cutting edge between the corner angle and the face cutting edge of a face mill. See Figure 78.

LIP—The material included between a relieved land and a tooth face. See Figure 75.

LIP ANGLE—The included angle between a tooth face and a relieved land. See Figure 75.

LOADING—Undesirable adherence of work material to the surfaces of the tool back of the cutting edges.

LOCKING TYPE SCREW—A special screw capable of securing an insert, wedge, or other device to a body by means of its design and/or assembled position.

MACHINABILITY RATING—A rating usually expressed as a percentage relating the difficulty of machining a given material to a standard. Usually

AISI B-1112 is the standard and a rating of 100 is assigned to it.

Margin—See preferred term Cylindrical Land.

Milling—A method of removing material in the form of chips by means of a rotating cutter.

Neck—The section of reduced diameter between the flutes and shank of a shank type cutter.

Negative Rake—See definition under Rake.

Nesting—The axial overlapping of cutters of different diameters without the use of interlocking projections. See Figure 82.

Nick—See preferred term Chip Breakers.

Non-Topping—Cutting in one operation only the root and sides of a form.

Normal—The condition which exists when a line or plane is perpendicular to another line or plane.

Normal Plane—A plane which is perpendicular to a given surface or a line in a given surface.

Normal Rake—See definitions under Rake.

Offset—1—The distance that a tooth face is positioned off center to obtain the desired rake. See Figure 77.—2—The distance a cutting edge is positioned from a reference plane containing the axis to obtain the desired relief in sharpening. See Figure 83.

Out-Cut or Out-Milling—See preferred term Conventional Milling.

Parallelism—The condition in which the radius to any point of a peripheral cutting edge is constant.

Peripheral Cutting-Edge Angle—See preferred term Corner Angle.

Peripheral Milling—Milling of a surface which is parallel to the cutter axis.

Peripheral Relief—See definition under Relief.

Peripheral Relief Angle—See preferred term Relief Angle under Relief.

Peripheral Speed—See preferred term Cutting Speed.

Periphery—The outside circumference of a cutter. See Figure 75.

Pitch

Milling Cutter—The distance between adjacent projections of a form produced on the work by the cutter.

Roughing End Mill—The distance from any point on the cutter o.d. form to a corresponding point on the next form, measured parallel to the axis and on the same side of the axis.

Plunge Cut—Refers to the relative movement between the work and the cutter directly in line with the center of the cutter when the cutter sinks directly into the work.

Positive Rake—See definition under Rake.

PRIMARY RELIEF—See definition under RELIEF.

PROFILING—See preferred term CONTOUR MILLING.

RADIAL RAKE—See definition under RAKE.

RADIAL RELIEF—See definition under RELIEF.

RADIAL RUNOUT—The total variation in a radial direction of all cutting edges in a plane of rotation.

RAKE—The angular relationship between the tooth face, or a tangent to the tooth face at a given point and a given reference plane or line.

Axial Rake—Applies to angular [not helical] flutes. The axial rake at a given point on the face of the flute is the angle between the tool axis and a tangent plane at the given point. See Figure 78.

Bi-Axial Rake—The combination of two different axial rakes on the same tooth. See Figure 84.

Bi-Negative Rake—The combination of two negative axial rakes on the same tooth. See Figure 84.

Effective Rake—The complement of the angle between the direction of motion of any point on the cutting edge and the direction of chip flow from the same point. Effective Rake is therefore the rake resulting from three factors—1—The cutter geometry.—2—The actual path of the cutting edge. —3—The actual direction of chip flow.

Helical Rake—Applies to helical teeth only [not angular]. The helical rake at a given point on the flute face is the angle between the tool axis and a tangent plane at the given point. See Figure 81.

Hook—A concave condition of a tooth face. The rake of a hooked tooth face must be determined at a given point. See Figure 85.

Negative Rake—Describes a tooth face in rotation whose cutting edge lags the surface of a tooth face. See Figure 80.

Normal Rake—The angle between a normal to the milled surface and a tangent to the tooth face at the same point on the cutting edge, measured in a plane perpendicular to the cutting edge at this same point. Normal Rake is therefore the dynamic rake value established by—1—The cutter geometry.—2—The cutter path; but disregarding the actual chip flow direction.

Positive Rake—Describes a tooth face in rotation whose cutting edge leads the surface of the tooth face. See Figure 77.

Radial Rake—The angle between the tooth face and a radial line passing through the cutting edge in a plane perpendicular to the cutter axis. See Figure 77.

Resultant Rake—The angle between a tangent to the tooth face at a given point on the cutting edge and a radial line to this point, measured in a plane perpendicular to the cutting edge. Resultant Rake is a term descriptive of cutter geometry alone without regard to operating factors and defines the resultant rake obtained on a cutter due to the combination of—1—Radial rake or hook.—2—Axial or helical rake.—3—A corner angle or corner radius.

True Rake—See *Effective Rake*.

Tangential Rake—The angle between a line tangent to a hooked tooth face at the peripheral cutting edge and a radial line passing through this point of tangency. See Figure 85.

RELIEF—The result of the removal of tool material behind or adjacent to the cutting edge to provide clearance and prevent rubbing [heel drag].

Axial Relief—The relief measured in the axial direction between a plane perpendicular to the axis and the relieved surface. It can be measured by the amount of indicator drop at a given radius in a given amount of angular rotation.

Back-Off—See preferred term *Relief*.

Back Taper—A slight reduction of the outside diameter [or an increase of the inside diameter] from front to back of the essentially cylindrical surface in which the cutting edges lie. See Figure 86.

Cam Relief—The relief from the cutting edge to the back of the tooth produced by a cam actuated cutting tool or grinding wheel on a relieving ["back-off"] machine. See Figure 86.

Concavity—The progressive decrease in cutter width from the periphery toward the center. See Figure 86.

Concave Relief—A concaved relieved surface behind the cutting edge.

Eccentric Relief—A convex relieved surface behind the cutting edge. See Figure 86.

End Relief—See preferred term *Axial Relief*.

Face Relief—See preferred term *Axial Relief*.

Flat Relief—A relieved surface behind the cutting edge which is essentially flat. See Figure 86.

Peripheral Relief—See preferred term *Radial Relief*.

Primary Relief—The relief immediately behind the cutting edge.

Relief Angle—The angle formed between a relieved surface and a given plane tangent to a cutting edge or to a point on a cutting edge. See Figure 86.

Radial Relief—Relief in a radial direction measured in the plane of rotation. It can be measured by the amount of indicator drop at a given radius in a given amount of angular rotation. See Figure 86.

Secondary Relief—See preferred term *Clearance*.

Side Relief—See preferred term *Axial Relief*.

RELIEF ANGLE—See definition under RELIEF.

RELIEVED LAND—See definition under LAND.

RESULTANT RAKE—See definition under RAKE.

Runout—See Axial Runout and Radial Runout.

Secondary Relief—See preferred term Clearance.

Set-Under—See preferred term Nesting.

Shank—The projecting portion of a cutter which locates and drives the cutter from the machine spindle or adapter.

Shear—1—An action or stress, resulting from applied forces, which causes or tends to cause two contiguous parts of a body to slide relatively to each other in a direction parallel to their plane of contact.—2—The type of cutting action produced by axial or helical rake when the direction of chip flow is other than at right angles to the cutting edge.

Side Relief—See preferred term Axial Relief under Relief.

Slab Milling—See preferred term Peripheral Milling.

Sleeve—See preferred term Adapter.

Spacing Collar—An arbor collar of specific length.

Sparking—The creation of sparks during milling due to combustion of small chips.

Speed—The rate of rotation of a cutter. It is usually expressed as revolutions per minute [rpm] or in peripheral surface feet per minute [sfm].

Spiral—See preferred term Helical.

Squealing—A high pitch noise resulting from a rubbing action of the cutter upon the work.

Stagger—The intentional misalignment of cutting edges or teeth in a cutter. See also Alternate.

Straight Shank—A cutter shank that is cylindrical.

Sub-Land—A one piece multiple diameter cutter in which the smaller diameter teeth overlap axially the larger diameter. See Figure 19.

Tang—The flatted end portion of a shank which mates with a slot in the driving member.

Tangential Rake—See definition under Rake.

Taper Shank—A cutter shank made to fit a specified [conical] taper socket.

Tip—A piece of tool material secured to a cutter tooth or blade.

Tooth—A projection on a cutter which carries a cutting edge. See Figure 75.

Tooth Face—The surface of the tooth on which the chip impinges. See Figure 75.

Tooth Rest—A guide to locate the cutting edge of a cutter in proper position for sharpening.

Tooth Spacing—See preferred term Circular Pitch.

Topping—A feature existing in a form cutter which cuts the top as well as the root and sides of a form, such as Threads and Gear Teeth.

Total Indicator Reading—See preferred term Total Indicator Variation [tiv].

Total Indicator Variation [tiv]—The difference between the maximum and minimum indicator readings during a checking cycle.

Tracer Milling—Duplication of a three dimensional form by means of a cutter controlled by a tracer which follows a master.

True Rake—See definition under Rake.

Undercut—See preferred term Positive Rake under Rake.

Unrelieved Land—See definition under Land.

Up-Milling—See Conventional Milling.

Wear Land—The cylindrical or flat land worn on the relieved portion of the cutter tooth. See Figure 87.

Wedge—A tapered block of hardened steel that is driven or held in place with a screw to secure a blade within the body or holder.

Weldon Shank—A straight shank with special flats for driving and locating the tool.

Wobble—See Axial Runout.

VI—Illustrations of Milling Cutter Elements

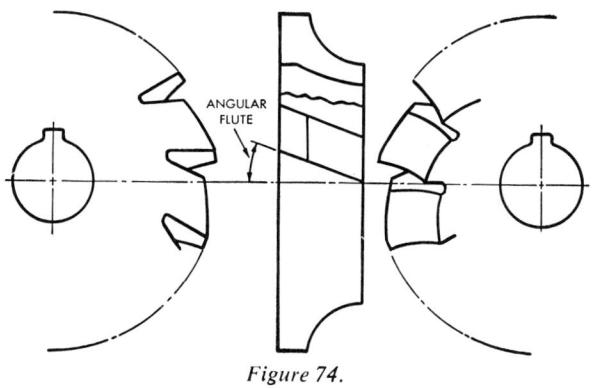

Figure 74.

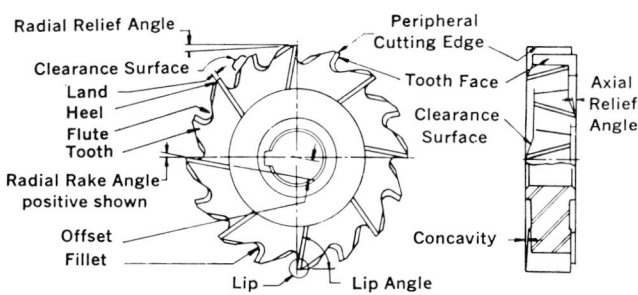

Figure 75.

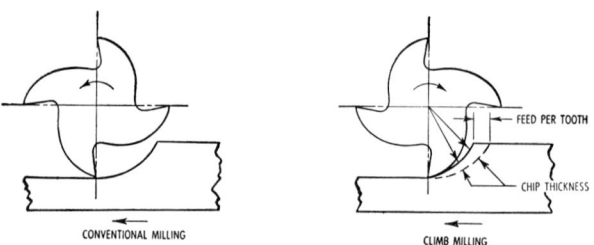

Figure 76.

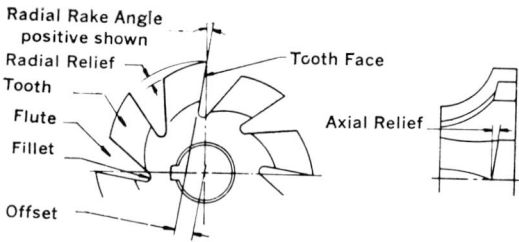

Figure 77.

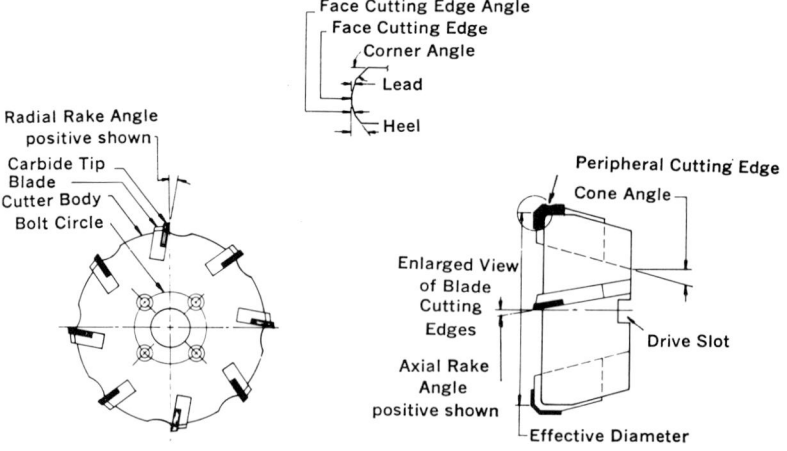

Figure 78.

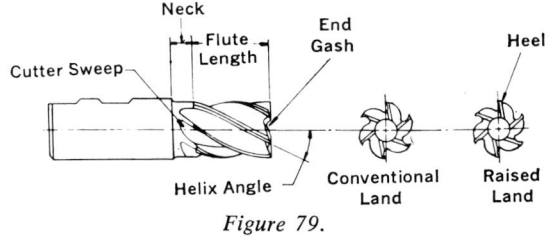

Figure 79.

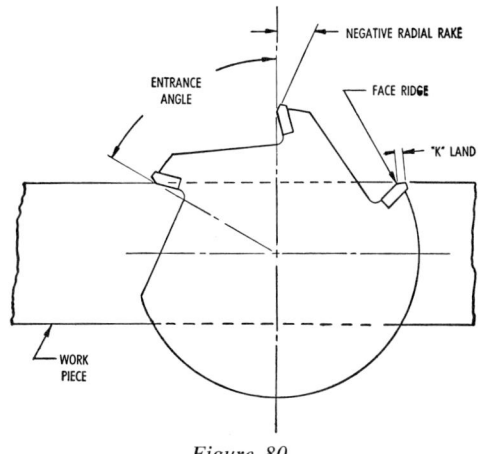

Figure 80.

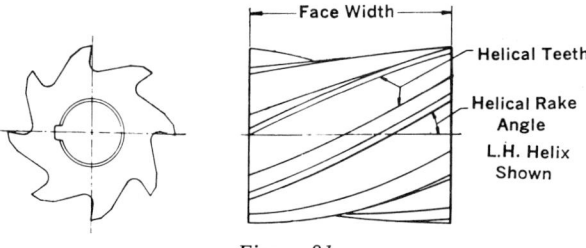

Figure 81.

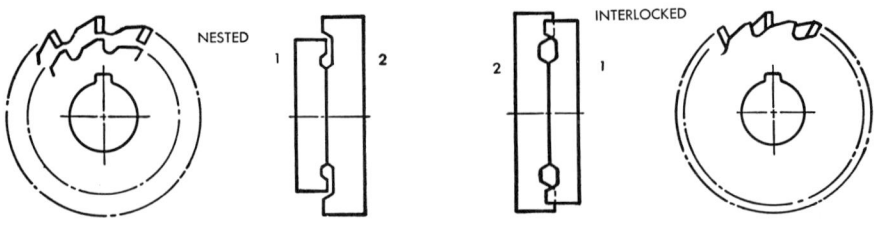

Figure 82.

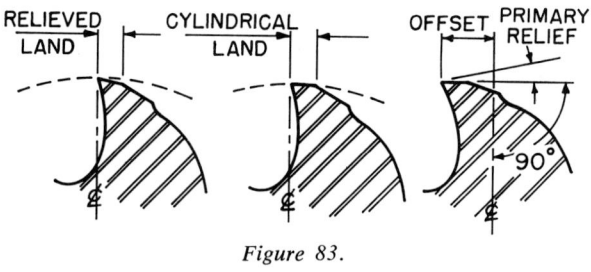

Figure 83.

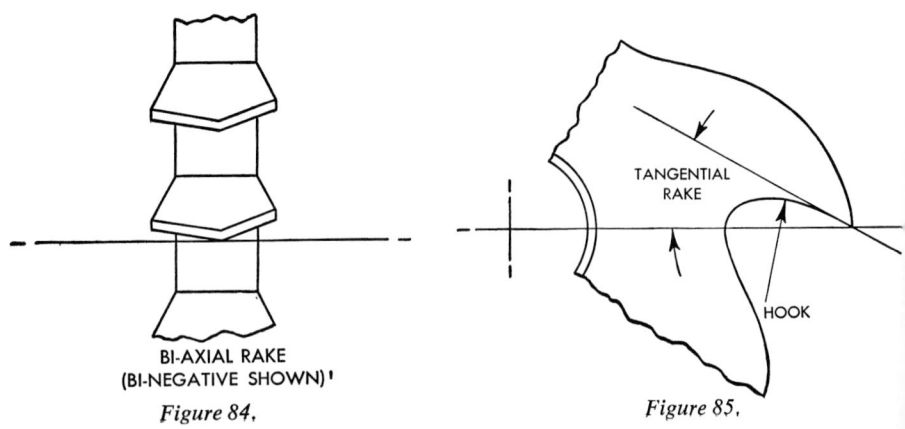

BI-AXIAL RAKE
(BI-NEGATIVE SHOWN)'

Figure 84.

Figure 85.

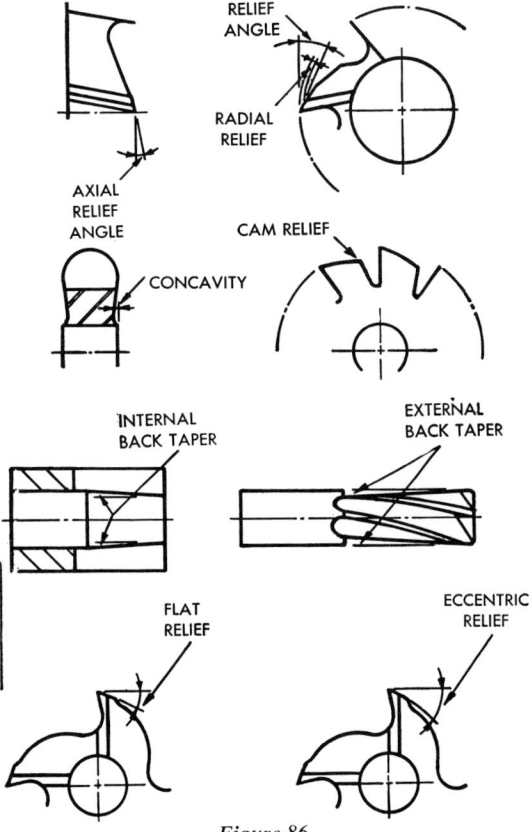

Figure 86.

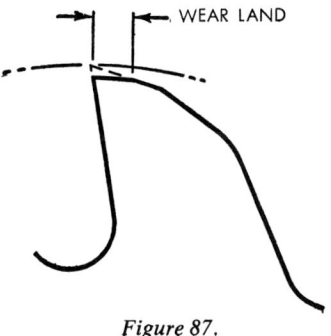

Figure 87.

MILLING CUTTER DESIGN

TYPE OF CUTTER CONSTRUCTION—Selection of the best type of of cutter construction will largely be determined by the kind of cut to be made, the work material to be machined and the amount of production required. Solid cutters of simple shape and reasonably small size are usually lower in first cost than inserted-blade cutters. Where intricately shaped cuts are involved, such as in multiple-thread milling, form-relieved cutters permit the lowest cost. For large production jobs, inserted-blade cutters with replaceable blades will probably be lowest in over-all cost. Tipped cutters, either solid or inserted, may be warranted by either the nature of the work material or high production requirements.

SIZE OF CUTTER AND TYPE OF DRIVE—The size of cutter will be determined by the depth and width of cut, reach required, rigidity requirements, piece part geometry, or approach and runby limitations of the application. The type of drive may be determined by the cutter size, type of machine or fixturing, or by the number and direction of cutting axes.

Arbor Type Cutters—Arbor type cutters are milling cutters with a bore and are mounted on an arbor. They must be large enough in diameter to be constructed with sufficient strength of cross section, with a bore which will accept an arbor large enough to provide reasonable rigidity during the milling operation, and with a large enough minimum cutting diameter to clear all fixturing. Also the diameter should be as small as possible to reduce the amount of torque on the arbor and to minimize the amount of side deflection while in use. This type of cutter is usually limited to slabbing, side milling, cut-off, or one axis contour milling. A two axis rise or fall motion may also be used with this type.

Shank Type Cutters—Shank type cutters and end mills are milling cutters with integral shanks and are usually supported on the shank end only. They should have a shank as large as possible and overhang between the cutting portion and spindle as short as possible to provide maximum rigidity to minimize deflection. This type may be restricted to one axis milling, such as keyway milling, but the major application is two or three axis milling such as profile, cavity, and contour end milling.

Face Milling Cutters—Face-milling cutters are large diameter milling cutters which are mounted directly to, and are supported on one side only, by the spindle nose or adapter. The diameter should be as small as possible for maximum rigidity, and large enough to span the width of cut. This type generally cuts on the face in a one axis direction perpendicular to the cutter axis.

CHIP SPACE—Cutters with as many teeth as possible around the outside diameter are desirable to promote smooth cutting. However, adequate chip space between teeth should never be sacrificed. Cutters should be designed so that one or more teeth will always be engaged in the cut, in order to promote smooth cutting operations. Tough, stringy materials require generous chip space. Materials which give a fragmentary chip, however, permit less chip clearance and a greater number of teeth. Cutters or end mills which plunge cut in a radial or axial direction should have coarse teeth and maximum chip space. This type of cut often produces chips which build up in quantity in the flutes before they can be ejected.

RAKE ANGLES—Rake angles will depend both on the material being cut and the material in the cutting tool. Most high-speed-steel cutters today are made with appreciable positive radial rake and axial rakes. A positive radial rake is desirable to part the chip from the work. However, the lip angle must be adequate for strength and absorption of heat. A high helix angle reduces impact and provides smooth cutting action, but is not necessarily most efficient with respect to horsepower consumption and heat generation.

Cemented carbides are used with lower rake angles, even negative angles in many cases. Because carbides have very high abrasion resistance and compressive strength, coupled with lower edge strength, the lip angle should be as great as possible, often exceeding 90 degrees.

The following table will serve as reference material in evaluating the effects of various rake angles.

Usual Ranges of Rake Angles for Typical Tool and Work Materials

Rake Angles	Tool Material	Steel	Cast Iron	Al., Mag., Non-Metallic
Radial Rake	H.S.S.	+5° — +20°	+5° — +20°	+10° — +20°
	T.C.	−15° — + 5°	0° — +10°	+5° — +15°
Axial Rake	H.S.S.	0° — +52°	0° — +52°	0° — +52°
	T.C.	−10° — +20°	−10° — +20°	0° — +30°

Softer work materials permit the use of the higher radial rake angles.

Harder work materials require the use of the lower radial rake angles.

Thin cutters permit the use of zero or low axial rake angles.

Wide cutters operate smoother with higher axial rake angles.

Cemented carbide is usually limited to more moderate axial rake angles because of added cost of fabricating this material into special shapes.

RELIEF ANGLES are influenced more by the material being milled than by the material in the cutting tool. They are determined also by the diameter of the tool and the feed per tooth expected. On the main cutting edge, the relief angles should be sufficient to avoid the heel of the land rubbing on the work, yet should be consistent with the edge strength required for the hardness of the work being milled.

Small-diameter cutters and end mills will require higher relief angles than those shown above and narrower lands than larger cutters.

Relief on side cutting edges is usually one-quarter to one-half that of the main cutting edge. In some cases, such as a plain saw, there is no side relief angle but the side of the saw is concaved to avoid rubbing in the cut except at the very corner of the tooth.

The width of the relieved land must be narrow enough so that the heel of the land will clear the work. This may vary from .005″ to .010″ on small end mills up to ⅛″ on large-diameter cutters. When this land becomes too wide as a result of repeated sharpenings, a clearance angle of a greater degree than the relief must be provided beyond the relief.

Excessive wear and heat may indicate too low a relief angle, and chatter in the cut may indicate too high a relief angle.

The following table indicates the usual ranges of relief angles for different tool and work materials.

Usual Ranges of Relief Angles for Typical Tool and Work Materials

Cutter	Tool Material	Work Material Steel	Cast Iron	Non-Ferrous and Non-Metallic Mat
Peripheral or End cutting edges	H.S.	5° — 10°	5° — 10°	7° — 12°
	T C.	4° — 6°	4° — 6°	5° — 10°
Side cutting edges	H.S.			
	T.C.	1° — 4°	1° — 4°	2° — 7°

Tables 9 to 12 in the resharpening section, following, give recommendations for specific cutter sizes.

In the cases of cutters which must mill to a given width, and where the maintenance of this width is important, two different practices are followed:

1. A very slight land or margin (sometimes called a hair line) without relief is left on the side teeth at the cutting edge. The relief angle of 2° to 4°

is ground back of this flat land, except on thin saws where the relief is increased to 3° to 5°.

2. The relief is ground to a sharp edge but to a reduced value of approximately 1° to 2°. This practice does not give sufficient relief for thin saws.

In both of the above cases, concavity (or dish) is ground on the side or end teeth in the same operation. This concavity on side teeth amounts to a .0015 to .003 per inch on cutters and as much as .005 to .0075 per inch on saws.

Multiple tooth end mills are usually relieved to a sharp edge on the end teeth with about 4° relief. Two flute end mills, because they are frequently fed endwise into solid stock, are usually relieved to the edge on the end teeth with about 7° relief. Shell mills are usually relieved 5° to 7° to a sharp edge on the end teeth.

SPECIAL FEATURES OF DESIGN—Special features may often be incorporated into milling cutters to improve their operation. Staggered, alternate or intermittent teeth are some of the ways to improve cutting action or finish on the work. Long cutting edges may be notched to produce short and readily disposable chips. On cemented-carbide-tip cutters, a face ridge or "K" land may be used to reduce chip friction and aid in the disposal of chips.

TOOL MATERIALS—Closely related to both design and application of milling cutters and end mills is the selection of the best tool material. Currently, two general classes of tool materials cover the requirements of most milling operations. These are the high speed steels and the sintered carbide tool materials. Other tool materials such as ceramics and compacted diamond are used in milling. Coatings and surface finishes are also in use to enhance tool life.

High Speed Steel—The high speed steels are tool steels capable of maintaining a useful cutting hardness at elevated temperatures. They do not lose hardness permanently unless exposed to temperatures higher than the tempering temperatures which are in excess of 1000°F. This property, often called red hardness, is due principally to two alloying elements, molybdenum and tungsten, separately or in combination. High speed steels are classified as "M" or "T" types depending upon which is the major alloying element. All high speed steels also contain carbon, vanadium and chromium while some grades also include cobalt as an alloying element.

Carbon contributes to hardness and is necessary for the formation of hard abrasion-resistant carbides. Vanadium forms exceptionally abrasion resistant carbides; however, high vanadium high speed steels are difficult to grind. Chromium is a stabilizing element. Cobalt additions yield higher red hardness and make the high speed steel resistant to tempering. Sulphur additions are sometimes used in high speed steels for form relieved cutters because the finish attainable is better.

As with most materials, high speed steels which are harder and more heavily alloyed tend to be more brittle than the standard general purpose types. However, the special purpose or "super" types are useful when coupled with rigid cutter designs and good milling setups.

Currently, the American Iron and Steel Institute lists thirty different standard types of high speed steels, and several non-standard types, as well as material produced by particle metallurgy rather than the common wrought process, are available. Thus, the selection of the best and most economical type can be a difficult problem. Fortunately, experienced cutter manufacturers have the background necessary to analyze milling operations and to recommend suitable high speed steel types.

Cemented Carbide Tool Materials—Cemented or sintered carbide tool materials, often called simply "carbides", are powder metallurgy products. The principal constituent is finely divided tungsten carbide which is bonded together with a smaller amount of cobalt by first forming under high pressure and then sintering at a high temperature. Some grades also contain tantalum carbide, titanium carbide and other elements. The composition and microstructure is controlled to vary the physical and cutting properties.

Carbides are characterized by high hardness and high abrasion resistance. They maintain hardness at elevated temperatures much in excess of those allowable with high speed steels and hence can be run at much higher cutting speeds.

Certain grades are designed for use at light feeds and where abrasion resistance is the principal requirement. Others are designed to be tougher and more temperature resistant for use on heavier roughing cuts.

As would be expected from their high effective cutting hardness, the carbide tool materials are somewhat brittle and not very shock resistant in comparison with high speed steel. To be successful, they must be used in well designed cutters and used on smooth operating rigid milling setups. Where these can be attained, carbide milling cutters and end mills are effective economical tools.

Carbide milling cutters are of two types: tipped or solid. Tipped cutters are made by fastening carbide cutting tips to a cutter body of steel or alloy cast iron. The tips may be either brazed or mechanically clamped to the cutter body. Solid cutters are completely fabricated from carbide. Their flutes are either formed prior to sintering or they may be ground into a solid piece of carbide. Small tools such as end mills are often made solid while the larger sized cutters are more often made in tipped construction.

As with the high speed steels, there are many carbide grades available and selection of the most suitable grade is often best handled by the cutter manufacturer after analysis of the proposed milling operation. Carbide grades are only loosely standardized so that it is not always possible to interchange corresponding grades made by different carbide manufacturers.

Ceramic and Diamond—Ceramic and diamond compacts are used as brazed or mechanically held insert materials. The diamond compact inserts machine almost all non-ferrous and abrasive materials.

Ceramics can be used in milling operations, primarily on cast iron and steel, if properly applied.

It is suggested that one contact the manufacturer of diamond compacts and ceramics to ensure their correct application and use.

Surface Treatment and Coatings—The surfaces of end mills have been modified in a number of ways over the years to enhance their performance. This trend is continuing as new processes and new techniques are developed. The chief mechanisms of these treatments have been to increase abrasion resistance, to increase lubricity, and to reduce built-up edge and related chip welding. The following treatments and coatings are the more common ones in use today:

1. *Nitride*

A very hard, shallow layer formed in the surface by the infusion of nitrogen to provide enhanced abrasion resistance.

2. *Steam Oxide*

An iron oxide surface formed in a furnace when steam is applied under moderately high heat. The oxide acts as a lubricant and reduces welding in ferrous applications. Nitride and steam oxide are often used as complementary treatments.

3. *Refractory Coatings*
   a. *Ceramic*

   A layer of material, normally aluminum oxide, often applied to carbide inserts to give longer tool life, particularly at high speeds, by increasing abrasion resistance.

b. *Titanium*

A family of various titanium compounds, most notably TiN and TiC, that are produced generally by vapor desposition as tight adhering coatings on the surfaces of the tool or inserts. The coating may be a single titanium compound such as TiN or a combination such as TiC + TiN. These titanium coatings reduce friction and chip welding as well as acting as a thermal insulator between the chip and the tool.

Typical Tool Applications

	Nitriding	Steam Oxide	Ceramic Coating	Titanium Coating
End Mill & Cutters	X	X		X
Inserts			X	X

For the most effective application of tool material, surface finish, and coating, one should consult with an experienced tool manufacturer.

SET UP AND OPERATION OF CUTTERS

In selecting the set up for a milling operation, factors to consider are the type, size and condition of the machine; work holding methods; selection and mounting of cutter; climb or conventional milling; and availability of coolant. These elements, some of which are discussed in this Section, contribute to the amount and quality of work produced.

WORK HOLDING METHODS—The quantity to be produced and the physical shape of the work piece influence the selection of a means to hold the work during the milling operation. Fixtures designed for a specific job are used where the production quantity justifies the tooling cost. Quick action hydraulic or pneumatic clamping devices facilitate loading and unloading. Manually operated fixtures are sometimes more economical, especially on short run jobs. See Fig. 1.

If a job does not warrant special design tooling, there are several inexpensive means of holding the work during the milling operation. Bolting directly to the table, gripping in a machine vise, clamping to angle plates or parallels are several such means. Be sure the work is held securely and safely. See Figures 1 and 2.

Figure 1—Manual type clamping fixture.   *Figure 2—Simple vise jaws.*

Some production parts are readily held in a simple machine vise where the removable jaws can be made to fit the contour of the part. In conventional milling the force of the cut is usually against the movable jaw of the vise, eliminating the hazardous practice of having the vise handle pass under the cutter. In climb milling the cutting forces will be against the fixed jaw of the vise. See Figures 2 and 3.

For long, or thin work, screw jacks or wedges help support the piece and prevent vibration. See Fig. 4.

Figure 3—Showing work clamped directly to the table.

Figure 4—Illustrating use of jack.

MOUNTING THE CUTTER—When mounted, the cutter should run true both radially and axially. An indicator setup, as illustrated in Figure 5, is generally used to check the accuracy of mounting.

Figure 5—Checking runout with dial indicator.

Arbor Type Cutters—Arbor type cutters usually have a straight hole with a keyway. In assembling the cutter to the arbor always use a key long enough to enter the shims and spacers on both sides of the cutter. Do not rely on friction to drive the cutter—always use a key.

Cutters, collars and shims should be clean before assembly. Dirty, out of parallel, or scored collars and cutters, assembled on the arbor, will cause it to spring out of line. Place the outer arbor support in position, before tightening the assembly to keep the arbor from springing. If two arbor supports are used, place one as near to the cutter as possible, and the other at the end of the arbor. On heavy cuts, clamp the arbor support to the knee or bed of the machine with arm braces. See Figure 7.

The size of the arbor selected depends on the size of cutter available, the area of the cut to be made, and the length of span necessary between spindle and arbor support. A 2″ arbor is sixteen times as rigid as 1″ arbor.

Before assembling be sure the taper on the arbor, and in the spindle, is clean. Dirt or chips may damage the spindle or allow the arbor to loosen.

Be sure there is ample working clearance between the work and stationary parts of the machine.

Arbor type cutters should be kept as small as practicable for a given arbor size, allowing for depth of cut or clearance over the work.

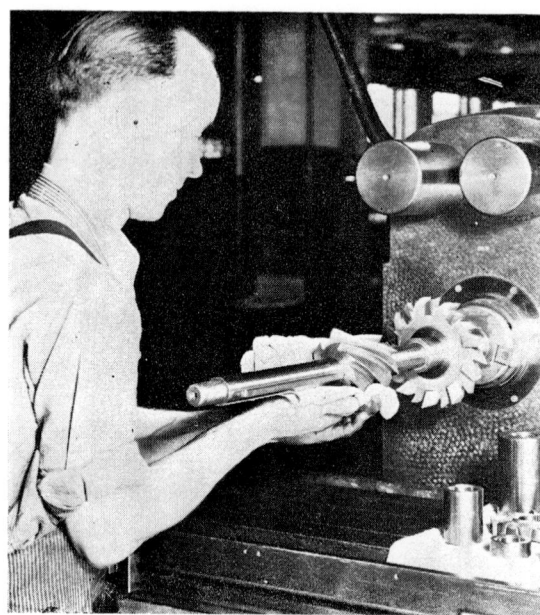

Figure 6—A cloth should be used to protect the hands when mounting cutters on the machine, otherwise cuts may result from slipping.

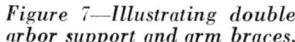

Figure 7—Illustrating double arbor support and arm braces.

Figure 8—*Cleaning out the taper.*

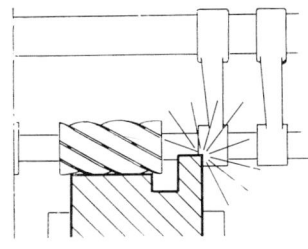

Figure 9 Showing work interference.

Shank Cutters—Shank type cutters have a straight or a taper shank mounted in an adapter or in the machine spindle taper. Straight shank types are sometimes held in a spring collet, but are usually fastened in their holder with one or more screws against flats on the shank. Taper shank cutters are generally made with a self-holding taper, such as Brown & Sharpe or Morse, or in some cases, with a steep taper, such as the No. 30, 40, 50 or 60 Milling Machine Standard Taper. Cutters with self-holding taper shanks do not require a draw-bar but steep taper shanks do unless some form of quick locking is provided. Exceptions for self-holding tapers may be cases where an unusually long length of cut exists, or with high helix cutters, where the hand of helix is the same as the hand of cut. When using high helix cutters with no draw-bar, particular care should be taken to have the hole and shank clean and the cutter securely seated in the socket. A soft faced hammer is often used to drive the shank into the taper. Care should be taken to avoid chipping cutter edges or damaging spindle bearings. With straight shank cutters chips in the hole are pushed away by the shank.

Adapters are produced in a range of combinations to mount shank cutters in various machine spindles.

End Mills—In selection and use of end mills, precautions should be observed for best results. A milling machine with ample power should be used. Select an end mill of proper design and mount it with the least possible overhang. The end mill must be sharp, and must run as concentric as possible. Scored, battered, or worn end mill shanks, holder/adaptor holes or shanks, as well as a worn or misaligned machine spindles each can contribute to a lack of concentricity. Non-concentric set-ups lead to premature wear, rough surface finish and excessive oversize cutting action. Emphasize rigidity of the individual job set up. The best aligned spindle-holder-end mill combination is ineffective if the set up is too light, work insecurely clamped, or improperly supported. Fig. 10 shows various end mill applications.

Figure 10—Various applications of end mills.

To determine selection of either a two-flute or a multiple-flute end mill, several basics must be considered:

1. Type of cut
2. Chip space required
3. Production rate desired
4. Surface finish required

Two-flute end mills have a greater chip handling capacity than multiple-flute end mills. Two-flute end mills are center cutting. Multiple-flute end mills are available in both center cutting and non-center cutting. An end mill must be center cutting in order to plunge cut.

When two-flute end mills and multiple-flute end mills are run at the same feed rate (inches per minute), multiple-flute end mills may produce finer finishes and longer tool life than two-flute end mills, owing to a lighter chip load per tooth. Too light a chip load can cause excessive wear.

Higher production rates may be achieved with multiple-flute end mills because feed per tooth is in direct proportion to the number of flutes, enabling an increase in feed rate.

Roughing end mills can be used in a wide variety of materials and will generally remove more material in less time than conventional heavy duty end mills. Roughing end mills permit the use of heavier feed rates, which will increase production rate and may extend tool life.

A common problem associated with slotting, using end mills, is the leaning slot. Non-perpendicular sides of the slot being milled are attributed to a combination of worn spindle, an end mill with excessive projection from the spindle, or excessive chip load.

The major cause of this condition is deflection of the end mill due to the cutting forces. (See Figure 11) The resulting slot is tipped so that its sides are not parallel to the machine spindle axis and its location is displaced laterally from the intended position. This condition can be minimized by reducing chip load, using shorter-fluted end mills with short overall lengths, or using end mills with straight or low helix angle flutes.

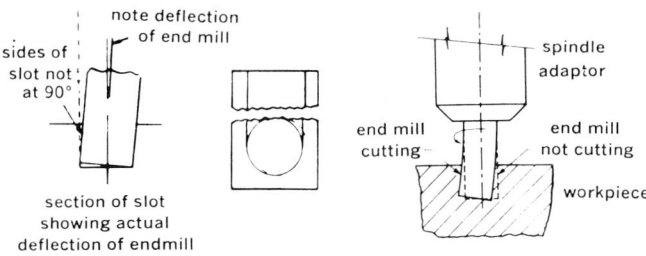

Figure 11—Illustrating End Mill Deflection.

Face Mills And Shell Mills—These cutters are mounted directly on the spindle nose of the machine or on an adapter. Face mills are bolted to the face of the spindle. A key positioned in the back of the cutter and in the spindle end, provides a positive drive. Concentricity is obtained by a centering plug, (See Figure 12) which centers the cutter with the spindle or by a ground counterbore on the back of the face mill which fits over the spindle end. On some small diameter face mills, where the cutter is not bolted directly to the spindle end, the centering plug is used as an arbor. All mounting surfaces of the spindle, plug and cutter must be clean.

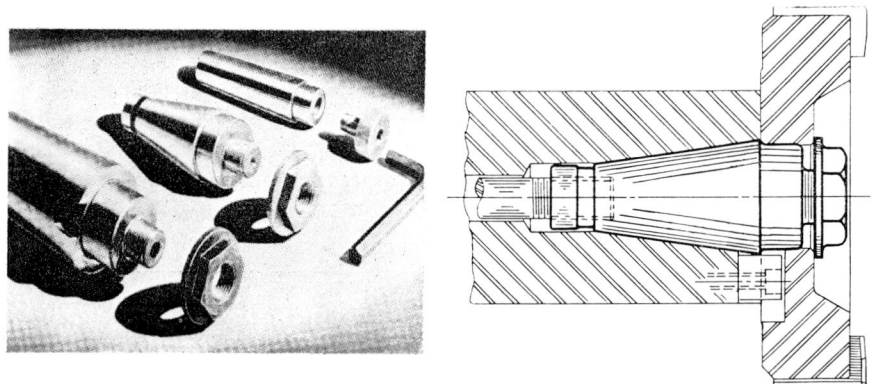

Figure 12—Various centering plugs.

Shell mills, a type of face mill, can be used for cutting on the outside diameter as well as on the face of the cutter. These are mounted on an adapter, similar to a centering plug, where the drive is obtained by a key on the adapter engaging in a slot on the back of the cutter. The cutter is locked in place with a retaining bolt or screw assembly. The adapter is mounted in the machine spindle. It is driven by a slotted flange engaging with driving keys mounted on the spindle end. See Figure 13. In assembling, be sure all surfaces are clean.

Figure 13—Shell mill and adapter.

CLIMB OR CONVENTIONAL MILLING—Material to be cut, shape of the workpiece, type and condition of the machine are factors that influence a decision to set up for climb milling. In climb milling, if the work is thin or difficult to hold securely, pressure of the cutter will force the work down rather than lift it up. It is particularly desirable to climb mill such materials as heat treated alloy steels and non-free machining stainless steels for better cutter life and to reduce the tendency to work harden.

In climb milling, a chip of definite thickness is formed at the start without the rubbing action resulting when conventional milling is used. However, in climb milling, care should be taken to equip the machine with some form of backlash eliminator, to prevent the cutter pulling the table into the cut resulting in cutter breakage. Occasionally arrangements to provide a load or drag to the table can be improvised.

General purpose cutters will produce a good finish when climb milling on medium to high carbon or alloy steels, even in a deep cut. On low carbon steels, however, they may produce a torn finish. Increasing the radial rake on the cutters or making a rough and finish cut will overcome the trouble.

Climb milling on one end of the table and conventional milling on the other, using the same cutters and holding devices, is sometimes used to speed up an operation. While the cut is made at one end, the operator can load at the other. See Figure 14.

Figure 14—Illustrating twin vises for climb and conventional milling.

Climb milling will be very helpful when using thin saws. In conventional milling, a thin saw may deflect sideways and break. The same saw, when climb milling, will very often cut straight and true.

CUTTER SELECTION—It is extremely important to select the proper cutter for a job. For milling large, flat surfaces a face mill large enough to cover the width of the cut in one pass, will be most practical. For a combination of flat surfaces and one or more vertical surfaces, a slab mill in combination with straddle cutters is often used. Whenever possible, the diameter of face mills should be one and one-half times the width of the cut so that the chip thickness is kept more uniform. On wide cuts, coarse tooth high helix slab mills are preferred, to reduce chatter and vibration.

For long, deep cuts, staggered tooth cutters are recommended. The alternate helix of the teeth prevents chatter and the design is such that chips do not stick in the cut. Chamfers on alternate corners of the teeth on saws, side mills, etc., will break up the chips and eliminate scoring the sides of the cut by chip drag.

The decision to use profile sharpened or form relieved cutters depends largely on the sharpening equipment available and to some extent on the cutter form. In general, profile sharpened cutters (or those sharpened by grinding on the lands only) are freer cutting than the form relieved type.

Form relieved cutters are made with either straight or helical flutes. Helical fluted form relieved cutters give a smoother cutting action but require special sharpening equipment. The profile sharpened type cutter can usually be made with more teeth than a corresponding form relieved cutter, an advantage at times, especially on small work. Profile sharpened form cutters often require a grinding machine that reproduces the cutter shape from a template.

Carbide tipped cutters are especially useful on cast iron, non-metallic and non-ferrous materials. They require a rigid machine with ample power. Carbides are more resistant to abrasion and permit much higher speeds and correspondingly greater feed rates than high speed steels. See Section on Feeds and Speeds. Carbide cutters are usually designed for specific jobs and work materials.

Climb milling is desirable for cutting steels with carbide. Carbide tipped cutters are not ordinarily recommended for milling highly alloyed materials such as stainless steels and heat resistant alloys.

CUTTING FLUIDS—A cutting fluid or coolant is required when using high speed steel cutters for milling steel. The coolant dissipates heat from the cut and minimizes possible damage to the cutter or work piece. Soluble oil and water is a common type of fluid, but various mineral and sulphur base oils are also quite popular. Water base solutions have good cooling qualities. Oil base solutions tend to give a good finish on the milled surface.

A cutting fluid is seldom used when milling steel or cast iron with carbide cutters, due to the tendency of the carbide to chip from intermittent cooling. On equipment which can provide a copious supply of cutting fluid under pressure, the fluid helps to clear the chips from the cut as well as cool the cutter and the work.

Cast iron, brass, plastics, etc. are generally cut dry. In confined areas, where the chips are apt to accumulate around the cutter, such as in milling T-slots, a blast of compressed air will keep the cutter cool and disperse the chips. When air is used, place safety guards around the operation to prevent chips and dust from injuring personnel.

SPEEDS AND FEEDS FOR MILLING—Speeds and feeds are the most important factors to consider for best results in milling. Improper feeds and speeds often cause low production, poor work quality and unnecessary damage to the cutter.

This section covers the basic principles of speed and feed selection for milling cutters and end mills. It will serve as a guide in setting-up new milling jobs.

Speed—In milling, SPEED is measured in peripheral feet per minute, (revolutions per minute times cutter circumference in feet). This is frequently referred to as "peripheral speed," "cutting speed," or "surface speed".

The figures given in Table 1 are suggested starting speeds only, a foundation on which to build. They will have to be tempered to suit the conditions ON THE JOB. For example:

Use Lower Speed Ranges For	Use Higher Speed Ranges For
Hard materials	Softer materials
Tough materials	Better finishes
Abrasive materials	Smaller diameter cutters
Heavy cuts	Light cuts
Minimum tool wear	Frail work pieces or set-ups
Maximum cutter life	Hand feed operations
	Maximum production rates
	Non-metallics

Table 1—Starting Peripheral Speeds

work material	tool material high speed steel	cemented carbide	work material	tool material high speed steel	cemented carbide
Magnesium	600 up	1000 up	Titanium		
Aluminum	600 up	1000 up	Under 100,000 PSI	35-55	150-180
Copper	300 up	1000 up	100,000 to		
Brass	300	800	135,000 PSI	25-35	120-150
Bronze	200	400	135,000 PSI & Over	15-25	80-120
Maleable Iron	100	350	High Tensile Steels		
Cast Iron	100	300	180,000—220,000 PSI	25-40	200-250
Cast Steel	70	200	220,000—260,000 PSI	10-25	110-200
Steel—100 Brinell	150	450	260,000—300,000 PSI	6-10	90-180
200 Brinell	70	350	High Temperature Alloys		
300 Brinell	40	200	Ferritic Low Alloys	40-60	150-300
400 Brinell	20	100	Austenitic Alloys	20-30	100-235
500 Brinell	10	75	Nickel Base Alloys	5-20	50-150
Stainless Steel:			Cobalt Base Alloys	5-10	50-100
Free Machining	70	400			
Other	40	300			

NON-METALLICS: These materials are so numerous and varied that specific recommendations cannot be made. In general, the use of cemented carbide at high peripheral speeds is indicated.

Chip Thickness—Chip thickness is one of the more important factors affecting tool life in milling operations. Very thin, or feather-edge chips dull cutting edges more rapidly than thick chips. Chip thickness is governed by the size and shape of the cutter and work piece. Chip thickness is affected by the feed per tooth. Figures 15-18 illustrate the relation of chip thickness to feed per tooth, for several conditions. It emphasizes that careful thought must be given to feed per tooth, to cutter size and design and relative location to the work.

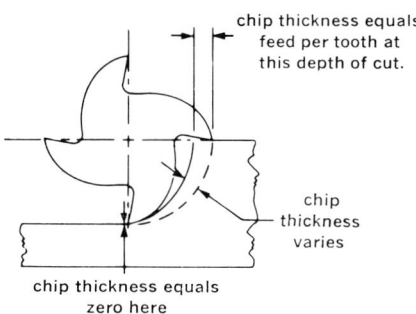

Figure 15—Deep peripheral cuts.

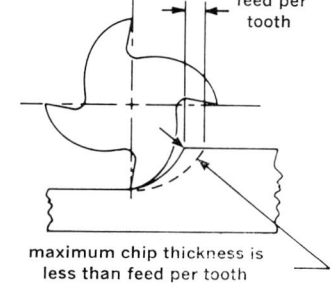

Figure 16—Shallow peripheral cuts.

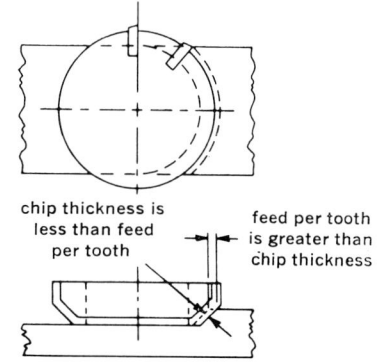

Figure 17—Face milling with beveled corner cutter.

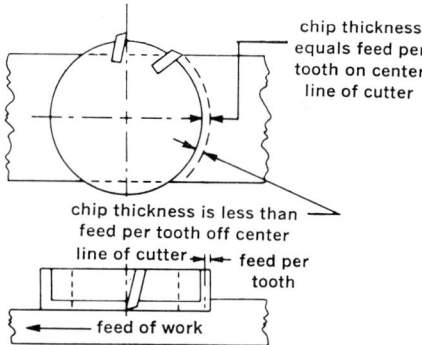

Figure 18—Face milling with square corner cutter.

Feed—Feed is usually measured in inches per minute. It is the product of feed per tooth times revolutions per minute times the number of teeth in the cutter. Due to variations in cutter sizes, numbers of teeth and revolutions per minute, all feed rates should be calcula'ed from feed per tooth. Feed per tooth is the basis of all feed rates per minute, whether the cutters are large or small, fine or coarse tooth, and are run at high or low peripheral speed. Because feed per tooth affects chip thickness, it is a very important factor in cutter life.

Highest possible feed per tooth will usually give longer cutter life between grinds and greater production per grind. Excessive feeds may overload the cutter teeth and cause breakage or chipping of the cutting edges.

The following factors should be kept in mind when using the recommended starting feed per tooth.

Use Higher Feeds For	Use Lower Feeds For
Heavy, roughing cuts	Light, and finishing cuts
Rigid set-ups	Frail set-ups
Easy to machine work materials	Hard to machine work materials
Rugged cutters	Frail and small cutters
Slab milling cuts	Deep slots
High tensile strength materials	Low tensile strength materials
Coarse tooth cutters	Fine tooth cutters
Abrasive materials	

Table 2—Starting Feeds

type of cut	feed per tooth
Face milling	.007
Straddle milling	.005
Slot milling (Side Mills)	.003
Slab Milling	
light duty	.004
heavy duty	.008
End Milling:	
½″ dia. and over	.002 to .003
Under ½″ dia.	.0002 to .002
Sawing	.0005 to .001
Thread milling	.0005 to .001

Note the feeds are listed as "Starting Feeds". Local conditions such as production rate, material, shape of workpiece, condition of machine, power supply, cutter life, etc., may require modification of feed or speed rates to achieve desired results.

Table 3—Feed (Inches Per Minute) For 1 R.P.M.

No. of Teeth	feed per tooth														
	.0002	.0004	.0006	.0008	.001	.002	.004	.006	.008	.010	.012	.014	.016	.018	.020
2	.0004	.0008	.0012	.0016	.002	.004	.008	.012	.016	.020	.024	.028	.032	.036	.040
3	.0006	.0012	.0018	.0024	.003	.006	.012	.018	.024	.030	.036	.042	.048	.054	.060
4	.0008	.0016	.0024	.0032	.004	.008	.016	.024	.032	.040	.048	.056	.064	.072	.080
6	.0012	.0024	.0036	.0048	.006	.012	.024	.036	.048	.060	.072	.084	.096	.108	.120
8	.0016	.0032	.0048	.0064	.008	.016	.032	.048	.064	.080	.096	.112	.128	.144	.160
10	.0020	.0040	.0060	.0080	.010	.020	.040	.060	.080	.100	.120	.140	.160	.180	.200
12	.0024	.0048	.0072	.0096	.012	.024	.048	.072	.096	.120	.144	.168	.192	.216	.240
14	.0028	.0056	.0084	.0112	.014	.028	.056	.084	.112	.140	.168	.196	.224	.252	.280
16	.0032	.0064	.0096	.0128	.016	.032	.064	.096	.128	.160	.192	.224	.256	.288	.320
18	.0036	.0072	.0108	.0144	.018	.036	.072	.108	.144	.180	.216	.252	.288	.324	.360
20	.0040	.0080	.0120	.0160	.020	.040	.080	.120	.160	.200	.240	.280	.320	.360	.400
22	.0044	.0088	.0132	.0176	.022	.044	.088	.132	.176	.220	.264	.308	.352	.396	.440
24	.0048	.0096	.0144	.0192	.024	.048	.096	.144	.192	.240	.288	.336	.384	.432	.480
26	.0052	.0104	.0156	.0208	.026	.052	.104	.156	.208	.260	.312	.364	.416	.468	.520
28	.0056	.0112	.0168	.0224	.028	.056	.112	.168	.224	.280	.336	.392	.448	.504	.560
30	.0060	.0120	.0180	.0240	.030	.060	.120	.180	.240	.300	.360	.420	.480	.540	.600
32	.0064	.0128	.0192	.0256	.032	.064	.128	.192	.256	.320	.384	.448	.512	.576	.640
36	.0072	.0144	.0216	.0288	.036	.072	.144	.216	.288	.360	.432	.504	.576	.648	.720
40	.0080	.0160	.0240	.0320	.040	.080	.160	.240	.320	.400	.480	.560	.640	.720	.800
44	.0088	.0176	.0264	.0352	.044	.088	.176	.264	.352	.440	.528	.616	.704	.792	.880
48	.0096	.0192	.0288	.0384	.048	.096	.192	.288	.384	.480	.576	.672	.768	.864	.960
52	.0104	.0208	.0312	.0416	.052	.104	.208	.312	.416	.520	.624	.728	.832	.936	1.040
56	.0112	.0224	.0336	.0448	.056	.112	.224	.336	.448	.560	.672	.784	.896	1.008	1.120
60	.0120	.0240	.0360	.0480	.060	.120	.240	.360	.480	.600	.720	.840	.960	1.080	1.200
72	.0144	.0288	.0432	.0576	.072	.144	.288	.432	.576	.720	.864	1.008	1.152	1.296	1.440
90	.0180	.0360	.0540	.0720	.090	.180	.360	.540	.720	.900	1.080	1.260	1.440	1.620	1.800

EXAMPLE: For .006 feed per tooth, 20 tooth cutter, 78 rpm
Feed for 1 rpm (from table) = .120
Feed for 78 rpm = .120 × 78 = 9.36 inches per minute

Table 4—Cutting Speeds

Cutter Diam. Inches	Surface Feet Per Minute (sfm)													
	10	15	20	25	30	35	40	45	50	60	70	80	90	100
	Revolutions Per Minute													
1/16	611	917	1222	1528	1834	2140	2445	2751	3057	3668	4280	4891	5502	6114
3/32	407	611	815	1019	1222	1426	1630	1834	2038	2446	2853	3261	3667	4075
1/8	306	458	611	764	917	1070	1222	1375	1528	1834	2139	2445	2750	3056
5/32	244	367	489	611	733	856	978	1100	1222	1466	1711	1955	2200	2444
3/16	204	306	407	509	611	713	815	917	1019	1222	1426	1630	1834	2038
1/4	153	229	306	382	458	535	611	688	764	917	1070	1222	1376	1528
5/16	122	183	244	306	367	428	489	550	611	733	856	978	1100	1222
3/8	102	153	204	255	306	357	408	458	509	611	713	815	916	1018
7/16	87	131	175	218	262	306	349	393	437	524	611	699	786	874
1/2	76	115	153	191	229	268	306	344	382	459	535	611	688	764
9/16	68	103	137	172	204	238	272	306	340	407	475	543	611	679
5/8	61	92	122	153	184	214	245	276	306	367	428	489	552	612
11/16	55	84	112	140	167	194	222	249	278	333	389	444	500	555
3/4	51	76	102	127	153	178	203	229	254	306	357	408	458	508
13/16	47	71	95	118	142	166	190	213	237	284	332	379	427	474
7/8	44	66	87	109	131	153	175	196	219	262	306	349	392	438
1	38	57	76	96	115	134	153	172	191	229	267	306	344	382
1 1/8	34	51	68	85	102	119	136	153	170	204	238	272	306	340
1 1/4	31	46	61	76	92	107	123	137	153	183	214	245	274	306
1 3/8	28	42	56	70	83	97	111	125	139	167	195	222	250	278
1 1/2	25	38	51	64	76	89	102	115	127	153	178	204	230	254
1 3/4	22	33	44	55	66	76	88	98	109	131	153	175	196	218
2	19	29	38	48	57	67	76	86	96	115	134	153	172	191
2 1/4	17	26	34	42	51	59	68	76	86	102	119	136	153	170
2 1/2	15	23	31	38	46	54	61	69	76	92	107	122	138	153
2 3/4	14	21	28	35	42	49	56	63	70	83	97	111	125	139
3	13	19	26	32	38	45	51	57	64	76	89	102	114	127
3 1/2	11	16	22	27	33	38	44	49	55	66	76	87	98	109
4	10	14	19	24	29	33	38	43	48	57	67	76	86	96
4 1/2	9	13	17	21	26	30	34	38	42	51	59	68	76	85
5	8	12	15	19	23	27	31	34	38	46	54	61	69	76
5 1/2	7	10	14	17	21	24	28	31	35	42	49	56	63	70
6	6	10	13	16	19	22	26	29	32	38	45	51	57	64
8	5	7	10	12	14	17	19	22	24	29	33	38	43	48
10	4	6	8	10	12	13	15	17	19	23	27	31	34	38
12	3	5	6	8	10	11	13	14	16	19	22	26	29	32
14	3	4	5	7	8	10	11	12	14	16	19	22	25	27
16	2	4	5	6	7	8	10	11	12	14	17	19	22	24
18	2	3	4	5	6	7	9	10	11	13	15	17	19	21

Table 5—Speed, Feed and Power Calculations

To Find	Having	Procedure	Formula
Peripheral cutting speed—V	Cutter diameter, d Rotational speed, N	Multiply cutter diameter by rotational speed and 0.262	$V = 0.262\ dN$
Rotational speed—N	Cutter diameter, d Peripheral cutting speed, V	Divide the product of 3.82 and cutting speed by cutter diameter	$N = \dfrac{3.82\ V}{d}$
Feed per tooth—f	Machine feed rate, F Rotational speed, N Number of teeth, I	Divide machine feed rate by product of rotational speed and number of teeth	$f = \dfrac{F}{NT}$
Machine feed rate—F	Feed per tooth, f Rotational speed, N Number of teeth, T	Multiply feed per tooth by rotational speed and number of teeth	$F = fNT$
Feed per revolution—f_r	Machine feed rate, F Rotational speed, N	Divide machine feed rate by rotational speed	$f_r = \dfrac{F}{N}$
Cutting power input—P	Width of cut, W Depth of cut, h Machine feed rate, F Workpiece material power constant, C	Divide the product of width of cut, depth of cut, and machine feed rate by the power constant	$P = \dfrac{WhF}{C}$

Definition of Symbols and Measurement Units

quantity	symbol	measurement unit
Cutting speed	V	Surface feet per minute, sfm.
Rotational speed	N	Revolutions per minute, rpm
Cutter diameter	d	Inches
Feed per tooth	f	Inches per tooth, ipt
Machine feed rate	F	Inches per minute, ipm
Feed per revolution	f_r	Inches per revolution, ipr
Cutting power input	P	Horsepower, hp
Power constant	C	Cubic inches per horsepower minute
Width of cut	W	Inches
Depth of cut	h	Inches
Number of teeth	T	—

POWER CONSTANTS FOR USE WITH TABLE 5	
work material	*C (constant)*
Magnesium	4.0
Aluminum	4.0
Copper	2.0
Brass	2.5
Bronze	2.0
Malleable Iron	1.0
Cast Iron	
Ferrite	1.5
Pearlitic	1.0
Chilled	.6
Steel	
Up to 150 Brinell	.7
300 Brinell	.6
400 Brinell	.5
500 Brinell	.4
Stainless Steel	
Free machining	1.0
Other	.6
Titanium	
Under 100.000 psi	.8
100,000-135,000 psi	.6
135,000 psi & over	.4
High Tensile Alloys	
180,000-220,000 psi	.5
220,000-260,000 psi	.4
260,000-300,000 psi	.3
High Temperature Alloys	
Ferritic Low Alloys	.6
Austenitic Alloys	.5
Nickel Base Alloys	.4
Cobalt Base Alloys	.4

Power constant values for various materials are the number of cubic inches of metal per minute capable of being removed per horsepower input with 60% power efficiency at the spindle nose and a 25% allowance for cutter dulling.

Troubles and Corrective Measures—Following are some of the more common troubles encountered and corrective measures involving variations in speeds and feeds, which may be taken to offset them:

Lack of rigidity: .... Increase speed, reduce feed

Excessive abrasion on the tool: .... Reduce speed, increase feed

Chipping of the cutting edge: .... Reduce feed per tooth

Burning of the cutting edge: .... Reduce speed

Cratering of cemented carbide: .... Reduce feed and speed

Chatter: .... Try other combinations of speeds and feeds

TYPES AND CHARACTERISTICS OF WORK MATERIALS

CAST IRON—The abrasive action of cast iron shows that cemented carbides generally are the most economical tool material for production runs. On chilled or very hard cast iron, cemented carbides may be the only choice.

STEELS—The hardness of the work material influences the choice of tool material for steel. For hardness below 300 Brinell, high speed steel tools made of regular high speed steel are generally satisfactory. Cemented carbides may be used, and cast alloy is a possible choice in some cases. For steels of 300 to 400 Brinell, regular grades of high speed steel operate at their extreme limit, but the high alloy grades are generally satisfactory. Cemented carbides may be the best choice, especially if production is high. For steels over 400 Brinell, the choice is usually limited to high alloy, high hardness, high speed steels and cemented carbides. For steels over 500 Brinell, the choice is usually limited to the cemented carbides.

STAINLESS STEELS, HEAT RESISTANT ALLOYS, and TITANIUM— High speed steel is generally a better tool material because of the tendency of these work materials to weld to and destroy the cutting edge of cemented carbides and cast cutting alloys. The high alloy high speed steels are generally the most satisfactory depending on production requirements.

NON-FERROUS MATERIALS—Regular grades of high speed steel are usually the most economical on short production runs. High alloy high speed steel may be economically justified in some cases due to longer life. Cemented carbides are usually the most economical on long production runs. The cast alloys are also a satisfactory tool material. Cast alloys and cemented carbides withstand the abrasive action of most of the non-ferrous materials better than high speed steel.

NON-METALLIC MATERIALS—Because most non-metallics are very abrasive and produce excessive tool wear, cemented carbides are usually the best tool material.

WORK HARDENING of the WORK MATERIALS—Work hardening is a condition which results from mechanical working of material during the cutting process. Work hardening increases as the cutter dulls. The effect upon the cutting tool is somewhat the same as if the material were hard to begin with. Work hardening is characteristic of most steels, particularly of stainless steels. To minimize its effect upon cutting edges, feeds and speeds should be selected to take a substantial chip load so that the work-hardening material is taken away in the form of chips. The cutting tool should not be permitted to dwell. Climb milling is sometimes advantageous in reducing the effects of work hardening.

CLOGGING of CUTTER FLUTES—This condition arises with cutting materials whose chips tend to stay in one long piece, rather than break up. It is true particularly of those materials having a high affinity for welding to the cutting edge. The long continuous chips jam up between the teeth with sufficient force to break out an entire tooth or to break the cutter in two. When milling materials of this type, cutters should be designed with plenty of chip space, or they should incorporate some type of chip breaker.

TYPES OF CUTTER BREAKDOWN

All work materials eventually cause wear on cutting tools in varying degrees. Their machining characteristics will give rise to one or more of the following cutter conditions as wear progresses:

ABRASION—All work materials abrade or wear cutting edges. The friction or rubbing action which takes place, as the cutter teeth rotate in the cut, continuously wears away small amounts of the cutting edges. This dulls the keen cutting edge and produces a "wear land." The greater the wear land the more force required to make the cutting edge penetrate the work and the more friction and heat are produced. These elements promote further abrasion. If the cutter is not removed at the proper time for resharpening to correct the abraded condition, the cutting edge will ultimately disintegrate.

CHIPPING OF EDGE—This condition occurs when the cutting edge is thin and weak from excessive relief and rake angles. It may also occur when the tool material is too brittle for the application. Chipping may also be caused by chatter, which occurs when the work piece is not rigidly supported, or when the cutter is not rigid enough.

In some cases, abrasion conceals chipping and it is difficult to determine which is the real cause of cutter wear. For example, if the cutter is permitted to continue in operation after it has become chipped, it wears rapidly because of increased pressure, and the chipped portions are obscured by wear.

BUILT-UP EDGE—This is the welding of portions of chips to the cutting face at or adjacent to the cutting edge. It is a result of the heat and pressure which exist in the cutting zone. If the built up edge become large enough it may actually function as a new cutting edge which will be irregular and which may periodically break loose from the tool. This can cause a rough work surface or a chipped cutting edge. Certain work materials have greater tendencies to form built up edges. Changes in cutting speeds and feeds and the application of cutting fluids can reduce build up edge formation.

CRATERING—This is the wearing of a crater in the tooth face some-what back from the cutting edge. As the crater grows and approaches the cutting edge it may cause chipping. Certain tool materials, both high speed steel and carbide, resist cratering.

OVERHEATING OF CUTTING EDGE—Overheating of cutting edges reduces their hardness and results in faster abrasion. Too much heat is caused in some work materials by the friction between the cutting edge and the work material, particularly if the work piece is relatively hard. In other cases, the dulling of a cutter causes a rise in temperature which ultimately causes breakdown of the cutting edge. Certain tool materials resist high temperatures better than others and may improve such op-erations. Your cutting-tool manufacturer, should be consulted on prob-lems of this kind.

CUTTER BREAKAGE—Cutter breakage occurs when the tool is subjected to stresses greater than it can withstand. Cutter breakage may result from hardness of the work material, too heavy a chip load per tooth, backlash and impact when climb milling, or to a poor machine or fixture setup.

SHARPENING CUTTERS AND END MILLS

WHEN TO SHARPEN—It is advisable to establish definite schedules for sharpening to obtain best results when using end mills and cutters. Several such schedules are described.

Sharpen At Predetermined Wear Land—Cutters should be sharpened as soon as the wear land (Fig. 19) reaches a predetermined width. This width should permit sharpening without excessive loss of tool life. It may vary from a few thousandths to $\frac{1}{16}$ inch, depending on the type of cutter and the finish required on the product.

This method is used on production runs where uneven amounts of stock is removed or where the material varies in machinability. It is also used on small quantity product lots.

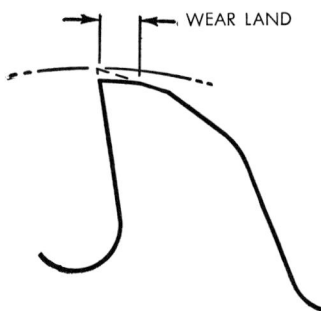

WEAR LAND

Figure 19—A cutter tooth with dimension lines showing wear land.

Sharpen After Predetermined Period Of Use—A "time period" to determine when to sharpen a cutter is often used on production runs where all conditions are repetitively uniform. The time period is the length of time it takes a sharp cutter to develop a width of wear land indicating that it should be sharpened.

A sharpening program based on a period of use means uniformly good sharpening as well as uniform machine and product conditions. A properly sharpened cutter can be expected to do approximately the same amount of work after each sharpening before the maximum width of wear land develops.

Sharpen When Product Quality Indicates—Sharpen cutters when product finish is unsatisfactory or when product size is not within dimensional tolerances. As a cutter dulls in use it cuts less freely than when sharp and has a tendency to produce rough surfaces affecting finish and size.

Sharpen When Power Increases—An increase in the amount of power used in milling operations is a good index of when to sharpen cutters if the amount of stock to be removed and product machinability are reasonably uniform. A visible indicator in the power line will show the operator when a power rise occurs and experience will indicate at what point cutters require sharpening.

FUNDAMENTALS OF SHARPENING

Mounting For Sharpening—All cutters should be sharpened to run true when in use. Arbor type cutters should be held on arbors that provide a good fit in the hole and avoid wobble. Shank type cutters may be held on centers or by the shank in a sleeve or socket that should be a good fit and concentric.

Use The Proper Grinding Wheel At Correct Speed—Many standard shapes of grinding wheels are readily available. The three most common for sharpening cutters are illustrated in Fig. 20.

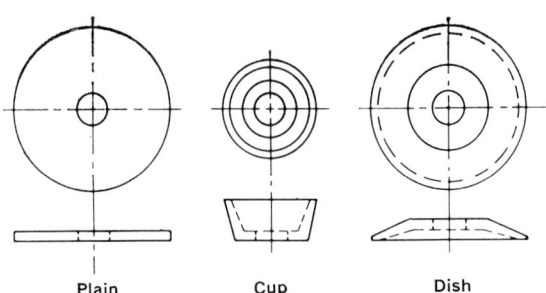

Plain   Cup   Dish

Figure 20—Common types of grinding wheels for sharpening cutters.

Table 6 is the standard marking for aluminum oxide and silicon carbide grinding wheels. A commonly used marking for diamond and CBN (cubic boron nitride) wheels is shown in Table 7.

When sharpening high speed steel cutters, the peripheral speed of the grinding wheel should be from 4500 to 6500 surface feet per minute. For sharpening cemented carbides, a speed of 5000 or 5500 sfm is recommended. Maximum safe operating speed, printed on the wheel blotter, should never be exceeded. Sometimes, the most efficient wheel speed for a specific operation or material may be less than the maximum safe speed. Too slow a speed means wastage of abrasive without too much useful work in return, whereas an excessive speed may result in overheating the cutting edge or wheel breakage.

Milling cutters are usually sharpened dry although wet grinding with a spray mist is becoming popular when sharpening some of the highly alloyed high speed steels. While most high speed steel cutters are sharpened with aluminum oxide wheels, CBN wheels are often suggested for sharpening cutters made of High Speed Steel containing more than 4% vanadium.

Grinding wheels used for sharpening the highly alloyed high speed steels should be one grade softer than for regular HSS. Slightly greater wheel wear must be accepted and the feed per pass should be less. CBN wheels allow for greater stock removal with greater form holding capabilities, but at higher initial expense per wheel. CBN wheels require more rigid set ups and extra requirements in their handling, trueing, and dressing. Contact your grinding wheel manufacturer for further information and specific recommendations. See Table 8.

Table 6—Standard Marking System Chart
(for aluminum oxide and silicon carbide grinding wheels)

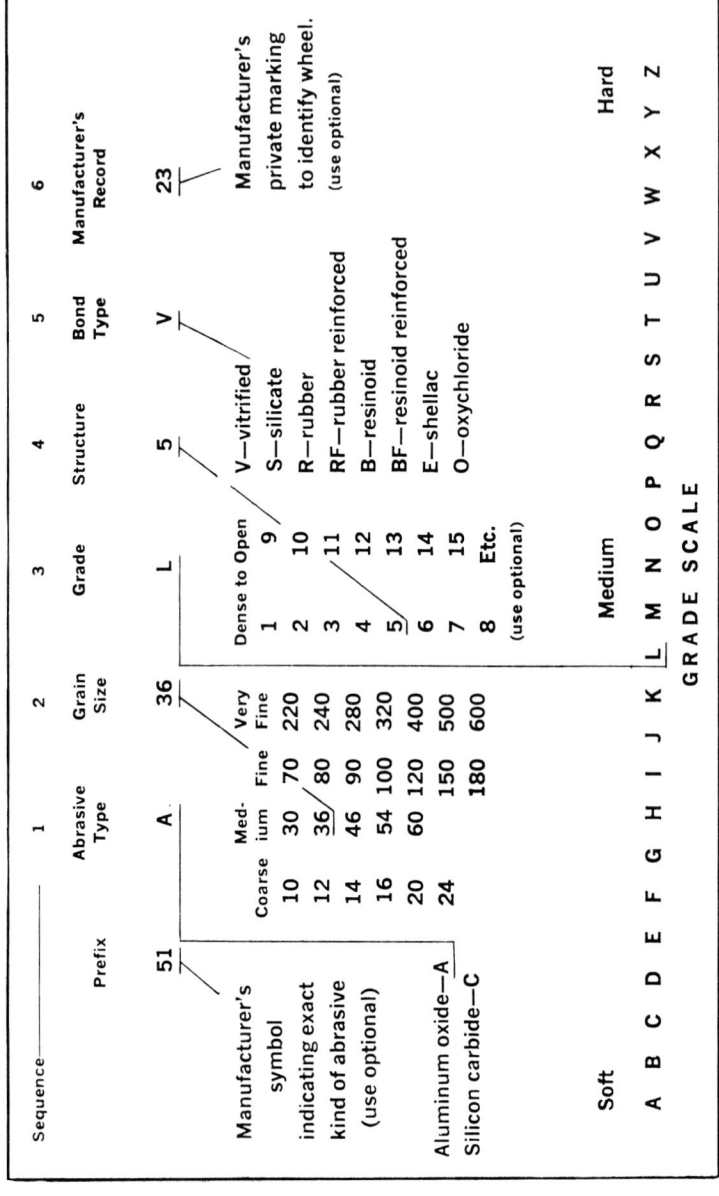

Sequence	1	2	3	4	5	6
Prefix	Abrasive Type	Grain Size	Grade	Structure	Bond Type	Manufacturer's Record
51	A	36	L	5	V	23

Prefix: Manufacturer's symbol indicating exact kind of abrasive (use optional)

Abrasive Type: Aluminum oxide—A, Silicon carbide—C

Grain Size:

Coarse	Med-ium	Fine	Very Fine
10	30	70	220
12	36	80	240
14	46	90	280
16	54	100	320
20	60	120	400
24		150	500
		180	600

Grade:

GRADE SCALE

Soft   A B C D E F G H I J K **L** M N O P Q R S T U V W X Y Z   Hard

Medium

Structure:
Dense to Open
1
2
3
4
5
6
7
8
Etc.
(use optional)

Bond Type:
V—vitrified
S—silicate
R—rubber
RF—rubber reinforced
B—resinoid
BF—resinoid reinforced
E—shellac
O—oxychloride

Manufacturer's Record: Manufacturer's private marking to identify wheel. (use optional)

Table 7—Identification Chart for Diamond and CBN Wheel Marking

Sequence							
1st	2nd	3rd	4th	5th	6th	7th	8th
D	100	N	100	B		1/8	*
Abrasive Type	Grit Size	Grade	Concentration	Bond	Bond Modification	Depth of Diamond Section (Inches)	Mfgrs. Identification
D-Diamond	24 120 600	A (Soft)	25 (Low)	B—Resinoid	Mfgrs. Specific Bond Record (optional)	1/32	Optional
	30 150 800					1/16	
	36 180 1000						
	46 220 1200		50	M—Metal		1/8	
MD)Man Made	54 240 1500	To	To			1/4	
SD)Diamond	60 280						
	80 320	Z (Hard)	75 (High)	V—Vitrified			
B-CBN	90 400						
	100 500		100				

Table 8—Grinding Wheel Specifications Commonly Used for Sharpening Cutters and End Mills

cutter material	operation	abrasive material	size grain	grade	bond	typical wheel markings	
High speed steel (regular)	Roughing	Aluminum oxide	36 to 46	H to K	Vitrified	AA46-JB-V40 77A46-J8V-6	32A46-I8VBE 9A46I8V22
	Finishing	Aluminum oxide	60 to 100	I to K	Vitrified	AA60-K8-V40 32A60-I8VBE	11A60J5V-22 7A60418V22
High alloy high speed steel	Roughing	Aluminum Oxide	36 to 60	G to J	Vitrified	32A60-G12VBEP	11A60H8V-6 7A-604H12V22
	Finishing	Aluminum Oxide	60 to 100	G to K	Vitrified	32A60-J8VG 77A60J5V-22	7A8048V22
Cemented carbides	Roughing	Silicon carbide	60	G to K	Vitrified	C60I7V 3C543I12V32FA	39C60-IVK GC60I8V-2
		Diamond	100	N	Resinoid	D100-N100B	D100-R4 BA 1/8 SD 100-N100B-6
	Finishing	Silicon carbide	100	G-H	Vitrified	C100 H 8	39C 100-IVK GC 100H8V-2
		Diamond	100-220	L-N	Resinoid	D220-L100B	SD 220- N100 B-6 D220N4 BA 1/8

If any grinding wheel acts too hard, decreasing the wheel SFM will make it act softer. Wheel grades with the same letter from different manufacturers often vary. It is best to experiment or consult a wheel maker for assistance if problems are encountered.

Maintain The Proper Rake And Relief Angle—When usage demonstrates that a cutter works well on a particular application, it should be resharpened with the same relief and rake angles.

Three common types of relief are used on profile sharpened cutters, namely, (a) concave, (b) flat, and (c) eccentric. The different land profiles are illustrated in Fig. 21. While (a) and (b) are actually different, the amount is so slight that tabulations for relief angles and indicator drops are generally accepted as satisfactory for either style. Because of the convex shape of the eccentric relief type land, the width of land is not as critical as the other two types.

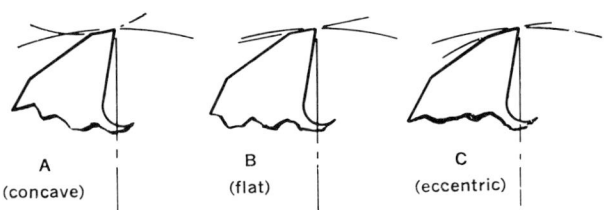

A	B	C
(concave)	(flat)	(eccentric)

Figure 21—Common types of relief.

Current practice of most cutter manufacturers is to furnish HSS cutters sharpened for general purpose milling of cast iron, ordinary steels, etc., with a definite amount of relief for the particular cutter diameter. This relief angle varies from approximately 22° for a $\frac{1}{16}$ diameter end mill to around 4° for cutters 4″ diameter and over.

The same general pattern is followed on carbide cutters, although the usual minimum is around 5° to 7° depending on the material the cutters are designed to cut.

For machining plastics, most stainless steels and non-ferrous materials, it is recommended that these amounts be increased up to 50% on small sizes with a minimum of 8° on the larger sizes to provide satisfactory cutting action.

Table 9 lists the primary flat or concave radial relief angles and table 12 the eccentric radial relief angles usually recommended for general purpose milling, measured by the indicator drop method.

In general, the relief angle on the outside diameter of milling cutters is ground to the edge. While there are several methods to check the amount of relief, such as projecting an image or specially designed commercial fixtures using angular edges coupled with optics, one simple inexpensive method is to measure by dial indicators while the cutter is mounted between bench centers or by its shank. Using two indicators, one with a sharp point to measure the drop in thousandths of an inch, the other to measure the exact amount of rotation, correct values can be checked. See Figure 22.

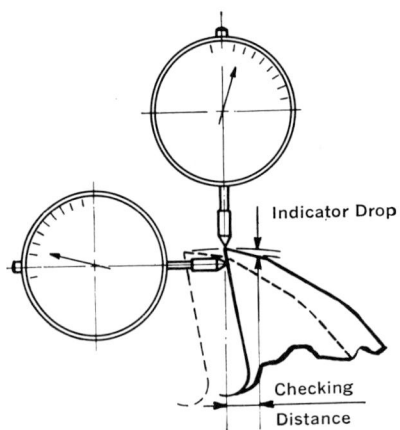

Figure 22—Indicator set-up for checking radial relief.

*Procedure To Check
Radial Relief Angles
With Indicators.*

1—Mount the cutter to rotate freely with no end movement.

2—Adjust the sharp pointed indicator to bear at the very tip of the cutting edge, pointing in a radial line, shown in Figure 22.

3—Roll the cutter the tabulated amount given under "checkiing distance" using the second indicator as a control.

4—Consult chart for amount of drop for the particular diameter and relief angle.

Note: After repeated sharpenings, the width of land will increase to the point where interference at the heel may occur if not reduced. A secondary clearance of increased angle is a means to control this condition. Do not increase the primary relief angle to get heel clearance.

Table 9—Measurement of Flat or Concave Radial Relief Angles by the Indicator-Drop Method.

Cutter Diam. Inches	Usual Range of Radial Angle (Degrees)	Flat or Concave Relief Indicator Drop for Relief Angles Shown		Checking Distance
		Min.	Max.	
1/16	20-25	.0018	.0027	.010
3/32	16-20	.0017	.0024	.010
1/8	15-19	.0021	.0032	.015
5/32	13-17	.0020	.0030	.015
3/16	12-16	.0020	.0034	.020
7/32	11-15	.0020	.0034	.020
1/4	10-14	.0019	.0033	.020
9/32	10-14	.0020	.0034	.020
5/16	10-13	.0020	.0033	.020
3/8	10-13	.0025	.0035	.020
7/16	9-12	.0025	.0038	.025
1/2	9-12	.0027	.0040	.025
9/16	9-12	.0028	.0041	.025
5/8	8-11	.0028	.0045	1/32
3/4	8-11	.0031	.0047	1/32
7/8	8-11	.0033	.0049	1/32
1	7-10	.0028	.0045	1/32
1 1/8	7-10	.0030	.0046	1/32
1 1/4	6-9	.0025	.0042	1/32
1 1/2	6-9	.0026	.0043	1/32
1 3/4	6-9	.0027	.0044	1/32
2	6-9	.0028	.0045	1/32
2 1/4	5-8	.0023	.0039	1/32
2 1/2	5-8	.0024	.0040	1/32
3	5-8	.0024	.0041	1/32
4	5-8	.0025	.0042	1/32
5	4-7	.0021	.0037	1/32
6	4-7	.0021	.0037	1/32
7	4-7	.0021	.0037	1/32
8	4-7	.0021	.0037	1/32
10	4-7	.0021	.0038	1/32
12	4-7	.0021	.0038	1/32

The actual value of the radial relief angle is normally kept within the range shown but may be varied to suit the cutter material, the work material, and the operating conditions.

Sharpening Side Or End Teeth—Different practices are followed when sharpening side teeth which must mill to a close tolerance on width and where maintenance of this width is important. Omit relief on the side teeth entirely, simply grind the sides concave; leave a narrow margin or land without relief at the cutting edge. Behind this margin the cutter is sharpened with 2° to 4° relief angle. Sharpen the side land with relief up to the edge but with less than the regular amount. Concavity rates vary from .001 to .008 per inch.

End teeth of End Mills and Shell Mills not intended to feed endwise are sharpened with 3° to 5° relief. End mills with teeth to center for end feeding, are generally relieved to the edge with from 6° to 8° relief. The diameter has no influence on relief angles for end teeth.

Face mills are sharpened on the face in various ways to suit different work materials and finish requirements.

Table 10 gives indicator measurements for flat relief on ends, side or face teeth for common relief angles.

Table 10—Indicator Drops for Relief on Side or End Teeth

measured travel of indicator point on the relieved surface	indicator drop of					
	1°	2°	3°	4°	5°	7°
$\frac{1}{64}$	.0003	.0005	.0008	.0011	.0014	.0019
$\frac{1}{32}$	.0005	.0011	.0016	.0022	.0027	.0038
$\frac{3}{64}$	.0008	.0016	.0025	.0033	.0041	.0057
$\frac{1}{16}$	.0011	.0022	.0033	.0044	.0055	.0077

Relief angles on form relieved cutters are built in by the manufacturer. Sharpening by grinding only the faces of the teeth in the same plane as when new, will maintain correct rake angles. This is discussed in succeeding pages.

Cutter Run-Out—A cutter performs best when the cutting edge of all teeth runs true with the axis. Then each tooth does its share of work. Radial and axial run-outs should be checked with an indicator after each sharpening.

NC Grinding Equipment—Numerically controlled (NC and CNC) grinding machines can economically sharpen end mills and shank-type milling cutters. The basic concepts of wheel selection, grinding speeds, and relief geometry apply to NC resharpening as well as conventional sharpening. NC tool and cutter grinders offer three advantages over conventional sharpening.

1. Simplified Operation. Once the NC machine is set up, the adjustments to the wheelhead are made through the program. Only the end mill needs to be positioned. This can provide substantial savings when sharpening a large volume of tools.

2. Consistency of Results. NC equipment eliminates the operator's "touch" from the quality of the end mill. Properly set up, an NC tool grinder makes nearly identical resharpened tools.

3. Versatility. The programmable feature of NC tool grinders allows the equipment to be used in many sharpening applications. From resharpening both peripheral and end teeth in one operation to altering tools with corner radii, each program can be customized to suit specific needs.

SHARPENING PROFILE CUTTERS
Radial Relief Sharpening

1. Relationship of Wheel To Work And Effect Upon Relief—There are several methods of grinding the relief on the periphery of profile cutters. Relief produced may vary with the method. The most common types are flat, concave, and eccentric, illustrated in Fig. 21.

Flat relief is produced by a cup wheel mounted with its axis approximately perpendicular to the cutter axis. The face of the wheel grinds a flat surface on the relieved area. The degree of relief can be changed by varying the offset, illustrated in Fig. 23 and accompanying Table 11.

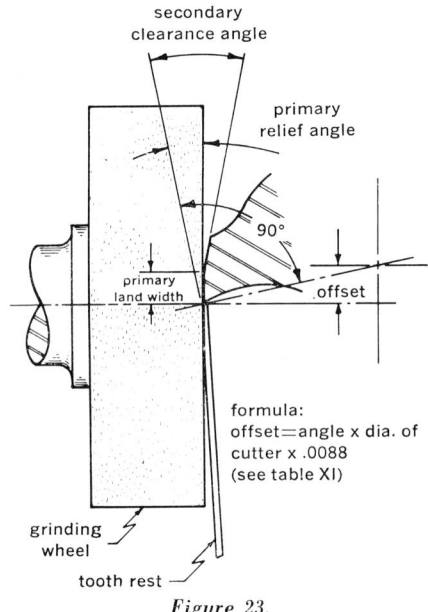

Figure 23.

Concave relief is produced with a plain wheel mounted with its axis approximately parallel with the cutter axis. One edge of the wheel periphery produces a concave surface on the relieved area. The amount of concavity is a function of wheel diameter. The degree of relief may be varied by adjusting the offset, or changing the grinding wheel diameter.

Table 11—Offset to Produce Various Relief and Clearance Angles

cutter diam.	primary relief		secondary clearance		primary land width
	angle	offset	angle	offset	
1/16	22°	.012	32°	.018	.005/.007
3/32	18°	.015	28°	.023	.005/.007
1/8	16°	.018	28°	.031	.005/.007
5/32	15°	.021	26°	.036	.007/.009
3/16	14°	.023	26°	.043	.007/.009
7/32	13°	.025	24°	.046	.007/.009
1/4	12°	.026	22°	.048	.007/.009
9/32	12°	.030	21°	.052	.009/.012
5/16	12°	.033	21°	.058	.009/.012
11/32	11°	.033	19°	.057	.009/.012
3/8	11°	.036	19°	.063	.009/.012
13/32	11°	.039	19°	.068	.012/.016
7/16	11°	.042	19°	.073	.012/.016
15/32	10°	.041	19°	.078	.012/.016
1/2	10°	.044	18°	.079	.012/.016
9/16	10°	.050	18°	.089	.016/.020
5/8	10°	.055	18°	.099	.016/.020
11/16	9°	.055	18°	.109	.016/.020
3/4	9°	.059	17°	.112	.016/.020
13/16	9°	.064	17°	.121	.020/.025
7/8	9°	.069	17°	.131	.020/.025
15/16	8°	.066	16°	.132	.020/.025
1	8°	.070	16°	.141	.020/.025
1 1/8	8°	.079	14°	.139	.025/.030
1 1/4	7°	.077	14°	.154	.025/.030
1 3/8	7°	.085	13°	.157	.025/.030
1 1/2	7°	.092	13°	.172	.025/.030
1 5/8	7°	.100	13°	.186	.030/.035
1 3/4	6°	.092	12°	.185	.030/.035
1 7/8	6°	.099	12°	.198	.030/.035
2	6°	.106	12°	.211	.030/.035
2 1/4	5°	.099	11°	.218	.035/.040
2 1/2	5°	.110	11°	.242	.035/.040
2 3/4	5°	.121	11°	.266	.035/.040
3	5°	.132	11°	.290	.035/.040
3 1/2	4°	.123	10°	.308	.040/.045
4	4°	.141	10°	.352	.040/.045
5	4°	.176	10°	.440	.040/.045
6	4°	.211	10°	.528	.040/.045

Eccentric relief is produced with a plain wheel positioned with its axis parallel or at a slight angle with the cutter axis. This setup differs from concave grinding in that the wheel periphery is dressed at an angle, or the wheel is tipped at an angle with the cutter axis. The wide surface on the periphery or face of the wheel will generate a convex or eccentric relief on a helical tooth, indicated in Fig. 24. Note that a helical tooth is required to generate eccentric relief, thus this method cannot be used to produce eccentric relief on straight fluted cutters. The degree of relief is varied by changing the angle of wheel inclination or the angle dressed on the wheel. For a given relief angle, the wheel angle will vary according to the helix angle. See table 12.

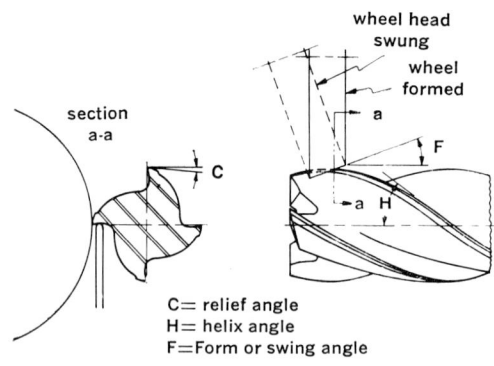

C= relief angle
H= helix angle
F=Form or swing angle

Formula: Tan (F) = Tan (C) x Tan (H)

Figure 24.

Note: The high point of the tooth rest must contact the tooth face at the high side of the wheel angle (O.D. of cutter) and be the same height as the wheel and work centers.

2. Tooth Rest And Its Function—Most profile sharpening set-ups involve three elements: cutter, grinding wheel and tooth rest. The tooth rest has many functions and each must be considered in selecting the proper design for a particular application.

Its main function is to support the cutter tooth being ground. It must be rigid enough to prevent deflection from grinding pressure or tangential movement of the cutter.

The tooth rest positions the cutter tooth for proper grinding and, to suit various conditions, must be adjustable. The degree of relief on the cutter

TABLE 12—Measurement of Eccentric Radial Relief Angles by the Indicator-Drop Method

Cutter Diam. Inches	Usual Range of Radial Angle (Degrees)	Eccentric Relief Indicator Drop for Relief Angles Shown		Check-ing Dis-tance	Wheel angles (See F, Figure 24)		
		Min.	Max.		20° Helix *Angle	30° Helix *Angle	40° Helix *Angle
1/16	20-25	.0036	.0047	.010	8° 34′	13° 27′	19° 10′
3/32	16-20	.0029	.0036	.010	6° 45′	10° 37′	15° 15′
1/8	15-19	.0040	.0052	.015	6° 21′	10° 1′	14° 24′
5/32	13-17	.0035	.0046	.015	5° 34′	8° 48′	12° 40′
3/16	12-16	.0042	.0057	.020	5° 11′	8° 12′	11° 49′
7/32	11-15	.0039	.0054	.020	4° 48′	7° 36′	10° 58′
1/4	10-14	.0035	.0050	.020	4° 26′	7°	10° 7′
9/32	10-14	.0035	.0050	.020	4° 26′	7°	10° 7′
5/16	10-13	.0035	.0046	.020	4° 13′	6° 42′	9° 41
3/8	10-13	.0035	.0046	.020	4° 13′	6° 42′	9° 41
7/16	9-12	.0040	.0053	.025	3° 52′	6° 6′	8° 50′
1/2	9-12	.0040	.0053	.025	3° 52′	6° 6′	8° 50′
9/16	9-12	.0040	.0053	.025	3° 52′	6° 6′	8° 50′
5/8	8-11	.0044	.0061	1/32	3° 29′	5° 31′	8°
3/4	8-11	.0044	.0061	1/32	3° 29′	5° 31′	8°
7/8	8-11	.0044	.0061	1/32	3° 29′	5° 31′	8°
1	7-10	.0038	.0055	1/32	3° 7′	4° 56′	7° 9′
1 1/8	7-10	.0038	.0055	1/32	3° 7′	4° 56′	7° 9′
1 1/4	6-9	.0033	.0050	1/32	2° 45′	4° 21′	6 °18′
1 1/2	6-9	.0033	.0050	1/32	2° 45′	4° 21′	6 °18′
1 3/4	6-9	.0033	.0050	1/32	2° 45′	4° 21′	6 °18′
2	6-9	.0033	.0050	1/32	2° 45′	4° 21′	6 °18′
2 1/4	5-8	.0027	.0044	1/32	2° 22′	3° 46′	5° 28′
2 1/2	5-8	.0027	.0044	1/32	2° 22′	3° 46′	5° 28′
3	5-8	.0027	.0044	1/32	2° 22′	3° 46′	5° 28′
4	5-8	.0027	.0044	1/32	2° 22′	3° 46′	5° 28′
5	4-7	.0022	.0038	1/32	2°	3° 11′	4° 37′
6	4-7	.0022	.0038	1/32	2°	3° 11′	4° 37′
7	4-7	.0022	.0038	1/32	2°	3° 11′	4° 37′
8	4-7	.0022	.0038	1/32	2°	3° 11′	4° 37′
10	4-7	.0022	.0038	1/32	2°	3° 11′	4° 37′
12	4-7	.0022	.0038	1/32	2°	3° 11′	4° 37′

The actual value of the radial relief angle is normally kept within the range shown but may be varied to suit the cutter material, the work material, and the operating conditions.

*Angle is calculated from the basic mean of the radical angle.

tooth is varied by adjusting the tooth rest up or down, or by mounting on a wheel head which can be elevated or lowered. Only the tooth being sharpened should contact the tooth rest.

The tooth rest can be used to index the cutter tooth into grinding position. It must be flexible enough to deflect out of the way as the back of an uncleared tooth is rotated into grinding position. It will snap into position when the cutter tooth rises above the tip of the rest. Use precaution to avoid damage to adjacent teeth by the snap of the tooth rest.

The tooth rest can maintain proper lead of the flutes when clearing helical fluted cutters. It must be offset from the wheel so that the leading edge of the tooth is supported before approaching the grind position. It should be formed at the same angle as the helix of the cutter and be wide enough so that the trailing edge of the tooth sparks out at the opposite side of the wheel in the proper grinding position. To assure a proper grinding pattern, it should contact the face of the tooth as close to the edge as possible.

3. *Suggested Equipment And Setup For Radial Relief Sharpening—* Straight gashed cutters may be sharpened with either a plain or cup wheel, indicated in Figs. 25 and 26. The tooth rest is mounted on the table slide. The cutter is located in a fixed position with the tooth rest. The cutter is then moved past the wheel.

Figure 25.

Figure 26.

Figure 27.

Figure 28.

Helical gashed cutters may be sharpened in a setup similar to that indicated in Figs. 27 and 28. It differs from the straight gash setup in that the tooth rest is mounted in fixed relation to the wheel head. This arrangement permits the helical cutting edge to rotate into grinding position as the cutter travels past the grinding wheel.

Staggered tooth cutters also may be ground in a similar manner except that a double bevelled tooth rest is generally used. See Figs. 29 and 30. This type makes it possible to grind consecutive teeth rapidly and accurately even though they have opposite hands of helix. When using this type of tooth rest care must be taken to position the contact point of the grinding wheel central with the high point of the tooth rest.

Figure 29.

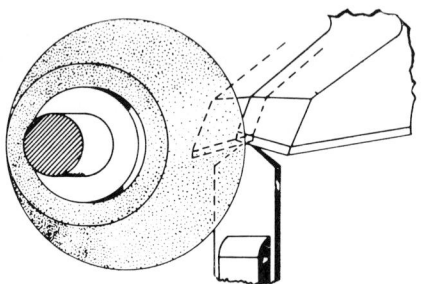

Figure 30.

SIDE TOOTH SHARPENING

1. *Side Concavity Or Dish*—A concave ground surface on the plain sides of a milling cutter or saw can provide an effective relief if the depth of cut is shallow or the finish requirements are not critical. If the depth of cut is appreciable and this type of relief is desired, the sides should be gashed, leaving narrow margins to reduce the contact surface and to allow coolant to enter this area.

Concavity can be ground into the side of a cutter by traversing a plain type or plunging a cup type wheel along the side of the cutter as it rotates. The degree of concavity is obtained by varying either the angle of the table slide or the inclination of the grinding wheel axis.

2. *Side Relief*—The most common way to provide relief on the side of a cutter is to grind axial relief along the entire length of the side tooth. This relief may be ground either to a sharp edge or to a narrow margin. In most cases a slight concavity is desirable.

When a cup wheel is used, as in Figure 31, the cutter tooth is oriented parallel with the direction of table travel. The cutter is tipped to provide axial clearance and swung to provide a slight concavity. This method is most commonly used for clearing end teeth on end mills.

When a plain wheel is used, as in Figure 32, it is fed sideways along the cutter tooth in a radial direction. The offset of the wheel produces the desired axial relief while the orientation of the cutter tooth gives the desired

Figure 31.   *Figure 32.*

concavity. Some sharpening machines also have a rise and fall mechanism to produce varying degrees of concavity or lead and chamfer angles.

Chamfer And Angle Sharpening—Chamfers on corners are usually sharpened with the cutter mounted in a fixed position, similar to that shown in Figure 33. In principle the setup is the same as for sharpening side teeth except the cutter is mounted at an angle to produce the chamfer.

Angle cutters can be ground in a similar manner provided the tooth is straight gashed and radial, as in Figure 34. However, if the cutter tooth is undercut or is helical gashed, the cutter cannot be held in a fixed position without distorting the angle being cleared. To maintain a true angle, the tooth must be mounted in a fixed relation with the wheelhead, and the cutter must be free to rotate as the tooth slides along the tooth rest during the grind.

 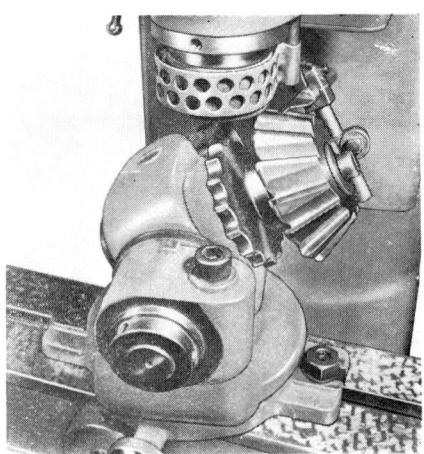

Figure 33.   *Figure 34.*

Radius Sharpening—Corner, convex or concave radii can be sharpened with the cutter mounted basically the same as for grinding chamfers or angles. The base of the cutter holding fixture, however, must be free to pivot about a point exactly below the center of the radius required on the cutter. Figure 35. In most cases the table slide remains fixed; however, table movement can be incorporated to combine grinding an angle tangent to a radius.

Figure 35.

Contour Sharpening—Profile cutters involving several interrelated complex forms cannot generally be sharpened economically on conventional grinding equipment, but should be sharpened on equipment designed to control the movement of the wheelhead or work table by a master template. Plain wheels dressed to correspond to the template follower shape are used to produce an undistorted cutter shape.

Complex forms may also be profile sharpened by cylindrical grinding the form on the periphery, then clearing each section individually to a hairline margin. This type of grinding is generally not economical and form relieved cutters are recommended if contour grinding equipment is not available.

The size of the radius is changed by relocating the cutter relative to the pivot point and by adjusting the wheel a corresponding amount. Plain wheels with an approximate convex dressed form are used. If the tooth rest is mounted on the wheelhead, be sure that the cutter edge of the grinding wheel is central with the high point of the tooth rest.

Recutting, Regashing Or Gumming—Cutters sharpened repeatedly end up eventually with wide lands and insufficient flute space. If these tools are salvaged, major reworking is required to deepen the flutes and thin the lands.

Soft, coarse grit wheels should be used in any recutting operation. Excessive heat during grinding can damage the tool beyond repair. It is essential to keep the grinding zone as cool as possible. Fillets at the root of gashes or flutes should be kept as large as possible. Sharp steps or notches, if required, should be added later with a hard wheel as a secondary operation. Maintain the original radial and axial rakes for consistent tool life.

Rework of this nature is a specialty and should not be attempted without thorough knowledge of grinding technics.

SHARPENING END MILLS—While most end mills are actually profile sharpened milling cutters, they have characteristics which require special attention in sharpening. End mills usually have a relatively small diameter in proportion to their length. The resulting flexibility may cause difficulty in holding a uniform diameter if grinding cuts are too heavy. Their smaller mass limits the amount of grinding heat which can be absorbed without metallurgical damage to the cutting edges. Since end mills are shank type cutters, concentricity must be maintained between the peripheral cutting edges and the shank. Most end mills have some form of end teeth. These are generally similar to the side teeth on some arbor-type cutters but sharpening complications are introduced by the fact that the end teeth of end mills extend to or very near to the center (axis) of the tool.

Sharpening Peripheral Teeth—As with other profile sharpened cutters, end mills are usually reconditioned by resharpening the relief and clearance surfaces of the peripheral teeth. Occasionally, however, there is a requirement of minimum diameter loss and it then may become economical to resharpen end mills by grinding the flute face and fillet to remove the wear land and restore the original flute shape. As pointed out earlier, such recutting or "gumming out" of flutes can generate a great deal of heat and may damage the tool material. Thus, such work is best left to specialists who have mastered the necessary techniques.

The setup used for resharpening the relief and clearance surfaces of end mill teeth is basically the same as that used for resharpening arbor type cutters. The differences are principally in the method of holding the cutter, in the stock removal rates, and sometimes in the finish requirements.

If an end mill has good center holes on each end it can be held between centers on a conventional cutter grinder as shown in Figures 36 and 37. Many end mills do not have centers on both ends so the tool must be held by its shank.

Figure 36—Tool support centers and their relation to grinding wheel and fixed tooth support finger.

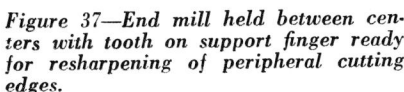

Figure 37—End mill held between centers with tooth on support finger ready for resharpening of peripheral cutting edges.

A simple rotating spindle head setup is shown mounted on a cutter grinder in Figures 38 and 39.

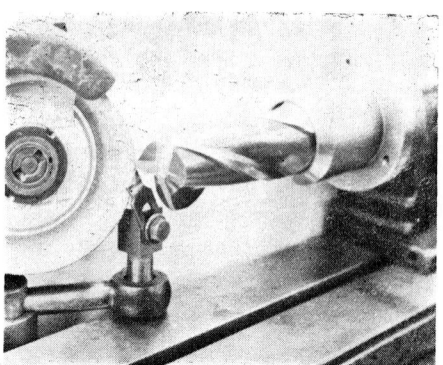

Figure 38—End mill held on shank in simple rotating spindle head with tool tooth on support finger ready for re-sharpening of peripheral cutting edges.

Figure 39—Tapered end mill held on shank in simple rotating spindle head with head swung to angle desired on end mill and tool tooth on support finger ready for resharpening.

Note that with both of these setups the tooth rest is fixed in relation to grinding wheel spindle and the work is rotated against the tooth rest as the cutter grinder table is traversed.

Another satisfactory setup involves the use of a special end mill sharpening fixture illustrated in Figures 40 and 41. Here, the end mill is held by its shank in a long spindle which can both rotate and slide axially. The tooth rest is mounted upon the fixture body. With this fixture, the machine table is locked and all the motions are incorporated in the fixture.

Figure 40—End mill held by shank in resharpening fixture spindle ready for radial cutting edge regrinding. Fixture provides for axial and rotational movement of tool past grinding wheel and has support finger and means for selecting proper relief and clearance angles built in.

Figure 41—Close-up of end mill in resharpening fixture spindle showing simple vertical, radial, and axial adjustment means for support finger providing quick and easy set-up.

A major factor in resharpening end mills is the attainment of the correct amount of relief *at the cutting edge*. The exact type of relief is relatively unimportant for most work. Several types of relief and the setups used for producing and checking them have been described previously.

The best resharpening procedure is to regrind the primary relief first until all of the wear has been removed, taking care to avoid excessive diameter loss. The secondary clearance is ground next to bring the primary relief lands to the desired widths. Table 11 shows general purpose relief and clearance angles and primary land widths along with the necessary setup offsets. After grinding the secondary clearance it is often desirable that the primary relief surfaces be given a light finish grind to refine the cutting edges. To minimize runout, this light finishing cut should be made at one machine setting, going completely around the end mill.

The amount of stock removed per pass is determined by end mill size, tool material, and finish requirements. In rough regrinding of end mills made of general purpose high speed steels, as much as 0.002 inches of stock can be removed per pass. For small flexible end mills this must be reduced to hold uniform diameters. Highly alloyed special purpose high speed steel may require cuts as light as 0.0005 inches of stock per pass. Light finishing cuts are also required on the primary relief surfaces to produce smooth sharp cutting edges. CBN wheels are recommended for minimum heat generation and may allow greater stock removal on roughing operations.

Sharpening End Teeth—In the resharpening of end teeth, the first step is always the removal of the wear on the end teeth and at the corner intersection of the end and peripheral teeth. Particular care must be taken that all of the corner wear is removed. This can be done either by hand or machine grinding but care must be taken to avoid overheating the end mills by too severe grinding. Machine grinding is preferred because stock removal is more uniform and resharpening is thus simplified. A common machine setup involves chucking the end mill and rotating it against a flat or cup wheel for a square end tool or against a concave radius dressed wheel for a ball end tool. Worn end mills and their appearance after this first operation are shown in Figures 42A and 42B.

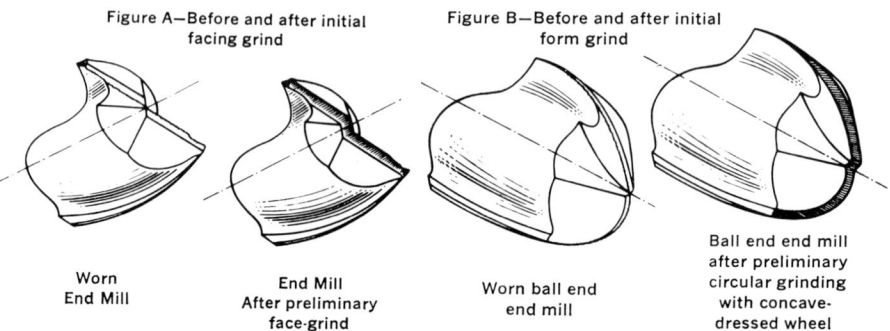

Figure A—Before and after initial facing grind

Figure B—Before and after initial form grind

Worn End Mill

End Mill After preliminary face-grind

Worn ball end end mill

Ball end end mill after preliminary circular grinding with concave-dressed wheel

Figure 42—2 Flute end mills. Before and after preliminary grind.

The end teeth of an ordinary two-flute end mill are similar to those of a square end drill. The three necessary operations and one optional feature, along with setup suggestions, are shown in Figures 43A to 43D. In each drawing, the shaded area indicates the surfaces being ground.

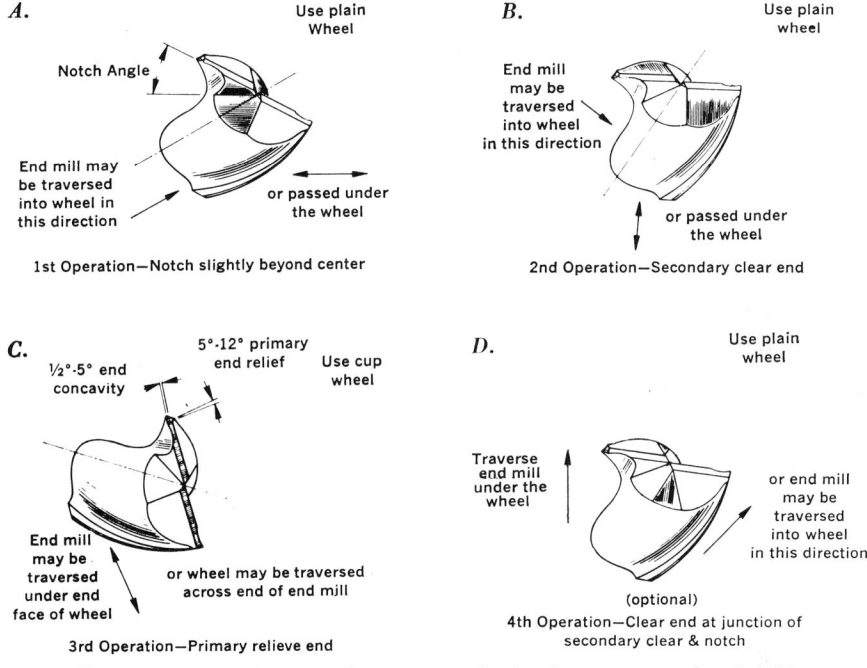

A.

Use plain Wheel

Notch Angle

End mill may be traversed into wheel in this direction

or passed under the wheel

1st Operation—Notch slightly beyond center

B.

Use plain wheel

End mill may be traversed into wheel in this direction

or passed under the wheel

2nd Operation—Secondary clear end

C.

½°-5° end concavity

5°-12° primary end relief

Use cup wheel

End mill may be traversed under end face of wheel

or wheel may be traversed across end of end mill

3rd Operation—Primary relieve end

D.

Use plain wheel

Traverse end mill under the wheel

or end mill may be traversed into wheel in this direction

(optional)

4th Operation—Clear end at junction of secondary clear & notch

Figure 43—Procedure for sharpening end of 2 flute square end end mills.

There are two ways to resharpen non center-cutting multi-flute end mills. The simplest is the cup or concave method shown in Figures 44A and 44B. Some end mills are made this way originally. If the original cup is deep enough, the first resharpening may require only the relieving operation of Figure 44B. This method of resharpening is not suitable for end mills which require large corner radii.

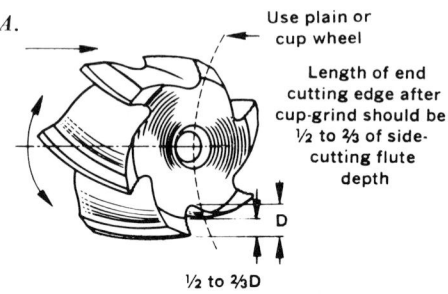

A.

Use plain or
cup wheel

Length of end
cutting edge after
cup-grind should be
½ to ⅔ of side-
cutting flute
depth

D

½ to ⅔D

1st Operation—Concave grind center on end

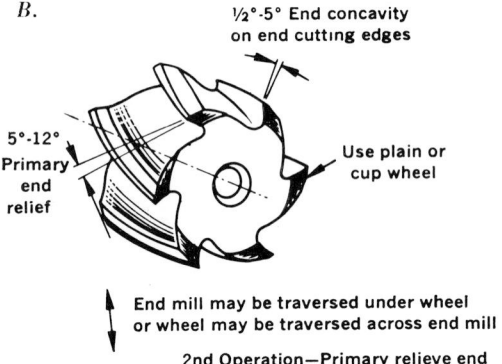

B.

½°-5° End concavity
on end cutting edges

5°-12°
Primary
end
relief

Use plain or
cup wheel

End mill may be traversed under wheel
or wheel may be traversed across end mill

2nd Operation—Primary relieve end

Figure 44 - Procedure for sharpening end of multi-flute cupped square end end mill.

The end gash method of resharpening non-center-cutting end mills is illustrated in Figures 45A through 45C. Note that the shallow end teeth of the heavy duty type of resharpening, shown in Figure 45B, are produced by a combined gashing and secondary clearing operation. While 4 flute end mills are shown in Figure 45, the procedure for end mills with 6, 8 or more flutes is similar but slight modification of grinding wheel angles may be necessary.

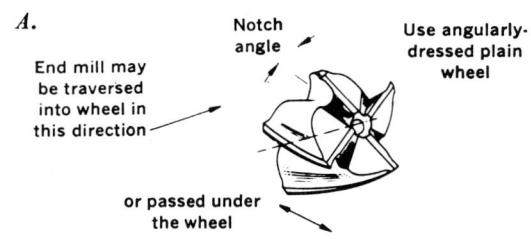

A.

Notch angle

Use angularly-dressed plain wheel

End mill may be traversed into wheel in this direction

or passed under the wheel

1st Operation—Notch gash end to center hole

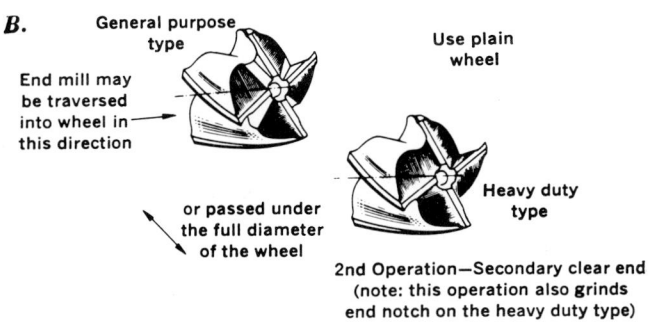

B.

General purpose type

Use plain wheel

End mill may be traversed into wheel in this direction

or passed under the full diameter of the wheel

Heavy duty type

2nd Operation—Secondary clear end (note: this operation also grinds end notch on the heavy duty type)

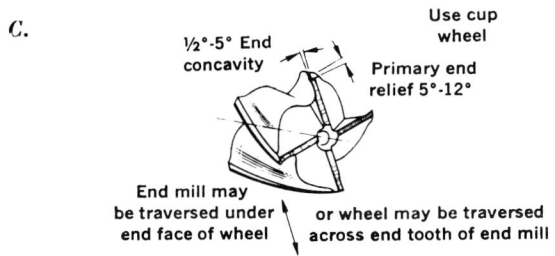

C.

$\frac{1}{2}$°-5° End concavity

Use cup wheel

Primary end relief 5°-12°

End mill may be traversed under end face of wheel

or wheel may be traversed across end tooth of end mill

3rd Operation—Primary relieve end

Figure 45— Procedure for sharpening end of 4-flute square end end mill with center hole.

Center cutting end mills with more than two flutes are more difficult to resharpen. The step by step procedure for 4-flute center cutting end mills is illustrated in Figures 46A through 46D. Here, two diametrically opposed teeth are made to cut to center in a manner similar to the end teeth of a two-flute end mill. All the other teeth are cut away slightly so that they do not extend to the center. Three-flute end mills are sharpened in a similar manner. The three types of end sharpening in common use are shown in Figures 47, 48 and 49. Figure 47 shows all three teeth extending to the center and the three end gashes meeting at the center point. Figure 48 shows two teeth extending to the center with the third tooth cut away at the center. Figure 49 shows one tooth extending to the center with the other two teeth cut away at the center point.

Ball end end mills also present resharpening problems due to the relieved radius form and the roughly spherical form of the secondary clearance. Some manufacturers grind both surfaces by machine, but most users will have to use a combination of machine and hand grinding to produce the ball end. A suggested procedure for two-flute ball end end mills is shown in Figures 50A through 50D. It is often helpful, in sharpening the primary relief on ball end end mills, to place the peripheral tangency point of the ball radius about .001″ to .002″ outside the peripheral diameter. This minimizes the chance of nicking the peripheral teeth at the tangency point. The effect on the form produced by the end mill is usually negligible.

Ball end end mills with more than two flutes require a combination of the procedures shown in Figures 46 and 50, since only two diametrically opposed teeth are carried to center. A certain amount of hand grinding may be required to blend the secondary clearance surfaces onto the end gashing cuts.

A.

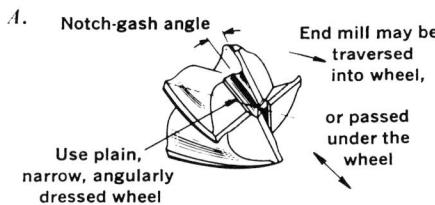

Notch-gash angle

End mill may be traversed into wheel,

or passed under the wheel

Use plain, narrow, angularly dressed wheel

1st Operation—Gash-notch two center-cutting end teeth past center

B.

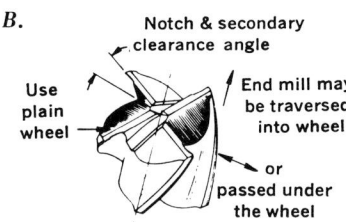

Notch & secondary clearance angle

Use plain wheel

End mill may be traversed into wheel

or passed under the wheel

2nd Operation—Notch-secondary-clear two end teeth not cutting to center

C.

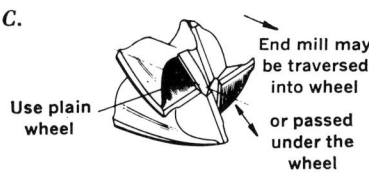

End mill may be traversed into wheel

Use plain wheel

or passed under the wheel

Conclusion of 2nd operation— Secondary clearing the two end teeth not cleared previously

D.

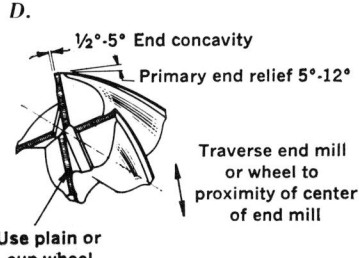

½°-5° End concavity

Primary end relief 5°-12°

Traverse end mill or wheel to proximity of center of end mill

Use plain or cup wheel

3rd Operation—Primary relieve end teeth

Three teeth to center
Figure 47.

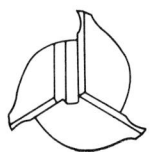

Two teeth to center
Figure 48.

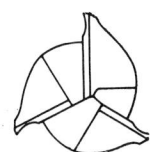

One tooth to center
Figure 49.

Three types of sharpening 3-flute end mills.

Figure 46—Procedure for sharpening end of 4-flute square end center-cutting end mill.

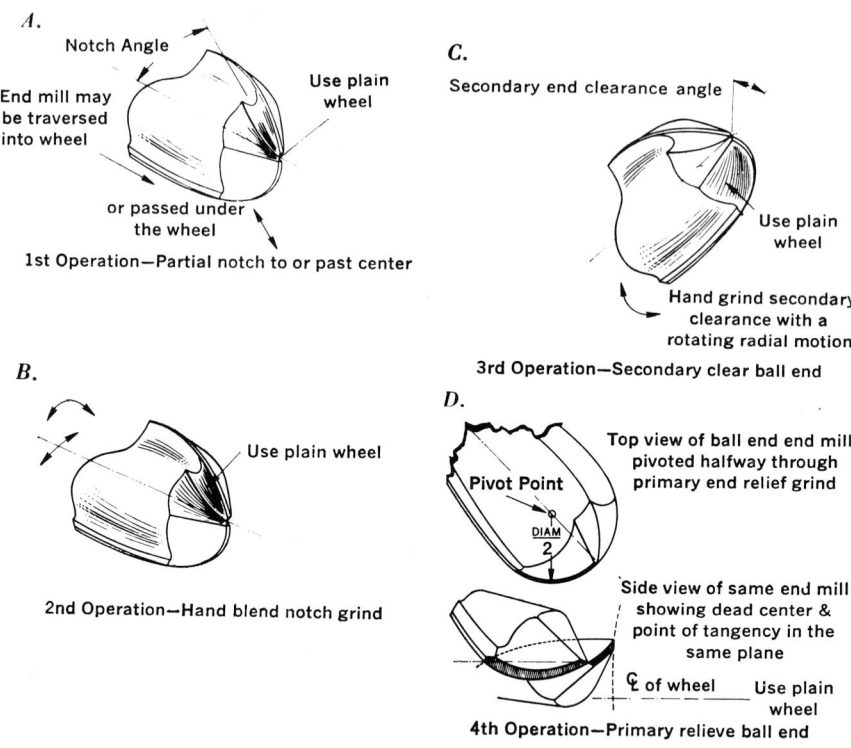

A.

Notch Angle

End mill may
be traversed
into wheel

Use plain
wheel

or passed under
the wheel

1st Operation—Partial notch to or past center

B.

Use plain wheel

2nd Operation—Hand blend notch grind

C.

Secondary end clearance angle

Use plain
wheel

Hand grind secondary
clearance with a
rotating radial motion

3rd Operation—Secondary clear ball end

D.

Pivot Point

DIAM
2

Top view of ball end end mill
pivoted halfway through
primary end relief grind

Side view of same end mill
showing dead center &
point of tangency in the
same plane

℄ of wheel   Use plain
wheel

4th Operation—Primary relieve ball end

Figure 50—Procedure for sharpening end of 2-flute ball end end mill.

Wherever possible, end notching or gashing cuts should be produced with grinding wheels which have corner radii to produce a fillet at the bottom portion of the notch grind. End tooth notch angles should produce about 0°-5° positive axial rake. Negative axial rake angles are seldom used on end mills made of high speed steel. An obvious exception is an end mill having hand of helix opposite hand of cut.

Primary relief land width of end teeth should be 1½ to 3 times that recommended for peripheral teeth. Primary end relief angles of 5°-12° are usually sufficient, with the lower values used for cutting the harder work materials. The primary end tooth relief angles should be increased on small diameter end mills used for plunge-cutting (drilling) applications.

Sharpening Corner Radii—Corner radii may be relieved on square end end mills by two methods. One method, used principally with small corner radii, is to dress the desired concave radius on the corner of a plain grinding wheel. With proper location of the wheel in relation to the end mill corner, the wheel is plunged into the end mill to a stop, thus forming and relieving the radius in a single operation.

The other method is to swing the end mill around the wheel (some equipment allows swinging the wheel about a fixed end mill) in a manner similar to that used for grinding the primary end relief on ball end end mills. If considerable stock must be removed, a rough form grinding operation may be necessary. Either method is acceptable providing the blending at the points of tangency with peripheral and end teeth are carefully controlled. It may be worthwhile to maintain a stock of concave dressed plain wheels to produce varying sizes of corner radii. If many end mills with different corner radii are sharpened, this practice can effect a substantial saving in both dressing time and in grinding wheel losses due to radius changes.

SHARPENING ROUGHING MILLS—Roughing mills with radius tooth and truncated tooth forms are form relieved on the OD. There are some general rules which apply to all roughing cutters.

It is important to remove the wear land when sharpening tools. Regrinding can generate considerable heat and care should be taken not to burn the cutting edge. See Table 8 for grinding wheel selection.

If a roughing end mill is center or end cutting, the teeth are resharpened the same way that the end teeth are sharpened on standard end mills.

Radius-tooth and truncated tooth roughing end mills are both sharpened similar to the method used for form relieved cutters. They are sharpened by grinding the radial rake face and fillet of the flute on a tool and cutter grinding machine.

A saucer-shaped grinding wheel is used. The wheel should be dressed to match the form of the flute face from the OD to the fillet. The grinding head is turned to an angle slightly greater than the helix angle of the end mill (usually 1° to 3° extra). This allows the leading edge of the wheel to hollow grind the rake face. Typically, the finished hook measures 8° to 12° (Figure 51).

A tooth support finger or lead-generating device may be used to ensure that proper lead is maintained. The support finger rests on the back of the tooth being sharpened, located below the form. Some manufacturers provide a step at the back of the tooth for this purpose (Figure 52).

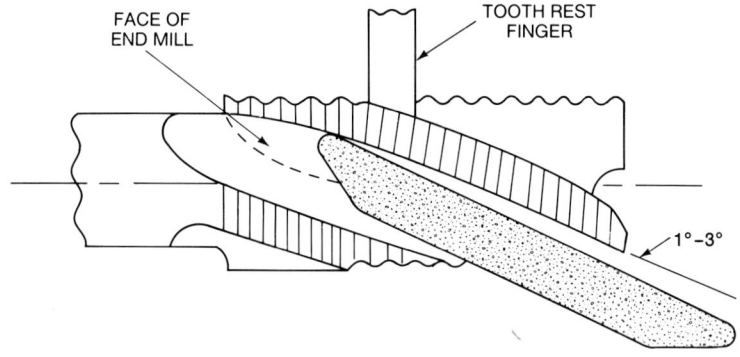

FACE OF
END MILL

TOOTH REST
FINGER

1°–3°

Figure 51.

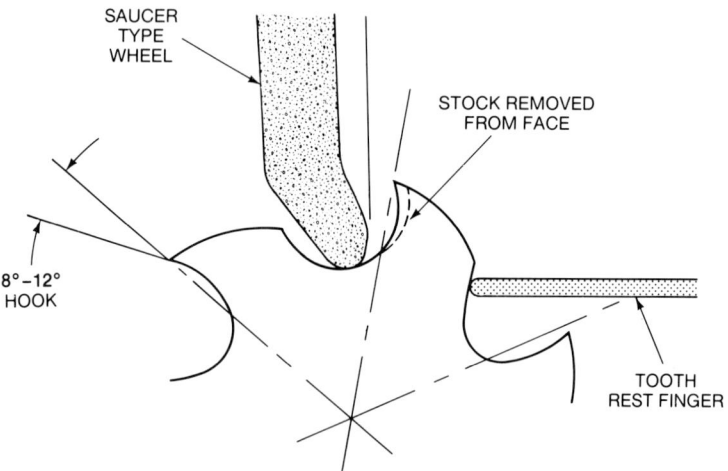

SAUCER
TYPE
WHEEL

STOCK REMOVED
FROM FACE

8°–12°
HOOK

TOOTH
REST FINGER

Figure 52.

SHARPENING FORM RELIEVED CUTTERS—Form relieved cutters are ground on the face of the teeth only when sharpened. The equipment and set-up required depends on the design of the gashes.

Straight Or Angular Gashed Cutters—Form relieved cutters with straight or angular gashes are usually sharpened with the face side of a dish wheel. However, if Hob sharpening equipment is available, some designs of straight gashed cutters may be sharpened using a conical wheel (See data on Helical Gashed Cutters).

Using the face side of a wheel, the cutter is mounted in a fixed position with its axis offset from the face of the wheel by the amount equal to the rake offset of the cutter tooth. If the cutter is angular gashed, its axis is also swung at an angle equal to the angle of the gash.

The cutter can be fed in a tangential or radial direction, or a combination of both, along the face of the wheel. Side gashes can be ground in the same set-up as to the top gash, however, a second operation is frequently used.

Since the face of the tooth is sharpened, a tooth rest cannot be used against the face for positioning as with profile cutters. Consequently, the cutter is usually positioned and indexed by one of the following methods:

First: Use of an index plate mounted on the cutter arbor or holder. Figure 53. This method produces uniformly spaced teeth.

Second: A fixed stop or indicator against which the relieved surface of each tooth is successively placed in contact. Figure 54. This method produces a true running cutter although the spacing of the teeth may not be uniform.

Third: A stop is positioned against the back of each tooth. Figure 55. A ground spot at the back of each tooth, duplicating spacing of the face of the tooth, is sometimes provided by the manufacturer. The stop should always be positioned on the back of the tooth being sharpened and located only on the ground spot.

Form relieved cutters are sometimes sharpened by first cylindrical grinding a spot on the cutting edge and then sharpening the face of each tooth to the point where the spot disappears.

Figure 53.

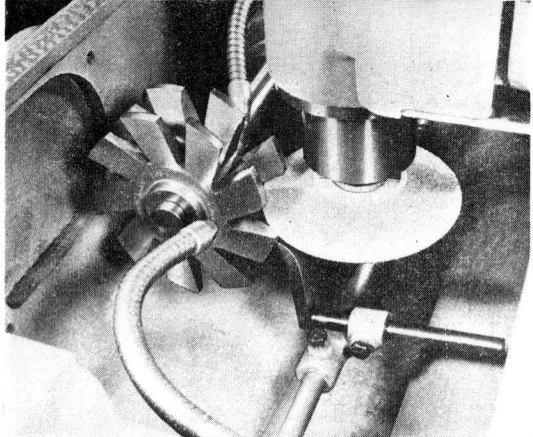

Figure 54.

Figure 55.

Helical Gashed Cutters—Form relieved cutters with helical gashes should always be sharpened with the cone side of a dish or conical wheel. The cutter is mounted to rotate in timed relation to the axial feed, so that an exact lead or helix can be generated on the face. The cutter axis is offset from the cone side of the wheel by the amount equal to the rake offset of the cutter tooth. The wheel axis should be swung at an angle equal to the helix angle of the cutter.

The cutter is fed in a tangential direction only past the wheel. Each end of the cutter face must pass by the high point of the grinding wheel to completely sharpen without distorting the form. If the cutter has side gashes or pockets, these should be ground as a secondary operation and should blend without grinding into the helical face.

The most common method of sharpening this type is with special sharpening equipment such as a Hob Sharpener with the mechanical means of reproducing the exact lead required, built into it. This machine incorporates an index plate for the indexing means. Figure 56.

A second method uses a sharpening guide or former mounted on the cutter arbor. Figure 57. Such a guide provides the required lead as well as the index corresponding with the number of teeth in the cutter. A follower is placed against the guide to rotate the cutter as it moves past the wheel.

A third method uses a tooth rest placed against the back of the cutter tooth on a ground spot sometimes provided by the manufacturer. Be sure the tooth rest is wide enough and formed at the correct angle to permit the wheel sufficient approach and run by to properly complete the grind.

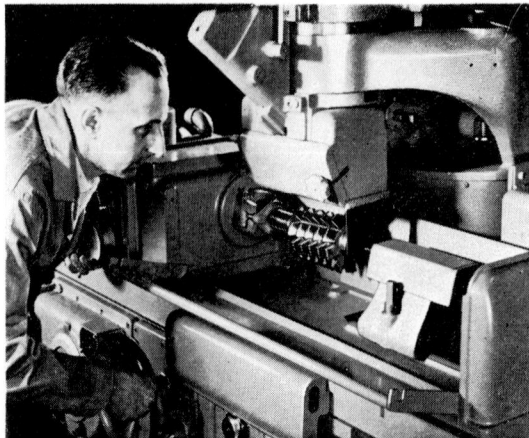

Figure 56.

Figure 57.

Maintaining Correct Offset And Lead—The form built into most form relieved cutters will be continued throughout the life of the cutter if the rake offset and lead marked on the cutter by the manufacturer is maintained during sharpening. Any variations in the form occurring after repeated sharpenings can generally be attributed to an incorrect offset or lead of gash.

If the depth of form changes, the offset ground on the cutter is too little or too great, Figures 58 and 59. If the form tapers, or if one end becomes too high or low, the lead or angle of gash used is incorrect. Figure 60.

Conversely, if the form varies a small amount, it can be modified by altering the rake offset or lead, indicated in these illustrations.

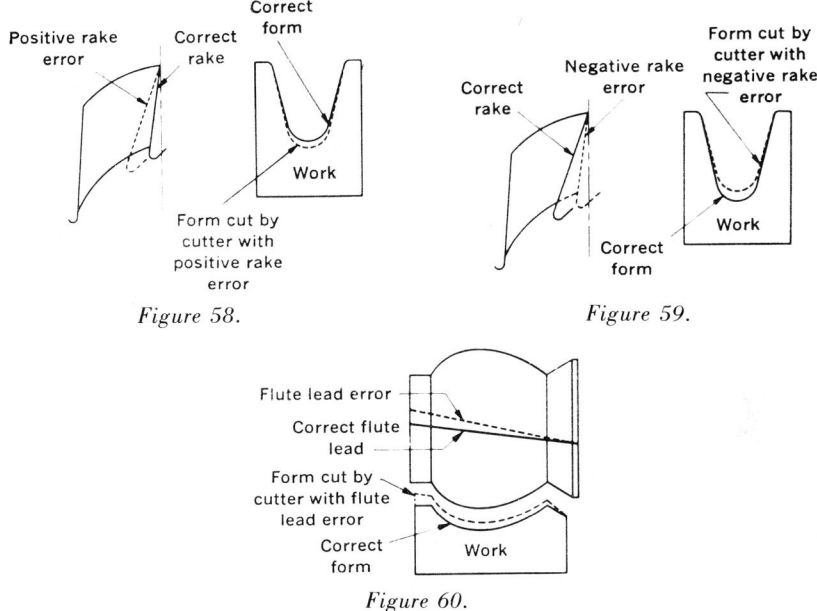

Figure 58.   Figure 59.

Figure 60.

HANDLING AND STORAGE—Precision tools merit careful handling and storage. Careless handling causes nicked cutting edges. Nicks are removed by additional costly grinding operations reducing the life of the tool.

Milling cutters should be protected against damage immediately after sharpening or removal from the milling machine. Standard procedure in plants equipped for plastic dipping, is to use this protective coating before the tool is stored. If this is not expedient, the cutter should be stored in a strong wooden box made especially for it. Care is recommended for all tools, but particularly in the case of carbide cutters which can be badly damaged by careless handling.

STANDARD MILLING CUTTERS and END MILLS

CONTENTS

MILLING CUTTERS

SAWS

END MILLS

HELICAL SERIES

STANDARD MILLING CUTTERS and END MILLS

CONTENTS

END MILLS (Continued)

(Continued on following page)

STANDARD MILLING CUTTERS and END MILLS

CONTENTS

END MILLS (Continued)

(Concluded on following page)

STANDARD MILLING CUTTERS and END MILLS

CONTENTS

(Concluded)

Single End (Non-Center Cutting)

Plain Milling Cutters

Light Duty

High Speed Steel

Cutters less than ¾" face have straight teeth.

Cutters ¾" face and wider have a helix angle of not less than 18° nor greater than 25°.

Standard Sizes and Dimensions

Diameter of Cutter	Width of Face	Diameter of Hole
2½	3/16	1
2½	¼	1
2½	5/16	1
2½	3/8	1
2½	½	1
2½	5/8	1
2½	¾	1
2½	1	1
2½	1½	1
2½	2	1
2½	3	1
3	3/16	1
3	¼	1
3	5/16	1
3	3/8	1
3	½	1¼
3	5/8	1
3	5/8	1¼
3	¾	1
3	¾	1¼
3	1	1¼
3	1½	1

(Concluded on following page)

Plain Milling Cutters

Light Duty

High Speed Steel

(Concluded)

Standard Sizes and Dimensions

Diameter of Cutter	Width of Face	Diameter of Hole
3	$1\frac{1}{4}$	$1\frac{1}{4}$
3	$1\frac{1}{2}$	$1\frac{1}{4}$
3	2	$1\frac{1}{4}$
3	3	$1\frac{1}{4}$
4	$\frac{1}{4}$	1
4	$\frac{5}{16}$	1
4	$\frac{3}{8}$	1
4	$\frac{3}{8}$	$1\frac{1}{4}$
4	$\frac{1}{2}$	$1\frac{1}{4}$
4	$\frac{5}{8}$	$1\frac{1}{4}$
4	$\frac{3}{4}$	$1\frac{1}{4}$
4	1	$1\frac{1}{4}$
4	$1\frac{1}{2}$	$1\frac{1}{4}$
4	2	$1\frac{1}{4}$
4	3	$1\frac{1}{4}$
4	4	$1\frac{1}{4}$

All dimensions are given in inches.
For keyway dimensions see Table 404.
For tolerances see Tables 401 and 402.

Plain Milling Cutters
Heavy Duty
High Speed Steel

Heavy duty helical tooth plain milling cutters have a helix angle of not less than 25° nor greater than 45°.

Standard Sizes and Dimensions

Diameter of Cutter	Width of Face	Diameter of Hole
2½	2	1
2½	4	1
3	2	1¼
3	2½	1¼
3	3	1¼
3	4	1¼
3	6	1¼
4	2	1½
4	3	1½
4	4	1½
4	6	1½

All dimensions are given in inches.
For keyway dimensions see Table 404.
For tolerances see Tables 401 and 402.

Plain Milling Cutters, High Helix
High Speed Steel

High helix plain milling cutters have a helix angle of not less than 45° nor greater than 52°.

Standard Sizes and Dimensions

Diameter of Cutter	Width of Face	Diameter of Hole
3	4	1¼
3	6	1¼
4	8	1½

All dimensions are given in inches.
For keyway dimensions see Table 404.
For tolerances see Tables 401 and 402.

Side Milling Cutters

High Speed Steel

Standard Sizes and Dimensions

Diameter of Cutter	Width of Face	Diameter of Hole
2	3/16	5/8
2	1/4	5/8
2	3/8	5/8
2½	1/4	7/8
2½	3/8	7/8
2½	1/2	7/8
3	1/4	1
3	5/16	1
3	3/8	1
3	7/16	1
3	1/2	1
4	1/4	1
4	3/8	1
4	1/2	1
4	1/2	1¼
4	5/8	1
4	5/8	1¼
4	3/4	1
4	3/4	1¼
4	7/8	1
5	1/2	1
5	1/2	1¼
5	5/8	1
5	5/8	1¼
5	3/4	1
5	3/4	1¼
5	1	1¼
6	1/2	1
6	1/2	1¼
6	5/8	1¼
6	3/4	1¼
6	1	1¼
7	3/4	1¼
7	3/4	1½
8	3/4	1¼
8	3/4	1½
8	1	1¼
8	1	1½

All dimensions are given in inches.
For keyway dimensions see Table 404.
For tolerances see Tables 401 and 402.

Staggered Tooth Side Milling Cutters
High Speed Steel

Standard Sizes and Dimensions

Diameter of Cutter	Width of Face	Diameter of Hole
2½	¼	⅞
2½	5/16	⅞
2½	⅜	⅞
2½	½	⅞
3	3/16	1
3	¼	1
3	5/16	1
3	⅜	1
3	½	1¼
3	⅝	1¼
3	¾	1¼
4	¼	1¼
4	5/16	1¼
4	⅜	1¼
4	7/16	1¼
4	½	1¼
4	⅝	1¼
4	¾	1¼
4	⅞	1¼
5	½	1¼
5	⅝	1¼
5	¾	1¼
6	⅜	1¼
6	½	1¼
6	⅝	1¼
6	¾	1¼
6	⅞	1¼
6	1	1¼
8	⅜	1½
8	½	1½
8	⅝	1½
8	¾	1½
8	1	1½

All dimensions are given in inches.
For keyway dimensions see Table 404.
For tolerances see Tables 401 and 402.

Half Side Milling Cutters
High Speed Steel

Right Hand Cutters have a Right Hand Helix.
Left Hand Cutters have a Left Hand Helix.

Standard Sizes and Dimensions

Diameter of Cutter	Width of Face	Diameter of Hole
4	¾	1¼
5	¾	1¼
6	¾	1¼

All dimensions are given in inches.
For keyway dimensions see Table 404.
For tolerances see Tables 401 and 402.

Woodruff Keyseat Cutters

Arbor Type

High Speed Steel

These cutters are standard with staggered teeth.

Standard Sizes and Dimensions

Cutter Number *	Nominal Diameter of Cutter †	Width of Face	Diameter of Hole
617	2⅛	3/16	¾
817	2⅛	¼	¾
1017	2⅛	5/16	¾
1217	2⅛	⅜	¾
822	2¾	¼	1
1022	2¾	5/16	1
1222	2¾	⅜	1
1422	2¾	7/16	1
1622	2¾	½	1
1228	3½	⅜	1
1628	3½	½	1
1828	3½	9/16	1
2028	3½	⅝	1
2428	3½	¾	1

All dimensions are given in inches.

* The cutter number indicates nominal key dimension or size cutter. The last two digits give the nominal diameter in 8ths of an inch, and the digits preceding the last two give the nominal width in 32nds of an inch. Thus, cutter No. 617 indicates a size 6/32 x 17/8 in. or 3/16 in. wide x 2⅛ in. diameter.

† Furnished 1/32 in. oversize to allow for sharpening.

For keyway dimensions see Table 404.

For tolerances see Tables 401 and 402. Tolerances are applied to the oversize diameter.

Woodruff Keyseat Cutters
Shank Type
High Speed Steel

These cutters are standard with right hand cut.

All sizes have 1/2″ diameter straight shank.

Standard Sizes and Dimensions

Cutter Number* American National Standard	Nominal Diameter of Cutter†	Width of Face	Length Overall
202	1/4	1/16	2 1/16
202½	5/16	1/16	2 1/16
302½	5/16	3/32	2 3/32
203	3/8	1/16	2 1/16
303	3/8	3/32	2 3/32
403	3/8	1/8	2 1/8
204	1/2	1/16	2 1/16
304	1/2	3/32	2 3/32
305	5/8	3/32	2 3/32
404	1/2	1/8	2 1/8
405	5/8	1/8	2 1/8
406	3/4	1/8	2 1/8
505	5/8	5/32	2 5/32
605	5/8	3/16	2 3/16
506	3/4	5/32	2 5/32
806	3/4	1/4	2 1/4
507	7/8	5/32	2 5/32
606	3/4	3/16	2 3/16
607	7/8	3/16	2 3/16
707	7/8	7/32	2 7/32
608	1	3/16	2 3/16
708	1	7/32	2 7/32
1208	1	3/8	2 3/8
609	1 1/8	3/16	2 3/16
807	7/8	1/4	2 1/4
808	1	1/4	2 1/4
709	1 1/8	7/32	2 7/32
809	1 1/8	1/4	2 1/4
610	1 1/4	3/16	2 3/16
710	1 1/4	7/32	2 7/32

(Continued on following page)

Woodruff Keyseat Cutters

Shank Type

(Concluded)

Standard Sizes and Dimensions

Cutter Number* American National Standard	Nominal Diameter of Cutter†	Width of Face	Length Overall
810	1¼	¼	2¼
811	1⅜	¼	2¼
812	1½	¼	2¼
1008	1	⁵⁄₁₆	2⁵⁄₁₆
1009	1⅛	⁵⁄₁₆	2⁵⁄₁₆
1010	1¼	⁵⁄₁₆	2⁵⁄₁₆
1011	1⅜	⁵⁄₁₆	2⁵⁄₁₆
1012	1½	⁵⁄₁₆	2⁵⁄₁₆
1210	1¼	⅜	2⅜
1211	1⅜	⅜	2⅜
1212	1½	⅜	2⅜

All dimensions are given in inches.

* The cutter number indicates the nominal key dimension or size cutter, that is, the last two digits give the nominal diameter in 8ths of an inch and the digits preceding the last two give the nominal width in 32nds of an inch. Thus, cutter No. 204 indicates a size ²⁄₃₂ × ⁴⁄₈ in. or ¹⁄₁₆ in. wide × ½ in. diameter.

† Furnished oversize to allow for sharpening.

The amount of oversize on diameter is incorporated in the tolerance.

For tolerances see Tables 401 and 402.

Woodruff Keyseat Cutters

Solid Carbide Head

Shank Type

These cutters are standard with straight flutes, right hand cut. All sizes have ½″ diameter, hardened steel, straight shanks with full length flat for set screws.

Standard Sizes and Dimensions

Cutter* Number	Nominal Diameter of Cutter**	Width of Face	Length Overall
303	3/8	3/32	2 3/32
403	3/8	1/8	2 1/8
304	1/2	3/32	2 3/32
404	1/2	1/8	2 1/8
305	5/8	3/32	2 3/32
405	5/8	1/8	2 1/8
505	5/8	5/32	2 5/32
605	5/8	3/16	2 3/16
406	3/4	1/8	2 1/8
506	3/4	5/32	2 5/32
606	3/4	3/16	2 3/16
806	3/4	1/4	2 1/4
507	7/8	5/32	2 5/32
607	7/8	3/16	2 3/16
707	7/8	7/32	2 7/32
807	7/8	1/4	2 1/4
608	1	3/16	2 3/16
708	1	7/32	2 7/32
808	1	1/4	2 1/4

* The cutter numbers indicate nominal key dimensions or cutter sizes. The last two digits give the nominal diameter in eighths of an inch, and the digits preceding the last two give the nominal width in thirty-seconds of an inch. Thus cutter No. 303 indicates a size of 3/32 × 3/8 inches or 3/32 wide × 3/8 in diameter.

** Furnished oversize to allow for sharpening.

All dimensions are given in inches.
For tolerances see Tables 401 and 402.

Plain Metal Slitting Saws
High Speed Steel

These saws are standard with concave sides.

Standard Sizes and Dimensions

Diameter of Cutter	Width of Face	Diameter of Hole
$2\frac{1}{2}$	$\frac{1}{32}$	$\frac{7}{8}$
$2\frac{1}{2}$	$\frac{3}{64}$	$\frac{7}{8}$
$2\frac{1}{2}$	$\frac{1}{16}$	$\frac{7}{8}$
$2\frac{1}{2}$	$\frac{3}{32}$	$\frac{7}{8}$
$2\frac{1}{2}$	$\frac{1}{8}$	$\frac{7}{8}$
3	$\frac{1}{32}$	1
3	$\frac{3}{64}$	1
3	$\frac{1}{16}$	1
3	$\frac{3}{32}$	1
3	$\frac{1}{8}$	1
3	$\frac{5}{32}$	1
4	$\frac{1}{32}$	1
4	$\frac{3}{64}$	1
4	$\frac{1}{16}$	1
4	$\frac{3}{32}$	1
4	$\frac{1}{8}$	1
4	$\frac{5}{32}$	1
4	$\frac{3}{16}$	1
5	$\frac{1}{16}$	1
5	$\frac{3}{32}$	1
5	$\frac{1}{8}$	1
5	$\frac{3}{8}$	$1\frac{1}{4}$
6	$\frac{1}{16}$	1
6	$\frac{3}{32}$	1
6	$\frac{1}{8}$	1
6	$\frac{1}{8}$	$1\frac{1}{4}$
6	$\frac{3}{16}$	$1\frac{1}{4}$
8	$\frac{1}{8}$	1
8	$\frac{1}{8}$	$1\frac{1}{4}$

All dimensions are given in inches.
For keyway dimensions see Table 404.
For tolerances see Tables 401 and 402.

Metal Slitting Saws

With Side Teeth
High Speed Steel

Standard Sizes and Dimensions

Diameter of Cutter	Width of Face	Diameter of Hole
2½	1/16	7/8
2½	3/32	7/8
2½	1/8	7/8
3	1/16	1
3	3/32	1
3	1/8	1
3	5/32	1
4	1/16	1
4	3/32	1
4	1/8	1
4	5/32	1
4	3/16	1
5	1/16	1
5	3/32	1
5	1/8	1
5	1/8	1¼
5	5/32	1
5	3/16	1
6	1/16	1
6	3/32	1
6	1/8	1
6	1/8	1¼
6	3/16	1
6	3/16	1¼
8	1/8	1
8	1/8	1¼
8	3/16	1¼

All dimensions are given in inches.
For keyway dimensions see Table 404
For tolerances see Tables 401 and 402

Metal Slitting Saws

With Staggered Peripheral and Side Teeth
High Speed Steel

Standard Sizes and Dimensions

Diameter of Cutter	Width of Face	Diameter of Hole
3	$\frac{3}{16}$	1
4	$\frac{3}{16}$	1
5	$\frac{3}{16}$	1
5	$\frac{1}{4}$	1
6	$\frac{3}{16}$	1
6	$\frac{3}{16}$	$1\frac{1}{4}$
6	$\frac{1}{4}$	1
6	$\frac{1}{4}$	$1\frac{1}{4}$
8	$\frac{3}{16}$	$1\frac{1}{4}$
8	$\frac{1}{4}$	$1\frac{1}{4}$
10	$\frac{3}{16}$	$1\frac{1}{4}$
10	$\frac{1}{4}$	$1\frac{1}{4}$
12	$\frac{1}{4}$	$1\frac{1}{2}$
12	$\frac{5}{16}$	$1\frac{1}{2}$

All dimensions are given in inches.
For keyway dimensions see Table 404
For tolerances see Tables 401 and 402

Screw Slotting Cutters

High Speed Steel

Standard Sizes and Dimensions

American National Standard Wire Gage Number	Width of Face	Diameter of Cutter	Diameter of Hole	Number of Teeth
7	0.144	2¾	1	72
8	0.128	2¾	1	72
9	0.114	2¾	1	72
10	0.102	2¾	1	72
11	0.091	2¾	1	72
12	0.081	2¾	1	72
13	0.072	2¾	1	72
14	0.064	2¾	1	72
15	0.057	2¾	1	72
16	0.051	2¾	1	72
17	0.045	2¾	1	72
18	0.040	2¾	1	72
19	0.036	2¾	1	72
20	0.032	2¾	1	72
21	0.028	2¾	1	72
22	0.025	2¾	1	72
23	0.023	2¾	1	72
24	0.020	2¾	1	72
14	0.064	2¼	⅝	60
15	0.057	2¼	⅝	60
16	0.051	2¼	⅝	60
17	0.045	2¼	⅝	60
18	0.040	2¼	⅝	60
19	0.036	2¼	⅝	60
20	0.032	2¼	⅝	60
21	0.028	2¼	⅝	60
22	0.025	2¼	⅝	60
23	0.023	2¼	⅝	60
24	0.020	2¼	⅝	60
14	0.064	1¾	⅝	90
15	0.057	1¾	⅝	90
16	0.051	1¾	⅝	90
17	0.045	1¾	⅝	90
18	0.040	1¾	⅝	90
19	0.036	1¾	⅝	90
20	0.032	1¾	⅝	90
21	0.028	1¾	⅝	90
22	0.025	1¾	⅝	90
23	0.023	1¾	⅝	90
24	0.020	1¾	⅝	90

All dimensions are given in inches.
For keyway dimensions See Table 404.
For tolerances See Tables 401 and 402.

Single Angle Milling Cutters
High Speed Steel

Right Hand Cutters

Standard Sizes and Dimensions

Diameter of Cutter	Angle Degrees	Width of Face	Diameter of Hole
2¾	45°	½	1
2¾	60°	½	1
3	45°	½	1¼
3	60°	½	1¼

Left Hand Cutters

Standard Sizes and Dimensions

Diameter of Cutter	Angle Degrees	Width of Face	Diameter of Hole
2¾	45°	½	1
2¾	60°	½	1
3	45°	½	1¼
3	60°	½	1¼

All dimensions are given in inches.
For keyway dimensions see Table 404.
For tolerances see Tables 401 and 402.

Double Angle Milling Cutters
High Speed Steel

Standard Sizes and Dimensions

Diameter of Cutter	Angle Degrees	Width of Face	Diameter of Hole
2¾	45°	½	1
2¾	60°	½	1
2¾	90°	½	1

All dimensions are given in inches.
For keyway dimensions see Table 404
For tolerances see Tables 401 and 402

60° Single Angle Milling Cutters
With Weldon Shanks
High Speed Steel

Right hand cutters are standard

Standard Sizes and Dimensions

Largest Diameter of Cutter	Width of Cutter	Diameter of Shank	Overall Length
¾	5⁄16	⅜	2⅛
1⅜	9⁄16	⅝	2⅞
1⅞	13⁄16	⅞	3¼
2¼	1 1⁄16	1	3¾

All dimensions are given in inches.
For tolerances see Tables 401 and 402
For shank dimensions see Table 1702

Single Angle Milling Cutters
With Threaded Hole
High Speed Steel

Right hand and left hand cutters with an included angle of 60° are standard. See drawing.

Standard Sizes and Dimensions

Diameter of Cutter	Width of Face	Threaded Hole	Hand of Rotation	Hand of Cutter
1¼	⁷⁄₁₆	⅜-24 UNF-2B RH	RH	RH
1¼	⁷⁄₁₆	⅜-24 UNF-2B LH	LH	LH
1⅝	⁹⁄₁₆	½-20 UNF-2B RH	RH	RH

All dimensions are given in inches.
For tolerances See Tables 401 and 402.

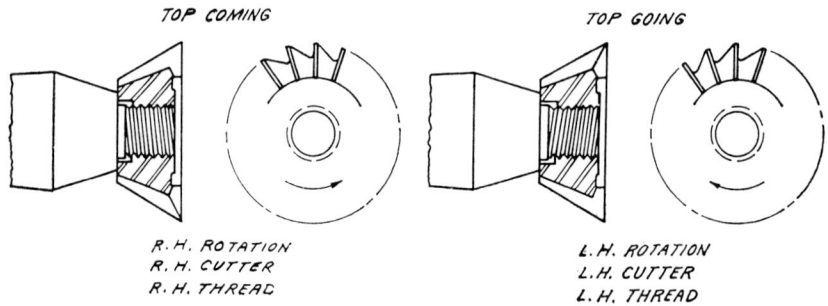

TOP COMING   TOP GOING

R.H. ROTATION   L.H. ROTATION
R.H. CUTTER   L.H. CUTTER
R.H. THREAD   L.H. THREAD

T-Slot Milling Cutters
With Weldon Shanks
High Speed Steel

Right hand cutters with staggered teeth are standard.

Standard Sizes and Dimensions

Bolt Size	Diameter of Cutter	Width of Cutter	Diameter of Neck	Length Overall	Diameter of Shank
$\frac{1}{4}$	$\frac{9}{16}$	$\frac{15}{64}$	$\frac{17}{64}$	$2\frac{19}{32}$	$\frac{1}{2}$
$\frac{5}{16}$	$\frac{21}{32}$	$\frac{17}{64}$	$\frac{21}{64}$	$2\frac{11}{16}$	$\frac{1}{2}$
$\frac{3}{8}$	$\frac{25}{32}$	$\frac{21}{64}$	$\frac{13}{32}$	$3\frac{1}{4}$	$\frac{3}{4}$
$\frac{1}{2}$	$\frac{31}{32}$	$\frac{25}{64}$	$\frac{17}{32}$	$3\frac{7}{16}$	$\frac{3}{4}$
$\frac{5}{8}$	$1\frac{1}{4}$	$\frac{31}{64}$	$\frac{21}{32}$	$3\frac{15}{16}$	1
$\frac{3}{4}$	$1\frac{15}{32}$	$\frac{5}{8}$	$\frac{25}{32}$	$4\frac{1}{16}$	1
1	$1\frac{27}{32}$	$\frac{53}{64}$	$1\frac{1}{32}$	$4\frac{13}{16}$	$1\frac{1}{4}$

All dimensions are given in inches.
For tolerances see Tables 401 and 402
For shank dimensions see Table 405

T-Slot Milling Cutters
Brown & Sharpe Taper Shanks
High Speed Steel

Right hand cutters with staggered teeth are standard.

Standard Sizes and Dimensions

Bolt Size	Diameter of Cutter	Width of Cutter	Diameter of Neck	Length Overall	Taper Number
$\frac{1}{2}$	$\frac{31}{32}$	$\frac{25}{64}$	$\frac{17}{32}$	5	7
$\frac{5}{8}$	$1\frac{1}{4}$	$\frac{31}{64}$	$\frac{21}{32}$	$5\frac{1}{4}$	7
$\frac{3}{4}$	$1\frac{15}{32}$	$\frac{5}{8}$	$\frac{25}{32}$	$6\frac{7}{8}$	9
1	$1\frac{27}{32}$	$\frac{53}{64}$	$1\frac{1}{32}$	$7\frac{1}{4}$	9

All dimensions are given in inches.
For tolerances see Tables 401 and 402.
For shank dimensions see Table 407.

Shell Mills

High Speed Steel

Right hand cutters with right hand helix are standard.

These mills are standard with square corners.

Standard Sizes and Dimensions

General Dimensions				Driving Slot			Counterbore		
Diam. of Cutter	Width of Cutter	Diam. of Hole	Length of Bearing	Width (Nom.)	Depth	Radius	Inside Diam. at Hub Face	Diam. of Hub	Angular Increase Degrees
1¼	1	½	⅝	¼	5/32	1/64	11/16	⅝	0
1½	1⅛	½	⅝	¼	5/32	1/64	11/16	⅝	0
1¾	1¼	¾	¾	5/16	3/16	1/32	15/16	⅞	0
2	1⅜	¾	¾	5/16	3/16	1/32	15/16	⅞	0
2¼	1½	1	¾	⅜	7/32	1/32	1 ¼	1 3/16	0
2½	1⅝	1	¾	⅜	7/32	1/32	1 ⅜	1 3/16	0
2¾	1⅝	1	¾	⅜	7/32	1/32	1 ½	1 3/16	5
3	1¾	1¼	¾	½	9/32	1/32	1 21/32	1½	5
3½	1⅞	1¼	¾	½	9/32	1/32	1 11/16	1½	5
4	2¼	1½	1	⅝	⅜	1/16	2 1/32	1⅞	5
4½	2¼	1½	1	⅝	⅜	1/16	2 1/16	1⅞	10
5	2¼	1½	1	⅝	⅜	1/16	2 9/16	1⅞	10
6	2¼	2	1	¾	7/16	1/16	2 13/16	2½	15

All dimensions are given in inches.

Tolerances

Element	Range	Direction	Tolerance
Diameter of Cutter	All Sizes	Plus	1/64
Width of Cutter	All Sizes	Plus or Minus	1/64
Diameter of Hole	All Sizes	Plus	.0005
Length of Bearing	All Sizes	Plus	1/64
Width, Driving Slot	All Sizes	Plus not less than	.008
		Plus not more than	.012
Depth, Driving Slot	All Sizes	Plus	1/64
Diameter, C'Bore	All Sizes	Plus or Minus	1/64
Diameter, Hub	All Sizes	Plus or Minus	1/64

For radial and axial runout see Table 402.

Convex Milling Cutters

Form Relieved

High Speed Steel

Standard Sizes and Dimensions

Diameter of Circle*	Diameter of Cutter	Width of Face	Diameter of Hole
1/8	2 1/4	1/8	1
3/16	2 1/4	3/16	1
1/4	2 1/2	1/4	1
5/16	2 3/4	5/16	1
3/8	2 3/4	3/8	1
7/16	3	7/16	1
1/2	3	1/2	1
5/8	3 1/2	5/8	1 1/4
3/4	3 3/4	3/4	1 1/4
7/8	4	7/8	1 1/4
1	4 1/4	1	1 1/4

All dimensions are given in inches.
* Cutters are designated by diameter of circle.
For tolerances see Tables 401 and 402
For keyway dimensions see Table 404

Concave Milling Cutters

Form Relieved

High Speed Steel

Standard Sizes and Dimensions

Diameter of Circle*	Diameter of Cutter	Width of Face	Diameter of Hole
1/8	2 1/4	1/4	1
3/16	2 1/4	3/8	1
1/4	2 1/2	7/16	1
5/16	2 3/4	9/16	1
3/8	2 3/4	5/8	1
7/16	3	3/4	1
1/2	3	13/16	1
5/8	3 1/2	1	1 1/4
3/4	3 3/4	1 3/16	1 1/4
7/8	4	1 3/8	1 1/4
1	4 1/4	1 9/16	1 1/4

All dimensions are given in inches.
* Cutters are designated by diameter of circle.
For tolerances see Tables 401 and 402
For keyway dimensions see Table 404

Corner Rounding Milling Cutters

Form Relieved

High Speed Steel

Right hand cutters are standard.

Standard Sizes and Dimensions

Radius of Circle*	Diameter of Cutter	Width of Face	Diameter of Hole
⅛	2½	¼	1
¼**	3	¹²⁄₃₂	1
⅜	3¾	⁹⁄₁₆	1¼
½	4¼	¾	1¼
⅝	4¼	¹⁵⁄₁₆	1¼

All dimensions are given in inches.
 * Cutters are designated by radius of circle. For tolerances see Tables 401 and 402. For keyway dimensions see Table 404.
** Right and left hand cutters are standard for this size.

Corner Rounding Cutters

With Weldon Shanks

Form Relieved

High Speed Steel

Right hand cutters are standard.

Standard Sizes and Dimensions

Radius of Circle	Outside Diameter of Cutter	End Diameter of Cutter	Diameter of Shank	Length Overall
¹⁄₁₆	⁷⁄₁₆	¼	⅜	2½
³⁄₃₂	½	¼	⅜	2½
⅛	⅝	¼	½	3
⁵⁄₃₂	¾	⁵⁄₁₆	½	3
³⁄₁₆	⅞	⁵⁄₁₆	½	3
¼	1	⅜	½	3
⁵⁄₁₆	1⅛	⅜	½	3¼
⅜	1¼	⅜	½	3½
³⁄₁₆	⅞	⁵⁄₁₆	¾	3⅛
¼	1	⅜	¾	3¼
⁵⁄₁₆	1⅛	⅜	⅞	3½
⅜	1¼	⅜	⅞	3¾
⁷⁄₁₆	1⅜	⅜	1	4
½	1½	⅜	1	4⅛

All dimensions are given in inches.
For tolerances see Tables 401 and 402. For shank dimensions see Table 405.

Finishing Gear Milling Cutters

14½ Degrees Pressure Angle

Form Relieved

High Speed Steel

Finishing Gear Milling Cutters are made with eight different cutter forms for each pitch depending on the number of teeth to be cut.

The range for each cutter is as follows:

Cutter Form Number	Range of Teeth	Cutter Form Number	Range of Teeth
1	135 thru a rack	5	21 thru 25
2	55 thru 134	6	17 thru 20
3	35 thru 54	7	14 thru 16
4	26 thru 34	8	12 and 13

Standard Sizes and Dimensions

Diametral Pitch	Diameter of Cutter	Diameter of Hole
1	8½	2
1¼	7¾	2
1½	7	1¾
1¾	6½	1¾
2	6½	1¾
2	5¾	1½
2½	6⅛	1¾
2½	5¾	1½
3	5⅝	1¾
3	5¼	1½
3	4¾	1¼
4	4¾	1¾
4	4½	1½
4	4¼	1¼
4	3⅝	1
5	4⅜	1¾
5	4¼	1½
5	3¾	1¼
5	3⅜	1

(Continued on following page)

Finishing Gear Milling Cutters

14½ Degree Pressure Angle

Form Relieved

High Speed Steel

(Concluded)

Standard Sizes and Dimensions

Diametral Pitch	Diameter of Cutter	Diameter of Hole
6	4¼	1¾
6	3⅞	1½
6	3½	1¼
6	3⅛	1
7	3⅝	1½
7	3⅜	1¼
7	2⅞	1
8	3½	1½
8	3¼	1¼
8	2⅞	1
9	3⅛	1¼
9	2¾	1
10	3	1¼
10	2¾	1
10	2⅜	⅞
11	2⅝	1
11	2⅜	⅞
12	2⅞	1¼
12	2⅝	1
12	2¼	⅞
14	2½	1
14	2⅛	⅞
16	2½	1
16	2⅛	⅞
18	2⅜	1
18	2	⅞
20	2⅜	1
20	2	⅞
22	2¼	1
22	2	⅞
24	2¼	1
24	1¾	⅞
26	1¾	⅞
28	1¾	⅞
30	1¾	⅞
32	1¾	⅞
36	1¾	⅞
40	1¾	⅞
48	1¾	⅞

All dimensions are given in inches.

For tolerances see Tables 401 and 402

For keyway dimensions see Table 404

Gear Milling Cutters
for Mitre and Bevel Gears

14½ Degree Pressure Angle

Form Relieved

High Speed Steel

Gear Milling Cutters for Mitre and Bevel Gears are made with eight different cutter forms for each pitch depending on the number of teeth to be cut.

To select the cutter form number for bevel gears with the axis at any angle, double the Back Cone Radius and multiply by the diametral pitch. This gives the number of teeth in the equivalent spur gear and is the basis for selecting the proper cutter form number from table.

Cutter Form Number	Range of Teeth	Cutter Form Number	Range of Teeth
1	135 thru a rack	5	21 thru 25
2	55 thru 134	6	17 thru 20
3	35 thru 54	7	14 thru 16
4	26 thru 34	8	12 and 13

Standard Sizes and Dimensions

Diametral Pitch	Diameter of Cutter	Diameter of Hole
3	4	1¼
4	3⅝	1¼
5	3⅜	1¼
6	3⅛	1
7	2⅞	1
8	2⅞	1
10	2⅜	⅞
12	2¼	⅞
14	2⅛	⅞
16	2⅛	⅞
20	2	⅞
24	1¾	⅞

All dimensions are given in inches.
For tolerances see Tables 401 and 402
For keyway dimensions see Table 404

Roughing Gear Milling Cutters

14½ Degree Pressure Angle

Form Relieved

High Speed Steel

Roughing cutters are made with No. 1 Cutter form only.

Standard Sizes and Dimensions

Diametral Pitch	Diameter of Cutter	Diameter of Hole
1	8½	2
1¼	7¾	2
1½	7	1¾
1¾	6½	1¾
2	6½	1¾
2	5¾	1½
2½	6⅛	1¾
2½	5¾	1½
3	5⅝	1¾
3	5¼	1½
3	4¾	1¼
4	4¾	1¾
4	4½	1½
4	4¼	1¼
4	3⅝	1
5	4⅜	1¾
5	4¼	1½
5	3¾	1¼
5	3⅜	1
6	3⅞	1½
6	3½	1¼
6	3⅛	1
7	3⅜	1¼
7	2⅞	1
8	3¼	1¼
8	2⅞	1

All dimensions are given in inches.
For tolerances see Tables 401 and 402
For keyway dimensions see Table 404

Roller Chain Sprocket Milling Cutters

American National Standard Tooth Form

Form Relieved

High Speed Steel

Roller Chain Sprocket Milling Cutters are made with six different forms for each pitch depending on the number of teeth to be cut.

Standard Sizes and Dimensions

Chain Pitch	Diameter of Roll	No. of Teeth in Sprocket	Diameter of Cutter	Width of Cutter	Diameter of Hole
$\frac{1}{4}$	0.130	6	$2\frac{3}{4}$	$\frac{5}{16}$	1
		7-8	$2\frac{3}{4}$	$\frac{5}{16}$	
		9-11	$2\frac{3}{4}$	$\frac{5}{16}$	
		12-17	$2\frac{3}{4}$	$\frac{5}{16}$	
		18-34	$2\frac{3}{4}$	$\frac{9}{32}$	
		35 & over	$2\frac{3}{4}$	$\frac{9}{32}$	
$\frac{3}{8}$	0.200	6	$2\frac{3}{4}$	$\frac{15}{32}$	1
		7-8	$2\frac{3}{4}$	$\frac{15}{32}$	
		9-11	$2\frac{3}{4}$	$\frac{15}{32}$	
		12-17	$2\frac{3}{4}$	$\frac{7}{16}$	
		18-34	$2\frac{3}{4}$	$\frac{7}{16}$	
		35 & over	$2\frac{3}{4}$	$\frac{13}{32}$	
$\frac{1}{2}$	0.313	6	3	$\frac{3}{4}$	1
		7-8	3	$\frac{3}{4}$	
		9-11	$3\frac{1}{8}$	$\frac{3}{4}$	
		12-17	$3\frac{1}{8}$	$\frac{3}{4}$	
		18-34	$3\frac{1}{8}$	$\frac{23}{32}$	
		35 & over	$3\frac{1}{8}$	$\frac{11}{16}$	
$\frac{5}{8}$	0.400	6	$3\frac{1}{8}$	$\frac{3}{4}$	1
		7-8	$3\frac{1}{8}$	$\frac{3}{4}$	
		9-11	$3\frac{1}{4}$	$\frac{3}{4}$	
		12-17	$3\frac{1}{4}$	$\frac{3}{4}$	
		18-34	$3\frac{1}{4}$	$\frac{23}{32}$	
		35 & over	$3\frac{1}{4}$	$\frac{11}{16}$	
$\frac{3}{4}$	0.469	6	$3\frac{1}{4}$	$\frac{29}{32}$	1
		7-8	$3\frac{1}{4}$	$\frac{29}{32}$	
		9-11	$3\frac{3}{8}$	$\frac{29}{32}$	
		12-17	$3\frac{3}{8}$	$\frac{7}{8}$	
		18-34	$3\frac{3}{8}$	$\frac{27}{32}$	
		35 & over	$3\frac{3}{8}$	$\frac{13}{16}$	
1	0.625	6	$3\frac{7}{8}$	$1\frac{1}{2}$	$1\frac{1}{4}$
		7-8	4	$1\frac{1}{2}$	
		9-11	$4\frac{1}{8}$	$1\frac{15}{32}$	
		12-17	$4\frac{1}{8}$	$1\frac{15}{32}$	
		18-34	$4\frac{1}{4}$	$1\frac{13}{32}$	
		35 & over	$4\frac{1}{4}$	$1\frac{11}{32}$	

(Continued on following page)

Roller Chain Sprocket Milling Cutters

American National Standard Tooth Form
Form Relieved

High Speed Steel

(Concluded)

Standard Sizes and Dimensions

Chain Pitch	Diameter of Roll	No. of Teeth in Sprocket	Diameter of Cutter	Width of Cutter	Diameter of Hole
1¼	0.750	6	4¼	1 13/16	1¼
		7-8	4⅜	1 13/16	
		9-11	4½	1 25/32	
		12-17	4½	1¾	
		18-34	4⅝	1 11/16	
		35 & over	4⅝	1⅝	
1½	0.875	6	4⅜	1 13/16	1¼
		7-8	4½	1 13/16	
		9-11	4⅝	1 25/32	
		12-17	4⅝	1¾	
		18-34	4¾	1 11/16	
		35 & over	4¾	1⅝	
1¾	1.000	6	5	2 3/32	1½
		7-8	5⅛	2 3/32	
		9-11	5¼	2 1/16	
		12-17	5⅜	2 1/32	
		18-34	5½	1 31/32	
		35 & over	5½	1⅞	
2	1.125	6	5⅜	2 13/32	1½
		7-8	5½	2 13/32	
		9-11	5⅝	2⅜	
		12-17	5¾	2 5/16	
		18-34	5⅞	2¼	
		35 & over	5⅞	2 5/32	
2¼	1.406	6	5⅞	2 11/16	1½
		7-8	6	2 11/16	
		9-11	6¼	2 21/32	
		12-17	6⅜	2 19/32	
		18-34	6½	2 15/32	
		35 & over	6½	2 13/32	
2½	1.563	6	6⅜	3	1¾
		7-8	6⅝	3	
		9-11	6¾	2 15/16	
		12-17	6⅞	2 29/32	
		18-34	7	2¾	
		35 & over	7⅛	2 11/16	
3	1.875	6	7½	3 19/32	2
		7-8	7¾	3 19/32	
		9-11	7⅞	3 17/32	
		12-17	8	3 15/32	
		18-34	8	3 11/32	
		35 & over	8¼	3 7/32	

All dimensions are given in inches.
For tolerances see Tables 401 and 402
For keyway dimensions see Table 404

Slitting Saws
Carbide Tipped

These saws are standard with full width hubs.

Standard Sizes and Dimensions

Diameter of Cutter	Width of Face	Diameter of Hole
4	$\frac{3}{32}$	1
4	$\frac{1}{8}$	1
4	$\frac{3}{16}$	1
6	$\frac{1}{8}$	$1\frac{1}{4}$
6	$\frac{3}{16}$	$1\frac{1}{4}$
6	$\frac{1}{4}$	$1\frac{1}{4}$

All dimensions are given in inches.
For keyway dimensions see Table 404.
For tolerances see Tables 401 and 402.

Plain Milling Cutters
Carbide Tipped

Standard Sizes and Dimensions

Diameter of Cutter	Width of Face	Diameter of Hole
3	2	$1\frac{1}{4}$
3	3	$1\frac{1}{4}$
3	4	$1\frac{1}{4}$
4	2	$1\frac{1}{2}$
4	3	$1\frac{1}{2}$
4	4	$1\frac{1}{2}$

All dimensions are given in inches.
For keyway dimensions see Table 404.
For tolerances see Tables 401 and 402.

Side Milling Cutters
Carbide Tipped
Standard Sizes and Dimensions

Diameter of Cutter	Width of Face	Diameter of Hole
3	1/4	1
3	5/16	1
3	3/8	1
3	1/2	1
4	1/4	1
4	3/8	1
4	3/8	1 1/4
4	1/2	1
4	1/2	1 1/4
4	5/8	1
4	5/8	1 1/4
4	3/4	1 1/4
5	3/8	1 1/4
5	1/2	1
5	1/2	1 1/4
5	3/4	1
5	3/4	1 1/4
6	3/8	1 1/4
6	1/2	1
6	1/2	1 1/4
6	5/8	1 1/4
6	3/4	1
6	3/4	1 1/4
6	1	1 1/4
8	1/2	1 1/2
8	3/4	1 1/4
8	3/4	1 1/2
8	1	1 1/4
8	1	1 1/2

All dimensions are given in inches.
For keyway dimensions see Table 404.
For tolerances see Tables 401 and 402.

Shell Mills
Carbide Tipped

Right hand cutters and left hand cutters are standard.
Square or chamfered corners optional with manufacturer.

Standard Sizes and Dimensions

General Dimensions					Driving Slot			Counter-bore
Diam. of Cutter	Width of Cutter	Length of Cut (Min.)	Diam. of Hole	Length of Bearing	Width (Nom.)	Depth	Radius	Diam. at Hub Face (Min.)
1¼	1	7/16	½	5/8	¼	5/32	1/64	11/16
1½	1⅛	7/16	½	5/8	¼	5/32	1/64	11/16
1¾	1¼	7/16	¾	3/4	5/16	3/16	1/32	15/16
2	1⅜	7/16	¾	3/4	5/16	3/16	1/32	15/16
2¼	1½	7/16	1	3/4	3/8	7/32	1/32	1¼
2½	1⅝	½	1	3/4	3/8	7/32	1/32	1⅜
2¾	1⅝	½	1	3/4	3/8	7/32	1/32	1½
3	1¾	9/16	1¼	3/4	½	9/32	1/32	1 21/32
3½	1⅞	11/16	1¼	3/4	½	9/32	1/32	1 11/16
4	2¼	11/16	1½	1	5/8	3/8	1/16	2 1/32
5	2¼	11/16	1½	1	5/8	3/8	1/16	2 9/16
6	2¼	11/16	2	1	3/4	7/16	1/16	2 13/16

All dimensions are given in inches.
For Radial and Axial Runout Tolerances see Table 402.

Tolerances

Element	Range	Direction	Tolerance
Diameter of Cutter	All Sizes	Plus	1/16
Width of Cutter	All Sizes	Plus or Minus	1/16
Diameter of Hole	All Sizes	Plus	.0010
Length of Bearing	All Sizes	Plus	1/32
Width, Driving Slot	All Sizes	Plus not less than	.008
		Plus not more than	.012
Depth, Driving Slot	All Sizes	Plus	1/64

Multiple Flute End Mills

With Plain Straight Shanks

High Speed Steel

Right hand cutters with right hand helix are standard. Helix angle not less than 10°.

Standard Sizes and Dimensions

Diameter of Cutter and Shank	Length of Cut	Length Overall
$\frac{1}{8}$	$\frac{5}{16}$	$1\frac{1}{4}$
$\frac{3}{16}$	$\frac{1}{2}$	$1\frac{3}{8}$
$\frac{1}{4}$	$\frac{5}{8}$	$1\frac{11}{16}$
$\frac{3}{8}$	$\frac{3}{4}$	$1\frac{13}{16}$
$\frac{1}{2}$	$\frac{15}{16}$	$2\frac{1}{4}$
$\frac{3}{4}$	$1\frac{1}{4}$	$2\frac{5}{8}$

All dimensions are given in inches.
For tolerances see Table 1701.

Two Flute Single-End End Mills
For Keyway Cutting
With Weldon Shanks
High Speed Steel

Right hand cutters with right hand helix are standard. Helix angle optional with manufacturer.

Standard Sizes and Dimensions

Diameter of Cutter	Diameter of Shank	Length of Cut	Length Overall
1/8	3/8	3/8	2 5/16
3/16	3/8	7/16	2 5/16
1/4	3/8	1/2	2 5/16
5/16	3/8	9/16	2 5/16
3/8	3/8	9/16	2 5/16
1/2	1/2	1	3
5/8	5/8	1 5/16	3 7/16
3/4	3/4	1 5/16	3 9/16
7/8	7/8	1 1/2	3 3/4
1	1	1 5/8	4 1/8
1 1/4	1 1/4	1 5/8	4 1/8
1 1/2	1 1/4	1 5/8	4 1/8

All dimensions are given in inches.
For tolerances see Table 1701.
For shank dimensions see Table 1702.

Stub Length, Four Flute, Medium Helix, Double-End Miniature End Mills

With 3/16″ Diameter Straight Shanks

High Speed Steel

Right hand cutters with right hand helix are standard. Helix angle greater than 19° but not more than 39°.

Standard Sizes and Dimensions

Diameter of Cutter	Diameter of Shank	Length of Cut	Length Overall
1/16	3/16	3/32	2
3/32	3/16	9/64	2
1/8	3/16	3/16	2
5/32	3/16	15/64	2
3/16	3/16	9/32	2

All dimensions are given in inches.
For tolerances see Table 1701.

Regular Length, Four Flute, Medium Helix, Double-End Miniature End Mills

With ³⁄₁₆″ Diameter Straight Shanks

High Speed Steel

Right hand cutters with right hand helix are standard. Helix angle greater than 19° but not more than 39°.

Standard Sizes and Dimensions

Diameter of Cutter	Diameter of Shank	Length of Cut	Length Overall
¹⁄₁₆	³⁄₁₆	³⁄₁₆	2¼
³⁄₃₂	³⁄₁₆	⁹⁄₃₂	2¼
⅛	³⁄₁₆	⅜	2¼
⁵⁄₃₂	³⁄₁₆	⁷⁄₁₆	2¼
³⁄₁₆	³⁄₁₆	½	2¼

All dimensions are given in inches.
For tolerances see Table 1701.

Long, Four Flute, Medium Helix, Double-End Miniature End Mills

With 3⁄16″ Diameter Straight Shanks

High Speed Steel

Right hand cutters with right hand helix are standard. Helix angle greater than 19° but not more than 39°.

Standard Sizes and Dimensions

Diameter of Cutter	Diameter of Shank	Length Below Shank	Length of Cut	Length Overall
1⁄16	3⁄16	3⁄8	7⁄32	2½
3⁄32	3⁄16	1⁄2	9⁄32	2⅝
1⁄8	3⁄16	3⁄4	3⁄4	3⅛
5⁄32	3⁄16	7⁄8	7⁄8	3¼
3⁄16	3⁄16	1	1	3⅜

All dimensions are given in inches.
For tolerances see Table 1701.

Stub Length, Two Flute, Medium Helix, Double-End Miniature End Mills

With $3/16''$ Diameter Straight Shanks

High Speed Steel

Right hand cutters with right hand helix are standard. Helix angle greater than 19° but not more than 39°.

Standard Sizes and Dimensions

Diameter of Cutter	Diameter of Shank	Length of Cut	Length Overall
1/32	3/16	3/64	2
3/64	3/16	1/16	2
1/16	3/16	3/32	2
5/64	3/16	1/8	2
3/32	3/16	9/64	2
7/64	3/16	5/32	2
1/8	3/16	3/16	2
9/64	3/16	7/32	2
5/32	3/16	15/64	2
11/64	3/16	1/4	2
3/16	3/16	9/32	2

All dimensions are given in inches.
For tolerances see Table 1701.

Ball End, Stub Length, Two Flute, Medium Helix, Double-End Miniature End Mills

With 3/16″ Diameter Straight Shanks

High Speed Steel

Right hand cutters with right hand helix are standard. Helix angle greater than 19° but not more than 39°.

Standard Sizes and Dimensions

Diameter of Cutter	Diameter of Shank	Length of Cut	Length Overall
1/16	3/16	3/32	2
3/32	3/16	9/64	2
1/8	3/16	3/16	2
5/32	3/16	15/64	2
3/16	3/16	9/32	2

All dimensions are given in inches.
For tolerances see Table 1701.

Regular Length, Two Flute, Medium Helix, Single-End Miniature End Mills

With 3/16″ Diameter Straight Shanks

High Speed Steel

Right hand cutters with right hand helix are standard. Helix angle greater than 19° but not more than 39°.

Standard Sizes and Dimensions

Diameter of Cutter	Diameter of Shank	Length of Cut	Length Overall
1/32	3/16	3/32	2 1/4
3/64	3/16	9/64	2 1/4
1/16	3/16	3/16	2 1/4
5/64	3/16	15/64	2 1/4
3/32	3/16	9/32	2 1/4
7/64	3/16	21/64	2 1/4
1/8	3/16	3/8	2 1/4
9/64	3/16	13/32	2 1/4
5/32	3/16	7/16	2 1/4
11/64	3/16	1/2	2 1/4
3/16	3/16	1/2	2 1/4

All dimensions are given in inches.
For tolerances see Table 1701.

Ball End, Regular Length, Two Flute, Medium Helix, Double-End Miniature End Mills

With 3/16″ Diameter Straight Shanks

High Speed Steel

Right hand cutters with right hand helix are standard. Helix angle greater than 19° but not more than 39°.

Standard Sizes and Dimensions

Diameter of Cutter	Diameter of Shank	Length of Cut	Length Overall
1/16	3/16	3/16	2 1/4
3/32	3/16	9/32	2 1/4
1/8	3/16	3/8	2 1/4
5/32	3/16	7/16	2 1/4
3/16	3/16	1/2	2 1/4

All dimensions are given in inches.
For tolerances see Table 1701.

Long, Two Flute, Medium Helix, Double-End Miniature End Mills

With 3/16″ Diameter Straight Shanks

High Speed Steel

Right hand cutters with right hand helix are standard. Helix angle greater than 19° but not more than 39°.

Standard Sizes and Dimensions

Diameter of Cutter	Diameter of Shank	Length Below Shank	Length of Cut	Length Overall
1/16	3/16	3/8	7/32	2 1/2
3/32	3/16	1/2	9/32	2 5/8
1/8	3/16	3/4	3/4	3 1/8
5/32	3/16	7/8	7/8	3 1/4
3/16	3/16	1	1	3 3/8

All dimensions are given in inches.
For tolerances see Table 1701.

Multiple Flute, Medium Helix, Single-End End Mills

With Weldon Shanks
High Speed Steel

Right hand cutters with right hand helix are standard. Starred sizes also standard with left hand cut, left hand helix. Helix angle greater than 19° but not more than 39°.

Standard Sizes and Dimensions

Diameter of Cutter	Diameter of Shank	Length of Cut	Length Overall	Number of Flutes
1/8	3/8	3/8	2 5/16	4
* 3/16	3/8	1/2	2 3/8	4
* 1/4	3/8	5/8	2 7/16	4
* 5/16	3/8	3/4	2 1/2	4
* 3/8	3/8	3/4	2 1/2	4
7/16	3/8	1	2 11/16	4
1/2	3/8	1	2 11/16	4
* 1/2	1/2	1 1/4	3 1/4	4
9/16	1/2	1 3/8	3 3/8	4
5/8	1/2	1 3/8	3 3/8	4
11/16	1/2	1 5/8	3 5/8	4
3/4	1/2	1 5/8	3 5/8	4
3/4	3/4	1 5/8	3 7/8	4
* 5/8	5/8	1 5/8	3 3/4	4
11/16	5/8	1 5/8	3 3/4	4
* 3/4	5/8	1 5/8	3 3/4	4
13/16	5/8	1 7/8	4	6
13/16	3/4	1 7/8	4 1/8	4
7/8	5/8	1 7/8	4	6
7/8	3/4	1 7/8	4 1/8	4
15/16	3/4	1 7/8	4 1/8	4
1	3/4	1 7/8	4 1/8	4
1	5/8	1 7/8	4	6
7/8	7/8	1 7/8	4 1/8	4
1	7/8	1 7/8	4 1/8	4
1 1/8	7/8	2	4 1/4	6
1 1/4	7/8	2	4 1/4	6
1	1	2	4 1/2	4
1 1/8	1	2	4 1/2	6
1 1/4	1	2	4 1/2	6
1 3/8	1	2	4 1/2	6
1 1/2	1	2	4 1/2	6
1 1/4	1 1/4	2	4 1/2	6
1 1/2	1 1/4	2	4 1/2	6
1 3/4	1 1/4	2	4 1/2	6
2	1 1/4	2	4 1/2	8

All dimensions are given in inches.
For tolerances see Table 1701.
For shank dimensions see Table 1702.

Long, Multiple Flute, Medium Helix, Single-End End Mills

With Weldon Shanks

High Speed Steel

Right hand cutters with right hand helix are standard. Helix angle greater than 19° but not more than 39°.

Standard Sizes and Dimensions

Diameter of Cutter	Diameter of Shank	Length of Cut	Length Overall	Number of Flutes
1/4	3/8	1 1/4	3 1/16	4
5/16	3/8	1 3/8	3 1/8	4
3/8	3/8	1 1/2	3 1/4	4
7/16	1/2	1 3/4	3 3/4	4
1/2	1/2	2	4	4
5/8	5/8	2 1/2	4 5/8	4
3/4	3/4	3	5 1/4	4
7/8	7/8	3 1/2	5 3/4	4
1	1	4	6 1/2	4
1 1/8	1	4	6 1/2	6
1 1/4	1	4	6 1/2	6
1 1/2	1	4	6 1/2	6
1 1/4	1 1/4	4	6 1/2	6
1 1/2	1 1/4	4	6 1/2	6
1 3/4	1 1/4	4	6 1/2	6
2	1 1/4	4	6 1/2	8

All dimensions are given in inches.
For tolerances see Table 1701.
For shank dimensions see Table 1702.

Extra Long, Multiple Flute, Medium Helix, Single-End End Mills

With Weldon Shanks

High Speed Steel

Right hand cutters with right hand helix are standard. Helix angle greater than 19° but not more than 39°.

Standard Sizes and Dimensions

Diameter of Cutter	Diameter of Shank	Length of Cut	Length Overall	Number of Flutes
1/4	3/8	1 3/4	3 9/16	4
5/16	3/8	2	3 3/4	4
3/8	3/8	2 1/2	4 1/4	4
1/2	1/2	3	5	4
5/8	5/8	4	6 1/8	4
3/4	3/4	4	6 1/4	4
7/8	7/8	5	7 1/4	4
1	1	6	8 1/2	4
1 1/4	1 1/4	6	8 1/2	6
1 1/2	1 1/4	8	10 1/2	6

All dimensions are given in inches.
For tolerances see Table 1701.
For shank dimensions see Table 1702.

Stub Length, Two Flute, Medium Helix, Single-End End Mills

With Weldon Shanks

High Speed Steel

Right hand cutters with right hand helix are standard. Helix angle greater than 19° but not more than 39°.

Standard Sizes and Dimensions

Diameter of Cutter	Diameter of Shank	Length of Cut	Length Overall
1/8	3/8	3/16	2 1/8
3/16	3/8	9/32	2 3/16
1/4	3/8	3/8	2 1/4

All dimensions are given in inches.
For tolerances see Table 1701.
For shank dimensions see Table 1702.

Two Flute, Medium Helix, Single-End End Mills

With Weldon Shanks

High Speed Steel

Right hand cutters with right hand helix are standard. Helix angle greater than 19° but not more than 39°.

Standard Sizes and Dimensions

Diameter of Cutter	Diameter of Shank	Length of Cut	Length Overall
1/8	3/8	3/8	2 5/16
3/16	3/8	7/16	2 5/16
1/4	3/8	1/2	2 5/16
5/16	3/8	9/16	2 5/16
3/8	3/8	9/16	2 5/16
7/16	3/8	13/16	2 1/2
1/2	3/8	13/16	2 1/2
1/2	1/2	1	3
9/16	1/2	1 1/8	3 1/8
5/8	1/2	1 1/8	3 1/8
11/16	1/2	1 5/16	3 5/16
3/4	1/2	1 5/16	3 5/16
5/8	5/8	1 5/16	3 7/16
11/16	5/8	1 5/16	3 7/16
3/4	5/8	1 5/16	3 7/16
13/16	5/8	1 1/2	3 5/8
7/8	5/8	1 1/2	3 5/8
1	5/8	1 1/2	3 5/8
7/8	7/8	1 1/2	3 3/4
1	7/8	1 1/2	3 3/4
1 1/8	7/8	1 5/8	3 7/8
1 1/4	7/8	1 5/8	3 7/8
1	1	1 5/8	4 1/8
1 1/8	1	1 5/8	4 1/8
1 1/4	1	1 5/8	4 1/8
1 3/8	1	1 5/8	4 1/8
1 1/2	1	1 5/8	4 1/8
1 1/4	1 1/4	1 5/8	4 1/8
1 1/2	1 1/4	1 5/8	4 1/8
1 3/4	1 1/4	1 5/8	4 1/8
2	1 1/4	1 5/8	4 1/8

All dimensions are given in inches.
For tolerances see Table 1701.
For shank dimensions see Table 1702.

Ball End, Two Flute, Medium Helix, Single-End End Mills

With Weldon Shanks

High Speed Steel

Right hand cutters with right hand helix are standard. Helix angle greater than 19° but not more than 39°.

Standard Sizes and Dimensions

Diameter of Cutter and End Circle	Diameter of Shank	Length of Cut	Length Overall
1/8	3/8	3/8	2 5/16
3/16	3/8	1/2	2 3/8
1/4	3/8	5/8	2 7/16
5/16	3/8	3/4	2 1/2
3/8	3/8	3/4	2 1/2
7/16	1/2	1	3
1/2	1/2	1	3
9/16	1/2	1 1/8	3 1/8
5/8	1/2	1 1/8	3 1/8
5/8	5/8	1 3/8	3 1/2
3/4	1/2	1 5/16	3 5/16
3/4	3/4	1 5/8	3 7/8
7/8	7/8	2	4 1/4
1	1	2 1/4	4 3/4
1 1/8	1	2 1/4	4 3/4
1 1/4	1 1/4	2 1/2	5
1 1/2	1 1/4	2 1/2	5

All dimensions are given in inches.
For tolerances see Table 1701.
For shank dimensions see Table 1702.

Long, Two Flute, Medium Helix, Single-End End Mills

With Weldon Shanks

High Speed Steel

Right hand cutters with right hand helix are standard. Helix angle greater than 19° but not more than 39°.

Standard Sizes and Dimensions

Diameter of Cutter	Diameter of Shank	Length Below Shank	Length of Cut	Length Overall
$\frac{1}{4}$	$\frac{3}{8}$	$1\frac{1}{2}$	$\frac{5}{8}$	$3\frac{1}{16}$
$\frac{5}{16}$	$\frac{3}{8}$	$1\frac{3}{4}$	$\frac{3}{4}$	$3\frac{5}{16}$
$\frac{3}{8}$	$\frac{3}{8}$	$1\frac{3}{4}$	$\frac{3}{4}$	$3\frac{5}{16}$
$\frac{1}{2}$	$\frac{1}{2}$	$2\frac{7}{32}$	1	4
$\frac{5}{8}$	$\frac{5}{8}$	$2\frac{23}{32}$	$1\frac{3}{8}$	$4\frac{5}{8}$
$\frac{3}{4}$	$\frac{3}{4}$	$3\frac{11}{32}$	$1\frac{5}{8}$	$5\frac{3}{8}$
1	1	$4\frac{31}{32}$	$2\frac{1}{2}$	$7\frac{1}{4}$
$1\frac{1}{4}$	$1\frac{1}{4}$	$4\frac{31}{32}$	3	$7\frac{1}{4}$

Ball End, Two Flute, Long, Medium Helix, Single-End End Mills

With Weldon Shanks

High Speed Steel

Right hand cutters with right hand helix are standard. Helix angle greater than 19° but not more than 39°.

Standard Sizes and Dimensions

Diameter of Cutter and End Circle	Diameter of Shank	Length Below Shank	Length of Cut	Length Overall
$\frac{1}{8}$	$\frac{3}{8}$	$\frac{13}{16}$	$\frac{3}{8}$	$2\frac{3}{8}$
$\frac{3}{16}$	$\frac{3}{8}$	$1\frac{1}{8}$	$\frac{1}{2}$	$2\frac{11}{16}$
$\frac{1}{4}$	$\frac{3}{8}$	$1\frac{1}{2}$	$\frac{5}{8}$	$3\frac{1}{16}$
$\frac{5}{16}$	$\frac{3}{8}$	$1\frac{3}{4}$	$\frac{3}{4}$	$3\frac{5}{16}$
$\frac{3}{8}$	$\frac{3}{8}$	$1\frac{3}{4}$	$\frac{3}{4}$	$3\frac{5}{16}$
$\frac{7}{16}$	$\frac{1}{2}$	$1\frac{7}{8}$	1	$3\frac{11}{16}$
$\frac{1}{2}$	$\frac{1}{2}$	$2\frac{1}{4}$	1	4
$\frac{5}{8}$	$\frac{5}{8}$	$2\frac{3}{4}$	$1\frac{3}{8}$	$4\frac{5}{8}$
$\frac{3}{4}$	$\frac{3}{4}$	$3\frac{3}{8}$	$1\frac{5}{8}$	$5\frac{3}{8}$
1	1	5	$2\frac{1}{2}$	$7\frac{1}{4}$

All dimensions are given in inches.
For tolerances see Table 1701.   } Applicable to both tables.
For shank dimensions see Table 1702.

Stub Length, Four Flute, Medium Helix, Double-End End Mills

With Weldon Shanks

High Speed Steel

Right hand cutters with right hand helix are standard. Helix angle greater than 19° but not more than 39°.

Standard Sizes and Dimensions

Diameter of Cutter	Diameter of Shank	Length of Cut	Length Overall
$\frac{1}{8}$	$\frac{3}{8}$	$\frac{3}{16}$	$2\frac{3}{4}$
$\frac{5}{32}$	$\frac{3}{8}$	$\frac{15}{64}$	$2\frac{3}{4}$
$\frac{3}{16}$	$\frac{3}{8}$	$\frac{9}{32}$	$2\frac{3}{4}$
$\frac{7}{32}$	$\frac{3}{8}$	$\frac{21}{64}$	$2\frac{7}{8}$
$\frac{1}{4}$	$\frac{3}{8}$	$\frac{3}{8}$	$2\frac{7}{8}$

All dimensions are given in inches.
For tolerances see Table 1701.
For shank dimensions see Table 1702.

Four Flute, Medium Helix, Double-End End Mills

With Weldon Shanks

High Speed Steel

Right hand cutters with right hand helix are standard. Starred sizes also standard with left hand cut, left hand helix. Helix angle greater than 19° but not more than 39°.

Standard Sizes and Dimensions

Diameter of Cutter	Diameter of Shank	Length of Cut	Length Overall
* 1/8	3/8	3/8	3 1/16
* 5/32	3/8	7/16	3 1/8
* 3/16	3/8	1/2	3 1/4
7/32	3/8	9/16	3 1/4
* 1/4	3/8	5/8	3 3/8
9/32	3/8	11/16	3 3/8
* 5/16	3/8	3/4	3 1/2
11/32	3/8	3/4	3 1/2
* 3/8	3/8	3/4	3 1/2
13/32	1/2	1	4 1/8
7/16	1/2	1	4 1/8
15/32	1/2	1	4 1/8
* 1/2	1/2	1	4 1/8
9/16	5/8	1 3/8	5
* 5/8	5/8	1 3/8	5
11/16	3/4	1 5/8	5 5/8
* 3/4	3/4	1 5/8	5 5/8
13/16	7/8	1 7/8	6 1/8
7/8	7/8	1 7/8	6 1/8
1	1	1 7/8	6 3/8

All dimensions are given in inches.
For tolerances see Table 1701.
For shank dimensions see Table 1702.

Stub Length, Two Flute, Medium Helix, Double-End End Mills

With Weldon Shanks

High Speed Steel

Right hand cutters with right hand helix are standard. Helix angle greater than 19° but not more than 39°.

Standard Sizes and Dimensions

Diameter of Cutter	Diameter of Shank	Length of Cut	Length Overall
$\frac{1}{8}$	$\frac{3}{8}$	$\frac{3}{16}$	$2\frac{3}{4}$
$\frac{5}{32}$	$\frac{3}{8}$	$\frac{15}{64}$	$2\frac{3}{4}$
$\frac{3}{16}$	$\frac{3}{8}$	$\frac{9}{32}$	$2\frac{3}{4}$
$\frac{7}{32}$	$\frac{3}{8}$	$\frac{21}{64}$	$2\frac{7}{8}$
$\frac{1}{4}$	$\frac{3}{8}$	$\frac{3}{8}$	$2\frac{7}{8}$

All dimensions are given in inches.
For tolerances see Table 1701.
For shank dimensions see Table 1702.

Two Flute, Medium Helix, Double-End End Mills

With Weldon Shanks

High Speed Steel

Right hand cutters with right hand helix are standard. Helix angle greater than 19° but not more than 39°.

Standard Sizes and Dimensions

Diameter of Cutter	Diameter of Shank	Length of Cut	Length Overall
1/8	3/8	3/8	3 1/16
5/32	3/8	7/16	3 1/8
3/16	3/8	7/16	3 1/8
7/32	3/8	1/2	3 1/8
1/4	3/8	1/2	3 1/8
9/32	3/8	9/16	3 1/8
5/16	3/8	9/16	3 1/8
11/32	3/8	9/16	3 1/8
3/8	3/8	9/16	3 1/8
13/32	1/2	13/16	3 3/4
7/16	1/2	13/16	3 3/4
15/32	1/2	13/16	3 3/4
1/2	1/2	13/16	3 3/4
9/16	5/8	1 1/8	4 1/2
5/8	5/8	1 1/8	4 1/2
11/16	3/4	1 5/16	5
3/4	3/4	1 5/16	5
7/8	7/8	1 9/16	5 1/2
1	1	1 5/8	5 7/8

All dimensions are given in inches.
For tolerances see Table 1701.
For shank dimensions see Table 1702.

Ball End, Two Flute, Medium Helix, Double-End End Mills

With Weldon Shanks

High Speed Steel

Right hand cutters with right hand helix are standard. Helix angle greater than 19° but not more than 39°.

Standard Sizes and Dimensions

Diameter of Cutter and End Circle	Diameter of Shank	Length of Cut	Length Overall
$\frac{1}{8}$	$\frac{3}{8}$	$\frac{3}{8}$	$3\frac{1}{16}$
$\frac{3}{16}$	$\frac{3}{8}$	$\frac{7}{16}$	$3\frac{1}{8}$
$\frac{1}{4}$	$\frac{3}{8}$	$\frac{1}{2}$	$3\frac{1}{8}$
$\frac{5}{16}$	$\frac{3}{8}$	$\frac{9}{16}$	$3\frac{1}{8}$
$\frac{3}{8}$	$\frac{3}{8}$	$\frac{9}{16}$	$3\frac{1}{8}$
$\frac{7}{16}$	$\frac{1}{2}$	$\frac{13}{16}$	$3\frac{3}{4}$
$\frac{1}{2}$	$\frac{1}{2}$	$\frac{13}{16}$	$3\frac{3}{4}$
$\frac{5}{8}$	$\frac{5}{8}$	$1\frac{1}{8}$	$4\frac{1}{2}$
$\frac{3}{4}$	$\frac{3}{4}$	$1\frac{5}{16}$	5
1	1	$1\frac{5}{8}$	$5\frac{7}{8}$

All dimensions are given in inches.
For tolerances see Table 1701.
For shank dimensions see Table 1702.

Four Flute, Center Cutting, Medium Helix, Single-End End Mills

With Weldon Shanks
High Speed Steel

Right hand cutters with right hand helix are standard. Helix angle greater than 19° but not more than 39°.

Standard Sizes and Dimensions

Diameter of Cutter	Diameter of Shank	Length of Cut	Length Overall
$\frac{1}{8}$	$\frac{3}{8}$	$\frac{3}{8}$	$2\frac{5}{16}$
$\frac{3}{16}$	$\frac{3}{8}$	$\frac{1}{2}$	$2\frac{3}{8}$
$\frac{1}{4}$	$\frac{3}{8}$	$\frac{5}{8}$	$2\frac{7}{16}$
$\frac{5}{16}$	$\frac{3}{8}$	$\frac{3}{4}$	$2\frac{1}{2}$
$\frac{3}{8}$	$\frac{3}{8}$	$\frac{3}{4}$	$2\frac{1}{2}$
$\frac{1}{2}$	$\frac{1}{2}$	$1\frac{1}{4}$	$3\frac{1}{4}$
$\frac{5}{8}$	$\frac{5}{8}$	$1\frac{5}{8}$	$3\frac{3}{4}$
$\frac{11}{16}$	$\frac{5}{8}$	$1\frac{5}{8}$	$3\frac{3}{4}$
$\frac{3}{4}$	$\frac{3}{4}$	$1\frac{5}{8}$	$3\frac{7}{8}$
$\frac{7}{8}$	$\frac{7}{8}$	$1\frac{7}{8}$	$4\frac{1}{8}$
1	1	2	$4\frac{1}{2}$
$1\frac{1}{8}$	1	2	$4\frac{1}{2}$
$1\frac{1}{4}$	$1\frac{1}{4}$	2	$4\frac{1}{2}$
$1\frac{1}{2}$	$1\frac{1}{4}$	2	$4\frac{1}{2}$

All dimensions are given in inches.
For tolerances see Table 1701.
For shank dimensions see Table 1702.

Long, Four Flute, Center Cutting, Medium Helix, Single-End End Mills

With Weldon Shanks

High Speed Steel

Right hand cutters with right hand helix are standard. Helix angle greater than 19° but not more than 39°.

Standard Sizes and Dimensions

Diameter of Cutter	Diameter of Shank	Length of Cut	Length Overall
$\frac{1}{4}$	$\frac{3}{8}$	$1\frac{1}{4}$	$3\frac{1}{16}$
$\frac{5}{16}$	$\frac{3}{8}$	$1\frac{3}{8}$	$3\frac{1}{8}$
$\frac{3}{8}$	$\frac{3}{8}$	$1\frac{1}{2}$	$3\frac{1}{4}$
$\frac{1}{2}$	$\frac{1}{2}$	2	4
$\frac{5}{8}$	$\frac{5}{8}$	$2\frac{1}{2}$	$4\frac{5}{8}$
$\frac{3}{4}$	$\frac{3}{4}$	3	$5\frac{1}{4}$
$\frac{7}{8}$	$\frac{7}{8}$	$3\frac{1}{2}$	$5\frac{3}{4}$
1	1	4	$6\frac{1}{2}$
$1\frac{1}{4}$	$1\frac{1}{4}$	4	$6\frac{1}{2}$

All dimensions are given in inches.
For tolerances see Table 1701.
For shank dimensions see Table 1702.

Extra Long, Four Flute, Center Cutting, Medium Helix, Single-End End Mills

With Weldon Shanks

High Speed Steel

Right hand cutters with right hand helix are standard. Helix angle greater than 19° but not more than 39°.

Standard Sizes and Dimensions

Diameter of Cutter	Diameter of Shank	Length of Cut	Length Overall
$\frac{1}{4}$	$\frac{3}{8}$	$1\frac{3}{4}$	$3\frac{9}{16}$
$\frac{5}{16}$	$\frac{3}{8}$	2	$3\frac{3}{4}$
$\frac{3}{8}$	$\frac{3}{8}$	$2\frac{1}{2}$	$4\frac{1}{4}$
$\frac{1}{2}$	$\frac{1}{2}$	3	5
$\frac{5}{8}$	$\frac{5}{8}$	4	$6\frac{1}{8}$
$\frac{3}{4}$	$\frac{3}{4}$	4	$6\frac{1}{4}$
$\frac{7}{8}$	$\frac{7}{8}$	5	$7\frac{1}{4}$
1	1	6	$8\frac{1}{2}$
$1\frac{1}{4}$	$1\frac{1}{4}$	6	$8\frac{1}{2}$

All dimensions are given in inches.
For tolerances see Table 1701.
For shank dimensions see Table 1702.

Three Flute, Medium Helix, Single-End End Mills

With Weldon Shanks

High Speed Steel

Right hand cutters with right hand helix are standard. Helix angle greater than 19° but not more than 39°.

Standard Sizes and Dimensions

Diameter of Cutter	Diameter of Shank	Length of Cut	Length Overall
1/8	3/8	3/8	2 5/16
3/16	3/8	1/2	2 3/8
1/4	3/8	5/8	2 7/16
5/16	3/8	3/4	2 1/2
3/8	3/8	3/4	2 1/2
7/16	3/8	1	2 11/16
1/2	3/8	1	2 11/16
1/2	1/2	1 1/4	3 1/4
9/16	1/2	1 3/8	3 3/8
5/8	1/2	1 3/8	3 3/8
3/4	1/2	1 5/8	3 5/8
5/8	5/8	1 5/8	3 3/4
3/4	5/8	1 5/8	3 3/4
7/8	5/8	1 7/8	4
1	5/8	1 7/8	4
3/4	3/4	1 5/8	3 7/8
7/8	3/4	1 7/8	4 1/8
1	3/4	1 7/8	4 1/8
1	7/8	1 7/8	4 1/8
1	1	2	4 1/2
1 1/8	1	2	4 1/2
1 1/4	1	2	4 1/2
1 1/2	1	2	4 1/2
1 1/4	1 1/4	2	4 1/2
1 1/2	1 1/4	2	4 1/2
1 3/4	1 1/4	2	4 1/2
2	1 1/4	2	4 1/2

All dimensions are given in inches.
For tolerances see Table 1701.
For shank dimensions see Table 1702.

Long, Three Flute, Medium Helix, Single-End End Mills

With Weldon Shanks

High Speed Steel

Right hand cutters with right hand helix are standard. Helix angle greater than 19° but not more than 39°.

Standard Sizes and Dimensions

Diameter of Cutter	Diameter of Shank	Length of Cut	Length Overall
$\frac{1}{4}$	$\frac{3}{8}$	$1\frac{1}{4}$	$3\frac{1}{16}$
$\frac{5}{16}$	$\frac{3}{8}$	$1\frac{3}{8}$	$3\frac{1}{8}$
$\frac{3}{8}$	$\frac{3}{8}$	$1\frac{1}{2}$	$3\frac{1}{4}$
$\frac{7}{16}$	$\frac{1}{2}$	$1\frac{3}{4}$	$3\frac{3}{4}$
$\frac{1}{2}$	$\frac{1}{2}$	2	4
$\frac{5}{8}$	$\frac{5}{8}$	$2\frac{1}{2}$	$4\frac{5}{8}$
$\frac{3}{4}$	$\frac{3}{4}$	3	$5\frac{1}{4}$
1	1	4	$6\frac{1}{2}$
$1\frac{1}{4}$	$1\frac{1}{4}$	4	$6\frac{1}{2}$
$1\frac{1}{2}$	$1\frac{1}{4}$	4	$6\frac{1}{2}$
$1\frac{3}{4}$	$1\frac{1}{4}$	4	$6\frac{1}{2}$
2	$1\frac{1}{4}$	4	$6\frac{1}{2}$

All dimensions are given in inches.
For tolerances see Table 1701.
For shank dimensions see Table 1702.

Four Flute, Center Cutting, Medium Helix, Double-End End Mills

With Weldon Shanks
High Speed Steel

Right hand cutters with right hand helix are standard. Helix angle greater than 19° but not more than 39°.

Standard Sizes and Dimensions

Diameter of Cutter	Diameter of Shank	Length of Cut	Length Overall
$\frac{1}{8}$	$\frac{3}{8}$	$\frac{3}{8}$	$3\frac{1}{16}$
$\frac{3}{16}$	$\frac{3}{8}$	$\frac{1}{2}$	$3\frac{1}{4}$
$\frac{1}{4}$	$\frac{3}{8}$	$\frac{5}{8}$	$3\frac{3}{8}$
$\frac{5}{16}$	$\frac{3}{8}$	$\frac{3}{4}$	$3\frac{1}{2}$
$\frac{3}{8}$	$\frac{3}{8}$	$\frac{3}{4}$	$3\frac{1}{2}$
$\frac{1}{2}$	$\frac{1}{2}$	1	$4\frac{1}{8}$
$\frac{5}{8}$	$\frac{5}{8}$	$1\frac{3}{8}$	5
$\frac{3}{4}$	$\frac{3}{4}$	$1\frac{5}{8}$	$5\frac{5}{8}$
$\frac{7}{8}$	$\frac{7}{8}$	$1\frac{7}{8}$	$6\frac{1}{8}$
1	1	$1\frac{7}{8}$	$6\frac{3}{8}$

All dimensions are given in inches.
For tolerances see Table 1701.
For shank dimensions see Table 1702.

Three Flute, Medium Helix, Double-End End Mills

With Weldon Shanks

High Speed Steel

Right hand cutters with right hand helix are standard. Helix angle greater than 19° but not more than 39°.

Standard Sizes and Dimensions

Diameter of Cutter	Diameter of Shank	Length of Cut	Length Overall
⅛	⅜	⅜	3 1/16
3/16	⅜	½	3¼
¼	⅜	⅝	3⅜
5/16	⅜	¾	3½
⅜	⅜	¾	3½
7/16	½	1	4⅛
½	½	1	4⅛
9/16	⅝	1⅜	5
⅝	⅝	1⅜	5
¾	¾	1⅝	5⅝
1	1	1⅞	6⅜

All dimensions are given in inches.
For tolerances see Table 1701.
For shank dimensions see Table 1702.

Heavy Duty, Medium Helix, Single-End End Mills

2″ Diameter Shank

High Speed Steel

Right hand cutters with right hand helix are standard. Helix angle greater than 19° but not more than 39°.

Standard Sizes and Dimensions

Number of Flutes	Diameter of Cutter	Length of Cut	Length Overall
Two	2	2	5¾
	2	3	6¾
	2	4	7¾
	2	6	9¾
	2½	4	7¾
	2½	6	9¾
Three	2	3	6¾
	2	4	7¾
	2	6	9¾
	2½	4	7¾
Four	2	2	5¾
	2	4	7¾
	2	6	9¾
	2½	4	7¾
	2½	6	9¾
Six	2	2	5¾
	2	4	7¾
	2	6	9¾
	2	8	11¾
	2½	4	7¾
	2½	6	9¾
	2½	8	11¾

All dimensions are given in inches.
For tolerances see Table 1701.
For shank dimensions see Tables 1702 and 1703.

Ball End, Heavy Duty, Medium Helix, Single-End End Mills

2″ Diameter Shank

High Speed Steel

Right hand cutters with right hand helix are standard. Helix angle greater than 19° but not more than 39°.

Standard Sizes and Dimensions

Number of Flutes	Diameter of Cutter	Length of Cut	Length Overall
Two	2	5	8¾
Four	2	5	8¾
	2½	5	8¾
Six	2	4	7¾
	2	6	9¾
	2	8	11¾

All dimensions are given in inches.
For tolerances see Table 1701.
For shank dimensions see Tables 1702 and 1703.

Heavy Duty, Medium Helix
Single-End End Mills
2½" Combination Shank
HIGH SPEED STEEL

Right hand cutters with right hand helix are standard. Helix angle greater than 19° but not more than 39°.

Standard Sizes and Dimensions

Diameter of Cutter	Number of Flutes	Length of Cut	Length Overall
2½	3	8	12
2½	3	10	14
2½	6	4	8
2½	6	6	10
2½	6	8	12
2½	6	10	14
2½	6	12	16
3	2	4	7¾
3	3	4	7¾
3	3	6	9¾
3	3	8	11¾
3	8	4	7¾
3	8	6	9¾
3	8	8	11¾
3	8	10	13¾
3	8	12	15¾

All dimensions are given in inches
For tolerances see Table 1701
For shank dimensions see Table 1703

Two Flute, High Helix, Single-End End Mills

With Weldon Shanks

High Speed Steel

Right hand cutters with right hand helix are standard. Helix angle greater than 39°.

Standard Sizes and Dimensions

Diameter of Cutter	Diameter of Shank	Length of Cut	Length Overall
$\frac{1}{4}$	$\frac{3}{8}$	$\frac{5}{8}$	$2 \frac{7}{16}$
$\frac{5}{16}$	$\frac{3}{8}$	$\frac{3}{4}$	$2 \frac{1}{2}$
$\frac{3}{8}$	$\frac{3}{8}$	$\frac{3}{4}$	$2 \frac{1}{2}$
$\frac{7}{16}$	$\frac{3}{8}$	1	$2\frac{11}{16}$
$\frac{1}{2}$	$\frac{1}{2}$	$1 \frac{1}{4}$	$3 \frac{1}{4}$
$\frac{5}{8}$	$\frac{5}{8}$	$1 \frac{5}{8}$	$3 \frac{3}{4}$
$\frac{3}{4}$	$\frac{3}{4}$	$1 \frac{5}{8}$	$3 \frac{7}{8}$
$\frac{7}{8}$	$\frac{7}{8}$	$1 \frac{7}{8}$	$4 \frac{1}{8}$
1	1	2	$4 \frac{1}{2}$
$1 \frac{1}{4}$	$1 \frac{1}{4}$	2	$4 \frac{1}{2}$
$1 \frac{1}{2}$	$1 \frac{1}{4}$	2	$4 \frac{1}{2}$
2	$1 \frac{1}{4}$	2	$4 \frac{1}{2}$

All dimensions are given in inches.
For tolerances see Table 1701.
For shank dimensions see Table 1702.

Long, Two Flute, High Helix, Single-End End Mills

With Weldon Shanks
High Speed Steel

Right hand cutters with right hand helix are standard. Helix angle greater than 39°.

Standard Sizes and Dimensions

Diameter of Cutter	Diameter of Shank	Length of Cut	Length Overall
$\frac{1}{4}$	$\frac{3}{8}$	$1\frac{1}{4}$	$3\frac{1}{16}$
$\frac{5}{16}$	$\frac{3}{8}$	$1\frac{3}{8}$	$3\frac{1}{8}$
$\frac{3}{8}$	$\frac{3}{8}$	$1\frac{1}{2}$	$3\frac{1}{4}$
$\frac{7}{16}$	$\frac{1}{2}$	$1\frac{3}{4}$	$3\frac{3}{4}$
$\frac{1}{2}$	$\frac{1}{2}$	2	4
$\frac{5}{8}$	$\frac{5}{8}$	$2\frac{1}{2}$	$4\frac{5}{8}$
$\frac{3}{4}$	$\frac{3}{4}$	3	$5\frac{1}{4}$
1	1	4	$6\frac{1}{2}$
$1\frac{1}{4}$	$1\frac{1}{4}$	4	$6\frac{1}{2}$
$1\frac{1}{2}$	$1\frac{1}{4}$	4	$6\frac{1}{2}$
2	$1\frac{1}{4}$	4	$6\frac{1}{2}$

All dimensions are given in inches.
For tolerances see Table 1701.
For shank dimensions see Table 1702.

Extra Long, Two Flute, High Helix, Single-End End Mills

With Weldon Shanks

High Speed Steel

Right hand cutters with right hand helix are standard. Helix angle greater than 39°.

Standard Sizes and Dimensions

Diameter of Cutter	Diameter of Shank	Length of Cut	Length Overall
$\frac{1}{4}$	$\frac{3}{8}$	$1\frac{3}{4}$	$3\frac{9}{16}$
$\frac{5}{16}$	$\frac{3}{8}$	2	$3\frac{3}{4}$
$\frac{3}{8}$	$\frac{3}{8}$	$2\frac{1}{2}$	$4\frac{1}{4}$
$\frac{1}{2}$	$\frac{1}{2}$	3	5
$\frac{5}{8}$	$\frac{5}{8}$	4	$6\frac{1}{8}$
$\frac{3}{4}$	$\frac{3}{4}$	4	$6\frac{1}{4}$
1	1	6	$8\frac{1}{2}$
$1\frac{1}{4}$	$1\frac{1}{4}$	6	$8\frac{1}{2}$
$1\frac{1}{2}$	$1\frac{1}{4}$	8	$10\frac{1}{2}$

All dimensions are given in inches.
For tolerances see Table 1701.
For shank dimensions see Table 1702.

Premium High Speed Steel

Two Flute, Medium Helix, Single-End End Mills

With Weldon Shanks

Right hand cutters with right hand helix are standard. Helix angle greater than 19° but not more than 39°.

Standard Sizes and Dimensions

Diameter of Cutter	Diameter of Shank	Length of Cut	Length Overall
$\frac{1}{8}$	$\frac{3}{8}$	$\frac{3}{8}$	$2\frac{5}{16}$
$\frac{3}{16}$	$\frac{3}{8}$	$\frac{7}{16}$	$2\frac{5}{16}$
$\frac{1}{4}$	$\frac{3}{8}$	$\frac{1}{2}$	$2\frac{5}{16}$
$\frac{5}{16}$	$\frac{3}{8}$	$\frac{9}{16}$	$2\frac{5}{16}$
$\frac{3}{8}$	$\frac{3}{8}$	$\frac{9}{16}$	$2\frac{5}{16}$
$\frac{1}{2}$	$\frac{1}{2}$	1	3
$\frac{5}{8}$	$\frac{5}{8}$	$1\frac{5}{16}$	$3\frac{7}{16}$
$\frac{3}{4}$	$\frac{3}{4}$	$1\frac{5}{16}$	$3\frac{9}{16}$
1	1	$1\frac{5}{8}$	$4\frac{1}{8}$
$1\frac{1}{4}$	$1\frac{1}{4}$	$1\frac{5}{8}$	$4\frac{1}{8}$
$1\frac{1}{2}$	$1\frac{1}{4}$	$1\frac{5}{8}$	$4\frac{1}{8}$

All dimensions are given in inches.
For tolerances see Table 1701.
For shank dimensions see Table 1702.

Premium High Speed Steel

Ball End, Two Flute, Medium Helix, Single-End End Mills

With Weldon Shanks

Right hand cutters with right hand helix are standard. Helix angle greater than 19° but not more than 39°.

Standard Sizes and Dimensions

Diameter of Cutter	Diameter of Shank	Length of Cut	Length Overall
$\frac{1}{8}$	$\frac{3}{8}$	$\frac{3}{8}$	$2\frac{5}{16}$
$\frac{3}{16}$	$\frac{3}{8}$	$\frac{1}{2}$	$2\frac{3}{8}$
$\frac{1}{4}$	$\frac{3}{8}$	$\frac{5}{8}$	$2\frac{7}{16}$
$\frac{5}{16}$	$\frac{3}{8}$	$\frac{3}{4}$	$2\frac{1}{2}$
$\frac{3}{8}$	$\frac{3}{8}$	$\frac{3}{4}$	$2\frac{1}{2}$
$\frac{1}{2}$	$\frac{1}{2}$	1	3
$\frac{5}{8}$	$\frac{5}{8}$	$1\frac{3}{8}$	$3\frac{1}{2}$
$\frac{3}{4}$	$\frac{3}{4}$	$1\frac{5}{8}$	$3\frac{7}{8}$
1	1	$2\frac{1}{4}$	$4\frac{3}{4}$
$1\frac{1}{4}$	$1\frac{1}{4}$	$2\frac{1}{2}$	5
$1\frac{1}{2}$	$1\frac{1}{4}$	$2\frac{1}{2}$	5

All dimensions are given in inches.
For tolerances see Table 1701.
For shank dimensions see Table 1702.

Premium High Speed Steel

Multiple Flute, Center Cutting, Medium Helix, Single-End End Mills

With Weldon Shanks

Right hand cutters with right hand helix are standard. Helix angle greater than 19° but not more than 39°.

Standard Sizes and Dimensions

Diameter of Cutter	Diameter of Shank	Length of Cut	Length Overall	Number of Flutes
$\frac{1}{8}$	$\frac{3}{8}$	$\frac{3}{8}$	$2\frac{5}{16}$	4
$\frac{3}{16}$	$\frac{3}{8}$	$\frac{1}{2}$	$2\frac{3}{8}$	4
$\frac{1}{4}$	$\frac{3}{8}$	$\frac{5}{8}$	$2\frac{7}{16}$	4
$\frac{5}{16}$	$\frac{3}{8}$	$\frac{3}{4}$	$2\frac{1}{2}$	4
$\frac{3}{8}$	$\frac{3}{8}$	$\frac{3}{4}$	$2\frac{1}{2}$	4
$\frac{1}{2}$	$\frac{1}{2}$	$1\frac{1}{4}$	$3\frac{1}{4}$	4
$\frac{5}{8}$	$\frac{5}{8}$	$1\frac{5}{8}$	$3\frac{3}{4}$	4
$\frac{3}{4}$	$\frac{3}{4}$	$1\frac{5}{8}$	$3\frac{7}{8}$	4
1	1	2	$4\frac{1}{2}$	4
$1\frac{1}{4}$	$1\frac{1}{4}$	2	$4\frac{1}{2}$	4, 6*
$1\frac{1}{2}$	$1\frac{1}{4}$	2	$4\frac{1}{2}$	4, 6*

*Optional with manufacturer.
All dimensions are given in inches.
For tolerances see Table 1701.
For shank dimensions see Table 1702.

Premium High Speed Steel

Ball End, Multiple Flute, Medium Helix, Single-End End Mills

With Weldon Shanks

Right hand cutters with right hand helix are standard. Helix angle greater than 19° but not more than 39°.

Standard Sizes and Dimensions

Diameter of Cutter	Diameter of Shank	Length of Cut	Length Overall	Number of Flutes
1/8	3/8	3/8	2 5/16	4
3/16	3/8	1/2	2 3/8	4
1/4	3/8	5/8	2 7/16	4
5/16	3/8	3/4	2 1/2	4
3/8	3/8	3/4	2 1/2	4
1/2	1/2	1 1/4	3 1/4	4
5/8	5/8	1 5/8	3 3/4	4
3/4	3/4	1 5/8	3 7/8	4
1	1	2	4 1/2	4
1 1/4	1 1/4	2	4 1/2	4, 6*
1 1/2	1 1/4	2	4 1/2	4, 6*

*Optional with manufacturer.
All dimensions are given in inches.
For tolerances see Table 1701.
For shank dimensions see Table 1702.

Premium High Speed Steel

Long, Multiple Flute, Center Cutting, Medium Helix, Single-End End Mills

With Weldon Shanks

Right hand cutters with right hand helix are standard. Helix angle greater than 19° but not more than 39°.

Standard Sizes and Dimensions

Diameter of Cutter	Diameter of Shank	Length of Cut	Length Overall	Number of Flutes
$\frac{3}{8}$	$\frac{3}{8}$	$1\frac{1}{2}$	$3\frac{1}{4}$	4
$\frac{1}{2}$	$\frac{1}{2}$	2	4	4
$\frac{5}{8}$	$\frac{5}{8}$	$2\frac{1}{2}$	$4\frac{5}{8}$	4
$\frac{3}{4}$	$\frac{3}{4}$	3	$5\frac{1}{4}$	4
1	1	4	$6\frac{1}{2}$	4
$1\frac{1}{4}$	$1\frac{1}{4}$	4	$6\frac{1}{2}$	4, 6*

Premium High Speed Steel

Extra Long, Multiple Flute, Center Cutting, Medium Helix, Single-End End Mills

With Weldon Shanks

Right hand cutters with right hand helix are standard. Helix angle greater than 19° but not more than 39°.

Standard Sizes and Dimensions

Diameter of Cutter	Diameter of Shank	Length of Cut	Length Overall	Number of Flutes
$\frac{3}{8}$	$\frac{3}{8}$	$2\frac{1}{2}$	$4\frac{1}{4}$	4
$\frac{1}{2}$	$\frac{1}{2}$	3	5	4
$\frac{5}{8}$	$\frac{5}{8}$	4	$6\frac{1}{8}$	4
$\frac{3}{4}$	$\frac{3}{4}$	4	$6\frac{1}{4}$	4
1	1	6	$8\frac{1}{2}$	4
$1\frac{1}{4}$	$1\frac{1}{4}$	6	$8\frac{1}{2}$	4, 6*

*Optional with manufacturer.
All dimensions are given in inches.
For tolerances see Table 1701.
For shank dimensions see Table 1702.

Applicable to both tables.

Roughing, Single-End End Mills
Non-Center Cutting
High Speed Steel

Right hand cutters with right hand helix are standard.

Standard Sizes and Dimensions

Diameter		Length		Diameter		Length	
Cutter	Shank	Cut	Overall	Cutter	Shank	Cut	Overall
1/2	1/2	1	3				
1/2	1/2	1 1/4	3 1/4	2	2	2	5 3/4
1/2	1/2	2	4	2	2	3	6 3/4
				2	2	4	7 3/4
				2	2	5	8 3/4
5/8	5/8	1 1/4	3 3/8				
5/8	5/8	1 5/8	3 3/4				
5/8	5/8	2 1/2	4 5/8	2	2	6	9 3/4
				2	2	7	10 3/4
3/4	3/4	1 1/2	3 3/4	2	2	8	11 3/4
3/4	3/4	1 5/8	3 7/8	2	2	10	13 3/4
3/4	3/4	3	5 1/4	2	2	12	15 3/4
1	1	2	4 1/2				
1	1	4	6 1/2	2 1/2	2	4	7 3/4
				2 1/2	2	6	9 3/4
1 1/4	1 1/4	2	4 1/2	2 1/2	2	8	11 3/4
1 1/4	1 1/4	4	6 1/2	2 1/2	2	10	13 3/4
1 1/2	1 1/4	2	4 1/2	3	2 1/2	4	7 3/4
1 1/2	1 1/4	4	6 1/2	3	2 1/2	6	9 3/4
				3	2 1/2	8	11 3/4
1 3/4	1 1/4	2	4 1/2	3	2 1/2	10	13 3/4
1 3/4	1 1/4	4	6 1/2				

Tolerances

Outside Diameter: .... +.025 −.005

Outside Diameter Taper:

Length of Cut	Up to 4 Incl.	Over 4 to 6 Incl.	Over 6 to 12 Incl.
Max. front taper	.002	.003	.004
Max. back taper	.005	.008	.010

Outside Diameter Form Runout: (measured on centers)

Length of Cut	Max. Runout
Up to 4 Inclusive	.005
Over 4 to 12 Inclusive	.006

Length of Cut: + 1/8   − 1/32

All dimensions are given in inches.
For all other element tolerances see Table 1701.
For shank dimensions see Table 1702.

Two Flute, Single-End End Mills
Center and Non-Center Cutting
With Plain Straight Shanks (not flatted)
Solid Carbide

Right hand cutters with right hand helix are standard in the listed diameters, and right hand cutters with straight flutes are standard in diameters through ½″. Center holes optional with manufacturer.

Standard Sizes and Dimensions

Diameter of Cutter	Diameter of Shank	Length of Cut	Length Overall
1⁄16	1⁄8	3⁄16	1½
3⁄32	1⁄8	9⁄32	1½
1⁄8	1⁄8	3⁄8	1½
5⁄32	3⁄16	1⁄2	2
3⁄16	3⁄16	9⁄16	2
7⁄32	1⁄4	5⁄8	2½
1⁄4	1⁄4	3⁄4	2½
9⁄32	5⁄16	3⁄4	2½
5⁄16	5⁄16	13⁄16	2½
3⁄8	3⁄8	7⁄8	2½
7⁄16	7⁄16	1	2¾
1⁄2	1⁄2	1	3
9⁄16	9⁄16	1⅛	3½
5⁄8	5⁄8	1¼	3½
11⁄16	5⁄8	1⅜	4
3⁄4	3⁄4	1½	4

All dimensions are given in inches.
For tolerances see Table 1701.

Ball End, Two Flute, Single-End End Mills

With Plain Straight Shanks (not flatted)

Solid Carbide

Right hand cutters with right hand helix and right hand cutters with straight flutes are standard.

Standard Sizes and Dimensions

Diameter of Cutter and End Circle	Diameter of Shank	Length of Cut	Length Overall
1/8	1/8	3/8	1½
5/32	3/16	1/2	2
3/16	3/16	9/16	2
7/32	1/4	5/8	2½
1/4	1/4	3/4	2½
9/32	5/16	3/4	2½
5/16	5/16	13/16	2½
3/8	3/8	7/8	2½
7/16	7/16	1	2¾
1/2	1/2	1	3

All dimensions are given in inches.
For tolerances see Table 1701.

Four Flute, Single-End End Mills

Center and Non-Center Cutting

With Plain Straight Shanks (not flatted)

Solid Carbide

Right hand cutters with right hand helix are standard in the listed diameters, and right hand cutters with straight flutes are standard in diameters through ½". Center holes optional with manufacturer.

Standard Sizes and Dimensions

Diameter of Cutter	Diameter of Shank	Length of Cut	Length Overall
1/16	1/8	3/16	1½
3/32	1/8	9/32	1½
1/8	1/8	3/8	1½
5/32	3/16	1/2	2
3/16	3/16	9/16	2
7/32	1/4	5/8	2½
1/4	1/4	3/4	2½
9/32	5/16	3/4	2½
5/16	5/16	13/16	2½
3/8	3/8	7/8	2½
7/16	7/16	1	2½
1/2	1/2	1	3
9/16	9/16	1 1/8	3½
5/8	5/8	1 1/4	3½
11/16	5/8	1 3/8	4
3/4	3/4	1 1/2	4

All dimensions are given in inches.
For tolerances see Table 1701.

Ball End, Four Flute, Single-End End Mills
With Plain Straight Shanks (not flatted)
Solid Carbide

Right hand cutters with right hand helix and right hand cutters with straight flutes are standard.

Standard Sizes and Dimensions

Diameter of Cutter and End Circle	Diameter of Shank	Length of Cut	Length Overall
1/8	1/8	3/8	1 1/2
5/32	3/16	1/2	2
3/16	3/16	9/16	2
7/32	1/4	5/8	2 1/2
1/4	1/4	3/4	2 1/2
9/32	5/16	3/4	2 1/2
5/16	5/16	13/16	2 1/2
3/8	3/8	7/8	2 1/2
7/16	7/16	1	2 3/4
1/2	1/2	1	3

All dimensions are given in inches.
For tolerances see Table 1701.

Two Flute, Double-End End Mills

With Weldon Shanks

Solid Carbide Cutting Lengths

Right hand cutters with right hand helix are standard. Center holes optional with manufacturer.

Standard Sizes and Dimensions

Diameter of Cutter	Diameter of Shank	Length of Cut	Length Overall
1/8	3/8	3/8	3 1/16
5/32	3/8	7/16	3 1/8
3/16	3/8	1/2	3 1/4
7/32	3/8	9/16	3 3/8
1/4	3/8	5/8	3 3/8
9/32	3/8	11/16	3 3/8
5/16	3/8	3/4	3 1/2
11/32	3/8	3/4	3 1/2
3/8	3/8	3/4	3 1/2
7/16	1/2	7/8	4
1/2	1/2	1	4

All dimensions are given in inches.
For tolerances see Table 1701.
For shank dimensions see Table 1702.

Four Flute, Double-End End Mills,

With Weldon Shanks

Solid Carbide Cutting Lengths

Right hand cutters with right hand helix are standard. Center holes optional with manufacturer.

Standard Sizes and Dimensions

Diameter of Cutter	Diameter of Shank	Length of Cut	Length Overall
1/8	3/8	3/8	3 1/16
5/32	3/8	7/16	3 1/8
3/16	3/8	1/2	3 1/4
7/32	3/8	9/16	3 3/8
1/4	3/8	5/8	3 3/8
9/32	3/8	11/16	3 3/8
5/16	3/8	3/4	3 1/2
11/32	3/8	3/4	3 1/2
3/8	3/8	3/4	3 1/2
7/16	1/2	7/8	4
1/2	1/2	1	4

All dimensions are given in inches.
For tolerances see Table 1701.
For shank dimensions see Table 1702.

Single-End End Mills

With Weldon Shanks

Helical Carbide Tipped
Non-Center Cutting

Standard Sizes and Dimensions

Diameter of Cutter	Diameter of Shank	Length of Cut (Nominal*)	Length Overall
½	½	1	3
9⁄16	½	1	3
5⁄8	5⁄8	1¼	3⅜
11⁄16	5⁄8	1¼	3⅜
¾	5⁄8	1¼	3⅜
13⁄16	5⁄8	1½	3⅝
7⁄8	7⁄8	1½	3¾
15⁄16	7⁄8	1½	3¾
1	1	1½	4
1 ⅛	1	1¾	4¼
1 ¼	1	1¾	4¼
1 ⅜	1	1¾	4¼
1 ½	1¼	2	4½
1 5⁄8	1¼	2	4½
1 ¾	1¼	2	4½
1 ⅞	1¼	2	4½
2	1¼	2	4½

*Minimum length of cut is 1/16" less than nominal.

All dimensions are given in inches.

For tolerances see Table 1701.

For shank dimensions see Table 1702.

Single-End End Mills

With Weldon Shanks
Carbide Tipped
Regular Length
Non-Center Cutting

Standard Sizes and Dimensions

Diameter of Cutter	Diameter of Shank	Length of Cut (Nominal*)	Length Overall
1/4	3/8	1/2	2 1/2
5/16	3/8	5/8	2 1/2
3/8	3/8	5/8	2 1/2
7/16	3/8	1	2 11/16
1/2	1/2	1	3
9/16	1/2	1	3 1/4
5/8	1/2	1	3 3/8
5/8	5/8	1	3 1/2
3/4	5/8	1	3 5/8
7/8	5/8	1 1/4	4
7/8	3/4	1 1/4	4
1	7/8	1 1/4	4
1	1	1 1/4	4
1 1/8	1	1 1/4	4 1/4
1 1/4	1	1 1/4	4 1/4
1 1/2	1 1/4	1 1/2	4 1/2
1 3/4	1 1/4	1 1/2	4 1/2
2	1 1/4	1 1/2	4 1/2

*Minimum length of cut is 1/32 less than nominal.

All dimensions are given in inches.

For tolerances see Table 1701.

For shank dimensions see Table 1702.

Single-End End Mills

With Weldon Shanks

Carbide Tipped

Three Straight Flutes

Center Cutting

Standard Sizes and Dimensions

Diameter of Cutter	Diameter of Shank	Minimum Length of Cut	Length Overall
$\frac{3}{8}$	$\frac{3}{8}$	$\frac{1}{2}$	$2\frac{1}{2}$
$\frac{7}{16}$	$\frac{3}{8}$	$\frac{11}{16}$	$2\frac{1}{2}$
$\frac{1}{2}$	$\frac{1}{2}$	$\frac{11}{16}$	3
$\frac{9}{16}$	$\frac{1}{2}$	$\frac{11}{16}$	3
$\frac{5}{8}$	$\frac{5}{8}$	$\frac{11}{16}$	$3\frac{1}{4}$
$\frac{3}{4}$	$\frac{5}{8}$	$\frac{11}{16}$	$3\frac{3}{8}$

All dimensions are given in inches.
For tolerances see Table 1701
For shank dimensions see Table 1702

TABLE   401

Standard Milling Cutter Tolerances[†]

Note: Additional tolerances may also be
shown on specific pages.

Outside Diameter Tolerances

Plain Milling Cutters (All Types) Side Milling Cutters Staggered Tooth Side Milling Cutters Half Side Milling Cutters H S S Metal Slitting Saws, All Types Screw Slotting Cutters Single and Double Angle Cutters	+.015	−.015
Plain Milling Cutters Carbide Side Milling Cutters Slitting Saws Tipped	+$\frac{1}{16}$	−$\frac{1}{16}$
Concave and Convex Cutters Corner Rounding Cutters—(Arbor Type) Gear and Sprocket Cutters	+$\frac{1}{16}$	−$\frac{1}{16}$
Corner Rounding Cutters — Shank Type........	+.010	−.010
Shell Mills.......... HSS................	+$\frac{1}{64}$	−0
Carbide Tipped........	+$\frac{1}{16}$	−0
T- Slot Cutters.............................	+.000	−.010
Woodruff Keyseat Cutters $\frac{1}{4}$ — $\frac{3}{4}$ incl. Shank Type, HSS $\frac{7}{8}$ — 1$\frac{1}{8}$ incl. 1$\frac{1}{4}$ — 1$\frac{1}{2}$ incl.	+.010 +.012 +.015	+.015 +.017 +.020
Shank Type, $\frac{3}{8}$ — $\frac{3}{4}$ incl. Solid Carbide $\frac{7}{8}$ — 1 incl.	+.010 +.012	+.015 +.017
Woodruff Keyseat Cutters, Arbor Type........	+.002	−.002
Grinding Tools, Solid Carbide................	+.000	−.005

TABLE 401 (Continued)

Shank Diameter Tolerances

All Cutters with Straight Shanks (For Weldon Shanks see Table 405)	−.0001	−.0005

Neck Diameter Tolerances

T Slot Cutters..............................	+.000	−.005

Length of Cut or Width of Cutting Face Tolerances

Plain Milling Cutters — HSS	Up to 1″ face	+.001	−.001
	Over 1″ to 2″ incl.	+.010	−.000
	Over 2″ to 8″ incl.	+.020	−.000
Plain Milling Cutters, Carbide Tipped		+½₂	−½₂
Side Milling Cutters.........................		+.002	−.001
Staggered Tooth Side Milling Cutters	Up to ¾″ face incl.	+.0000	−.0005
	Over ¾″ to 1″ incl.	+.0000	−.0010
Half Side Milling Cutters.....................		+.015	−.000
Metal Slitting Saws and Screw Slotting Cutters HSS		+.001	−.001
Slitting Saws, Carbide Tipped		+.001	−.001
Single and Double Angle Cutters..............		+.015	−.015
T- Slot Cutters.............................		+.000	−.005
Woodruff Keyseat Cutters, Shank Type	⅟₁₆ to ⁵⁄₃₂ face incl.	+.0000	−.0005
	³⁄₁₆ to ⁷⁄₃₂ face incl.	−.0002	−.0007
	¼	−.0003	−.0008
	⁵⁄₁₆	−.0004	−.0009
	⅜	−.0005	−.0010
Woodruff Keyseat Cutters, Arbor Type	³⁄₁₆ face	−.0002	−.0007
	¼ face	−.0003	−.0008
	⁵⁄₁₆ face	−.0004	−.0009
	⅜ face and over	−.0005	−.0010
Roller Chain Sprocket Milling Cutters.........		+.010	−.000
Concave and Corner Rounding Cutters, Arbor Type.................................		+.010	−.010
Grinding Tools, Solid Carbide Shank Type		+ ⅟₆₄	− ⅟₆₄
Arbor Type................................		+.005	−.005

(Concluded on following page)

TABLE 401 (Concluded)

Diameter of Circle, Tolerances

Convex Cutters.............................		+.002	−.002
Concave Cutters {Up to $\frac{1}{16}$″ dia. of circle incl.		+.002	−.001
{Over $\frac{1}{16}$″ dia. of circle		+.004	−.002
Corner Rounding {Up to $\frac{1}{8}$″ radius of circle incl.		+.001	−.001
Cutters {Over $\frac{1}{8}$″ radius of circle		+.002	−.001

Angle Tolerances

Single Angle Cutters..........................	+10 min.	−10 min.
Double Angle Cutters, Half Angle..............	+10 min.	−10 min.

Length Overall Tolerances

Shank Type	+$\frac{1}{16}$	−$\frac{1}{16}$

Length of Neck Tolerances

Grinding Tools, Shank Type, Solid Carbide..............................	+$\frac{1}{64}$	−$\frac{1}{64}$

Hole Diameter Tolerances

All cutters except Shell Mills & Screw Slotting Cutters		
Thru 1″ hole diameter, { HSS..............	+.00075	−.0000
if not over 3″ long { Carbide Tipped....	+.0010	−.0000
Thru 1″ hole diameter, if over 3″ long.........	+.0010	−.0000
Over 1″ thru 2″ hole diameter................	+.0010	−.0000
Shell Mills...............{ HSS..............	+.0005	−.0000
{ Carbide Tipped....	+.0010	−.0000
Screw Slotting Cutters.......................	+.0010	−.0000
Grinding Tools, Solid Carbide	+.0005	−.0000

†This table is not applicable to Roughers. See page 638 for Rougher tolerances.

TABLE 402

Radial and Axial Runout for
American National Standard Milling Cutters

All figures are Total Indicator Variation (tiv)

Section I

Type of Cutter	Nominal Diameter		MAX. RADIAL RUNOUT (when rotated on a stationary shoulder test arbor)		MAX. AXIAL RUNOUT (when rotated on a stationary shoulder test arbor)	
	Over	Thru	Form Relieved	Profile Sharpened	Form Relieved	Profile Sharpened
Arbor Type Cutters: Side Mills, Shell Mills, Keyseat, Angular, Gear, Convex, Concave, Corner Rounding	2, 5	2, 5, 8½	.003 .004 .005	.002 .003 .004	.003 .004 .005	.002 .003 .004
Plain Mills	5	5 8		.003 .004		.002 .003
Metal Slitting Saws	5	5 12		.003 .004	(see Section II)	(see Section II)
Screw Slotting Cutters	All Sizes			.005	(see Section II)	(see Section II)

(Concluded on following page)

TABLE 402 (Concluded)

Section II
Axial Runout of Metal Slitting Saws and Screw Slotting Cutters

Type of Cutter	Nominal Diameter		MAX. AXIAL RUNOUT (When clamped between Standard Diameter Collars on test arbor) NOMINAL FACE WIDTH					
	Over	Thru	thru 1/32	over 1/32 thru 1/16	over 1/16 thru 1/8	over 1/8 thru 3/16	over 3/16 thru 1/4	over 1/4 thru 5/16
Metal Slitting Saws and Screw Slotting Cutters		3	.004	.003	.003	.002		
	3	4	.005	.005	.004	.003		
	4	6		.007	.006	.005	.004	
	6	8			.008	.006	.005	
	8	12				.007	.006	.005

Section III
Radial Runout of Shank Type Cutters

Type of Cutter	Flute Length		MAX. RADIAL RUNOUT OF PERIPHERY WITH RESPECT TO SHANK (Shank rotated in Vee Block)
	Over	Thru	
Shank Type Cutters		2	.002
	2	4	.003
	4	8	.004

SECTION IV
Axial Runout of Shank Type Cutters

Type of Cutter	Diameter		MAX. AXIAL RUNOUT
	Over	Thru	
Keyseat Cutters	All	All	.001
All Others		1	.002
	1	2½	.003

All dimensions are given in inches.

Table 403

Multiple Thread Milling Cutter Tolerances

4 to 80 Threads per Inch. Topping and Non Topping.

Classes of Cutters

AT — PRECISION GROUND
BT — COMMERCIAL GROUND
CT — ACCURATE UNGROUND
DT — COMMERCIAL UNGROUND

NOTE 1 — The finest tolerance in items 2, 3, 4, 5 or 6 determines the cutter's classification.

NOTE 2 — A tolerance value for items 2, 3, 4, 5 or 6 which falls between any two classes must be reduced to the tolerance of the more accurate class for purposes of classification.

NOTE 3 — For arbor type cutters total indicator variation (tiv) when rotated on a stationary shoulder mandrel.
For shank type cutters the total indicator variation (tiv) when rotated between centers.

	HOLE SIZE	Class	Under 1"	1" to 2" inc.	Over 2" to 3" inc.
1	HOLE DIAMETER TOLERANCE (Plus only)	AT	.0002	.0003	.0004
		BT	.0003	.0004	.0005
		CT	.0004	.0005	.0006
		DT	.0005	.0006	.0007

	CUTTER DIAMETER	Class	Under 2"	2" to 4" inc.	Over 4" to 6" inc.
2	OUTSIDE DIAMETER RADIAL RUNOUT (See Note 3)	AT	.0006	.0008	.0010
		BT	.0012	.0013	.0015
		CT	.0015	.0020	.0025
		DT	.0020	.0025	.0030
3	AXIAL RUNOUT OF FORM (See Note 3)	AT	.0003	.0004	.0005
		BT	.0004	.0006	.0008
		CT	.0008	.0012	.0016
		DT	.0012	.0016	.0020

(Concluded on following page)

Table 403 (Concluded)

Multiple Thread Milling Cutter Tolerances

4 to 80 Threads per Inch. Topping and Non Topping.

Threads per Inch (American National and Unified Forms only)	Class	4.000 to & inc. 6	6.001 to & inc. 12	12.001 to & inc. 24	24.001 to & inc. 36	36.001 to & inc. 48	48.001 to & inc. 60	60.001 to & inc. 80
Flank Length for Special Forms — Inches		.187 to .1250 inc.	.1249 to .0625	.0624 to .0312	.0311 to .0208	.0207 to .0156	.0155 to .0125	.0124 to .0094 inc.

4 — Maximum Error in Form Half Angle, Minutes (plus or minus)

Class	4.000 to & inc. 6	6.001 to & inc. 12	12.001 to & inc. 24	24.001 to & inc. 36	36.001 to & inc. 48	48.001 to & inc. 60	60.001 to & inc. 80
AT	5	8	12	18	25	30	35
BT	8	10	15	25	35	40	50
CT	10	15	20	30	40	··	··
DT	15	20	30	45	55	··	··

5 — Total Diametral Variation from Parallelism or Desired Taper — In Face Width (Axial Length of Relieved Portion) of

Class	In Any Inch	1" or less	Over 1" to & inc. 2"	Over 2" to & inc. 3"	Over 3" to & inc. 4"	Over 4" to & inc. 5"	Over 5" to 6" inc.
AT	.0003	.0003	.0005	.0007	.0009	.0011	.0013
BT	.0005	.0005	.0008	.0011	.0014	.0017	.0020
CT	.0010	.0010	.0015	.0020	.0025	.0030	.0035
DT	.0015	.0015	.0020	.0025	.0030	.0035	.0040

6 — Maximum Variation from Basic Thread Pitch (Plus or minus) — In Axial Length of Thread Section

Class	In Any Inch	1" or less	Over 1" to & inc. 2"	Over 2" to & inc. 3"	Over 3" to & inc. 4"	Over 4" to & inc. 5"	Over 5" to 6" inc.
AT	.0003	.0003	.0005	.0007	.0009	.0011	.0013
BT	.0004	.0004	.0008	.0010	.0012	.0014	.0016
CT	.0005	.0005	.0010	.0015	.0020	.0025	.0030
DT	.0010	.0010	.0020	.0025	.0030	.0035	.0040

TABLE 404

Standard Keyways for Cutters

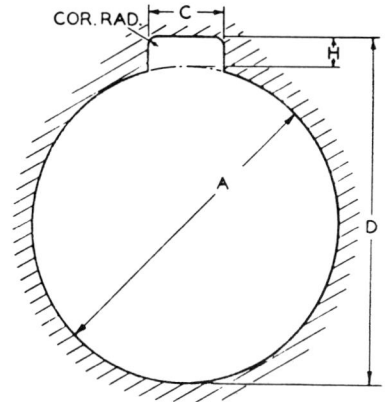

COR. RAD.

Cutter Hole and Keyway

Cutter Hole "A"	Nominal Size Key (Square)	C		D		H Nominal	Corner Radius
		Maximum	Minimum	Maximum	Minimum		
$\frac{1}{2}$	$\frac{3}{32}$	.106	.099	.5678	.5578	$\frac{3}{64}$	.020
$\frac{5}{8}$	$\frac{1}{8}$	.137	.130	.7085	.6985	$\frac{1}{16}$	$\frac{1}{32}$
$\frac{3}{4}$	$\frac{1}{8}$	.137	.130	.8325	.8225	$\frac{1}{16}$	$\frac{1}{32}$
$\frac{7}{8}$	$\frac{1}{8}$	.137	.130	.9575	.9475	$\frac{1}{16}$	$\frac{1}{32}$
1	$\frac{1}{4}$	.262	.255	1.1140	1.1040	$\frac{3}{32}$	$\frac{3}{64}$
$1\frac{1}{4}$	$\frac{5}{16}$	.343	.318	1.3950	1.3850	$\frac{1}{8}$	$\frac{1}{16}$
$1\frac{1}{2}$	$\frac{3}{8}$	.410	.385	1.6760	1.6660	$\frac{5}{32}$	$\frac{1}{16}$
$1\frac{3}{4}$	$\frac{7}{16}$	.473	.448	1.9580	1.9480	$\frac{3}{16}$	$\frac{1}{16}$
2	$\frac{1}{2}$	.535	.510	2.2080	2.1980	$\frac{3}{16}$	$\frac{1}{16}$
$2\frac{1}{2}$	$\frac{5}{8}$	.660	.635	2.7430	2.7330	$\frac{7}{32}$	$\frac{1}{16}$
3	$\frac{3}{4}$	.785	.760	3.2750	3.2650	$\frac{1}{4}$	$\frac{3}{32}$
$3\frac{1}{2}$	$\frac{7}{8}$	.910	.885	3.9000	3.8900	$\frac{3}{8}$	$\frac{3}{32}$
4	1	1.035	1.010	4.4000	4.3900	$\frac{3}{8}$	$\frac{3}{32}$
$4\frac{1}{2}$	$1\frac{1}{8}$	1.160	1.135	4.9630	4.9530	$\frac{7}{16}$	$\frac{1}{8}$
5	$1\frac{1}{4}$	1.285	1.260	5.5250	5.5150	$\frac{1}{2}$	$\frac{1}{8}$

All dimensions are given in inches.
For intermediate size hole use the keyway for the next larger size hole listed.

TABLE 1701
Standard End Mill & Router Tolerances

Note: Additional tolerances may also be shown
on specific pages.

Outside Diameter Tolerances HSS

HELICAL SERIES

Tolerances

Multiple Flute................................	+.005	−.000
2 Flute, Keyway Cutting......................	+.000	−.0015

MEDIUM HELIX SERIES

Double End, 3/16 S.S.
*4 Flute....................................	+.003	−.000
2 Flute, Stub and Regular Length.............	+.000	−.0015
*2 Flute, Long Length........................	+.003	−.000

Single End, Weldon Shank
Multiple Flute.............................	+.003	−.000
2 Flute, Stub Length........................	+.000	−.0015
2 Flute, Regular and Long Length.............	+.003	−.000

Double End, Weldon Shank
*4 Flute....................................	+.003	−.000
2 Flute.....................................	+.000	−.0015

Center Cutting, Single End, Weldon Shank.......
3 & 4 Flute.................................	+.003	−.000

Center Cutting, Double End, Weldon Shank
3 & 4 Flute.................................	+.000	−.0015

Heavy Duty, Single End, 2″, 2½″ Shank
2, 3, 4, 6 & 8 Flute........................	+.005	−.000

HIGH HELIX SERIES

Single End Weldon Shank
2 Flute.....................................	+.003	−.000

PREMIUM HIGH SPEED STEEL SERIES

Center Cutting, Single End, Weldon Shank
2 Flute and Multiple Flute..................	+.003	−.000

---

* If the shank is the same diameter as the cutting portion on Double End, HSS straight shank end mills, and where the tolerance is normally positive, the tolerance on the cutting diameter becomes +.0000 −.0025.

(Continued)

TABLE 1701 (Continued)

Outside Diameter Tolerances—CARBIDE

CARBIDE SERIES		
Solid Carbide		
Up to 1/4″ Dia., Incl.	+.000	−.002
Over 1/4″ to 3/4″ Dia. Incl.	+.000	−.003
Carbide Tipped	+.005	−.0015
ROUTERS, SOLID CARBIDE		
Single End, 1-2-3 Flutes	+.000	−.003
Single End, Diamond Pattern	+.000	−.005

Length of Cut Tolerances—HSS

END MILLS, HSS	Tolerances	
All except Miniature and Heavy Duty	+1/32	−1/32
Miniature	+1/32	−1/64
Heavy Duty, 2″, 2 1/2″ Shank	+1/16	−1/16

Length of Cut Tolerances—CARBIDE

END MILLS, CARBIDE		
Solid Carbide		
Up to 3/16″ Dia. Incl.	+1/32	−1/32
Over 3/16″ to 3/4″ Dia. Incl.	+1/16	−1/16
Carbide Tipped	see individual specification pages	
ROUTERS, SOLID CARBIDE		
Up to 3/16″ Dia. Incl.	+1/32	−1/32
Over 3/16″ to 1/2″ Dia. Incl.	+1/16	−1/16

(Continued)

TABLE 1701 (Concluded)

Shank Diameter Tolerances HSS and CARBIDE

STRAIGHT SHANKS (Including Weldon Shanks) —.0001 —.0005

(See Tables 1702, 1703, 1704 for Design features and
tolerances for Weldon and Combination Shanks)

Length Overall Tolerances HSS and CARBIDE

End Mills and Routers

HSS, Except Heavy Duty 3″ O.D.	$+\frac{1}{16}$	$-\frac{1}{16}$
HSS, Heavy Duty, 3″ O. D.	$+\frac{1}{8}$	$-\frac{1}{8}$
Carbide, Solid.............................	$+\frac{1}{16}$	$-\frac{1}{16}$
Carbide Tipped............................	$+\frac{1}{8}$	$-\frac{1}{8}$

End Mill Radial Runout Tolerances HSS and CARBIDE

Length of Cut		Maximum Radial Runout of Periphery with respect to Shank
Over	Thru	
—	2	.002
2	4	.003
4	12	.004

Axial Runout of End Mill Teeth HSS and CARBIDE

Diameter		Maximum Axial Runout
Over	Thru	
—	1	.002
1	3	.003

TABLE 1702
Weldon Shanks

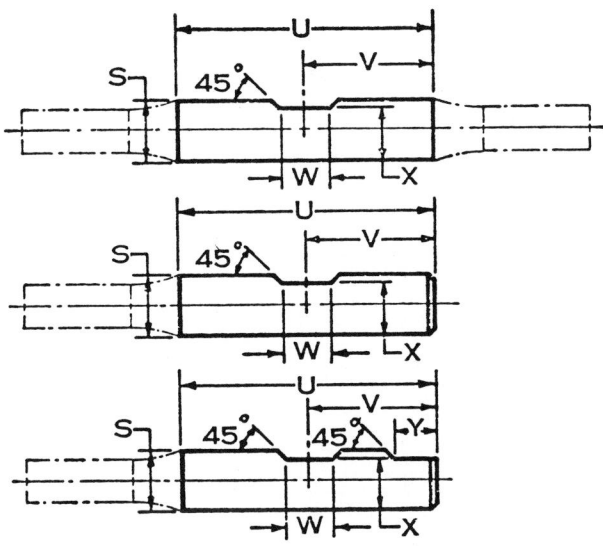

Dimensions

Diameter of Shank S	Length of Shank U	V	W		X	Y
			minimum	maximum		
$\frac{3}{8}$	$1\frac{9}{16}$	$\frac{25}{32}$	0.280	0.282	0.325	..
$\frac{1}{2}$	$1\frac{25}{32}$	$\frac{57}{64}$	0.330	0.332	0.440	..
$\frac{5}{8}$	$1\frac{29}{32}$	$\frac{61}{64}$	0.400	0.402	0.560	..
$\frac{3}{4}$	$2\frac{1}{32}$	$1\frac{1}{64}$	0.455	0.457	0.675	..
$\frac{7}{8}$	$2\frac{1}{32}$	$1\frac{1}{64}$	0.455	0.457	0.810	$\frac{1}{2}$
1	$2\frac{9}{32}$	$1\frac{9}{64}$	0.515	0.517	0.925	$\frac{1}{2}$
$1\frac{1}{4}$	$2\frac{9}{32}$	$1\frac{9}{64}$	0.515	0.517	1.156	$\frac{1}{2}$
$1\frac{1}{2}$	$2\frac{11}{16}$	$1\frac{3}{16}$	0.515	0.517	1.406	$\frac{9}{16}$
2	$3\frac{1}{4}$	$1\frac{27}{32}$	0.700	0.702	1.900	$\frac{27}{32}$
$2\frac{1}{2}$	$3\frac{1}{2}$	$1\frac{15}{16}$	0.700	0.702	2.400	$\frac{27}{32}$

All dimensions are given in inches.

Tolerances

Element	Range	Direction	Tolerance
Diameter of Shank, S	All Sizes	Minus	.0001 to .0005
Length of Shank, U	All Sizes	Plus or Minus	$\frac{1}{32}$
Dimension, V	All Sizes	Plus or Minus	$\frac{1}{64}$
Dimension, X	All Sizes	Minus	$\frac{1}{64}$
Dimension, Y	$\frac{7}{8}$" to $2\frac{1}{2}$" incl.	Plus or Minus	$\frac{1}{32}$

All dimensions are given in inches.

TABLE 1703

Combination Shanks*
For End Mills, Right Hand Cut

*Modified for use as a Weldon or a Pin Drive Shank.

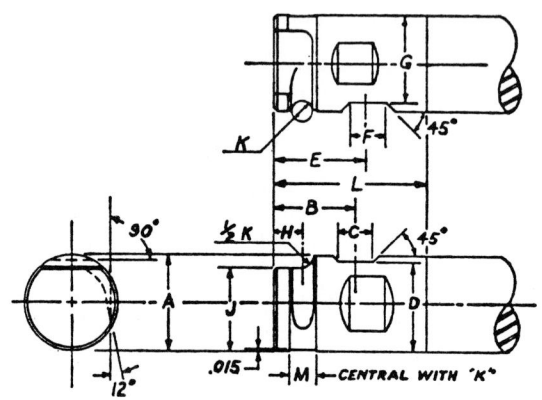

Dimensions

Dia. of Shank A	Length of Shank L	B	C	D	E	F	G	H	J	K	M
1½	2¹¹⁄₁₆	1 ³⁄₁₆	.515	1.406	1½	.515	1.371	⁹⁄₁₆	1.302	.377	⁷⁄₁₆
2	3 ¼	1²³⁄₃₂	.700	1.900	1¾	.700	1.809	⅝	1.772	.440	½
2½	3 ½	1¹⁵⁄₁₆	.700	2.400	2	.700	2.312	¾	2.245	.503	⁹⁄₁₆

Tolerances

Element	Direction	Tolerance
Diameter of Shank, A	Minus	.0001 to .0005
Length of Shank, L	Plus or Minus	¹⁄₃₂
Dimension, B	Plus or Minus	¹⁄₆₄
Dimension, C	Plus	.002
Dimension, D	Minus	¹⁄₆₄
Dimension, E	Plus or Minus	¹⁄₆₄
Dimension, F	Plus or Minus	.005
Dimension, G	Minus	¹⁄₆₄
Dimension, H	Plus	¹⁄₆₄
Dimension, J	Plus or Minus	.002
Dimension, K	Plus	.003

All dimensions are given in inches.

TABLE 1704
Combination Shanks*
For End Mills, Left Hand Cut

*Modified for use as a Weldon or a Pin Drive Shank.

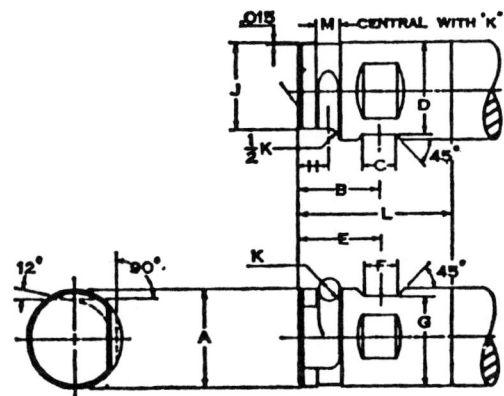

Dimensions

Dia. of Shank A	Length of Shank L	B	C	D	E	F	G	H	J	K	M
1½	2¹¹⁄₁₆	1 ³⁄₁₆	.515	1.406	1½	.515	1.371	⁹⁄₁₆	1.302	.377	⁷⁄₁₆
2	3 ¼	1²³⁄₃₂	.700	1.900	1¾	.700	1.809	⅝	1.772	.440	½
2½	3 ½	1¹⁵⁄₁₆	.700	2.400	2	.700	2.312	¾	2.245	.503	⁹⁄₁₆

Tolerances

Element	Direction	Tolerance
Diameter of Shank, A	Minus	.0001 to .0005
Length of Shank, L	Plus or Minus	¹⁄₃₂
Dimension, B	Plus or Minus	¹⁄₆₄
Dimension, C	Plus	.002
Dimension, D	Minus	¹⁄₆₄
Dimension, E	Plus or Minus	¹⁄₆₄
Dimension, F	Plus or Minus	.005
Dimension, G	Minus	¹⁄₆₄
Dimension, H	Plus	¹⁄₆₄
Dimension, J	Plus or Minus	.002
Dimension, K	Plus	.003

All dimensions are given in inches.

TABLE 1705

Morse Tapers

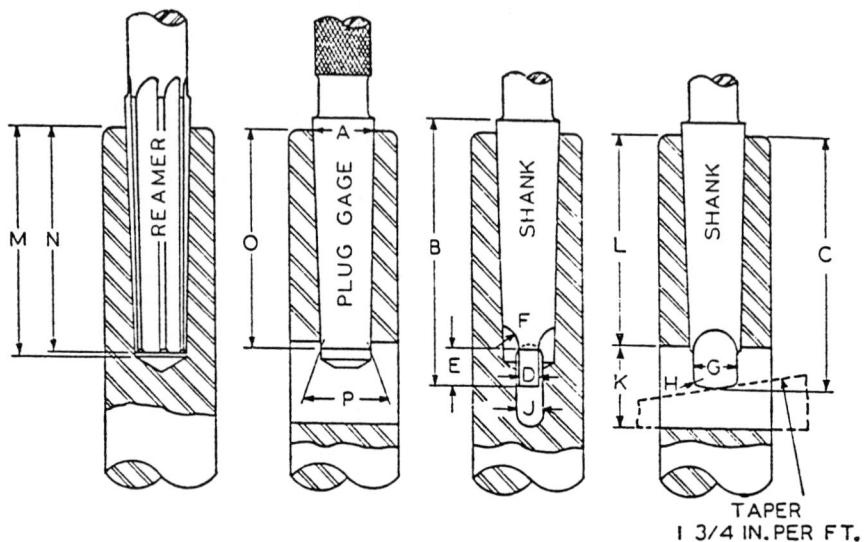

TAPER
I 3/4 IN. PER FT.

(Concluded on following page)

TABLE 1705
Morse Tapers

Number of Taper	P — Diam. of Plug at Small End	A — Diam. at End of Socket	SHANK B — Whole Length	SHANK C — Depth	SOCKET L — End of Socket to Tang Slot	SOCKET M — Depth of Drilled Hole	SOCKET N — Depth of Reamed Hole	SOCKET O — Standard Plug Depth	TANG D — Thickness	TANG E — Length	TANG F — Radius	TANG G — Diameter	TANG H — Radius	TANG SLOT J — Width	TANG SLOT K — Length	Taper per Inch	Taper per Foot
0	.25200	.35610	2 11/32	2 7/32	1 15/16	2 1/16	2 1/2	2	.156	1/4	5/32	15/64	3/64	.172	9/16	.052050	.62460
1	.36900	.47500	2 9/16	2 7/16	2 1/16	2 3/16	2 5/32	2 1/8	.203	3/8	3/16	11/32	3/64	.218	3/4	.049882	.59858
2	.57200	.70000	3 1/8	2 15/16	2 1/2	2 21/32	2 29/64	2 9/16	.250	7/16	1/4	17/32	1/16	.266	7/8	.049951	.59941
3	.77800	.93800	3 7/8	3 11/16	3 1/16	3 5/16	3 1/4	3 3/16	.312	9/16	9/32	23/32	5/64	.328	1 3/16	.050196	.60235
4	1.02000	1.23100	4 7/8	4 5/8	3 7/8	4 3/16	4 1/8	4 1/16	.469	5/8	5/16	31/32	3/32	.484	1 1/4	.051938	.62326
4½	1.26600	1.50000	5 3/8	5 1/8	4 5/16	4 5/8	4 9/16	4 1/2	.562	11/16	3/8	1 13/64	1/8	.578	1 3/8	.052000	.62400
5	1.47500	1.74800	6 1/8	5 7/8	4 15/16	5 5/16	5 1/4	5 3/16	.625	3/4	3/8	1 15/32	1/8	.656	1 1/2	.052626	.63151
6	2.11600	2.49400	8 9/16	8 1/4	7	7 13/16	7 21/64	7 1/4	.750	1 1/8	1/2	2	5/32	.781	1 3/4	.052138	.62565
7	2.75000	3.27000	11 5/8	11 1/4	9 1/2	10 5/32	10 5/64	10	1.125	1 3/8	3/4	2 5/8	3/16	1.156	2 5/8	.052000	.62400

The undercut shown on the tang having diameter G, and length E, may be eliminated at the option of the manufacturer provided the tang is heat-treated to a minimum hardness of Rockwell C30.

TOLERANCES ON RATE OF TAPER, all sizes 0.002 per foot. This tolerance may be applied on shanks only in the direction which increases the rate of taper and on sockets only in the direction which decreases the rate of taper.

TABLE 1706

Brown & Sharpe Tapers

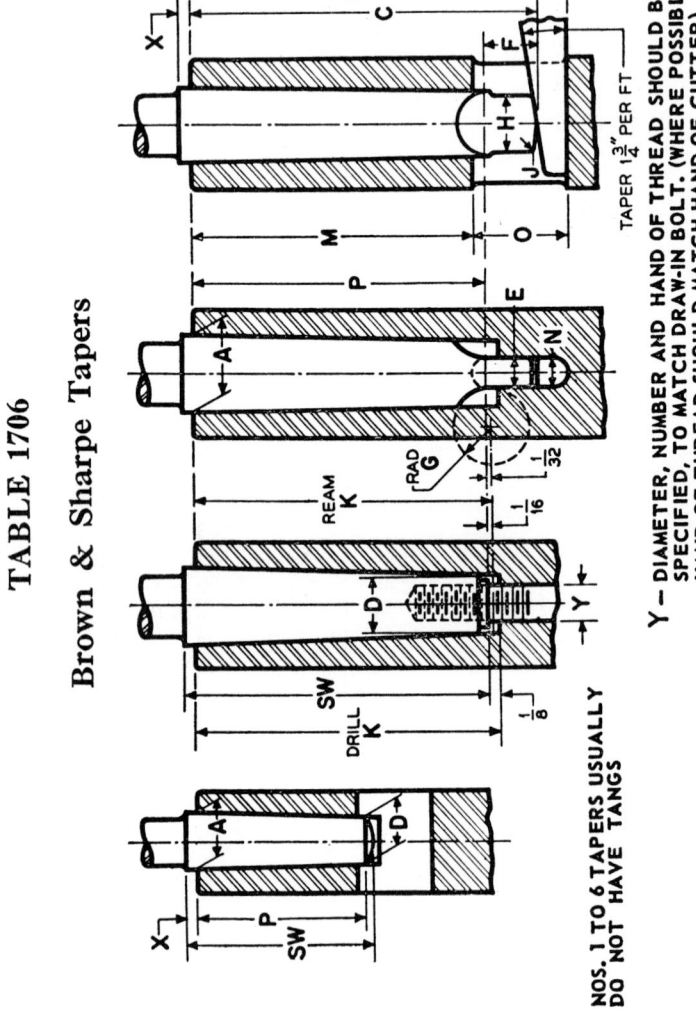

TAPER $1\frac{3}{4}''$ PER FT

Y — DIAMETER, NUMBER AND HAND OF THREAD SHOULD BE SPECIFIED, TO MATCH DRAW-IN BOLT. (WHERE POSSIBLE, HAND OF THREAD SHOULD MATCH HAND OF CUTTER)

NOS. 1 TO 6 TAPERS USUALLY DO NOT HAVE TANGS

TABLE 1706
Brown & Sharpe Tapers

Number of Taper	D Dia. of Plug At Small End	A Dia. at End of Socket	SHANK — B Total Length of Shank With Tang	SHANK — SW Shank Length Without Tang	SHANK — C Shank Depth	SHANK — X Shank Projects From End of Socket	SOCKET — M End of Socket To Tang Slot	SOCKET — K Min. Depth of Tapered Hole (Drilled)	SOCKET — K Min. Depth of Tapered Hole (Reamed)	P Plug Depth	TANG — E Thickness	TANG — F Length	TANG — G Radius of Mill	TANG — H Diameter	TANG — J Radius	TANG SLOT — N Width	TANG SLOT — O Length	TANG SLOT — R Shank End To Back of Tang Slot	Taper Per Foot	Taper Per Inch
1	.20000	.23922	1 9/16	1 1/16	1 3/16	3/32	15/16	1 1/16	1 1/4	15/16	.125	3/16	3/16	.170	.030	.141	3/8	1/8	.50200	.041833
2	.25000	.29968	1 19/32	1 11/32	1 1/2	3/32	1 11/64	1	1 1/4	1 3/16	.156	1/4	3/16	.220	.030	.172	1/2	1/64	.50200	.041833
3	.31250	.37525	1 31/32	1 41/64	1 13/32	3/32	1 15/32	1	1 1/4	1	.188	1/4	5/16	.282	.040	.203	5/8	7/32	.50200	.041833
4	.3500	.4023	1 7/8	1 29/64	1 21/32	3/32	1 19/64	1	1 13/16	1 1/2	.218	5/16	5/16	.320	.050	.228	11/16	15/64	.50240	.041867
5	.4500	.5232	2 7/32	1 25/64	1 15/16	3/32	2 1/64	2	2 1/8	1 3/4	.250	3/8	5/16	.420	.060	.260	3/4	1/4	.50160	.041800
6	.5000	.5996	2 3/16	2 17/32	2 1/8	3/32	2 1/64	2 1/2	2 7/8	2 1/8	.281	3/8	5/16	.460	.060	.291	7/8	19/64	.50329	.041941
7	.6000	.7254	3 5/8	3 3/8	3 17/32	1/8	2 29/64	3 3/16	3 3/16	3	.312	1/2	3/8	.560	.070	.322	15/16	5/16	.50147	.041789
8	.7500	.8987	4 1/4	3 11/16	4 1/8	1/8	3 29/64	3 11/16	3 5/8	3 1/16	.343	1/2	3/8	.710	.080	.353	1	21/64	.50100	.041750
9	.9001	1.0671	4 1/2	4 1/8	4 5/8	1/8	4 1/64	4 1/8	4 1/16	4	.375	9/16	7/16	.860	.100	.385	1 1/16	3/8	.50085	.041738
10	1.04465	1.2893	6 17/32	5 13/16	6 13/32	1/8	5 17/64	5 13/16	5 3/4	5 11/16	.437	5/8	1/2	1.010	.110	.447	1 5/16	7/16	.51612	.043010
11	1.24995	1.5318	7 19/64	6 7/8	7 15/16	1/8	6 19/64	7	6 3/4	6 1/4	.437	3/4	1/2	1.210	.130	.510	1 5/16	7/16	.50100	.041750
12	1.5001	1.7968	8 1/16	7 1/4	7 15/16	1/8	6 15/16	7 1/4	7 5/8	7 5/8	.500	3/4	1/2	1.460	.150	.510	1 7/8	1/2	.49973	.041644
13	1.75005	2.0731	8 11/16	7 7/8	8 9/16	1/8	7 9/16	7 7/8	7 13/16	7 3/4	.500	3/4	5/8	1.710	.170	.510	1 1/4	3/8	.50020	.041683
14	2.0000	2.3438	9 9/32	8 3/8	9 9/32	1/8	8 1/32	8 3/8	8 5/16	8	.562	27/32	3/4	1.960	.190	.572	1 11/16	9/16	.50000	.041666
15	2.2500	2.6146	9 9/32	8 7/8	9 25/32	1/8	8 17/32	8 7/8	9 5/16	9	.562	27/32	7/8	2.210	.210	.572	1 11/16	9/16	.50000	.041666
16	2.5000	2.8854	10 3/8	9 3/8	10 1/4	1/8	9	9 7/8	9 5/16	9	.625	15/16	1	2.450	.230	.635	1 7/8	7/8	.50000	.041666
17	2.7500	3.1563						10 3/8	10 5/16	10 1/4									.50000	.041666
18	3.0000	3.4271																	.5000	.041666

HOB SECTION

HOB SECTION CONTENTS

THE HOBBING PROCESS

The hobbing process involves a rotating workpiece, a rotating cutting tool with teeth arranged in a helical thread and a machine which maintains a timed relationship between the tool and workpiece and which feeds the tool through, or into, the workpiece. The cutting tool is called a hob.

The generating cutting action that is employed requires that the cutting edges on the tool have a form that will produce the desired tooth form on the workpiece. Generally, the form produced is unlike the form on the cutting edges of the hob.

For analysis of the generating action the hob is often considered to be a rack. Fig. A illustrates a gear having involute tooth profiles in mesh with a hob having straight-sided teeth.

Figure B illustrates the generating action of the hobbing process. As shown, the teeth of the hob cut into the workpiece in successive order, and each in slightly different position. Each hob tooth which comes in contact with the work will remove a small chip to help form the complete gear tooth profile. The cutting action is continuous, and several hob teeth will be cutting at the same time, each working on different portions of adjacent teeth. Thus in the course of this uninterrupted working process, the teeth of the hob progressively envelop the teeth of the workpiece with a network of cuts. The number of hob teeth involved in generating the workpiece

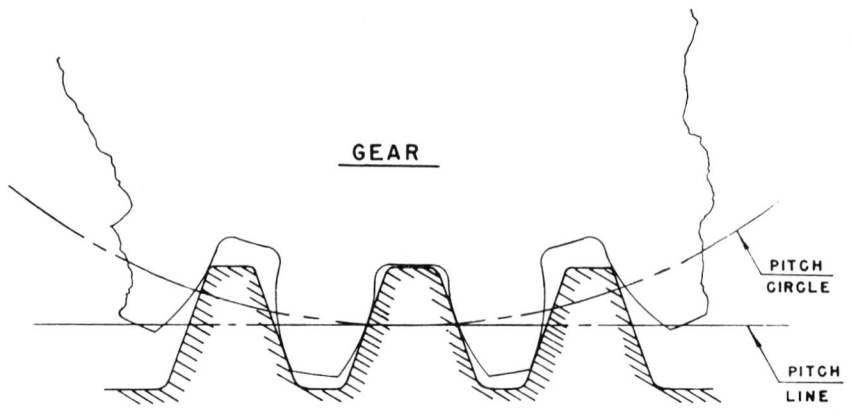

GEAR

PITCH CIRCLE

PITCH LINE

HOB

Fig. A

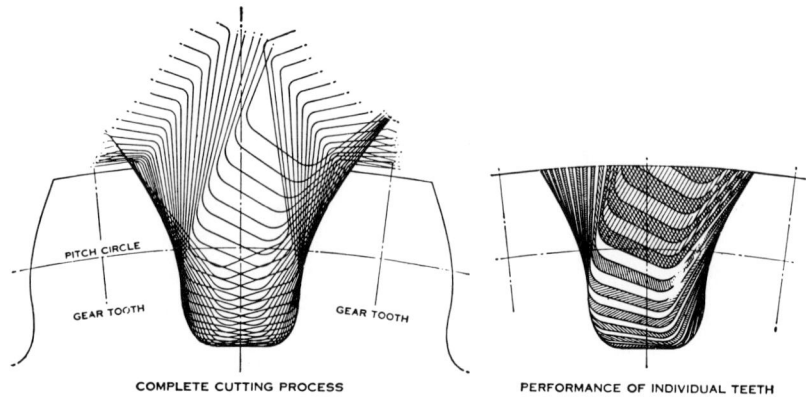

PITCH CIRCLE

GEAR TOOTH   GEAR TOOTH

COMPLETE CUTTING PROCESS   PERFORMANCE OF INDIVIDUAL TEETH

Fig. B

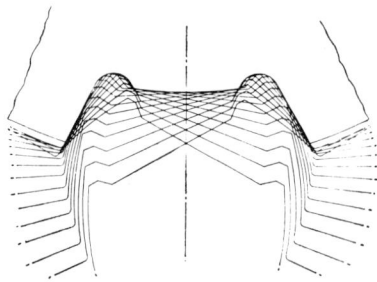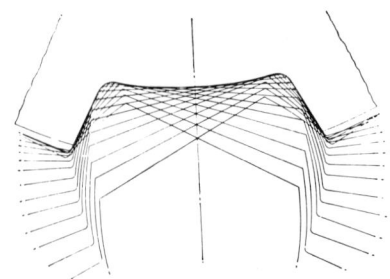

Spline Hob with Lugs   Spline Hob without Lugs

Fig. C

tooth profile increases with the number of flutes in the hob and the number of teeth in the workpiece.

Figures B & C, which are schematic, show that the workpiece remains fixed and the hob, in addition to the feed movement and rotation is moving around the circumference of the workpiece. In actual operation on a hobbing machine, the workpiece and hob rotate in timed relation about their respective axes while the hob is fed the desired distance.

NOMENCLATURE OF HOBS

I—DEFINITION

Hob—A rotary cutting tool with its teeth arranged along a helical thread, used for generating gear teeth or other evenly spaced forms on the periphery of a cylindrical workpiece. The hob and the workpiece are rotated in timed relationship to each other while the hob is fed axially or tangentially across or radially into the workpiece. The cutting edges of the hob teeth lie in a helicoid which is usually conjugate to the form produced on the workpiece. CAUTION: Not to be confused with multiple thread milling cutters, rack cutters, etc., wherein the teeth are not arranged along a helical thread. See Figure 1.

II—GENERAL CLASSIFICATIONS

A—Classification Based Upon Method of Mounting.

1—*Arbor-Type Hobs*—Those which have a hole for mounting on an arbor. They usually have a keyway to receive an arbor driving key. These are sometimes called Shell-Type Hobs. See Figure 3.

2—*Shank-Type Hobs*—Those having a straight or tapered shank to fit into the hobbing machine spindle or adapter. They also usually have a center or extended bearing pilot for supporting the end opposite the shank. See Figure 12.

B—Classification Based Upon Hand of Hob.

The "hand" of a hob is based upon the "hand" of the thread helix of the hob teeth.

1—*Right Hand Hob*—One whose thread helix corresponds to that of a right hand screw.

2—*Left Hand Hob*—One whose thread helix corresponds to that of a left hand screw. *Caution*—When a hob is equipped with unsymmetric features such as tapered holes, shanks, or clutch drives, the hand of hob specification must be supplemented with direction of rotation data. The recommended way is to specify that when the hob is viewed perpendicular to the axis with "Top Coming" or "Top Going" [state which], the desired feature is on the right or left end [state which] of the hob.

C—Classification Based Upon Topping Feature.

1—*Non-Topping Hob*—One which produces only the sides and root diameter of the workpiece profile, but does not finish the outside diameter of the workpiece and does not produce an obvious chamfer at the tip of the workpiece form. Such a hob may produce a small amount of gear tooth tip modification. See Figure 2.

2—*Semi-Topping Hob*—One which produces an obvious modification such as a chamfer or a radius at the intersection of the side of the workpiece profile with the workpiece outside diameter. See Figure 2.

3—*Topping Hob*—One which produces the entire workpiece profile including the complete outside diameter of the workpiece. See Figure 2.

D—Classification Based Upon Number of Threads.

1—*Single Thread Hob*—One whose teeth are arranged along a single helical thread so that the lead of the thread is equal to the axial pitch of the hob.

2—*Multiple Thread Hob*—One whose teeth are arranged along two or more parallel helical threads which are so spaced that the lead of the thread is equal to the axial pitch of the hob multiplied by the *Number of Threads*.

E—Classification Based Upon Type of Flute.

1—*Straight Flute Hob*—One whose flutes are parallel to the hob axis. Sometimes called a *Straight Gash Hob*.

2—*Helical Flute Hob*—One whose flutes are helical. The hand of the helix is usually opposite that of the hob thread and of such lead that the flutes are normal to the hob thread at the hob pitch diameter. Sometimes called *Spiral Flute, Helical Gash,* or *Spiral Gash Hobs*.

F—Classification Based Upon Type of Operation.

1—*Finishing Hob*—One which will produce the workpiece tooth profile to final finished dimensions.

2—*Roughing Hob*—One which produces a workpiece tooth profile with *Stock Allowance* for a subsequent finishing operation. Most often, finishing stock is allowed only on the sides of the workpiece tooth profile so the teeth on a roughing hob are thinner than those on a finishing hob.

3—*Pre-Shave Hob*—One which leaves a small amount of stock on the sides of the workpiece tooth profile to allow for final finishing by shaving.

Such a hob may have a modified tooth profile so that stock left on the workpiece is best distributed for finishing by shaving.

4—*Pre-Grind Hob*—One which leaves a specified amount of stock on the sides, and sometimes the bottom, of the workpiece tooth profile to allow for final finishing by grinding. Such a hob may have a modified profile to distribute the stock left on the workpiece to best suit the grinding process used.

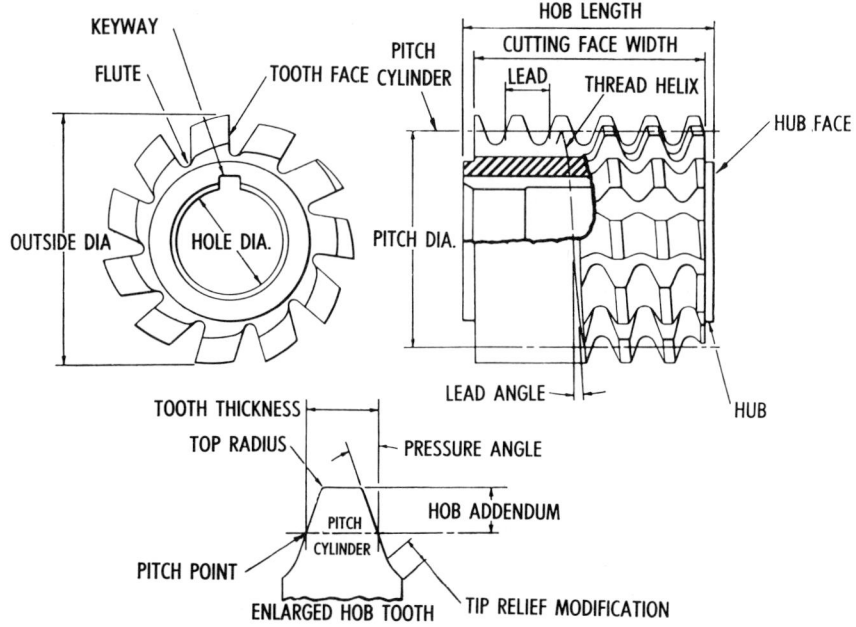

Figure 1—*A typical right hand single thread gear hob.*

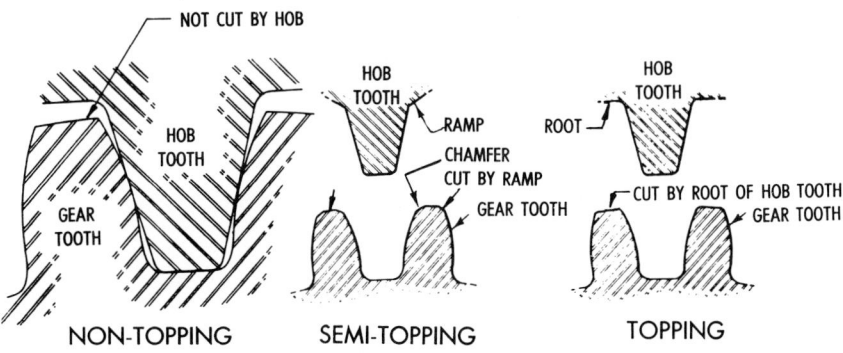

Figure 2—*Characteristics of non-topping, semi-topping and topping hobs.*

G—Classification Based Upon Accuracy of Hob Elements.

1—*Class*—In reference to hobs, classifies them according to the degree of accuracy of hob elements. They are designated as Classes AA, A, B, C or D. Class AA hobs have the closest tolerances and Class D the widest.

III—Various Types of Hobs

Figure 3—Arbor Type Gear Hob—Has essentially straight sided teeth for cutting involute gears.

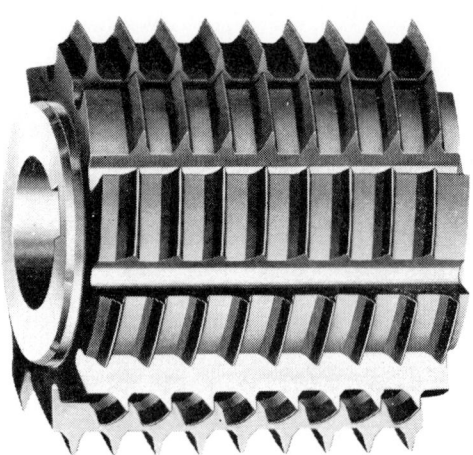

Figure 4—Alternate Tooth Hob—Has alternate teeth removed along the hob thread. Usually used to produce pointed teeth or other special features. Sometimes called Interrupted Thread or Staggered Tooth Hob.

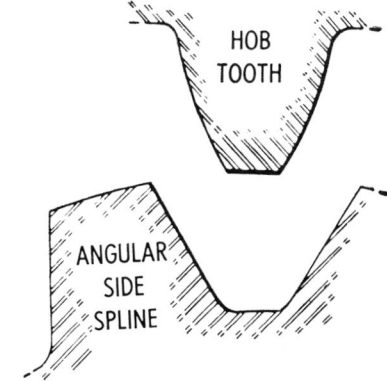

Figure 5——Angular Side Spline and Hob——Has curved tooth profiles to generate angular side splines.

Figure 6——Involute Spline Hob——Similar to Gear Hob, but with shorter teeth and usually of a greater pressure angle.

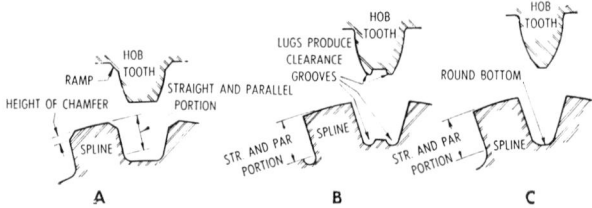

Figure 7——Characteristics of Parallel Side Spline Hobs with curved tooth profiles to generate straight side forms.——(a) Hob tooth shown without lugs. Note generated Fillet at base of spline.——(b) Hob tooth with lugs extending beyond the parallel sides of spline.——(c) With relativly deep narrow teeth. Such hobs produce large or full fillets at root of spline.

Figure 8——Ratchet Hob——Has special tooth form to produce ratchet teeth. A common type illustrated.

Figure 9—Serration Hob—Has straight sided teeth for cutting involute serrations or curved tooth profiles for cutting straight sided serrations.

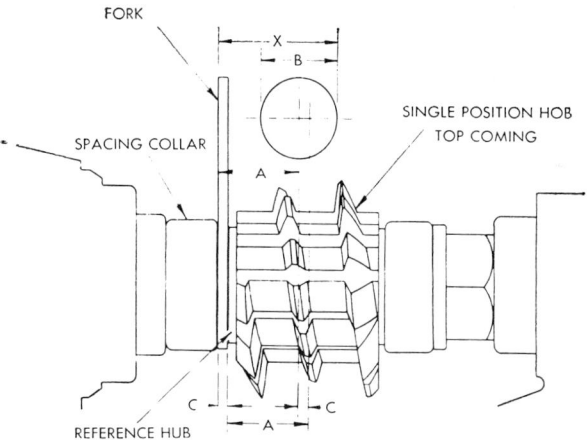

FORK

X

B

SINGLE POSITION HOB
TOP COMING

SPACING COLLAR

A

C

C

A

REFERENCE HUB

Figure 10—Single Position Hob—Must be used in a specified position relative to the work to produce the desired form.

Figure 11——Sprocket Hob——Has curved tooth profiles to generate non-involute tooth forms.

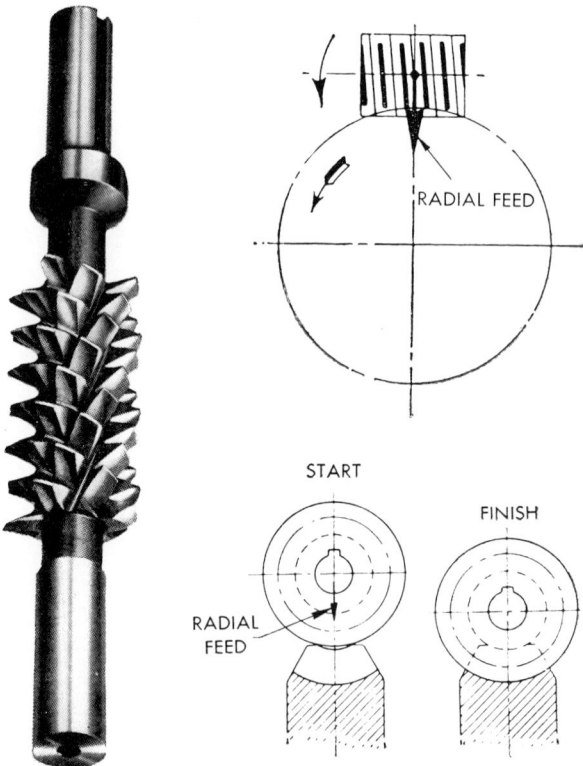

Figure 12——Shank Type Worm Gear Hob——Infeed Type—— Usually similar in appearance to a Gear Hob. Its diameter approximates that of the worm meshing with the gear being cut. This hob is fed radially to depth.

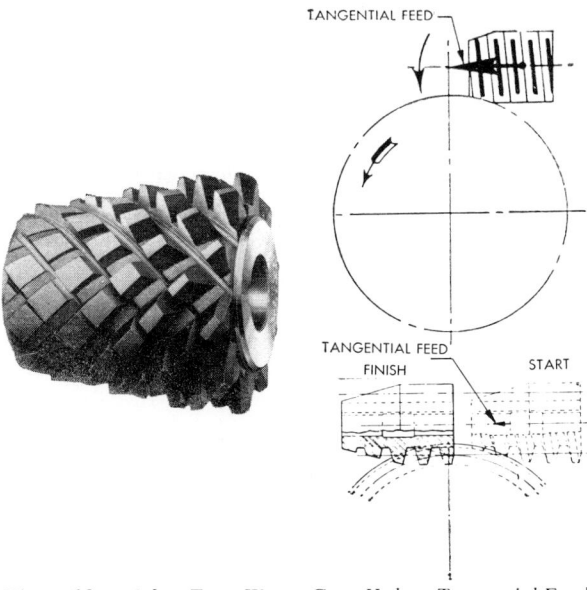

Figure 13—Arbor Type Worm Gear Hob—Tangential Feed Type—Hob is set to depth and fed tangent to the blank. The entering end is usually tapered with the roughing teeth frequently made thinner. The diameter of the finishing end approximates that of the mating worm.

IV—NOMENCLATURE OF HOB ELEMENTS AND OTHER TERMS RELATING TO HOBBING

ADJACENT FLUTE SPACING—The variation from the desired angle between adjacent tooth faces measured in the plane of rotation.

ADJACENT THREAD SPACING—The difference in the average variations obtained by traversing along the desired helical path of one thread, indexing and traversing in a similar manner on an adjacent thread. See Figure 14.

APPROACH—See preferred term TIP RELIEF MODIFICATION.

APPROACH DISTANCE—The linear distance in the direction of feed between the point of initial hob contact and the point of full hob contact. See Figure 15.

ARBOR COLLAR—A hollow cylinder which fits an arbor, and is used to position the hob.

AUXILIARY LEADS—A feature employed on some hobs, especially worm gear hobs, wherein both sides of the hob thread have leads different from the nominal hob lead: one side longer, the other side shorter. This results in the tooth thickness being successively less toward the roughing end of the hob.

Axial Feed—The rate of change of hob position parallel to the workpiece axis usually specified in inches per revolution of the workpiece. See Figure 15.

Axial Plane—A plane containing the axis of rotation.

Axial Pressure Angle—See definition under Pressure Angle.

Back-Off—See preferred term Cam Relief, under Relief.

Cam—The radial drop of the form in the angular distance between adjacent tooth faces.

Centering Device—A ground locating pin used to center a tooth or space of the hob on the centerline of the workpiece.

Chamfer—A beveled surface to eliminate an otherwise sharp corner.

Climb Hobbing—Rotation of a hob in the opposite direction to the feed of the hob relative to the workpiece at the point of contact. See Figure 16.

Clutch Keyway—See Keyway.

Common Factor Ratio—In multiple thread hobs, the condition wherein the Gear Tooth-Hob Thread Ratio is not a whole number, but there is a common factor of the number of gear teeth and the number of hob threads.

Conventional Hobbing—Rotation of a hob in the same direction as the feed of the hob relative to the workpiece at the point of contact. See Figure 17.

Cutting Face Width—The axial length of the relieved portion of the hob. See Figure 1.

Cutting Speed—The peripheral lineal speed resulting from rotation, usually expressed as surface feet per minute [sfm].

Depth of Cut—The radial depth to which the hob is sunk into the workpiece. See Whole Depth.

Drawbar—A rod which retains the arbor, adapter or hob shank in the spindle.

Even Ratio—In multiple thread hobs, the condition wherein the Gear Tooth-Hob Thread Ratio is a whole number.

Feed—The rate of change of hob position while cutting. See Axial Feed, Tangential Feed and Infeed.

Fillet—1—A curved line joining two lines to eliminate a sharp internal corner.—2—A curved surface joining two surfaces to eliminate a sharp internal corner.

Flute—A longitudinal groove either straight or helical that forms the tooth face of one row of hob teeth and the back of the preceding row. It also provides chip space. See Figure 1.

Flute Helix Angle—The angle which a helical tooth face makes with an axial plane, measured on the hob pitch cylinder.

Flute Lead—The axial advance of a helical tooth face in one turn around the axis of a hob.

FLUTE LEAD VARIATION—The deviation of a hob tooth face from the desired helical surface.

FORMER—See preferred term SHARPENING GUIDE.

FULL TOP RADIUS—Continuous radius tangent to top and side cutting edges.

GEAR TOOTH-HOB THREAD RATIO—The ratio of the number of teeth in the workpiece to the number of threads in the hob. See COMMON FACTOR RATIO, EVEN RATIO, and PRIME RATIO.

GENERATED FILLET—At the bottom of the hobbed form a fillet joining the root diameter with the desired generated form. This fillet is not a true radius. See Figure 18.

GENERATED FILLET HEIGHT—On the hobbed workpiece, the radial distance from the root diameter to the point where the generated fillet joins the desired generated form. See Figure 18.

GRINDING CRACKS—Fractures in the hob caused by improper grinding techniques in sharpening. See Figure 19.

HIGH POINT—See preferred term PROTUBERANCE.

HOB ADDENDUM—Radial distance between the top of the hob tooth and the PITCH CYLINDER. Do not confuse with gear addendum. See Figure 1.

HOB ARBOR—A device to mount in or on the spindle of a hobbing machine, which is designed to carry and drive an arbor-type hob.

HOB DEDENDUM—In TOPPING HOBS, the radial distance between the bottom of the hob tooth profile and the PITCH CYLINDER. Do not confuse with gear dedendum.

HOB LENGTH—Overall length of hob. See Figure 1.

HOB RUNOUT—The runout of hob when mounted in a hobbing machine, measured radially on hub diameter, and axially on hub face.

HOB SHIFT—The axial movement of a hob along its axis to engage a different section with the workpiece, to distribute the wear.

HUB—A qualifying surface at each end of an arbor type hob which is provided for checking diameter and face runout. See Figure 1.

HUB DIAMETER RUNOUT—The total variation in radial distance of the hub periphery from the axis.

HUB FACE—The side surface of the HUB. See Figure 1.

HUB FACE RUNOUT—The total axial variation of the hub face from a true plane of rotation.

HUNTING RATIO—See preferred term PRIME RATIO.

INFEED—The radial rate of change of hob position, relative to the workpiece axis, usually specified in inches per revolution of the workpiece. Figure 12.

KEY—A mechanical member through which the turning force is transmitted to the hob.

Keyseat—The pocket, usually in the driving element, in which the key is retained.

Keyway—A slot through which the turning force is transmitted to the hob. May be either a longitudinal slot through the hole or a transverse slot across the hub face. If the latter, it is called a Clutch Keyway. See Figure 1.

Lead—The axial advance of a thread for one complete turn, or convolution. See Figure 1.

Lead Angle—The angle between a tangent to a helix and a plane of rotation. In a hob, lead angle usually refers specifically to the angle of thread helix measured on the pitch cylinder. (See Fig. 1)

Lead Variation—The axial deviation of the hob teeth from the correct thread lead.

Leader—See preferred term Sharpening Guide.

Linear Pitch—See preferred term Axial Pitch, under Pitch.

Linear Pressure Angle—See preferred term Axial Pressure Angle, under Pressure Angle.

Lug—An extension of hob tooth profile above the nominal top cutting edge. Sometimes called Spurs or Prongs. See Figure 7.

Non-Adjacent Flute Spacing—The variation from the desired angle between any two non-adjacent tooth faces measured in the plane of rotation.

Normal Circular Pitch—See definition under Pitch.

Normal Diametral Pitch—See definition under Pitch.

Normal Plane—A plane perpendicular to a pitch cylinder helix.

Normal Pressure Angle—See definition under Pressure Angle.

Number of Threads—In multiple thread hobs, the number of parallel helical paths along which hob teeth are arranged, sometimes referred to as Number of Starts. Should not be confused with the term, Number of Threads per Inch, which is commonly used in designating the axial pitch of screw threads.

Offset—See preferred term Rake Offset.

Outside Diameter—The diameter of the cylinder which contains the tops of the cutting edges of the hob teeth. See Figure 1.

Outside Diameter Runout—The total variation in the radial distance from the axis to the tops of the hob teeth.

Overtravel—The linear distance in the direction of feed of the hob beyond the last point of contact of the hob with the workpiece.

Pilot End—On shank type hobs, a cylindrical or conical bearing surface opposite the driving end.

Pitch—The distance between corresponding, equally spaced hob Thread elements along a given line or curve. The use of the single word Pitch without qualification may be confusing. Specific terms such as NORMAL DIAMETRAL PITCH, NORMAL CIRCULAR PITCH, or AXIAL PITCH are preferred.

Axial Pitch—The pitch parallel to the axis in an axial plane between corresponding elements of adjacent hob thread sections. The term *Axial Pitch* is preferred to the term *Linear Pitch*. See Figure 1.—LEAD.

Linear Pitch—See preferred term *Axial Pitch*.

Normal Circular Pitch—The distance between corresponding elements on adjacent hob thread sections measured along a helix that is normal to the *Thread Helix* in the *Pitch Cylinder*.

Normal Diametral Pitch—π [3.1416] divided by the *Normal Circular Pitch*.

PITCH CIRCLE—A transverse section of the hob PITCH CYLINDER. See Figure 1.

PITCH CYLINDER—A reference cylinder in a hob from which design elements, such as lead, lead angle, profile, and tooth thickness are derived. Figure 1.

PITCH DIAMETER—The diameter of the PITCH CYLINDER.

PITCH POINT—The point at which a TOOTH PROFILE intersects the PITCH CYLINDER. See Figure 1.

PRESSURE ANGLE—The angle between a tooth profile and a line perpendicular to the PITCH CYLINDER at the PITCH POINT. In hobs, the PRESSURE ANGLE is usually specified in the NORMAL PLANE or in the AXIAL PLANE.

Axial Pressure Angle—The *Pressure Angle* as measured in an *Axial Plane*. The term *Axial Pressure Angle* is preferred to the term *Linear Pressure Angle*.

Normal Pressure Angle—The *Pressure Angle* as measured in a *Normal Plane*.

PRIME RATIO—In multiple thread hobs, the condition wherein the GEAR TOOTH-HOB THREAD RATIO is not a whole number and there is no common factor of the number of gear teeth and the number of hob threads.

PROTUBERANCE—A modification near the top of the hob tooth which produces UNDERCUT at the bottom of the tooth of the workpiece. See Figure 20.

RAKE—The angular relationship between the tooth face and a radial line intersecting the tooth face at the hob outside diameter measured in a plane perpendicular to the axis.

Negative Rake—The condition wherein the peripheral cutting edge lags the tooth face in rotation. See Figure 21.

Positive Rake—The condition wherein the peripheral cutting edge leads the tooth face in rotation. See Figure 22.

Zero Rake—The condition wherein the tooth face coincides with a radial line. See Figure 23.

RAKE OFFSET—The distance between the tooth face and a radial line parallel to the tooth face. Used for checking rake. See Figures 21 and 22.

RAMP—A modification at the bottom of the hob tooth which produces a chamfer at the top corners of the tooth of the workpiece. See Figure 2.

RELIEF—The result of the removal of tool material behind or adjacent to a cutting edge to provide clearance and to prevent rubbing [heel drag].

Cam Relief—The relief from the cutting edges to the back of the tooth produced by a cam actuated cutting tool or grinding wheel on a relieving [back-off] machine.

Side Relief—The relief provided at the sides of the teeth behind the cutting edges. The amount depends upon the radial cam, the axial cam, and the nature of the tooth profile.

SCALLOPS—The shallow depressions on the generated form produced by hob tooth action.

SETTING ANGLE—The angle used for setting hob swivel to align the hob thread with the workpiece teeth.

SHANK—That projecting portion of a hob which locates and drives the hob in the machine spindle or adapter.

SHARPENING ALLOWANCE—The amount by which the pitch diameter of a worm gear hob exceeds that of the worm, to allow for the reduction in diameter by sharpening.

SHARPENING GUIDE—A cylindrical part with flutes, having the same lead as the hob flutes, used for guiding the hob along the correct lead when sharpening.

SHORT LEAD—A feature employed on some hobs wherein the hob lead is made shorter than the nominal or theoretical lead, in order to generate lower on the workpiece, to meet a particular fillet or undercut requirement.

SIDE RELIEF—See definition under RELIEF.

STOCK ALLOWANCE—The amount of a modification of the hob tooth to leave material on the workpiece tooth form for subsequent finishing.

TANGENTIAL FEED—The rate of change of hob position along its own axis, usually specified in inches per revolution of the workpiece. See Figure 13.

THREAD—A helical ridge, generally of constant form or profile. In a hob, unlike a worm or screw, the thread is not continuous and exists only at the cutting edges of the hob teeth. Therefore, it is sometimes referred to as the THREAD ENVELOPE.

THREAD ENVELOPE—See preferred term THREAD.

THREAD HELIX—The helix of the hob thread in the PITCH CYLINDER. See Figure 1.

TIP RELIEF—A gear tooth modification in which a small amount of material is removed from the basic profile near the tip of the gear tooth. See Figure 24.

TIP RELIEF MODIFICATION—A modification on the sides of the hob tooth near the bottom which produces a small amount of TIP RELIEF on the gear tooth. Such modification is usually incorporated in full depth gear Finishing Hobs except in the finer pitches. See Figure 1.

TOOTH—A projection on a hob which carries a cutting edge.

TOOTH FACE—The tooth surface against which the chips impinge. See Figure 1.

TOOTH PROFILE—Outline or contour of hob tooth cutting edges.

TOOTH THICKNESS—The actual width or thickness of the hob thread at the PITCH CYLINDER. The use of the single term TOOTH THICKNESS without qualification may be confusing. The specific term NORMAL TOOTH THICKNESS and AXIAL TOOTH THICKNESS are preferred.

Axial Tooth Thickness—The tooth thickness as measured in an axial plane. See Figure 1.

Normal Tooth Thickness—The tooth thickness as measured along a helix normal to the thread helix.

TOP RADIUS—Radius of the arc joining the top and a side cutting edge of a hob tooth. See Figure 1.

TOTAL INDICATOR READING [tir]—See preferred term TOTAL INDICATOR VARIATION.

TOTAL INDICATOR VARIATION [tiv]—The difference between maximum and minimum indicator readings during a checking cycle.

UNDERCUT—The condition at the base of a hobbed workpiece form wherein additional material beyond the basic form is removed. Under certain conditions this may occur naturally, while in other cases it may be produced by intentional modification of the hob tooth. See Figure 25.

WEAR LAND—A cylindrical or flat land worn on the relieved portion of the hob tooth behind the cutting edge.

WHOLE DEPTH—The radial depth (exclusive of the effect of lugs) which the hob is designed to produce on the workpiece.

WOBBLE—The motion of a hob when the radial runout varies along the hob length.

WORM GEAR HOB OVERSIZE—See preferred term SHARPENING ALLOWANCE.

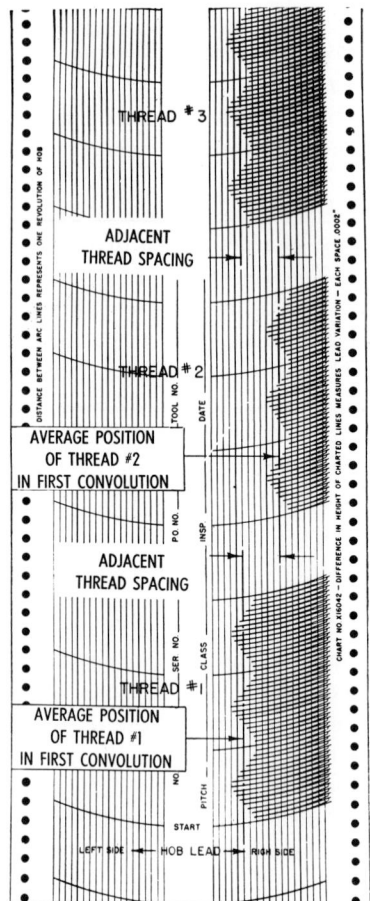

Figure 14—Adjacent Thread Spacing.

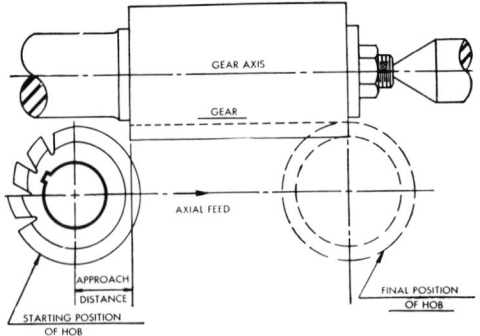

Figure 15——Approach Distance.

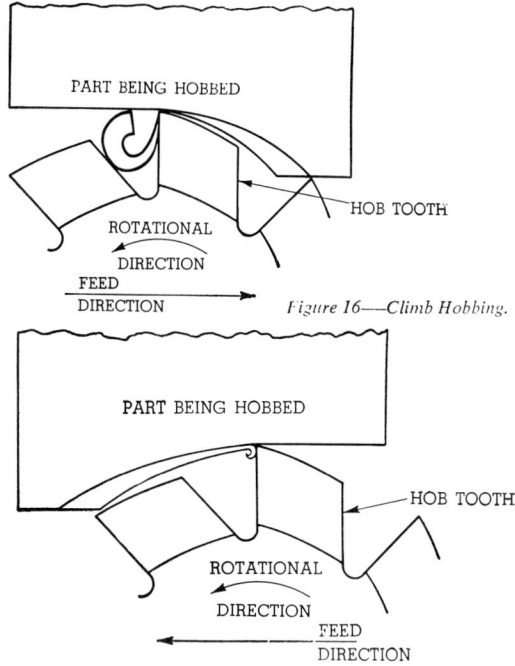

Figure 16—Climb Hobbing.

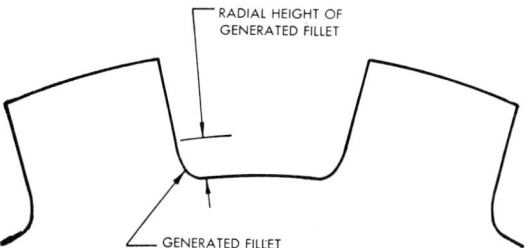

Figure 17—Conventional Hobbing.

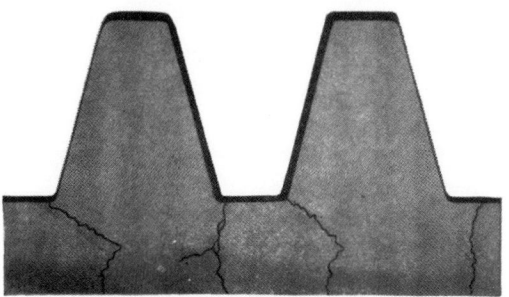

Figure 18—Generated Fillet.

Figure 19—Grinding Cracks.

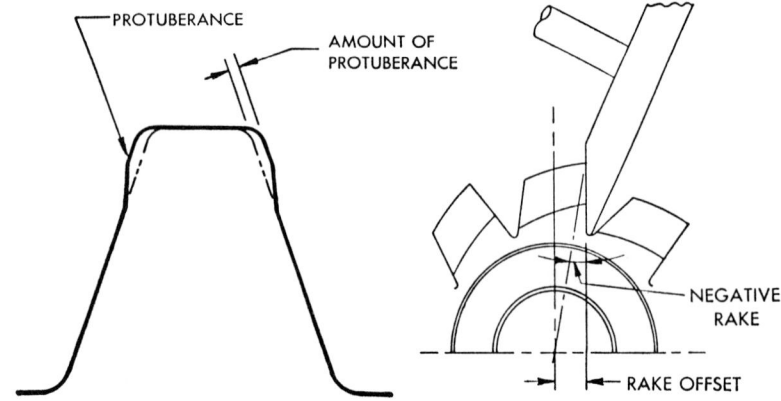

Figure 20——Protuberance on a Gear Hob Tooth. Figure 21——Negative Rake.

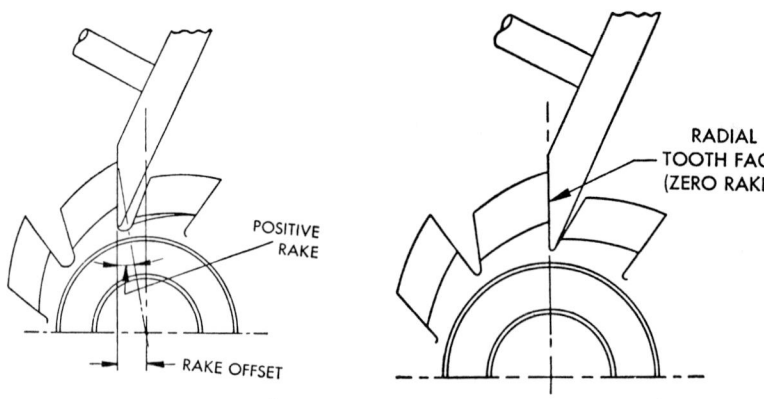

Figure 22——Positive Rake. Figure 23——Zero Rake.

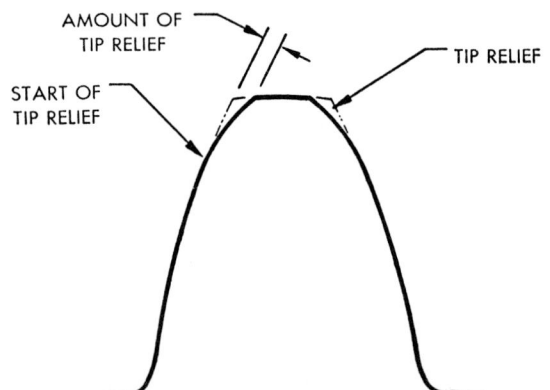

Figure 24——Tip Relief on a Gear Tooth.

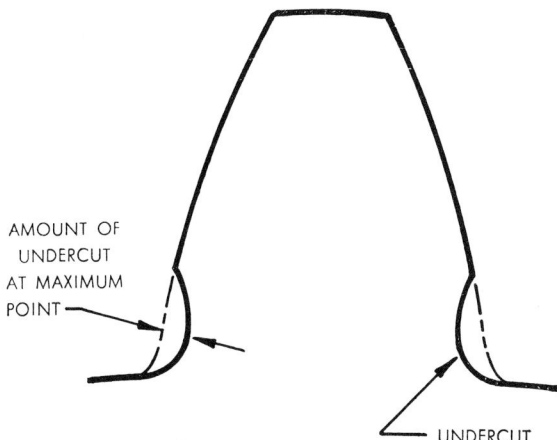

AMOUNT OF
UNDERCUT
AT MAXIMUM
POINT

UNDERCUT

Figure 25——Undercut on a Gear Tooth.

HOB FEATURES

The hobbing process can be used to produce nearly all profiles which are equally spaced about a cylindrical workpiece. The projections or indentations need not be symmetrical about a radial line, nor does the form need to be of any special curvature. The only limiting factor is that the projections or indentations must be of such proportion that they will roll in and out of the hob without interference. Common hob applications include the production of involute spur and helical gears; involute and parallel-side splines; worm and worm gears; roller chain, silent chain, and film sprockets; involute and straight-side serrations; and square and hexagon shapes.

There are several important hob features which must be correctly specified if the hobs are to give satisfactory service. Several of the more important features which are common to most hobs will be discussed here. These include: (1) Size and drive, (2) accuracy, (3) Tooth form, (4) Number of Threads, and (5) Hob material.

Size and Drive

The size of a hob is determined by consideration of the hob thread pitch, the hob shank, the hobbing machine to be used, and the size and design of the workpiece. Through the Metal Cutting Tool Institute, hob manufacturers have recommended standard hob sizes for the production of involute forms of various pitches and depths. Tables of these standard hob sizes are included later in this section. These hob sizes have been found satisfactory for most applications. Larger diameter hobs are sometimes used to reduce the size of generating flats through incorporation of an increased number of flutes. Occasionally, the hob size must be reduced to avoid interference with portions of the workpiece.

Most hobs are made in arbor-type construction with a hole for mounting on an arbor. Shank-type hobs are normally used when a hole of sufficient diameter would excessively weaken the hob. In some cases, shank type hobs are used to eliminate mounting inaccuracies.

Arbor-type hobs are usually driven by means of a key through the hole. A clutch keyway across the end of the hob is sometimes employed when a keyway through the hole would excessively weaken the hob. Shank-type hobs are specially made to suit the user's hobbing equipment. The driving shank and the pilot end may be made either straight or tapered.

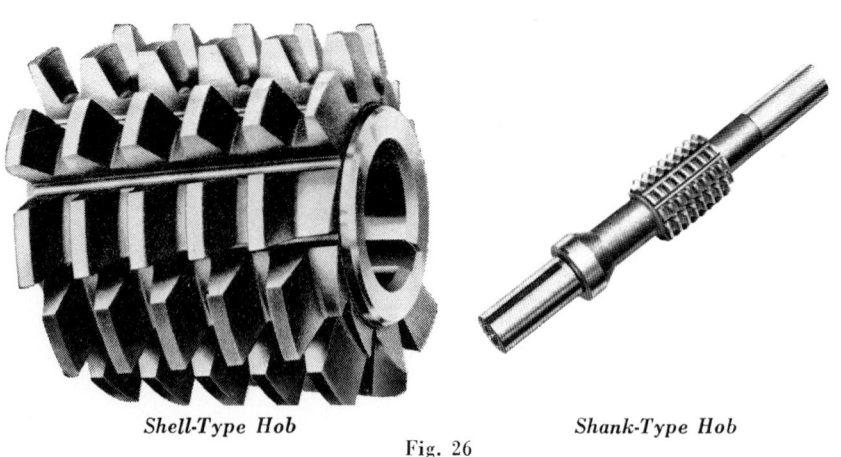

Shell-Type Hob   Shank-Type Hob

Fig. 26

Accuracy

Hobs are classified according to the tolerances maintained in their manufacture. Most hobs can be made with either ground or unground form. The selection of the class of hob for any specific application depends principally upon the accuracy requirements of the part. However, actual results obtained upon the workpiece may vary considerably depending upon the condition of the machine and tooling, the care taken in setting up the machine, the preparation and design of the workpiece blank, the accuracy of hob sharpening and the inspection methods.

The Metal Cutting Tool Institute has adopted five standard classifications for hobs used in generating involute tooth forms. The standard includes multiple-thread hobs to four threads as well as single-thread hobs. The classes are designated as:

Class AA
Class A
Class B
Class C
Class D

The tolerances are shown on pages 704 through 713. These are applicable to hobs for involute spur and helical gears, and involute splines. Note that the tolerances apply only to hobs of the recommended standard diameters or smaller and that the tolerances on hobs of larger diameter must be increased in proportion to the increased diameter.

Tooth Forms

Most hobs are of the full generating type which can be used in any position across their cutting face width. However, some special tooth forms require the use of semi-generating or non-generating hobs which must be accurately positioned in relation to the workpiece; these are designated as single-position hobs.

Forms Produced by
Full Generating Hobs

Special Forms Produced
by Single-Position Hobs

Fig. 27

Hobs are used for finishing, semi-finishing, and roughing operations. Finishing hobs are made to produce the finished part tooth form without further machining. Semi-finishing hobs include pre-shave and pre-grind hobs which are made to produce a part tooth form with a stock allowance for the finishing process. Roughing hobs usually allow more stock for finishing than corresponding semi-finishing hobs.

In order to generate the desired part tooth form, the hob must have a tooth profile that is conjugate to the tooth profile on the part. The conjugate hob tooth form is one which would transmit uniform angular velocity to the form on the tooth of the workpiece. In hobbing, both the work and the hob rotate at a uniform angular velocity on the hobbing machine. The hob tooth is usually different from the part tooth profile with its shape primarily determined by the generating pitch diameter.

The total depth of the hob tooth is determined by the depth of the work-piece tooth and whether the hob is topping or non-topping. A topping hob cuts the outside diameter of the blank to a specified size. When the total depth of the hob tooth is equal to the whole depth of the part tooth, the hob is topping. When the total depth of the hob tooth is greater than the whole depth of the part tooth the hob is either non-topping or semi-topping.

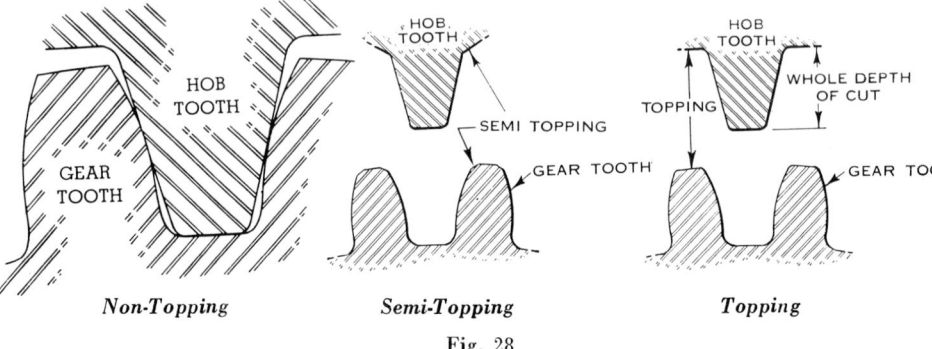

Non-Topping   Semi-Topping   Topping

Fig. 28

The hob tooth form can be modified to meet special design requirements. The most common of these modifications include profiles with protuber-ances, lugs and ramps. A protuberance at the top of the hob tooth produces undercut to provide clearance for the shaving cutter or grinding wheel. Lugs produce clearance grooves to eliminate interference between mating spline members. A ramp at the base of the hob tooth will produce a cham-fer at the top of the workpiece tooth. Most coarse pitch, full depth, finishing gear hobs incorporate a modification of the profile to produce a small amount of tip relief on the gear teeth.

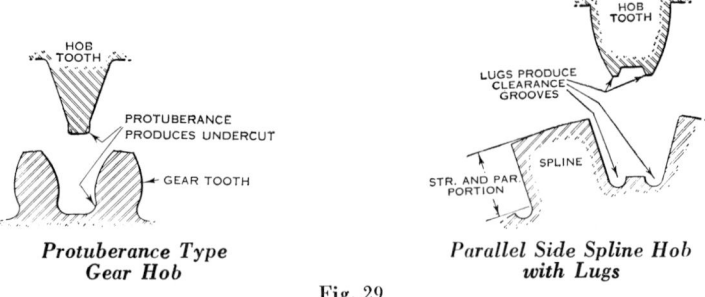

Protuberance Type   Parallel Side Spline Hob
Gear Hob   with Lugs

Fig. 29

Number of Threads

The number of threads on a hob are determined primarily by the pro-duction rate required and the accuracy and finish requirements. Single thread hobs are recommended where accuracy and finish requirements are of major importance. The accuracy of workpiece tooth spacing is independent of hob accuracy with single thread hobs since all workpiece teeth are generated by the same group of hob teeth along a single thread. The part tooth profile will be smoother since it will be made up of a maximum number of generating flats.

While multiple thread hobs can often improve production rates, their application must be carefully considered. The incorporation of the additional thread or threads in a multiple thread hob make necessary a greater tolerance on the hob lead.

Performance of multiple thread hobs can depend greatly upon the gear tooth—hob thread ratio. The illustration indicates three types of ratios which can exist between the number of teeth on the part and number of threads on the hob. These are designated as even, common factor, and prime. The type of ratio can affect spacing and profile accuracy on the part.

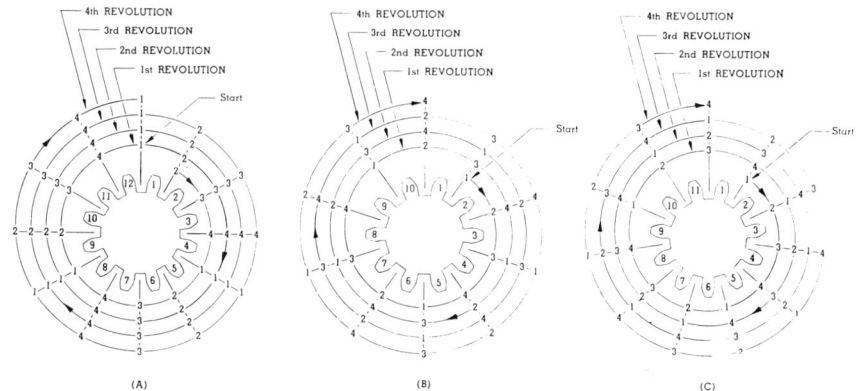

Fig. 30. *Sketch A represents the "even" ratio, where the number of teeth, 12, is divisible by the number of threads, 4, an even number of times, 3. Note that the same threads mesh with the same teeth on each revolution. Sketch B indicates the "common-factor" ratio, where the number of teeth, 10, is not divisible by the number of threads, 4, an even number of times, 2-½, but they are both divisible by a common multiple, 2. Groups of teeth equal to the multiple, 2, mesh with pairs of threads, 1-3 and 2-4. Sketch C shows the "prime" ratio, where the number of teeth, 11, is not evenly divisible by the number of threads, 4, an even number of times, 2-¾, and there is no common factor. Each tooth meshes with all threads.*

With an even ratio, each tooth space is acted upon by the teeth of one particular thread and any error between the threads on the hob is transferred to the part in the form of a tooth spacing error. Since all threads do not act upon each tooth, maximum profile accuracy is obtained.

When the ratio is prime, each thread acts upon each tooth space, cancelling thread spacing errors in such a way as to produce maximum tooth spacing accuracy. However, in cancelling the tooth spacing error, the lead and profile errors on the hob may combine to produce either more or less profile error on the workpiece tooth than that obtained with an even gear tooth—hob thread ratio.

The results obtained with a common factor ratio may be intermediate to those obtained with the even and prime ratios in that groups of teeth are cut by teeth in certain groups of threads.

Another consideration with multiple thread hobs is the ratio between the number of threads and the number of flutes on the hob. This ratio affects the size and position of the generating flats.

Hob Material

Most hobs are made of high speed steel, and standard analyses are suitable for many hob applications. However, the trend is toward higher Brinell hardness for gear materials, and some modern gear materials are very abrasive. As a result, the use of special high speed steels for hobs is increasing. There are also cases where hobs made from special high speed steels can be justified for average materials in order to extent hob life. However, the use of such steels should be carefully considered because they are more costly, sometimes more difficult to sharpen, and when improperly applied, the performance will not improve in proportion to the cost of the hob.

The use of carbide tipped hobs has thus far been primarily limited to low strength, non-ferrous or non-metallic workpiece materials. For such applications the high abrasion resistance of carbide has produced excellent results. However, most attempts to apply carbide tipped hobs to workpieces of steel or cast iron have been either unsuccessful or uneconomic.

The trend of applying hard thin coatings over the high speed steel has extended hob life in certain applications.

GEAR HOBS

Although the involute tooth profile is not the only tooth profile applicable to gears, it is universally understood to be the tooth profile whenever gears are discussed. This section will consider gear hobs as those used to generate gear teeth having involute profiles. With the exception of special features, one hob will generate all numbers of teeth of the desired pitch.

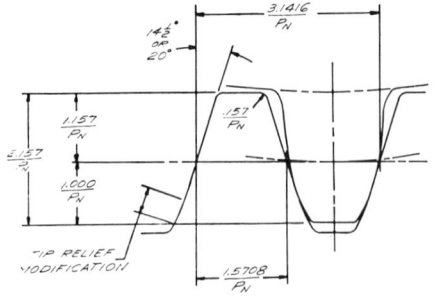

American Standard 14½° or 20°
Full Depth Involute System
*Pn = Normal Diametral Pitch

Fig. 31

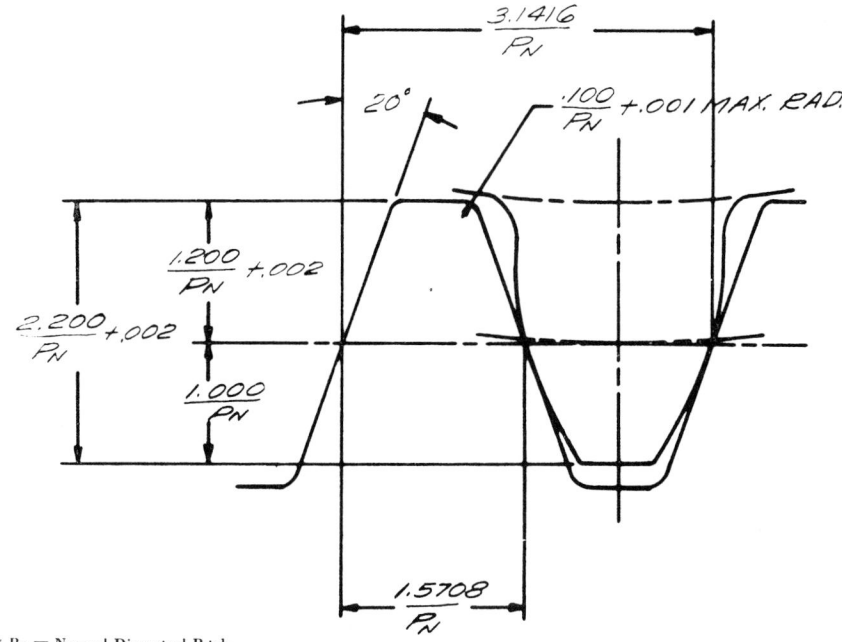

* Pn = Normal Diametral Pitch.

Fig. 32. *American Standard 20° Full Depth, Fine-Pitch Involute System*

There are several systems of gear tooth proportions in common use. The most popular systems are based on a 20° Normal Pressure Angle, although higher and lower pressure angle systems are occasionally used. Figures 31 and 32 show hob tooth proportions for the American Standard full depth systems. Hobs can be made for any one of the standard and most special systems.

Several stub tooth gear systems are also used and hobs can be made to produce gears based upon such systems.

Tables of tooth proportions for American Standard full depth system are found in the Engineering Data Section.

The method of gear finishing employed will influence the proportions of the hobbed teeth. Gears finished by shaving should have sufficient depth to provide clearance for the shaving cutter. The additional depth of the hob tooth produces undercut on the small numbers of gear teeth providing clearance for the tip of the shaving cutter tooth. The results are a smooth blend between the root fillet and the finished profile. To obtain undercut on the greater numbers of teeth, a hob tooth having protuberance is employed.

A pre-shave system that has been generally accepted and applied requires a depth of $\dfrac{2.35}{Pn}$ for pitches 1 through 19.99 Normal Diametral Pitch and $\dfrac{2.350}{Pn} + .002$ for 20 Normal Diametral Pitch and finer.

The tables on pages 695–696 include recommended shaving stock allowances and a standard protuberance design. For a given protuberance, the amount of undercut varies with the numbers of teeth hobbed. Where specific undercut conditions are desired, special design is necessary.

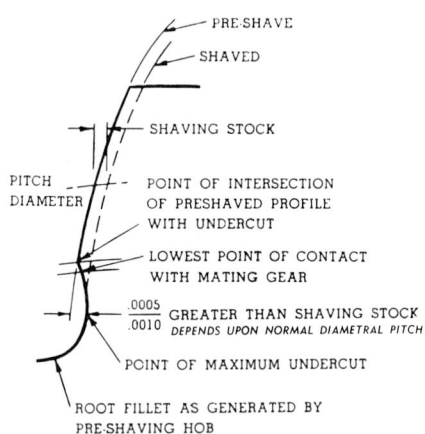

Fig. 23. *Recommended pre-shave gear tooth profile for coarser than 20 NDP*

PRE-SHAVE SYSTEM FOR
14½ N.P.A. FULL-DEPTH GEAR TEETH
19.99 Normal Diametral Pitch and Coarser

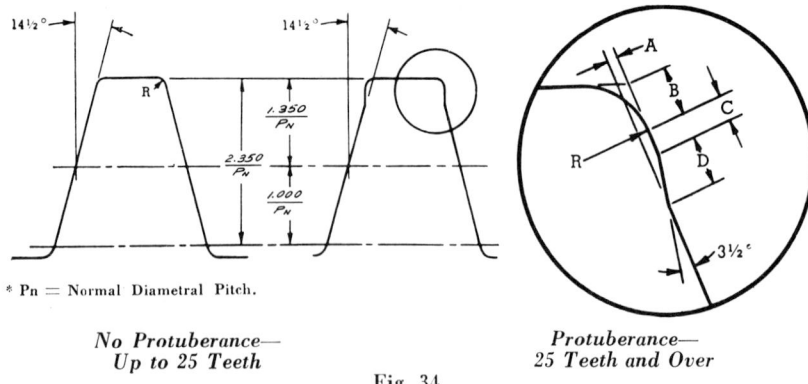

* Pn = Normal Diametral Pitch.

No Protuberance—
Up to 25 Teeth

Protuberance—
25 Teeth and Over

Fig. 34

Recommended Pre-Shave Hob Tooth Design

(All readings in thousandths of an inch)

Normal Diametral Pitch	Stock Allowance on Tooth Thickness	A	B	C	D	R
3	3–4	2.5–3	77	33–49	45	100
4	3–4	2.5–3	58	25–39	45	75
5	2.5–3.5	2–2.5	43	20–32	45	55
6	2.5–3.5	2–2.5	35	17–27	37	45
7	2–3	2–2.5	31	14–22	37	40
8	2–3	2–2.5	27	13–20	37	35
9	2–3	1.5–2	27	11–17	29	35
10	2–3	1.5–2	23	10–15.5	29	30
11	1.5–2.5	1.5–2	23	9–14	29	30
12	1.5–2.5	1.5–2	19	8–13	29	25
13	1.5–2.5	1.5–2	19	8–12	29	25
14	1.5–2.5	1–1.5	15	7–11	20	20
15	1–2	1–1.5	15	7–11	20	20
16	1–2	1–1.5	15	6–9.5	20	20
17	1–2	1–1.5	15	6–9.5	20	20
18	1–2	1–1.5	12	6–9	20	15
19	1–2	1–1.5	12	5–8	20	15

These proportions are satisfactory for most conditions. In some cases different protuberance design may be required to meet workpiece requirements.

Pre-Shave System For
20° N.P.A. Full-Depth Gear Teeth

19.99 Diametral Pitch and Coarser

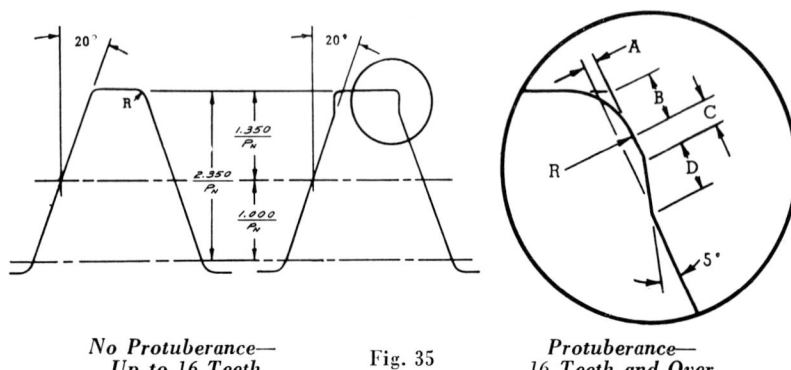

| No Protuberance—
Up to 16 Teeth | Fig. 35 | Protuberance—
16 Teeth and Over |

Pre-Shave System For
20° N.P.A. Full-Depth Gear Teeth
20 Normal Diametral Pitch and Finer

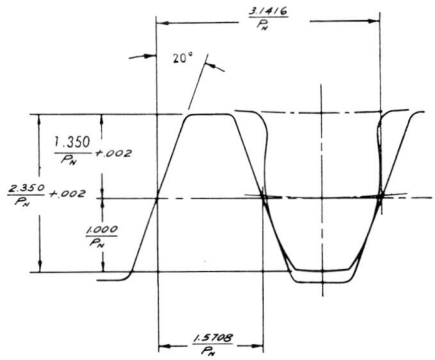

Fig. 36. *No Protuberance—All Numbers of Teeth*

Normal Diametral Pitch	Stock Allowance on Tooth Thickness
20 – 48	.0005 – .0015
48 – 72	.0003 – .0007

STANDARD HOB SPECIFICATIONS & TOLERANCES
RECOMMENDED HOB SIZES
Single Thread Coarse Pitch Hob Sizes for Ground and Unground Hobs

(1-19.99 Normal Diametral Pitch)

For Spur and Helical Gears

Normal Diametral Pitch	Nominal Hole Diameter	Outside Diameter	Overall Length
1	2½″	10¾″	15″
1¼	2″	8¾″	12″
1½	2″	8″	10″
1¾	2″	7¼″	9″
2.00–2.24	1½″	5¾″	8″
2.25–2.49	1½″	5½″	7½″
2.50–2.74	1½″	5″	7″
2.75–2.99	1½″	5″	6″
3.00–3.49	1¼″	4½″	5″
3.50–3.99	1¼″	4¼″	4¾″
4.00–4.99	1¼″	4″	4″
5.00–6.99	1¼″	3½″	3½″
7.00–7.99	1¼″	3¼″	3¼″
8.00–11.99	1¼″	3″	3″
12.00–13.99	1¼″	2¾″	2¾″
	¾″	2″	2″
14.00–19.99	1¼″	2½″	2½″
	¾″	1⅞″	1⅞″

Multiple Thread Coarse Pitch Hob Sizes for Ground and Unground Hobs

(2-19.99 Normal Diametral Pitch)

For Spur and Helical Gears

Normal Diametral Pitch	Number of Threads	Nominal Hole Diameter	Outside Diameter	Overall Length
2–2.99	2	1½″	6½″	8″
3–3.99	2	1½″	5½″	5½″
4–4.99	2 or 3	1½″	5½″	5½″
5–6.99	2, 3 or 4	1½″	5″	5″
7–7.99	2, 3 or 4	1¼″	4″	4″
8–8.99	2, 3 or 4	1¼″	3¾″	3¾″
9–11.99	2, 3 or 4	1¼″	3½″	3½″
12–13.99	2, 3 or 4	1¼″	3¼″	3¼″
14–15.99	2, 3 or 4	1¼″	3″	3″
16–19.99	2, 3 or 4	1¼″	2¾″	2¾″

Single Thread Fine Pitch Hob Sizes
for Ground and Unground Hobs

(20 Normal Diametral Pitch and Finer)

For Spur and Helical Gears

Normal Diametral Pitch	Nominal Hole Diameter	Outside Diameter	Overall Length
20–21.99	$1\frac{1}{4}''$	$2\frac{1}{2}''$	$2\frac{1}{2}''$
	$\frac{3}{4}''$	$1\frac{7}{8}''$	$1\frac{7}{8}''$
22–23.99	$1\frac{1}{4}''$	$2\frac{1}{2}''$	$2''$
	$\frac{3}{4}''$	$1\frac{7}{8}''$	$1\frac{7}{8}''$
24–29.99	$1\frac{1}{4}''$	$2\frac{1}{2}''$	$2''$
	$\frac{3}{4}''$	$1\frac{7}{8}''$	$1\frac{7}{8}''$
	$\frac{1}{2}''$	$1\frac{1}{4}''$	$1\frac{1}{4}''$
30–55.99	$\frac{3}{4}''$	$1\frac{7}{8}''$	$1\frac{1}{2}''$
	$\frac{1}{2}''$	$1\frac{1}{8}''$	$1\frac{1}{8}''$
	.3937 (10 mm)	$1\frac{1}{8}''$	$\frac{3}{4}''$
	.315 (8 mm)	$\frac{3}{4}''$	$\frac{1}{2}''$
56–85.99	$\frac{3}{4}''$	$1\frac{5}{8}''$	$1\frac{1}{2}''$
	$\frac{1}{2}''$	$1\frac{1}{8}''$	$1\frac{1}{8}''$
	.3937 (10 mm)	$1\frac{1}{8}''$	$\frac{3}{4}''$
	.315 (8 mm)	$\frac{3}{4}''$	$\frac{1}{2}''$
86–130.99	$\frac{3}{4}''$	$1\frac{5}{8}''$	$*1\frac{1}{2}''$
	$\frac{1}{2}''$	$1\frac{1}{8}''$	$*\frac{7}{8}''$
	.3937 (10 mm)	$1\frac{1}{8}''$	$\frac{3}{4}''$
	.315 (8 mm)	$\frac{3}{4}''$	$\frac{1}{2}''$
131–200	$\frac{3}{4}''$	$1\frac{5}{8}''$	$**1\frac{1}{2}''$
	$\frac{1}{2}''$	$1\frac{1}{8}''$	$**\frac{7}{8}''$
	.3937 (10 mm)	$1\frac{1}{8}''$	$**\frac{3}{4}''$
	.315 (8 mm)	$\frac{3}{4}''$	$\frac{1}{2}''$

* $\frac{3}{4}''$ Active face width

** $\frac{5}{8}''$ Active face width

Multiple Thread Fine Pitch Hob Sizes
for Ground and Unground Hobs

(20 Normal Diametral Pitch and Finer)

For Spur and Helical Gears

Diametral Pitch	Number of Threads	Nominal Hole Diameter	Outside Diameter	Overall Length
20–29.99	2, 3 or 4	$1\frac{1}{4}''$	$2\frac{1}{2}''$	$2\frac{1}{2}''$
30-50	2, 3 or 4	$\frac{3}{4}''$	$1\frac{7}{8}''$	$1\frac{1}{2}''$

Single Thread Combination Pitch Hob Sizes
for Ground and Unground Hobs

(3/4 Thru 12/14 Pitch)

For Spur and Helical Gears

Combination Pitch[1]	Nominal Hole Diameter	Outside Diameter	Overall Length
3/4	1¼″	4″	4″
4/5	1¼″	3½″	3½″
5/7	1¼″	3¼″	3¼″
6/8	1¼″	3″	3″
7/9	1¼″	3″	3″
8/10	1¼″	3″	3″
9/11	1¼″	2¾″	2¾″
10/12	1¼″ 3/4″	2¾″ 2″	2¾″ 2″
12/14	1¼″ 3/4″	2½″ 2″	2½″ 2″

NOTE: 1. The numerator of the fraction denotes normal diametral pitch. The denominator of the fraction denotes the diametral pitch for calculating addendum and whole depth.

Roller Chain Sprocket Hob Sizes

Single Thread, for Chain Pitches ¼ thru 3″

Based on sprocket profiles of American National Standard B29.1 Transmission Roller Chains and Sprocket Teeth

Standard Sizes and Dimensions

CHAIN			HOB		
No.	Pitch	Roller Diameter	Outside Diameter	Overall Length	Hole
25	¼	.130	2½	2½	1¼
35	⅜	.200	3	2½	1¼
40	½	.313	3⅛	2½	1¼
50	⅝	.400	3½	2½	1¼
60	¾	.469	3½	2⅞	1¼
80	1	.625	4¼	3½	1¼
100	1¼	.750	4¾	4½	1¼
120	1½	.875	5¼	5¼	1½
140	1¾	1.000	5½	6	1½
160	2	1.125	6⅞	6¾	2
180	2¼	1.406	8	8	2
200	2½	1.563	8½	8½	2
240	3	1.875	9¾	11	2

Minimum Marking:
Chain Pitch   Type of Tooth
Lead Angle — L.A.   Roller Diameter
Flute Lead — F.L.

HOB TOLERANCES FOR INVOLUTE GEARS AND SPLINES

PURPOSE
The purpose is to provide dimensional tolerances applicable to the manufacture of single and multiple-thread hobs for generating spur and helical gears, and splines having involute tooth form.

SCOPE
Tolerances are given for hobs having 1, 2, 3 and 4 threads. They are based upon the hob sizes as shown in the tables of "Recommended Hob Sizes." Larger than recommended hob diameters are subject to tolerance increases proportional to one-half increase in diameter. Tolerances are not decreased for smaller than recommended hob diameters.

CLASSES
Hobs are classified with reference to dimensional tolerances as follows:

Class AA
Class A
Class B
Class C
Class D

TOOTH PROFILE
Hobs for generating involute tooth forms have essentially straight-sided tooth profiles in the normal plane section. Standard full depth, coarse pitch, finishing gear hobs have a profile modification beginning near the base of the tooth to produce a slight amount of tip relief on the gear tooth.

TOLERANCE DEFINITIONS

HOLE

Diameter—The basic diameter of the hole in the hob.

Tolerance—The amount that a hole may be oversize from the basic diameter of the hole.

Bearing contact—The area of contact obtained using a plug gage that makes contact over the full length of the hob.

RUNOUT

Hub face—The total indicator variation on the end face in one revolution of the hob.

Hub diameter—The total indicator variation on the hub diameter in one revolution of the hob.

Outside diameter—The total indicator variation on the tops of the hob teeth in one revolution of the hob.

SHARPENING

Spacing between adjacent flutes—The total indicator variation obtained between any two successive flutes when a hob is indexed.

Spacing between non-adjacent flutes—The total indicator variation obtained between any two flutes when the hob is indexed through one complete revolution.

Rake to cutting depth—The total indicator variation when traversing the tooth face from the top to the cutting depth.

Flute lead—The total indicator variation when traversing the faces of all of the teeth in any one flute following the specified lead.

LEAD VARIATION

Tooth-to-tooth—The total indicator variation on successive teeth when traversing along the true helical path.

Any one axial pitch—The total indicator variation in one complete revolution along the true helical path (360°) on a single thread hob; one-half revolution (180°) on a 2-thread hob; one-third revolution (120°) on a 3-thread hob; one-fourth revolution (90°) on a 4-thread hob; one-fifth revolution (72°) on a 5-thread hob; and one-sixth revolution (60°) on a 6-thread hob.

Any three axial pitches—The total indicator variation in three revolutions along the true helical path (1080°) on a single thread hob; one-and-a-half revolutions along the true helical path (540°) on a 2-thread hob; one revolution (360°) on a 3-thread hob; three-fourths revolution (270°) on a 4-thread hob; three-fifths revolution (216°) on a 5-thread hob; and one-half revolution (180°) on a 6-thread hob.

Figure 37, following these definitions, shows the development of a 3-thread hob having 12 flutes. In cases where the number of flutes is not evenly divisible by the number of threads, the indicator variation of one axial pitch and three axial pitches may be interpolated.

Adjacent thread to thread spacing—The difference in the average variations obtained by traversing along the true helical path of one thread, indexing and traversing in a similar manner on the adjacent thread.

Another method used for hobs having straight flutes is the axial pitch check along the flute, at or in back of the cutting edge. An indicator and Jo blocks or a comparator are often applied.

TOOTH PROFILE

Pressure angle or profile—The departure of the actual tooth profile from the correct tooth profile as denoted by total indicator variation or by magnified layout comparison.

Tolerances shown apply to straight side profiles in the axial or normal section. Hobs with curved profiles are special and subject to individual consideration.

Tooth thickness—The difference between the measured thickness and the specified thickness at the hob pitch cylinder.

Start of Tip Relief Modification—The tolerance permitted in locating the point on the hob tooth, plus or minus, at which a profile modification begins.

Symmetry in start of Tip Relief Modification—The radial tolerance for the start of the modification with reference to the start of the modification on the opposite hob tooth profile.

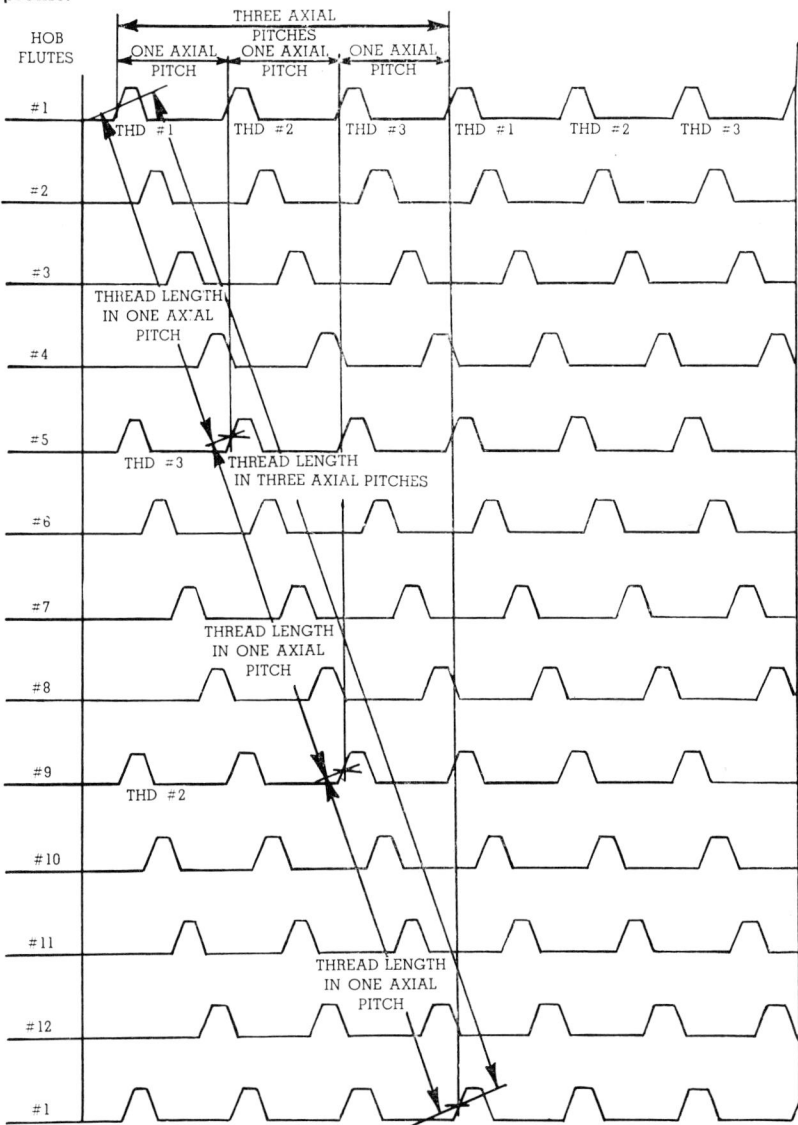

Fig. 37. *Development of a 3-Thread, 12-Flute Hob Through 360 Degrees*

Single Thread Coarse Pitch Gear Hob Tolerances[1]
(In ten thousandths of an inch)

Hob Element	CLASS	NORMAL DIAMETRAL PITCH[2]							
		1 Thru 1.99	2 Thru 2.99	3 Thru 3.99	4 Thru 4.99	5 Thru 5.99	6 Thru 8.99	9 Thru 12.99	13 Thru 19.99
RUNOUT[3]									
Hub Face	AA	–	–	2	2	2	1	1	1
	A	8	5	2	2	2	2	2	2
	B	10	8	4	4	3	3	2	2
	C	10	8	4	4	3	3	2	2
	D	10	8	5	5	4	4	3	3
Hub Diameter	AA		–	2	2	2	1	1	1
	A	10	5	4	3	3	3	2	2
	B	12	8	6	5	4	4	3	2
	C	12	8	6	5	4	4	3	2
	D	15	10	8	8	6	6	5	5
Outside Diameter	AA	–	–	5	4	3	3	3	3
	A	30	20	15	15	10	10	10	10
	B	40	30	25	20	15	15	15	10
	C	50	45	40	25	20	17	17	12
	D	60	55	50	45	35	35	30	25
LEAD[3]									
Tooth to Tooth	AA	–	–	4	3	2	1.7	1.7	1.7
	A	7	5	4	3	2	2	2	2
	B	10	8	6	4	3	3	3	3
	C	15	12	8	6	5	4	4	4
	D	25	20	16	14	12	10	10	8
In Any One Turn of Helix	AA	–	–	8	6	4	3	3	2
	A	25	18	10	8	6	5	5	4
	B	35	25	17	11	9	7	7	6
	C	45	35	22	14	11	9	9	8
	D	60	50	40	30	25	20	20	18
In Any Three Turns of Helix	AA	–	–	12	9	6	5	5	4
	A	38	26	15	12	9	8	8	7
	B	53	38	22	16	12	11	10	9
	C	70	50	30	21	16	14	13	12
	D	120	100	80	60	50	40	35	25
TOOTH									
Pressure Angle [3, 4]	AA	–	–	2	2	1.7	1.7	1.7	1.7
	A	10	5	3	3	2	2	2	2
	B	16	8	5	5	4	3	3	3
	C	25	15	10	5	4	3	3	3
	D	80	55	30	18	12	8	8	6
Thickness (minus only)	AA	–	–	15	15	10	10	10	10
	A	30	20	15	15	10	10	10	10
	B	30	20	15	15	10	10	10	10
	C	35	25	20	20	15	15	15	15
	D	40	35	30	25	20	20	20	20
Start of Tip Relief Modification (plus or minus)	AA	–	–	100	80	70	60	60	40
	A	200	180	160	140	120	100	80	60
	B	220	200	180	160	140	120	100	80
	C	220	200	180	160	140	120	100	80
	D	260	240	220	200	180	160	140	120
Symmetry in Start of Tip Relief Modification	AA	–	–	70	60	50	40	40	25
	A	150	130	120	100	90	80	60	50
	B	180	150	130	120	100	90	80	70
	C	180	150	130	120	100	90	80	70
	D	200	180	160	140	120	110	100	90

(Continued)

Single Thread Coarse Pitch Gear Hob Tolerances[1]

(In ten thousandths of an inch)

(Concluded)

Hob Element	CLASS	NORMAL DIAMETRAL PITCH[2]							
		1 Thru 1.99	2 Thru 2.99	3 Thru 3.99	4 Thru 4.99	5 Thru 5.99	6 Thru 8.99	9 Thru 12.99	13 Thru 19.99
FLUTES									
Adjacent Flute Spacing[5]	AA	–	–	20	15	10	8	8	6
	A	40	30	25	20	15	10	10	10
	B	50	45	40	30	20	15	15	10
	C	50	45	40	30	20	15	15	10
	D	60	60	50	50	30	25	25	20
Non-Adjacent Flute Spacing[5]	AA	–	–	40	35	25	15	15	15
	A	80	60	50	40	30	30	30	25
	B	100	90	80	60	50	50	50	40
	C	100	90	80	60	50	50	50	40
	D	120	120	100	100	80	80	70	60
Rake To Cutting Depth[6]	AA	–	–	10	8	6	5	5	3
	A	30	15	10	8	6	5	5	3
	B	50	25	15	10	8	7	7	5
	C	50	25	15	10	8	7	7	5
	D	100	75	50	40	30	20	20	15

		CUTTING FACE WIDTH				
Flute Lead Over Cutting Face Width		Up to 1	1.001 to 2	2.001 to 4	4.001 to 7	7.001 & Up
	AA	8	10	15	20	20
	A	10	15	25	30	50
	B	10	15	25	30	50
	C	10	15	25	30	50
	D	15	23	38	45	75

HOLE		HOLE DIAMETER					
		2.500	2.000	1.500	1.250	.750	.500 & Smaller
Hole Diameter (plus only)	AA	–	–	–	2	2	2
	A	8	8	5	2	2	2
	B	10	10	8	3	2	2
	C	10	10	8	3	2	2
	D	10	10	8	5	4	3

NOTE: 1. Tolerances apply only to standard hob sizes.

2. For combination pitch hobs, the coarser of the two pitches shall apply.

3. Total indicator variation.

4. Exclusive of Tip Relief Modification.

5. Compared against master index plate.

6. Radial (zero rake) tooth faces are standard.

Multiple Thread Coarse Pitch Gear Hob Tolerances[1]

(In ten thousands of an inch)

Hob Elements	CLASS	NORMAL DIAMETRAL PITCH[2]							
		1 Thru 1.99	2 Thru 2.99	3 Thru 3.99	4 Thru 4.99	5 Thru 5.99	6 Thru 8.99	9 Thru 12.99	13 Thru 19.99
RUNOUT[3]									
Hub Face	A	8	5	2	2	2	2	2	2
	B	10	8	4	4	3	3	2	2
	C	10	8	4	4	3	3	2	2
	D	10	8	5	5	4	4	3	3
Hub Diameter	A	10	5	4	3	3	3	2	2
	B	12	8	6	5	4	4	3	2
	C	12	8	6	5	4	4	3	2
	D	15	10	8	8	6	6	5	5
Outside Diameter	A	30	20	15	15	10	10	10	10
	B	40	30	25	20	15	15	15	10
	C	50	45	40	25	20	17	17	12
	D	60	55	50	45	35	35	30	25
LEAD[3]									
In One Axial Pitch	A	25	20	10	8	6	5	5	4
	B	35	30	17	12	10	8	8	7
	C	45	35	22	18	15	12	12	10
	D	60	50	40	30	25	20	20	18
In Three Axial Pitches	A	38	30	15	12	9	8	8	7
	B	53	38	22	20	15	12	12	10
	C	70	50	30	28	20	18	16	14
	D	120	100	80	60	50	40	35	25
Tooth to Tooth 2 Thread Hob	A	8	6	5	4	3	3	3	3
	B	12	10	7	6	5	5	5	4
	C	18	14	10	9	7	6	6	5
	D	27	22	18	16	14	12	11	9
Tooth to Tooth 3 Thread Hob	A	9	7	6	4	4	4	3	3
	B	14	12	8	7	6	6	5	5
	C	21	16	12	10	8	7	6	5
	D	29	24	20	18	16	14	12	10
Tooth to Tooth 4 Thread Hob	A	10	7	6	5	4	4	4	3
	B	16	13	9	8	7	6	6	5
	C	24	18	13	11	9	7	7	6
	D	31	26	22	20	18	16	13	11

(Continued)

Multiple Thread Coarse Pitch Gear Hob Tolerances[1]

(In ten thousandths of an inch)

(Continued)

Hob Element	CLASS	NORMAL DIAMETRAL PITCH[2]							
		1 Thru 1.99	2 Thru 2.99	3 Thru 3.99	4 Thru 4.99	5 Thru 5.99	6 Thru 8.99	9 Thru 12.99	13 Thru 19.99
ADJACENT THREAD SPACING									
2 Thread Hob	A	11	9	8	7	6	5	4	3
	B	14	12	11	10	9	8	6	5
	C	20	17	15	13	11	10	9	8
	D	26	22	19	17	15	13	12	11
3 Thread Hob	A	13	11	10	8	7	6	5	4
	B	16	14	12	11	10	9	7	6
	C	22	19	16	14	13	11	10	9
	D	28	24	20	18	16	15	13	12
4 Thread Hob	A	15	13	12	9	8	7	6	5
	B	18	16	14	12	11	10	8	7
	C	24	21	18	15	14	12	11	10
	D	30	26	22	20	18	16	14	13
TOOTH									
Pressure Angle[3, 4] 2 Thread Hob	A	12	7	5	4	3	3	2	2
	B	18	10	7	5	5	4	3	3
	C	27	16	11	7	5	4	3	3
	D	80	55	30	18	12	8	8	7
Pressure Angle[3, 4] 3 Thread Hob	A	15	8	5	4	3	3	3	2
	B	20	10	7	5	5	4	4	3
	C	27	16	11	7	5	4	4	3
	D	80	55	30	18	12	8	8	7
Pressure Angle[3, 4] 4 Thread Hob	A	15	8	5	4	3	3	3	2
	B	20	10	7	5	5	4	4	3
	C	27	16	11	7	5	4	4	3
	D	80	55	30	18	12	8	8	7
Thickness (minus only)	A	30	20	15	15	10	10	10	10
	B	30	20	15	15	10	10	10	10
	C	35	25	20	20	15	15	15	15
	D	40	35	30	25	20	20	20	20
Start of Tip Relief Modification (plus or minus)	A	200	180	160	140	120	100	80	60
	B	220	200	180	160	140	120	100	80
	C	220	200	180	160	140	120	100	80
	D	260	240	220	200	180	160	140	120
Symmetry in Start of Tip Relief Modification	A	150	130	120	100	90	80	60	50
	B	180	150	130	120	100	90	80	70
	C	180	150	130	120	100	90	80	70
	D	200	180	160	140	120	110	100	90

(Continued)

Multiple Thread Coarse Pitch Gear Hob Tolerances[1]

(In ten thousandths of an inch)

(Concluded)

Hob Element	CLASS	NORMAL DIAMETRAL PITCH[2]							
		1 Thru 1.99	2 Thru 2.99	3 Thru 3.99	4 Thru 4.99	5 Thru 5.99	6 Thru 8.99	9 Thru 12.99	13 Thru 19.99
FLUTES									
Adjacent Flute Spacing[5]	A	40	30	25	20	15	10	10	10
	B	50	45	40	30	20	15	15	10
	C	50	45	40	30	20	15	15	10
	D	60	60	50	50	30	25	25	20
Non-Adjacent Flute Spacing[5]	A	80	60	50	40	30	30	30	25
	B	100	90	80	60	50	50	50	40
	C	100	90	80	60	50	50	50	40
	D	120	120	100	100	80	80	70	60
Rake to Cutting Depth[6]	A	30	15	10	8	6	5	5	3
	B	50	25	15	10	8	7	7	5
	C	50	25	15	10	8	7	7	5
	D	100	75	50	40	30	20	20	15

Flute Lead Over Cutting Face Width		CUTTING FACE WIDTH				
		Up to 1	1.001 to 2	2.001 to 4	4.001 to 7	7.001 & Up
	A	10	15	25	30	50
	B	10	15	25	30	50
	C	10	15	25	30	50
	D	15	23	38	45	75

HOLE		HOLE DIAMETER					
		2.500	2.000	1.500	1.250	.750	.500 & Smaller
Hole Diameter (plus only)	A	8	8	5	2	2	2
	B	10	10	8	3	2	2
	C	10	10	8	3	2	2
	D	10	10	8	5	4	3

NOTE: 1. Tolerances apply only to standard hob sizes.
2. For combination pitch hobs, the coarser of the two pitches shall apply.
3. Total indicator variation.
4. Exclusive of Tip Relief Modification.
5. Compared against master index plate.
6. Radial (zero rake) tooth faces are standard.

Single Thread Fine Pitch Gear Hob Tolerances[1]
(In ten thousandths of an inch)

Hob Element	Class	NORMAL DIAMETRAL PITCH[2]		
		20 Thru 29.99	30 Thru 50.99	51 and Finer
RUNOUT[3]				
Hub Face	AA	1	1	1
	A	2	2	2
	B	2	2	–
	C	2	2	2
	D	3	3	–
Hub Diameter	AA	1	1	1
	A	2	2	2
	B	2	2	–
	C	2	2	2
	D	4	3	–
Outside Diameter	AA	2	2	2
	A	10	7	5
	B	10	7	–
	C	12	10	8
	D	20	15	–
LEAD[3]				
Tooth to Tooth	AA	1.7	1.5	1.5
	A	2	2	2
	B	3	2	–
	C	4	3	3
	D	6	5	–
In Any One Turn of Helix	AA	2	1.5	1.5
	A	4	3	3
	B	6	4	–
	C	8	8	6
	D	16	14	–
In Any Three Turns of Helix	AA	4	3	3
	A	7	5	5
	B	9	7	–
	C	12	12	8
	D	20	16	–
TOOTH				
Pressure Angle[3, 4]	AA	1.7	1.5	1.5
	A	2	2	2
	B	3	2	–
	C	3	3	3
	D	5	4	–
Thickness (minus only)	AA	10	5	5
	A	10	5	5
	B	10	5	–
	C	15	10	10
	D	20	15	–
Start of Tip Relief Modification (plus or minus)	AA	40	30	–
	A	40	30	–
	B	50	40	–
	C	60	50	–
	D	100	80	–
Symmetry in Start of Tip Relief Modification	AA	25	25	–
	A	35	25	–
	B	45	35	–
	C	55	45	–
	D	80	60	–

(Continued)

Single Thread Fine Pitch Gear Hob Tolerances[1]

(In ten thousandths of an inch)
(Concluded)

Hob Element	Class	NORMAL DIAMETRAL PITCH[2]		
		20 Thru 29.99	30 Thru 50.99	51 and Finer
FLUTES				
Adjacent Flute Spacing[5]	AA	6	6	6
	A	10	10	10
	B	10	10	–
	C	10	10	10
	D	17	17	–
Non-Adjacent Flute Spacing[5]	AA	15	15	15
	A	25	20	20
	B	35	30	–
	C	35	30	30
	D	50	40	–
Rake to Cutting Depth[6]	AA	3	3	3
	A	3	3	3
	B	5	5	–
	C	5	5	5
	D	15	10	–

		CUTTING FACE WIDTH		
		Up to 1	1.001 to 2	2.001 to 4
Flute Lead Over Cutting Face Width	AA	8	10	15
	A	10	15	25
	B	10	15	25
	C	10	15	25
	D	15	23	38

HOLE		HOLE DIAMETER		
		1.250	.750	.500 and Smaller
Hole Diameter (plus only)	AA	2	2	2
	A	2	2	2
	B	3	2	2
	C	3	2	2
	D	5	4	3

NOTE: 1. Tolerances apply only to standard hob sizes.
2. For combination pitch hobs, the coarser of the two pitches shall apply.
3. Total indicator variation.
4. Exclusive of Tip Relief Modification.
5. Compared against master index plate.
6. Radial (zero rake) tooth faces are standard.

Multiple Thread Fine Pitch Gear Hob Tolerances[1]

(In ten thousandths of an inch)

Hob Element	Class	NORMAL DIAMETRAL PITCH[2]	
		20 Thru 29.99	30 Thru 50.99
RUNOUT[3]			
Hub Face	A	2	2
	B	2	2
	C	2	2
Hub Diameter	A	2	2
	B	2	2
	C	2	2
Outside Diameter	A	10	7
	B	10	7
	C	12	10
LEAD[3]			
In One Axial Pitch	A	4	3
	B	7	4
	C	10	8
In Three Axial Pitches	A	7	5
	B	10	7
	C	14	12
Tooth to Tooth 2 Thread Hob	A	2	2
	B	3	2
	C	5	3
Tooth to Tooth 3 Thread Hob	A	3	2
	B	4	3
	C	5	4
Tooth to Tooth 4 Thread Hob	A	3	3
	B	4	4
	C	5	4
ADJACENT THREAD SPACING			
2 Thread Hob	A	3	3
	B	5	5
	C	7	6
3 Thread Hob	A	4	4
	B	6	6
	C	8	7
4 Thread Hob	A	4	4
	B	7	6
	C	9	8

(Continued)

Multiple Thread Fine Pitch Gear Hob Tolerances[1]

(In ten thousandths of an inch)

Hob Element	Class	NORMAL DIAMETRAL PITCH[2]	
		20 Thru 29.99	30 Thru 50.99
TOOTH			
Pressure Angle[3, 4] 2 Thread Hob	A	2	2
	B	3	2
	C	3	3
Pressure Angle[3, 4] 3 Thread Hob	A	2	2
	B	3	2
	C	3	3
Pressure Angle [3, 4] 4 Thread Hob	A	2	2
	B	3	2
	C	3	3
Thickness (minus only)	A	10	5
	B	10	5
	C	15	10
Start of Tip Relief Modification (plus or minus)	A	50	40
	B	60	50
	C	60	50
Symmetry in Start of Tip Relief Modification	A	40	30
	B	60	50
	C	60	50
FLUTES			
Adjacent Flute Spacing[5]	A	10	10
	B	10	10
	C	10	10
Non-Adjacent Flute Spacing[5]	A	25	20
	B	35	30
	C	35	30
Rake to Cutting Depth[6]	A	3	3
	B	5	5
	C	5	5

		CUTTING FACE WIDTH		
Flute Lead Over Cutting Face Width		Up to 1	1.001 to 2	2.001 to 4
	A	10	15	25
	B	10	15	25
	C	10	15	25

(Continued)

Multiple Thread Fine Pitch Gear Hob Tolerances[1]

(In ten thousandths of an inch)
(Concluded)

Hob Element	Class	HOLE DIAMETER		
		1.250	.750	.500 and Smaller
HOLE				
Hole Diameter (plus only)	A	2	2	2
	B	3	2	2
	C	3	2	2

NOTE: 1. Tolerances apply only to standard hob sizes.
2. For combination pitch hobs, the coarser of the two pitches shall apply.
3. Total indicator variation.
4. Exclusive of Tip Relief Modification.
5. Compared against master index plate.
6. Radial (zero rake) tooth faces are standard.

MARKING SYMBOLS

Hob Element or Feature	Standard Marking Symbol
Normal Circular Pitch	x.xxNCP
Normal Diametral Pitch	x NDP
Normal Pressure Angle	x° NPA
Number of Threads	x THD.
Hand of Thread	x H
Whole Depth of Gear Tooth	x.xxx WD
Lead Angle	x° xx' LA
Flute Lead	xxx.xx FL
Positive Rake	x°-.xxx
Negative Rake	NEG x°-.xxx
Pre-Shave	PRE-S
Pre-Grind	PRE-GR
Topping	TOP
Semi-Topping	S-TOP
Class	Class x
Roughing	RGH
Finishing	FIN
Number of Teeth In Spline	x-N
Spline O.D.	x.xx OD
Spline Tooth Thickness	.xxxT

Standard Keyways for Hobs

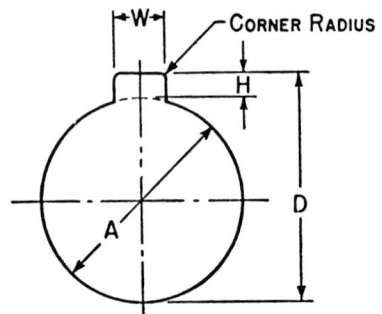

Nominal Hole Diameter A	Nominal Keyway W x H	W Max.	W Min.	D Max.	D Min.	Max. Corner Radius
*.315 (8 mm)	1/16" x 1/32"	.068	.065	.351	.347	.008
.3937 (10 mm)	1/16" x 1/32"	.068	.065	.431	.427	.008
1/2"	1/8" x 1/16"	.137	.130	.583	.573	1/32"
3/4"	1/8" x 1/16"	.137	.130	.833	.823	1/32"
1 1/4"	1/4" x 1/8"	.262	.255	1.395	1.385	3/64"
1 1/2"	3/8" x 3/16"	.400	.385	1.708	1.698	1/16"
2"	1/2" x 1/4"	.525	.510	2.270	2.260	1/16"
2 1/2"	5/8" x 5/16"	.650	.635	2.838	2.828	1/16"

*Keyway is optional.

WORM AND WORM GEAR HOBS

Worms

Worms are produced by turning, rolling, milling, grinding, hobbing, or with a thread generating cutter. The thread profiles or helicoids produced by each process are different. When lead angles and thread depth are small, the differences are negligible, but when lead angles are large or threads deep, the thread profiles differ appreciably. These differences are a most important consideration in the design of worm gear hobs.

Worm Hobs

When hobbed, the worm threads are involute helicoids and the worm is actually a helical gear having a rather large helix angle and a number of teeth equal to the number of threads in the worm.

Worms with three or more threads can be hobbed in most hobbing machines, although special equipment or attachments may be needed. Under some conditions even a two-thread worm can be hobbed, but it is generally considered impractical to hob single thread worms.

Worm hobs are usually made single thread. The same worm hob can be used to cut a worm with any number of threads, but all worms cut by that hob would have the same normal circular pitch and the same normal pressure angle.

For better finish, or to avoid machine interferences, worm hobs are usually larger in diameter and have more flutes than a gear hob of equivalent pitch.

Worm hobs are tapered on the entering end to divide and reduce the initial cutting loads, and are narrower than gear hobs of equivalent pitch to avoid the drag on the finished worm threads that would result from an excessive number of finishing hob teeth.

Worm hobs are made in the same tolerance classes as gear hobs, but the tolerances for a worm hob that is larger in diameter than a gear hob of corresponding pitch are increased in proportion to the increase in diameter.

Worm Gear Hobs

Worm gears may be produced by infeed or tangential hobbing.

In the infeed method the workpiece or the hob is fed radially to the desired depth. See Fig. 12. Usually the infeed hob can be used in several settings across its cutting face width.

Under certain conditions (principally a small hob diameter, small number of teeth in the worm gear, deep thread form or an Even Gear Tooth-Hob Thread Ratio) an infeed hob will produce worm gear teeth with undesirably large generating flats.

In the tangential feed method the entire length of hob is fed along its axis through the gear. See Fig. 13. The tangential feed hob is tapered. The small end contains short, and thin roughing teeth, the large end the full depth finishing teeth.

Tangential hobbing, though slower than infeed hobbing, tends to produce more accurate worm gears and, if the proper feed is selected, there need be no visible generating flats.

The number of threads in a worm gear hob is usually the same as the number of threads in the worm.

For desirable contact conditions, the thread form of worm gear hobs should be a close duplicate of the worm. Since differences in manufacturing methods result in appreciable differences in worm thread profile when lead angles are large or threads deep, the hob maker must have sufficiently detailed specifications of the worm thread form, a sample worm, or pertinent information on the method of manufacture.

Worm gear hobs are made in three classes:

> Class A
> Class B
> Class D

The tolerances for each class are tabulated in the table on page 718.

Worm gear hob pitch diameters are made slightly larger than the worm pitch diameter. This enlargement is the sharpening allowance. Sharpening allowance on worm gear hobs tends to alter and reduce the contact area between the worm and worm gear. This effect is accentuated when the lead angles are large and the thread form is deep. Therfore, the sharpening allowance should be less under these conditions than when worm lead angles are small and thread form shallow.

For best contact conditions the sharpening allowance in a worm gear hob requires an angular adjustment of the hob axis from the 90° position with the axis of the worm gear. The angle of adjustment, or tipping angle, is approximately the difference between the worm lead angle and the worm gear hob lead angle calculated at the enlarged hob pitch diameter.

Since each sharpening reduces the hob pitch diameter, a change in the tipping angle is generally required after each sharpening.

Sharpening the hob to a point where its pitch diameter is less than that of the worm usually results in undesirable contact which cannot be corrected by any angular adjustment of the hob.

TOLERANCES FOR WORM GEAR HOBS

PURPOSE

The purpose is to provide dimensional tolerances applicable to the manufacture of single and multiple-thread hobs for generating worm gear tooth forms.

SCOPE

Tolerances are given for hobs having 1 through 6 threads. Since worm gear hob diameters are special and depend primarily upon the worm diameter, the tolerances are based upon corresponding recommended hob sizes for spur and helical gears. Diameters larger than these are subject to tolerance increases. Tolerances are not decreased for smaller than recommended hob diameters.

CLASSES

Hobs are classified with reference to dimensional tolerances as:

Class A
Class B
Class D

TOOTH PROFILE

Worm gear hobs often require curved tooth profiles. The tolerances as shown apply only to straight-sided profiles. Hobs with curved tooth profiles are special and are subject to individual consideration.

REMARKS

The profile of worm threads depends largely upon the manufacturing process employed. Since the hob tooth profile is based upon the worm thread profile, the hob manufacturer should be supplied with the pertinent worm processing data. For worms having high lead angles and deep forms, the hob manufacturer should be supplied with a sample worm.

Tolerances for Single- and Multiple-Thread Worm-Gear Hobs

Class "A"
Class "B"
Class "D"

All readings in tenths of a thousandth of an inch

AXIAL PITCH OF WORM	CLASS	3.000 Thru 1.501	1.500 Thru 1.001	1.000 Thru .751	.750 Thru .626	.625 Thru .501	.500 Thru .351	.350 Thru .251	.250 Thru .161	.160 Thru .101	.100 Thru .061	.060 and Finer
RUNOUT (1 - 6 Threads)												
Hub Face*	A	8	5	2	2	2	2	2	2	2	2	2
	B	10	8	4	4	3	3	2	2	2	2	
	D	10	8	5	5	4	4	3	3	3	3	3
Hub Diameter*	A	10	5	4	3	3	3	2	2	2	2	2
	B	12	8	6	5	4	4	3	2	2	2	
	D	15	10	8	8	6	6	5	5	4	4	4
Pilot and Taper or Straight Shank						For all classes and pitches — 2						
LEAD VARIATION												
Tooth to Tooth* 1 Thread	A	7	5	4	3	2	2	2	2	2	2	2
	B	10	8	6	4	3	3	3	3	3	2	
	D	25	20	16	14	12	10	10	8	5	4	4
2 Thread	A	8	6	5	4	3	3	3	3	2	2	2
	B	12	10	7	6	5	5	5	4	3	2	
	D	27	22	18	16	14	12	11	9	5	4	4
3 Thread	A	9	7	6	4	4	4	3	3	3	2	2
	B	14	12	8	7	6	6	5	5	4	3	
	D	29	24	20	18	16	14	12	10	6	5	5
4 Thread	A	10	7	6	5	4	4	4	3	3	3	2
	B	16	13	9	8	7	6	6	5	4	4	
	D	31	26	22	20	18	16	13	11	7	5	5

5 Thread	A	12	8	7	5	4	4	4	4	3	3	3
	B	18	14	10	9	8	7	6	6	5	4	.
	D	33	28	24	22	20	17	14	12	8	6	6
6 Thread	A	13	9	7	6	5	4	4	4	4	3	3
	B	19	15	11	9	8	7	7	7	6	5	.
	D	35	30	26	24	21	18	15	13	9	7	6
Any one Axial Pitch* 1 Thread	A	25	18	10	8	6	5	5	4	4	3	3
	B	35	25	17	11	9	7	7	6	6	4	.
	D	60	50	40	30	25	20	20	18	16	12	8
2 - 6 Threads	A	25	20	10	8	6	5	5	4	4	3	3
	B	35	30	17	12	10	8	8	7	7	4	.
	D	60	50	40	30	25	20	20	18	16	12	8
Any Three Axial Pitches* 1 Thread	A	38	26	15	12	9	8	8	7	7	5	5
	B	53	38	22	16	12	11	10	9	9	7	.
	D	120	100	80	60	50	40	35	25	20	16	12
2 - 6 Threads	A	38	30	15	12	9	8	8	7	7	5	5
	B	53	38	22	20	15	12	10	10	10	7	.
	D	120	100	80	60	50	40	35	25	20	16	12
Adjacent Thread to Thread Spacing* 2 Thread	A	11	9	8	7	6	5	4	3	3	3	3
	B	14	12	11	10	9	8	6	5	5	5	.
	D	26	22	19	17	15	13	12	11	7	6	6
3 Thread	A	13	11	10	8	7	6	5	4	4	4	3
	B	16	14	12	11	10	9	7	6	6	6	.
	D	28	24	20	18	16	15	13	12	8	7	7
4 Thread	A	15	12	9	8	7	6	5	4	4	4	3
	B	18	14	12	11	10	8	7	7	6	6	.
	D	30	22	20	18	16	14	13	9	7	7	7

(Continued on following page)

* Total Indicator Variation.

(Continued from preceding page)

CIRCULAR PITCH	CLASS	3.000 Thru 1.501	1.500 Thru 1.001	1.000 Thru .751	.750 Thru .626	.625 Thru .501	.500 Thru .351	.350 Thru .251	.250 Thru .161	.160 Thru .101	.100 Thru .061	.060 and Finer
5 Thread	A	17	15	14	10	9	8	7	6	5	1	3
	B	21	18	16	13	12	11	9	8	7	6	·
	D	32	28	24	22	20	17	15	14	10	8	8
6 Thread	A	19	17	16	12	10	9	8	7	6	5	4
	B	23	20	18	15	13	11	10	9	8	7	·
	D	34	30	26	24	22	18	16	15	11	9	8
TOOTH PROFILE												
Pressure Angle or Profile*(1) 1 Thread	A	10	5	3	3	2	2	2	2	2	2	2
	B	16	8	5	5	4	3	3	3	3	2	·
	D	80	55	30	18	12	8	8	6	6	5	4
2 Thread	A	12	7	5	4	3	3	2	2	2	2	2
	B	18	10	7	5	5	4	3	3	3	2	·
	D	80	55	30	18	12	8	8	7	7	6	5
3 and 4 Threads	A	15	8	5	4	3	3	3	2	2	2	2
	B	20	10	7	5	5	4	4	3	3	2	·
	D	80	55	30	18	12	8	8	7	7	6	5
5 and 6 Threads	A	15	8	5	5	4	4	3	3	3	2	2
	B	20	10	7	6	6	5	4	3	3	3	·
	D	80	55	30	18	12	9	8	7	7	6	5
Tooth Thickness (minus only) 1 - 6 Threads	A	30	20	15	15	10	10	10	10	10	5	5
	B	30	20	15	15	10	10	10	10	10	5	·
	D	40	35	30	25	20	20	20	20	15	10	10

SHARPENING (1 - 6 Threads)

Adjacent Flute Spacing*	A	40	30	25	20	15	10	10	10	10	10	10
	B	50	45	40	30	20	15	15	10	10	10	··
	D	60	60	50	50	30	25	25	20	15	10	10
Non-Adjacent Flute Spacing*	A	80	60	50	40	30	30	30	25	25	20	20
	B	100	90	80	60	50	50	50	35	30	30	··
	D	120	120	100	100	80	80	70	50	40	40	35
Rake to Cutting Depth*	A	30	15	10	8	6	5	5	3	3	3	3
	B	50	25	15	10	8	7	7	5	5	5	··
	D	100	75	50	40	30	20	20	15	12	10	8

CUTTING FACE WIDTH

		0 to 1"	1" to 2"	2" to 4"	4" to 7"	7" and up
Flute Lead	A	10	15	25	30	50
	B	10	15	25	30	50
	D	15	23	38	45	75

HOLE (1 - 6 Threads)

HOLE DIAMETER

		2.500"	2.000"	1.500"	1.250"	.750"	.500" & smaller
Diameter, Straight Hole (Plus Only)	A	8	8	5	2	2	2
	B	10	10	8	3	2	2
	D	10	10	8	5	4	3

ALL DIAMETERS

LENGTH

Percent of Bearing Contact, Straight Hole	A	75
	B	75
	D	50

* Total Indicator Variation.

(a) Tolerances apply only to straight side profiles in axial or normal section. Hobs with curved profiles are special and are subject to individual consideration.

SPECIAL HOBS

There are many types of special hobs. The discussion that follows includes some of the more common ones.

Single-Position Hobs

Single-position hobs are single-purpose hobs which can be used in one position only. Although they are applicable to many special forms, they have been extensively used for the production of various types of ratchet wheels. When the tooth form desired is sharp at the root and has a radial, straight, or undercut profile, a single-position hob will usually be required. The single-position hob is used when the desired tooth form can not be produced by a full generating hob. Single-position hobs can be made in either ground or unground form, depending upon tooth form and accuracy requirements. Single-position hob sizes are specially determined for each application. Such hobs are sharpened in the normal manner. As shown in the illustrations, the hob must have a specific location relative to the workpiece. Only the correct location will permit the hob to produce the desired profile upon the workpiece.

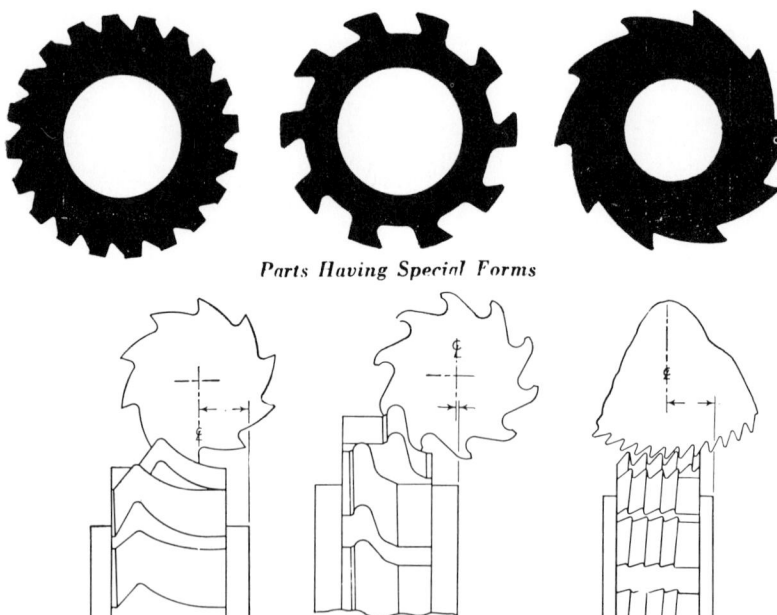

Parts Having Special Forms

Fig. 38. *Single-Position Hobs for Special Forms*

Alternate-Tooth Hobs

Alternate-tooth hobs are a special type of hob in which the cutting teeth are staggered from row to row. This is accomplished by removing or eliminating alternate tooth forms along each flute. Alternate tooth forms along the hob thread are eliminated so that each flute appears to be staggered in relation to the one preceding it. Hobs of this type are used for hobbing particularly long and narrow teeth, and teeth which have pointed tops.

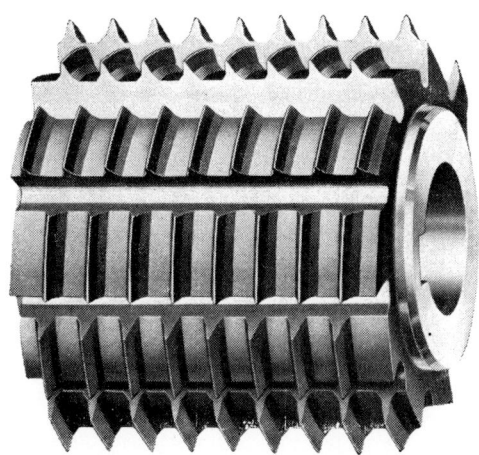

Fig. 39. *Alternate-Tooth Hob*

Gear-Chamfering Hobs

Gear-chamfering hobs are generally used for chamfering one side of a gear tooth profile at one end of a gear. Such hobs were primarily designed for chamfering flywheel ring gears. They are generally of single purpose design for use on a particular application. Gear-chamfering hobs can also be made of such design that they will simultaneously chamfer both sides of a gear tooth profile. See Fig. 40.

HOBBING FEEDS AND SPEEDS

As with most machining operations, a number of factors influence the optimum speed-feed combination. In hobbing, these factors include (1) material being hobbed, (2) finish, (3) accuracy, (4) machine capacity, (5) hob specifications, and (6) tool life. In most cases, finish and accuracy are established items. The problem is to determine a hob speed and feed for a given hob and machine capacity that will produce the established finish and accuracy at a given production rate with justifiable tool life.

Available information for determining hob speed and feed for a specific job can usually be found in past performance records, test data, and published articles.

Fig. 40. *Chamfering Hob and Chamfered Gear*

HOBBING PRACTICE

Hobs can produce accurate work only if the hobbing setup is such that the hob and workpiece are accurately rotated and fed together in the proper timed relationship. A poor, loose, non-rigid setup will yield poor work even with the most accurate hob.

The hobbing machine and its spindle must be kept in good condition. The spindle must run true and its axial play must be reduced to a minimum. Any wobble or runout of the hob can cause workpiece profile errors and thus nullify the inherent accuracy of the hob. To maintain accurate indexing, the index worm and worm gear must be adjusted to the minimum amount of backlash permitted by the operating speed. When single thread hobs are used, any angular spacing errors on the workpiece can be attributed to faulty indexing by the hobbing machine.

Tooling arrangements for supporting, locating and clamping the workpiece must be accurate and rigid. The fixturing must be rigid enough to resist deflection by the cutting forces, but care must be taken that the clamping arrangements do not distort the workpiece.

HOB SHARPENING

To maintain its initial accuracy, a hob must be sharpened to tolerances corresponding to its over-all accuracy. The hob user should be aware of this responsibility. Inspection of all hobs after each sharpening will assure best results. In many instances, hobs are inspected before and after sharpening so that the amount of hob tooth removed by sharpening is based upon the amount of wear obtained from hobbing.

Good grinding practice must be used in sharpening hobs to produce a satisfactory surface finish and to avoid residual stresses which may cause grinding cracks.

Hob sharpening errors influence the accuracy produced in the parts being hobbed. Figs. 41-44 show some of the more common sharpening errors and their effect upon involute tooth profiles.

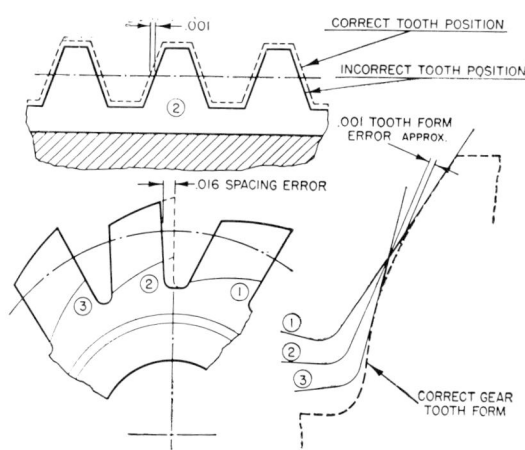

Fig. 41. *Serious hob lead errors result from sharpening with unequal spacing of cutting faces*

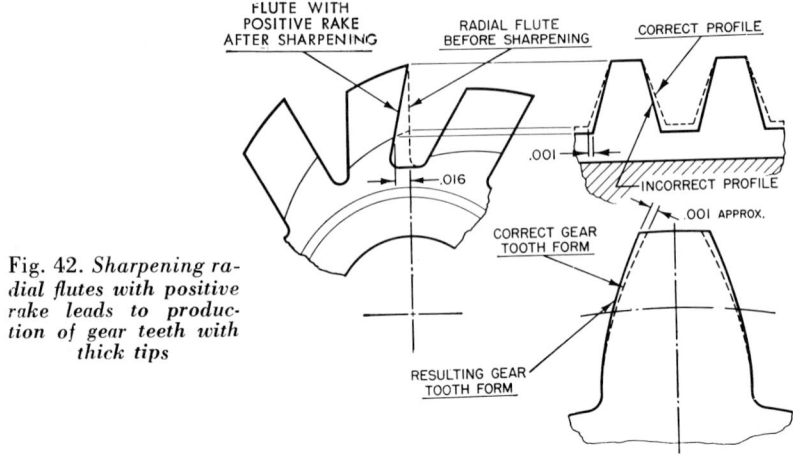

Fig. 42. *Sharpening radial flutes with positive rake leads to production of gear teeth with thick tips*

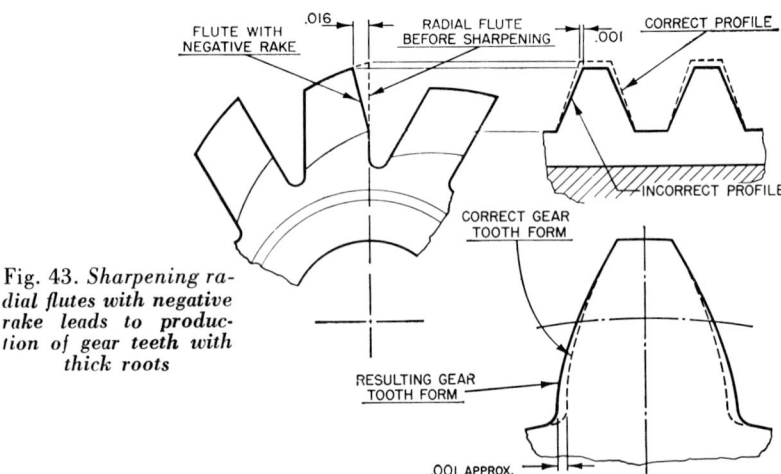

Fig. 43. *Sharpening radial flutes with negative rake leads to production of gear teeth with thick roots*

Fig. 44. *Flute lead er-rors in sharpening tend to produce distorted "leaning" teeth on the finished gear*

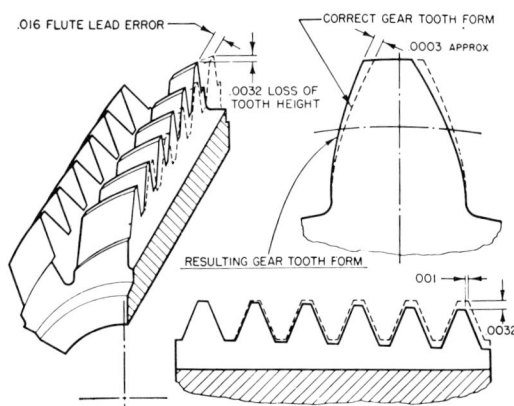

GEAR SHAVING CUTTER SECTION

GEAR SHAVING SECTION CONTENTS

FOREWORD

The gear shaving process is an accepted method for producing accurate gears. Performed after the teeth have been generated by hobbing, shaping or broaching methods, the shaving process removes small amounts of material from the gear tooth profiles to provide high quality gears with excellent surface finish characteristics.

The continuing trend to higher speeds and more compact designs in gear drives for instruments, machinery, automobiles, aircraft engines and ships has meant that more and more accuracy is required in the shaving tools, as well as in the methods of shaving.

Continuing research programs in the laboratories of our tool companies as well as similar work by tool material producers enables the accuracy of gear shaving cutters to keep pace with the requirements of industry.

Shaving cutter design is a complex process that relies on a broad background of experience. As a result, the intricate design details of the cutters are usually specified by the tool manufacturers.

This section is compiled to familiarize design engineers and tool engineers with the types of shaving cutters in general usage, the standard tolerances to which they are manufactured and basic cutter application information. Many of the details contributing to the ultimate success of the gear shaving process such as material, gear blank accuracy, and preshaving production equipment are necessarily omitted.

The gear shaving process can be applied to involute internal or external gears having either spur or helical teeth (see Figs. 5 and 6). Gear quality improvement by shaving is a result of producing accurate profiles, tooth spacing, helix angle, eccentricity and improving surface finish. There are two basic methods of shaving external spur or helical gears. One uses a serrated-tooth rotary tool in the form of a helical gear, and the other uses a serrated tooth tool in the form of a rack. Internal spur or helical gears are shaved by the rotary shaving method.

Rotary gear shaving is accomplished by a shaving cutter that has the same normal diametral pitch and normal pressure angle as the gear being shaved. The helix angle of the cutter normally differs from that of the work gear by from 5-deg to 18 degrees. The work and the cutter are rotated in tight mesh under pressure with a relatively slow feed per revolution. Axial sliding between the mating teeth resulting from the crossed axes relationship of the cutter and work gear causes the edges of the serrated cutter profiles to remove small amounts of stock from the work gear in the form of curled hair-like chips.

Shaving of the full face is accomplished by reciprocating either the work gear or the cutter. Rotation of the cutter in mesh with the work is accomplished by driving either the work gear or the cutter.

The teeth may be shaved to a constant tooth thickness (straight) by the rotary shaving process, or they may be made to taper by tilting the cutter or work gear. When gear configuration precludes taper shaving by tilting work gear or cutter, the cutter can be ground with a counterpart taper. Crowned gear teeth, in which the teeth are thinner at the ends than in the middle, can be accomplished either by rocking the work gear during the shaving operation or by using a cutter with reverse-crowned teeth.

On some gears both taper and crown are required. These features can be shaved simultaneously when the machine is equipped with both taper and crowning attachment.

NOMENCLATURE OF GEAR SHAVING CUTTERS

I—General Types of Gear Shaving Cutters

A—Rotary Gear Shaving Cutters.

A rotary gear shaving cutter is a helical gear-like tool having tooth forms conjugate to the forms of the teeth to be produced on the shaved gear. The faces of each of the teeth on the shaving cutter have a plurality of narrow grooves separated by narrow lands. These grooves are called serrations. The edges of the lands form parallel cutting edges. See Figure 4.

B—Rack Type Gear Shaving Cutters.

A rack type shaving cutter is a rack shaped tool on which the faces of each of the teeth are provided with a plurality of narrow parallel grooves separated by narrow lands. These grooves are called serrations. The edges of the lands form parallel cutting edges. See Figure 11.

II—General Classifications

A—Classification Based on Construction.

1—Rotary shaving cutters are solid cutters made of one piece of tool material such as high speed steel.

2—Rack shaving cutters are of solid flat tool material such as high speed steel. Each rack blade is made of a separate piece.

B—Classification Based on Type.

1—*Rotary Gear Shaving Cutters, Specific Types*—The exact detailed specifications as well as overall dimensions for rotary gear shaving cutters depend upon the type of shaving process in which they will be used. See Figures 1, 2 and 3.

When specifying shaving cutters it is necessary to indicate the type of shaving process to be used, and whether the gear teeth are to be crowned or tapered.

[a] *Axial Traverse Cutters* [also referred to as "Conventional" or "Transverse" shaving cutters].—In Axial-Traverse Shaving the workgear is fed along a path parallel to its axis, with the center line (point pivot) of the tool passing from one edge of the gear face to the opposite edge. For this type of shaving, it is necessary to apply an upfeed, gradually reducing the size of the gear to be shaved by controlled increments of feed, shortening the center distance between the tool and the work. Crowned gear teeth are produced by rocking the table during the shaving cycle.

Although the process is particularly adaptable to shaving extremely wide-faced gears and long involute splines, narrow-faced gears can be shaved equally well.

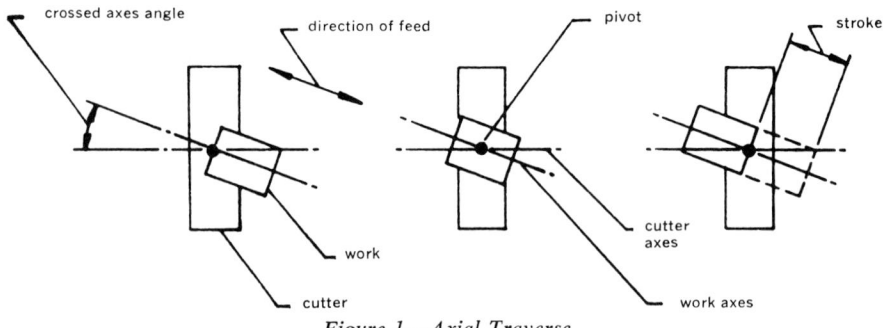

Figure 1—Axial Traverse

[b] *Angular Traverse Cutters* [also referred to as "Diagonal" or "Modified Underpass" shaving cutters].—With these cutters, the work gear is reciprocated across the cutter in a path between zero and 90 degrees to the work gear axis. Normally this angle is from 30 to 60 degrees. This angle setting range avoids having specially designed serrations in the cutter. The direction of rotation is reversed at each end of the stroke.

A two-stroke cycle can be used with angular-traverse shaving.

Often a multi-stroke cycle may be used in conjunction with an automatic upfeed mechanism to provide longer cutter life and greater stock removal.

The cutter may be narrower than the work gear. Crowned gear teeth can be produced by rocking-table action with constant thickness shaving cutters. If desired, reverse crowned cutters can also be used to produce crowned gear teeth without table rocking.

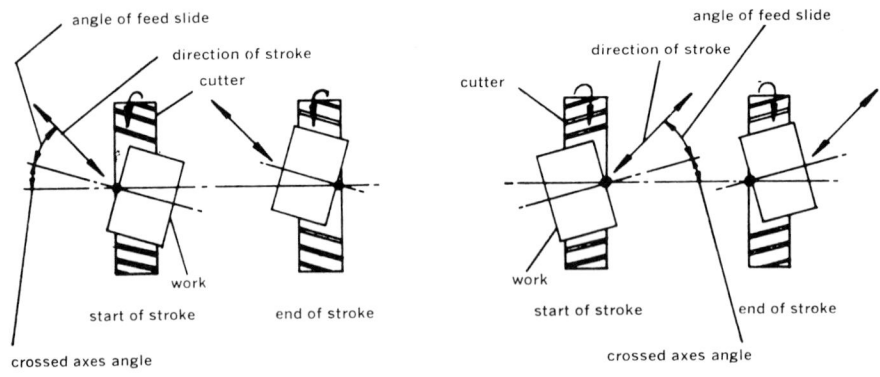

Figure 2—Angular Traverse

[c] *Right Angle Traverse Cutters* [also referred to as "Tangential" or "Underpass" shaving cutters]—Here the work-gear is reciprocated across the cutter at an angle of 90 degrees to the work gear axis. Usually one pass and return of the work to starting position will finish the operation. The direction of rotation is reversed at the end of the stroke.

The cutter is wider than the work gear and is provided with a staggered-serration design. The crossed axes angle used when shaving shoulder gears with this process may be decreased to prevent cutter interference with the shoulder. Low crossed axes angles require the removal of minimum shaving stock. Tooth crowning is produced by a reverse-crowned cutter in this process.

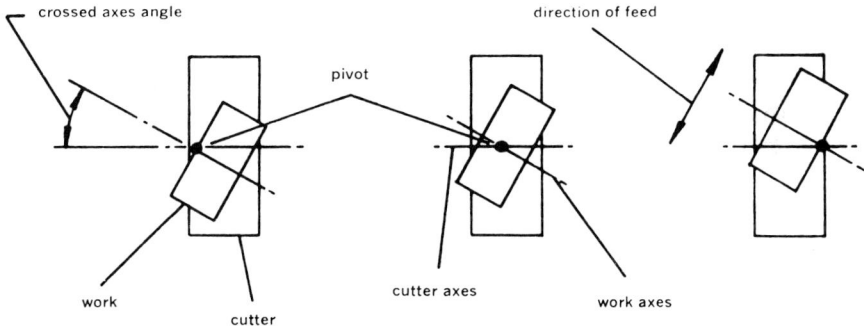

Figure 3—Right Angle Traverse

[d] *Plunge Feed Cutters*—With these cutters, the work meshes with the cutter and is fed to depth without traverse. Work and cutter centerlines are in crossed axes relationship. See Figure 7.

[e] *Parallel Axes Cutters*—With these cutters, the work and cutter centerlines are on parallel axes. The work is traversed back and forth in mesh with the cutter, and fed to depth. See Figure 8.

[f] *Crown Shaving Cutters*—These shaving cutters have constant thickness teeth and produce a crowned tooth surface by rocking the work axis while using either axial traverse or parallel axes shaving methods. See Figure 9.

[g] *Reversed Crown Shaving Cutters*—These shaving cutters have concave tooth surfaces and produce a crown in gear teeth when angular traverse, right angle traverse, or plunge feed methods are used. See Figure 10.

Note—Axial traverse, angular traverse, or parallel axes shaving cutters can be nearly identical, but may have different serration designs. Right angle traverse cutters are wider than the gear and have modified serration designs. Plunge feed cutters are also wider than the gear and have a modified lead, and modified serration design.

2—[a] *Hand of Cutter*—A right hand shaving cutter has helical teeth that twist away from the observer in a clockwise direction when viewed from either side of the cutter.—Left hand shaving cutters twist away from the observer in a counter-clockwise direction when viewed from either side of the cutter.—When external helical gears are shaved by rotary gear shaving processes, the hand of the cutter is opposite from that of the work. Internal helical gear shaving cutters have the same hand as the work. When shaving internal or external spur gears the hand of the cutter may be either right hand or left hand, depending on the application.

[b] *Serrations*—Serrations are the narrow grooves on the cutter teeth between the lands that form the cutting edges.

(Differential or Staggered)—Serrations on successive cutter teeth located in a pre-determined staggered relationship. Normally used only when shaving method is right-angle-traverse or angular-traverse with an angular path greater than 60 degrees from the axial plane.

(Normal)—Serrations with sides perpendicular to the tooth faces. Sides of serrations may be parallel or in angular relationship to form a buttressed land section.

(Transverse)—Serrations with sides perpendicular to the cutter axis.
(Width)—The width of the grooves in the cutter tooth faces.
Land—(center)—The tooth face section between two serrations,
(End)—the tooth face section between the edge of the tooth and the
first serration.
Clearance Holes—Holes at the roots of the teeth, used for manufac-
turing purposes.

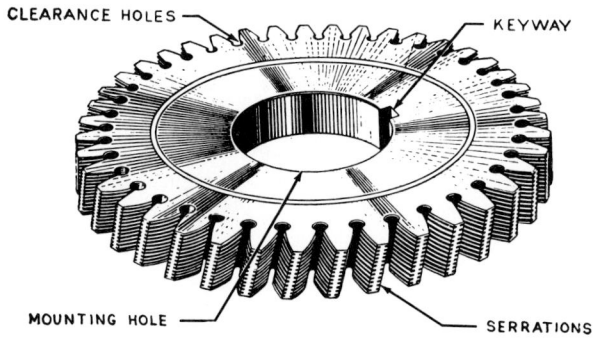

(RIGHT HAND SHAVING CUTTER)

*Figure 4—Shaving cutter details, i.e., serrations,
clearance holes, etc.*

III—RACK SHAVING CUTTERS

A—General Nomenclature.

1—*Rack Blade*—A single tooth in a rack made from one piece of tool
material such as high speed steel. See Figure 11.

2—*Box*—The box type fixture in which individual rack blades are assembled. See Figure 11.

3—*Straight Racks*—These have blades set at right angles to the centerline of the box.

4—*Angular Racks*—These have blades set at an angle to the box centerline. See Figure 12.

5—*End Blocks*—Precision-ground blocks used in each end of the rack
box to assemble the blades. See Figure 11.

6—*Draw Bar*—The bar that is inserted through the assembled blades to
hold them in position. See Figure 11.

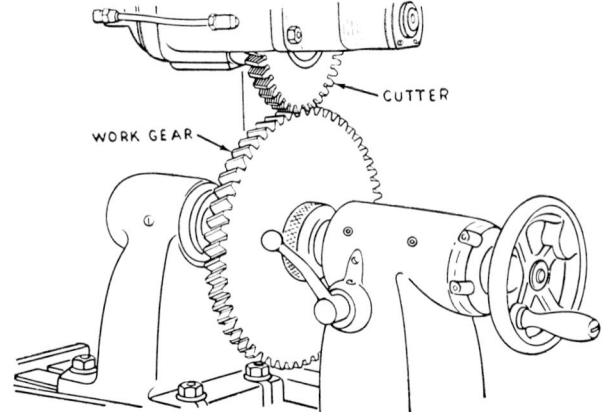

Figure 5—External helical gear shaving setup.

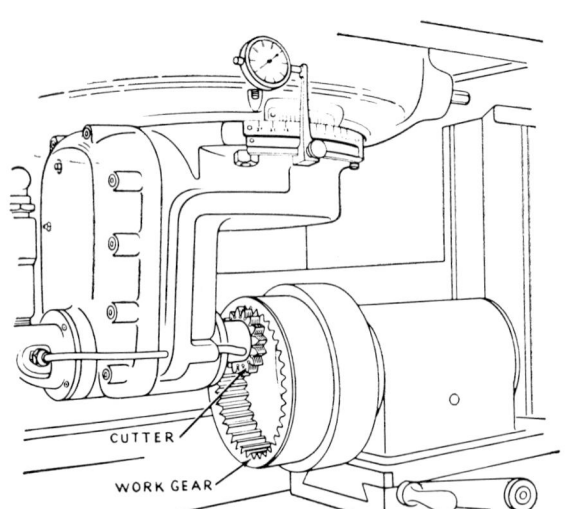

Figure 6—Internal gear shaving setup.

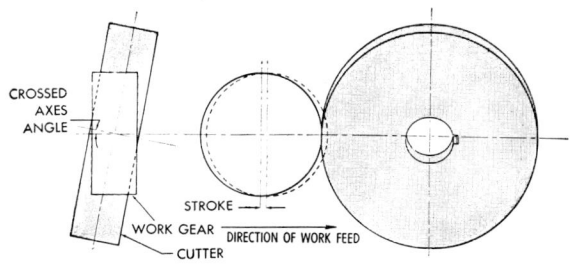

Figure 7—Plunge feed shaving.

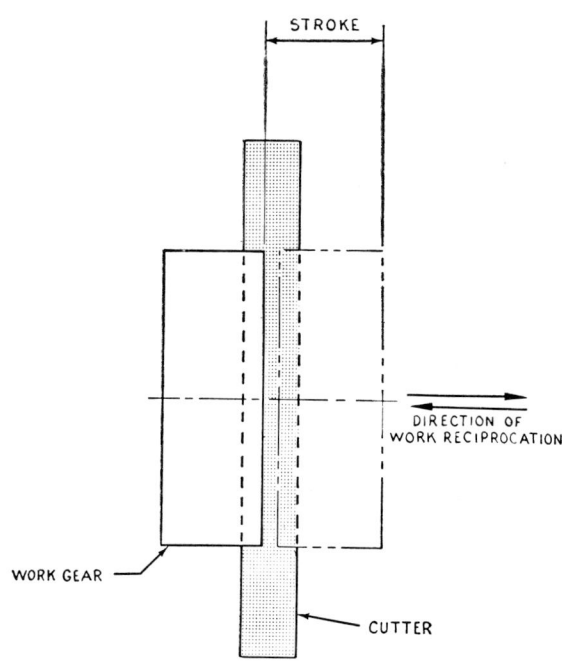

Figure 8—Parallel axes shaving.

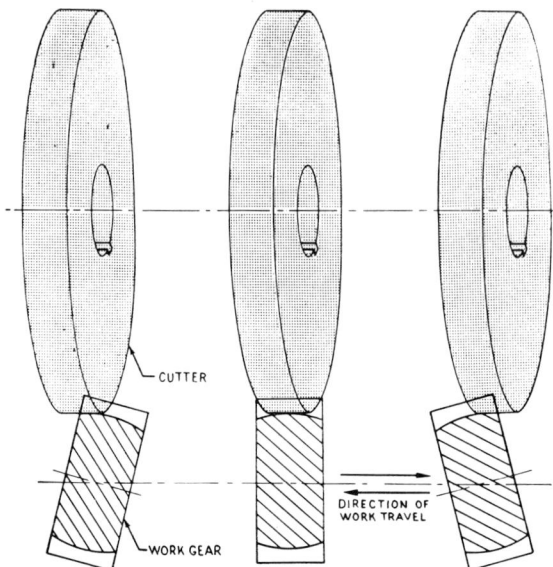

Figure 9——Rocking table crown shaving.

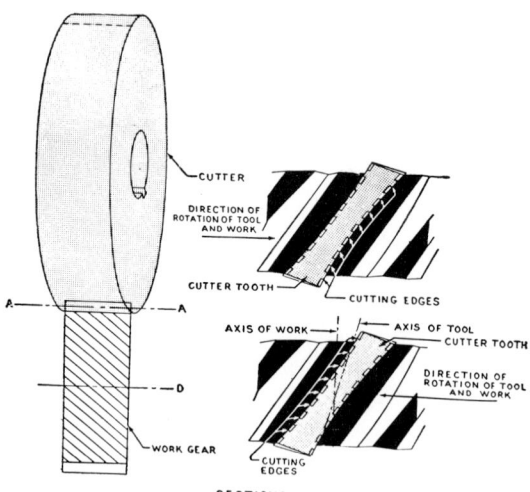

SECTIONS AA

Figure 10——Tooth shapes of reversed crown cutters.

V—RACK SHAVING CUTTERS

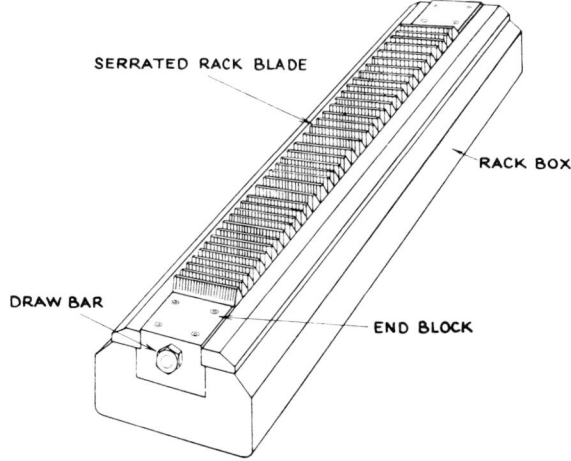

Figure 11—Typical spur type rack for shaving helical gears.

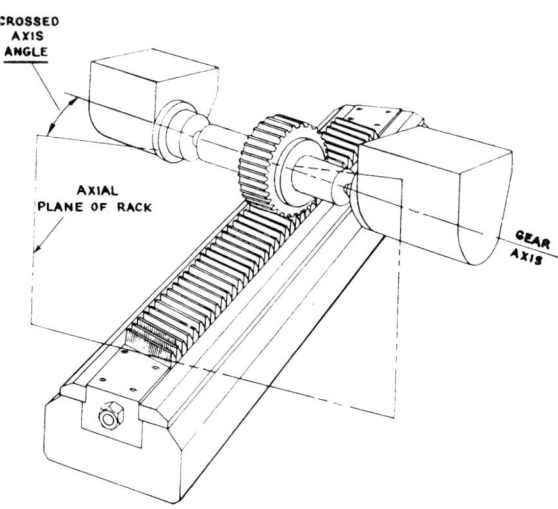

Figure 12—Typical spur gear with helical shaving rack.

BASIC DESIGN CONSIDERATIONS FOR GEAR SHAVING

Fillets

The tips of the shaving cutter teeth must not contact the gear root fillet during the shaving operation. If such contact does occur, excessive wear of the cutter results and the accuracy of the involute profile is affected. See Fig. 13.

The shaving cutter must finish the gear tooth below its active profile. Thus, the height of the fillet should not exceed the lowest point of contact between the shaving cutter teeth and the teeth on the work gear.

Shaper cutters generate higher fillets than hobs for the same depth of cut. As a result, shaper-cut gears are cut deeper than hobbed gears. Tooth depth for 1 to 19 diametral pitch shaper-cut gears inclusive should be $\frac{2.35}{Pd}$. For 20 diametral pitch and finer, a depth of $\frac{2.40}{Pd}$ + .002-in. allows ample clearance for the shaving cutter under ordinary conditions.

When full-round root fillets are used, additional whole depth must be provided. It is advisable to maintain the same length of involute (straight portion) on the preshaving tool and determine the size of tip radius and extra depth from that point as shown in Fig. 14. The shape of the fillets is a generated form determined by the type of preshave tool. It is usually specified as the form produced by a hob or shaper cutter with a given tip radius or shape.

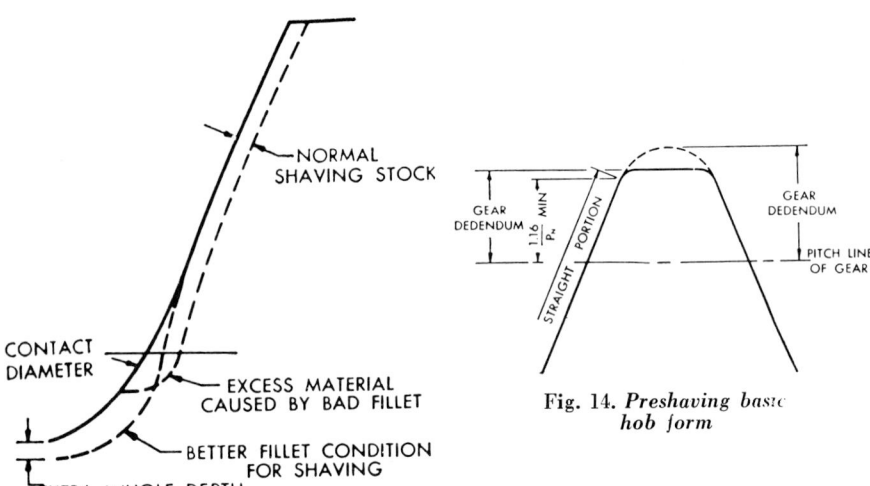

Fig. 13. *How to avoid contact between shaving cutter and root fillet*

Fig. 14. *Preshaving basic hob form*

Protuberance

Protuberance type (high-point) hobs and shaper cutters are often used prior to shaving to produce a slight undercut or relief near the base of the tooth. This method assures a smooth blending of the shaved tooth profile and the unshaved tooth fillet, as well as reducing shaving cutter tooth tip wear.

Standard hob and shaper cutter designs usually generate a natural undercut on gears having small numbers of teeth. In many cases, however, this undercut is excessive from the standpoint of good gear design as shown in Fig. 15. For this reason, protuberance-type tools are seldom recommended for gears having fewer than 20 teeth with a 20-deg pressure angle, or 30-teeth with a 14½-deg pressure angle.

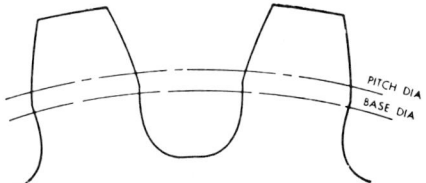

Fig. 15. *Natural undercut produced by a standard hob or gear shaper cutter on gears having small numbers of teeth*

The following rules should be followed in designing protuberance-type tools:

1. The amount of undercut should be determined from mating-gear conditions.

2. The amount of undercut should not be excessive to a point where the teeth are weakened.

3. The undercut must be properly positioned. A hob or shaper cutter designed to produce the proper undercut on a 30-tooth gear, will produce too high an undercut on the active profile of a 20-tooth gear. The undercut produced by the same tool on a 100-tooth gear will be too low for any useful purpose. A preshave protuberance-type hob or shaper cutter for crowned gear teeth may require some modification in protuberance design to avoid fillet interference with the cutter at both extremes of the gear face.

Table No. 1 shows the relationship between protuberance and number of teeth for different pressure angles.

On page 776 of the Engineering Data Section is a table that shows the

TABLE 1

RELATIONSHIP BETWEEN PROTUBERANCE AND NUMBER OF TEETH FOR DIFFERENT PRESSURE ANGLES

Protuberance Application (Tooth Pressure Angle)	No Protuberance (Number of Teeth)	Protuberance (Number of Teeth)
14½	Up to 30	25 and up
17½	Up to 25	20 and up
20-Stub Depth	Up to 25	20 and up
20 and up, Full Depth	Up to 20	16 and up

standard tooth proportions for the Full Depth System, Gear Shaper System and the Preshave Systems.

Protuberance-type tools serve a useful purpose on shaved gears that have restricted tooth clearances (shallow tooth depth) as well as on large gears. In both cases, the fillet is cut by the pre-shave tool to avoid interference with the path of the shaving tool.

Indiscriminate use of one protuberance-type preshave tool on all gears regardless of the number of teeth is a questionable practice. Theoretically, an individual protuberance-type tool is required for each gear in accordance with its number of teeth. This is not a practical method, however, and one tool is used for a range of gears with varying numbers of teeth. Often, a hob or shaper cutter with no protuberance can be used for gears with small numbers of teeth; a second protuberance type tool being used for the gears with larger numbers of teeth.

On long and short addendum gears, the amount and position of undercut must be carefully specified, so that the final tooth design is not impaired. Enlarged tooth form and mating gear layouts, Figure 16, can be used to advantage in studying undercut problems.

The amount of undercut produced by the protuberance-type tool should be from 0.0005-in. to 0.001-in. greater than the amount of shaving stock for each side of the gear tooth. On crowned gears this calculation should be made for the thin end. The position of the undercut should be such that its upper margin meets the involute profile at a point below its contact diameter.

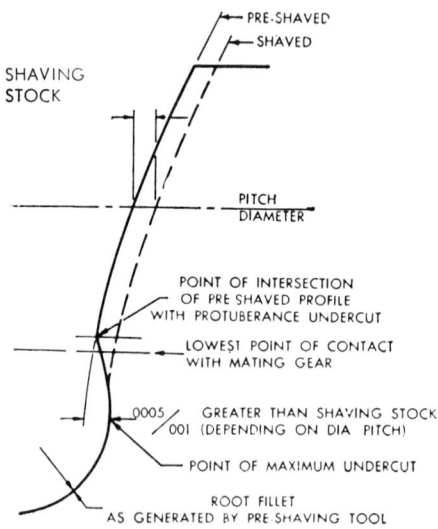

Fig. 16. Undercut produced by a protuberance-type preshaving tool

On small pinions, it is practically impossible to maintain this objective. Under these conditions, the junction point may come slightly above the contact diameter. The undercut should be of a shape to allow at least 0.0005-in. of stock at the mating gear contact diameter to be removed at the shaving operation.

Root fillets can often be increased slightly on protuberance-cut teeth if they remain sufficiently low to clear the tips of the shaving cutter teeth. Properly designed undercut forms are essential to obtain optimum use of shaving process. This relates to the height of the point of intersection of preshaved profile and undercut as well as the configuration of fillet and amount of undercut.

Shaving Stock

The amount of stock removed by the shaving process is a key to its successful application. Sufficient stock should be removed to permit correction of errors in the pre-shaved teeth. However, if too much stock is removed, cutter life and work accuracy are effectively reduced. Table No. 2 shows the recommended amount of stock to be left for shaving. The tables make

TABLE 2

RECOMMENDED AMOUNT OF STOCK TO LEAVE FOR SHAVING

Normal Diametral Pitch	Shaving Stock on Tooth Thickness	Stock Allowed Over Two Pins		
		$14\frac{1}{2}$	$20°$	$25°$
2 to 3.999	.0045	.013	.011	.009
4 to 5.999	.0040	.011	.009	.007
6 to 7.999	.0035	.009	.007	.006
8 to 12.999	.0030	.007	.006	.005
13 to 16.999	.0025	.005	.004	.004
17 to 25.999	.0020	.004	.003	.003
26 to 32.999	.0015	.002	.002	.002
33 to 47.999	.0010	.002	.001	.001
48 to 95.999	.0005	.001	.001	.001
96 and Up	.0004 or less	.0005	.0005	.0005

Note: These stock allowances should be gradually reduced as the crossed axes angle goes below 10°

no allowance for manufacturing tolerances. For example, on 8-pitch, 20-deg pressure angle gears, the table shows a stock allowance of 0.006-in. measured over pins. The production preshave stock tolerance could be plus or minus 0.001-in. from this amount.

Shoulder Gears

A shoulder gear is one where an adjacent gear or other element of larger diameter constitues a shoulder that limits the angle at which the axes of the cutter and smaller gear can be crossed during the shaving operation. Factors limiting this amount of crossed axes angle include: height of shoulder, diameter of shoulder, width between both gear faces, and shaving cutter diameter. Thus, only the cutter diameter and cutter helix angle can be altered by the tool manufacturer to provide a crossed-axes design that will avoid interference. It is most beneficial if this factor is taken into consideration at the time of the design so that the minimum 5-deg crossed-axes setting can be achieved.

The shoulder gear shaving cutter layout shown in Fig. 17 illustrates how the interference problem can be studied. Although a minimum crossed-axes setting of 5-deg is recommended, this can be reduced to 3-deg with some sacrifice in cutting efficiency. Such a layout can be used by gear designers to determine minimum gear face spacing and gear diameters that will provide for optimum shaving conditions.

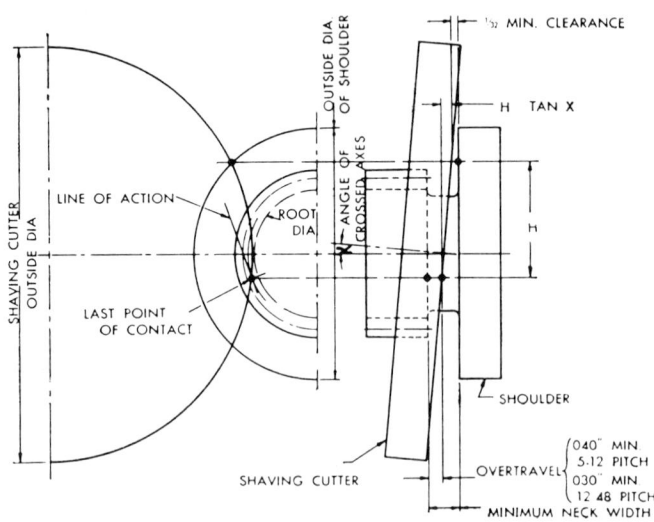

Fig. 17. *Limitation of crossed-axes angle on shoulder gears*

STANDARD TOLERANCES FOR ROTARY TYPE GEAR SHAVING CUTTERS

Gear shaving cutters are made to standard tolerances by the manufacturer. Tables, No's 3, 4 and 5 gives standard blank sizes and tolerances for rotary shaving cutters for involute spur and helical gears, coarse and fine pitch series.

TABLE 3

Rotary Shaving Cutters

For
Involute Spur and Helical Gears
Ground Circular Groove Type

Fine Pitch Series

20 thru 120 Normal Diametral Pitch
2″ thru 4″ Pitch Diameter
20° Maximum Helix Angle
½″ Nominal Blank Width

Tolerances

	Class A	Class AA
Involute Profile — (Active Length — tiv From True Involute)	.00020	.00015
Lead — (Uniformity — tiv Per Half-inch of Face)	.0003	.0002
Parallelism — (Opposite Sides of Same Tooth Alike Within)	.0003	.0002
Helix Angle — (Deviation from True Angle — Per Half-inch of Face)	.0005	.0003
Tooth Spacing — (Adjacent Teeth at PD)	.0002	.00015
Circular Pitch — (Variation — tiv)	.0004	.0002
Spacing Accumulation — (Over 3 Consecutive Teeth)	.0003	.0002
Runout — (tiv at PD) Over 19° NPA 13° thru 19° NPA	.0007 .0009	.0005 .0006
Face Runout — (tiv below Teeth)	.0002	.0002
Tooth Thickness	−.0010	−.0010
Hole	+.0002	+.0002
Outside Diameter	+.010 −.040	+.005 −.020

All dimensions are given in inches

Rotary Shaving Cutters

For

Involute Spur and Helical Gears

Coarse Pitch Series
4 thru 19.999 Normal Diametral Pitch
13° and Over Normal Pressure Angle
thru 9.499 Pitch Diameter

Blank Size

Pitch Diameter	Face Width (approx.)
Thru 4.499	½″
4.500 thru 6.499	⅝″
6.500 thru 9.499	¾ thru 1″

Tolerances

	Class A	Class AA
Involute Profile — (Active Length — tiv From True Involute)		
Thru .177 Working Depth	.00020	.00015
.178 thru .395 Working Depth	.00025	.00020
.396 thru .610 Working Depth	.00030	.00025
Lead — (Uniformity — tiv Per Inch of Face)	.0004	.0003
Parallelism — (Opposite Sides of Same Tooth Alike Within)	.0003	.0002
Helix Angle — (Deviation from True Angle — Per Inch of Face)	.0010	.0005
Tooth Spacing — (Adjacent Teeth at PD)	.0002	.00015
Circular Pitch — (Variation — tiv)	.0004	.0002
Spacing Accumulation — (Over 3 Consecutive Teeth)	.0004	.0003
Runout — (tiv at PD)		
Thru 4.499″ PD		
Over 19° NPA	.0007	.0005
13° thru 19° NPA	.0011	.0007
4.500 thru 9.499″ PD		
Over 19° NPA	.0009	.0006
13° thru 19° NPA	.0012	.0008
Face Runout — (tiv below Teeth)	.0002	.0002
Tooth Thickness	−.0010	−.0010
Hole	+.0002	+.0002
Outside Diameter	+.010 −.040	+.005 −.020

All dimensions are given in inches

TABLE 5

Rotary Shaving Cutters

For
Involute Spur and Helical Gears

Coarse Pitch Series

4 thru 19.999 Normal Diametral Pitch
13° and Over Normal Pressure Angle
9.500″ thru 13.499″ Pitch Diameter
1″ Face Width (Approx.)

Tolerances

	Class A	Class AA
Involute Profile — (Active Length — tiv From True Involute)		
Thru .177 Working Depth	.00020	.00015
.178 thru .395 Working Depth	.00025	.00020
.396 thru .610 Working Depth	.00030	.00025
Lead — (Uniformity — tiv Per Inch of Face)	.0004	.0003
Parallelism — (Opposite Sides of Same Tooth Alike Within)	.0003	.0002
Helix Angle — (Deviation from True Angle —Per Inch of Face)	.0010	.0005
Tooth Spacing — (Adjacent Teeth at PD)	.0002	.00015
Circular Pitch — (Variation — tiv)	.0004	.0002
Spacing Accumulation — (Over 3 Consecutive Teeth)	.0004	.00025
Runout — (tiv at PD)		
Over 19° NPA	.0010	.0007
13° thru 19° NPA	.0014	.0010
Face Runout — (tiv below Teeth)	.0003	.0002
Tooth Thickness	−.0010	−.0010
Hole	+.0002	+.0002
Outside Diameter	+.010 −.040	+.010 −.040

All dimensions are given in inches

ENGINEERING DATA SECTION

ENGINEERING DATA SECTION CONTENTS

GENERAL TABLES
Hardness Conversion Table

approximate equivalent hardness numbers for
rockwell "c" hardness numbers for steel

rockwell "c" scale hardness no.	diamond pyramid hardness no.	brinell hardness no. 10 m/m ball 3000 KG load std. ball	tensile strength (approx.) 1000 psi*	rockwell "c" scale hardness no.	diamond pyramid no. 10	brinell hardness no. 10 m/m ball 3000 KG load std. ball	tensile strength (approx.) 1000 psi*
80	1865	—	—	49	498	464	239
79	1787	—	—	48	484	451	232
78	1710	—	—	47	471	442	225
77	1633	—	—	46	458	432	219
76	1556	—	—	45	446	421	212
75	1478	—	—	44	434	409	206
74	1400	—	—	43	423	400	201
73	1323	—	—	42	412	390	196
72	1245	—	—	41	402	381	191
71	1160	—	—	40	392	371	186
70	1076	—	—	39	382	362	181
69	1004	—	—	38	372	353	176
68	940	—	—	37	363	344	172
67	900	—	—	36	354	336	168
66	865	—	—	35	345	327	163
65	832	—	—	34	336	319	159
64	800	—	—	33	327	311	154
63	772	—	—	32	318	301	150
62	746	—	—	31	310	294	146
61	720	—	—	30	302	286	142
60	697	—	—	29	294	279	138
59	674	—	326	28	286	271	134
58	653	—	315	27	279	264	131
57	633	—	305	26	272	258	127
56	613	—	295	25	266	253	124
55	595	—	287	24	260	247	121
54	577	—	278	23	254	243	118
53	560	—	269	22	248	237	115
52	544	500	262	21	248	231	113
51	528	487	253	20	238	226	110
50	513	475	245	—	—	—	—

*Apply to steel and alloys of steel only
Values for R"C" 68 and less, from American Standard Z76.4-1961
Values over R"C" 68 courtesy Wilson Mechanical Instrument Division

Machinability Rating

This data is based on turning tests with single point tools. They are an indication, but are not necessarily accurate for other types of metal cutting, such as milling, drilling or reaming.

The metals commonly used by industry are divided into six classes and are assigned a machinability rating as follows:

(Ferrous metals as compared with A.I.S.I. B-1112 cold-drawn steel)

Class 1—70% or over (easy)
Class 2—50% to 65%
Class 3—40% to 50%
Class 4—below 40% (severe)

(Nonferrous metals as compared with A.I.S.I. B-1112 cold-drawn steel)

Class 5—100% or over
Class 6—below 100%

Prefix-Letter Designations

C—Basic open-hearth carbon steel
B—Acid Bessemer carbon steel
CB—Either acid Bessemer or basic open-hearth carbon steel at the option of the manufacturer
A—Basic open-hearth alloy steels
E—Electric-furnace steels of both carbon and alloy types
NE—National Emergency series steels

Class 1—Ferrous (70% and higher)

A.I.S.I.	Rating, per cent	Brinell
C 1110	85	137–166
C 1115	85	143–179
C 1117	85	143–179
C 1118	80	143–179
C 1120	80	143–179
C 1132	75	187–229
C 1137	70	187–229
C 1022	70	159–192
C 1016	70	137–174
B 1111	95	179–229
B 1112	100	179–229
B 1113	135	179–229
A 4023	70	156–207
A 4027*	70	166–212
A 4119	70	170–217
Malleable iron		
(Ferritic)	100-250	120–200
(Pearlitic)	70	190–240
Cast steel		
(0.35% C)	70	170–212
Stainless iron		
(Free cutting)	70	163–207

Class 2—Ferrous (50% to 65%)

A.I.S.I.	Rating, per cent	Brinell
C 1141	65	183–241
C 1020	65	137–174
C 1030	65	170–212
C 1035	65	174–217
C 1040*	60	179–229
C 1045*	60	179–229
NE 1330*	65	179–235
NE 1340*	60	187–241
NE 1350*	55	187–241

A.I.S.I.	Rating, per cent	Brinell
A 2317	55	174–217
A 3045*	60	179–229
A 3120	60	163–207
A 3130*	55	179–217
A 3140*	55	187–229
A 3145*	50	187–235
A 4032*	65	170–229
A 4037*	65	179–229
A 4042*	60	183–235
A 4047*	55	183–235
A 4130*	65	187–229
A 4137*	60	187–229
A 4145*	55	187–229
A 4150*	50	187–235
A 4615	65	174–217
A 4640*	55	187–235
A 4815	50	187–229
A 5120	65	170–212
A 5140*	60	174–229
A 5150*	55	179–235
A 5045*	65	179–229
NE 8024 (a)	60	174–217
NE 8124 (a)	55	174–217
NE 8233 (a)	60	179–229
NE 8339* (a)	60	179–229
NE 8620	60	170–217
NE 8630*	65	179–229
NE 8720	65	179–229
NE 8739* (a)	60	179–229
NE 8744* (a)	55	183–235
NE 8749* (a)	50	183–241
NE 8817 (a)	60	170–228
NE 9415	55	170–217
NE 9425 (b)	50	170–217
NE 9430*	65	179–228
NE 9440*	60	187–235
NE 9450*	45	187–235

(a) Discontinued.
(b) New steel, rating estimated.

Class 3—Ferrous (40% to 50%)

A.I.S.I.	Rating, per cent	Brinell
C 1008	50	126–163
C 1010	50	131–170
C 1015	50	131–170
C 1050*	50	179–229
C 1070*	45	183–241
A 1320	50	170–229
A 1330*	50	179–235
A 1335*	50	179–235
A 1340*	45	179–235
A 2330*	50	179–229
A 2340*	45	170–235
A 3240*	45	183–235
A 4340*	45	187–241
A 6120	50	179–217
A 6145*	50	179–235
A 6152*	45	183–241
NE 9250*	50	179–217

Class 3—(Con't)

NE 9260*	45	187–255
NE 9261* (b)	50	179–217
NE 9262*	45	187–241
NE 8442* (a)	45	187–255
NE 8447* (a)	40	187–255
NE 8949* (a)	50	187–255
Ingot iron	50	101–131
Wrought iron	50	101–131
Stainless 18–8 F.M.	45	179–212
Cast iron	50	160–193

Class 4—Ferrous (40% and below)

A. I. S. I.	Rating, per cent	Brinell
A 2515*	30	170–229
E 3310*	40	170–229
E 52100**	30	183–229
Ni-Resist*	30	... ...
Stainless 18–8 (Austenitic) *	25	150–160
Manganese oil-hardening steel**	30	... ...
Tool steel, low-tungsten chromium, and carbon**	30	... ...
High-speed steel**	30	200–218
High-carbon, high-chrome tool steel**	25	... ...

Class 5—Nonferrous (above 100%)

	Rating, per cent	Brinell
Magnesium alloys Dow "J" (wrought) 92.3 Mg, 6.5 Al, 0.2 Mn, 1 Zn	500–2000	58
Dow "H" (cast) 90.8 Mg, 6 Al, 0.2 Mn, 3 Zn	500–2000	50
Aluminum 11S-T 3 (5.5 Cu, 0.5 Pb, 0.5 Bi, 93.5 Al)	500–2000	95 (500 kg)
2S (100 Al) O to H temper	300–1500	23–44 (500 kg)
17S-T (4 Cu, 0.5 Mn, 0.5 Mg, 95 Al)	300–1500	100 (500 kg)
Brass, leaded F. C., C. D. (62 Cu, 35 Zn, 3 Pb)	200–400	50 (500 kg)
Red, leaded (78.5 Cu, 20 Zn, 1.5 Pb)	180	55 (500 kg)
Bronze, phos., leaded (94 Cu, 1 Pb, 5 Sn)	100	75 (500 kg)
Zinc	200	

Class 6—Nonferrous (below 100%)

	Rating, per cent	Brinell
Aluminum bronze (cast) 89.7 Cu, 10 Al, 0.2 Fe, 0.5 Sn	60	140–160
Brass Yellow (63 Cu, 37 Zn)	80	50–70 (500 kg)
Red (80 Cu, 20 Zn)	60	55 (500 kg)
Bronze, mang. (59 Cu, 39 Zn, 0.7 Sn, 0.5 Mn, 0.8 Fe)	60	80–95 (500 kg)
Bronze, phos. (95 Cu, 5 Sn)	40	140
Copper, cast	70	30
Copper, rolled (¼ hard)	60	80
Everdur (Cu-Si) 95.80 Cu, 1.10 Mn, 3.10 Si	60	200
Everdur (Cu-Si) 95.60 Cu, 1.0 Mn, 3.0 Si, 0.4 Pb	120	200
Gun Metal (cast) 88 Cu, 10 Sn, 2 Zn	60	65
Nickel (hot-rolled)	20	110–150
Nickel (cold-drawn) *	30	100–140
Monel metal, *regular (68 Ni, 29 Cu)	40	125–150
As cast, "H" (65 Ni, 29 Cu, 2 Fe, 2.75 Si, 0.2 C, 0.7 Mn)	35	175–250
As cast, "S" (64 Ni, 29 Cu, 2.5 Fe, 3.75 Si, 0.1 C, 0.5 Mn)	20	280–325
Rolled (67 Ni, 31 Cu, 1.16 Fe, 0.7 Si, 0.1 C, 0.8 Mn)	45	207–224
"K" (66 Ni, 29 Cu, 0.9 Fe, 0.5 Si, 0.15 C, 2.75 Al)	50	215–265
Inconel, temper B, cold-drawn* (78 Ni, 14 Cr, 6 Fe)	45	130–170

* Annealed

** Spheroidized anneal

NOTE: The terms "annealed" and "spheroidized anneal" refer specifically to the commercial practice in steel mills, prior to cold-drawing or cold-rolling, in the production of the steels specifically mentioned.

TABLE 405

Weldon Shanks

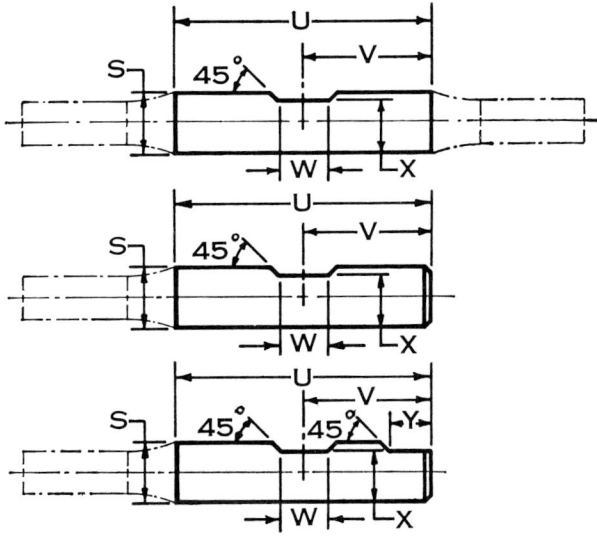

Dimensions

Diameter of Shank S	Length of Shank U	V	W		X	Y
			minimum	maximum		
$3/8$	$1\,9/16$	$25/32$	0.280	0.282	0.325	..
$1/2$	$1\,25/32$	$57/64$	0.330	0.332	0.440	..
$5/8$	$1\,29/32$	$61/64$	0.400	0.402	0.560	..
$3/4$	$2\,1/32$	$1\,1/64$	0.455	0.457	0.675	..
$7/8$	$2\,1/32$	$1\,1/64$	0.455	0.457	0.810	$1/2$
1	$2\,9/32$	$1\,9/64$	0.515	0.517	0.925	$1/2$
$1\,1/4$	$2\,9/32$	$1\,9/64$	0.515	0.517	1.156	$1/2$
$1\,1/2$	$2\,11/16$	$1\,3/16$	0.515	0.517	1.406	$9/16$
2	$3\,1/4$	$1\,27/32$	0.700	0.702	1.900	$27/32$
$2\,1/2$	$3\,1/2$	$1\,15/16$	0.700	0.702	2.400	$27/32$

All dimensions are given in inches.

Tolerances

Element	Range	Direction	Tolerance
Diameter of Shank, S	All Sizes	Minus	.0001 to .0005
Length of Shank, U	All Sizes	Plus or Minus	$1/32$
Dimension, V	All Sizes	Plus or Minus	$1/64$
Dimension, X	All Sizes	Minus	$1/64$
Dimension, Y	$7/8''$ to $2\,1/2''$ incl.	Plus or Minus	$1/32$

All dimensions are given in inches.

TABLE 1703

Combination Shanks*
For End Mills, Right Hand Cut

*Modified for use as a Weldon or a Pin Drive Shank.

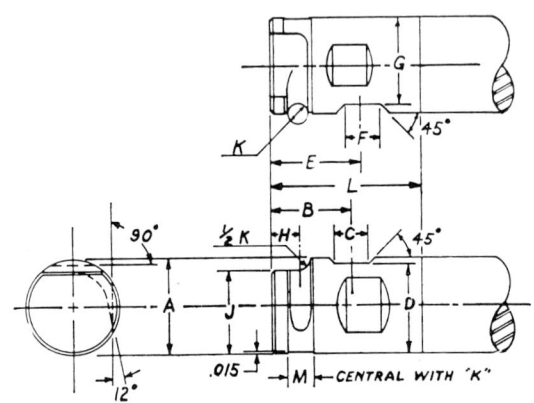

Dimensions

Dia. of Shank	Length of Shank	B	C	D	E	F	G	H	J	K	M
A	L										
$1\frac{1}{2}$	$2^{11}\!/_{16}$	$1\,^{3}\!/_{16}$	.515	1.406	$1\frac{1}{2}$	.515	1.371	$^{9}\!/_{16}$	1.302	.377	$^{7}\!/_{16}$
2	$3\frac{1}{4}$	$1^{23}\!/_{32}$	.700	1.900	$1\frac{3}{4}$	.700	1.809	$^{5}\!/_{8}$	1.772	.440	$\frac{1}{2}$
$2\frac{1}{2}$	$3\frac{1}{2}$	$1^{15}\!/_{16}$	.700	2.400	2	.700	2.312	$^{3}\!/_{4}$	2.245	.503	$^{9}\!/_{16}$

Tolerances

Element	Direction	Tolerance
Diameter of Shank, A	Minus	.0001 to .0005
Length of Shank, L	Plus or Minus	$\frac{1}{32}$
Dimension, B	Plus or Minus	$\frac{1}{64}$
Dimension, C	Plus	.002
Dimension, D	Minus	$\frac{1}{64}$
Dimension, E	Plus or Minus	$\frac{1}{64}$
Dimension, F	Plus or Minus	.005
Dimension, G	Minus	$\frac{1}{64}$
Dimension, H	Plus	$\frac{1}{64}$
Dimension, J	Plus or Minus	.002
Dimension, K	Plus	.003

All dimensions are given in inches.

TABLE 1704

Combination Shanks*
For End Mills, Left Hand Cut

*Modified for use as a Weldon or a Pin Drive Shank.

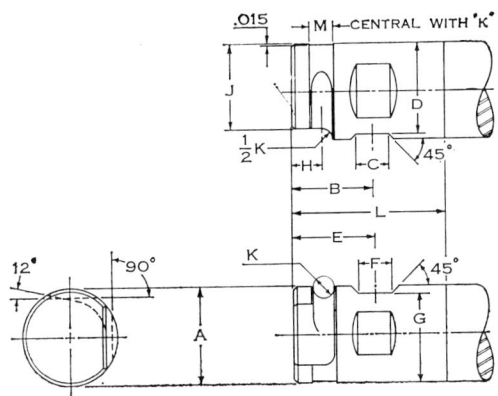

Dimensions

Dia. of Shank	Length of Shank	B	C	D	E	F	G	H	J	K	M
A	L										
1½	2¹¹⁄₁₆	1 ³⁄₁₆	.515	1.406	1½	.515	1.371	⁹⁄₁₆	1.302	.377	⁷⁄₁₆
2	3 ¼	1²³⁄₃₂	.700	1.900	1¾	.700	1.809	⅝	1.772	.440	½
2½	3 ½	1¹⁵⁄₁₆	.700	2.400	2	.700	2.312	¾	2.245	.503	⁹⁄₁₆

Tolerances

Element	Direction	Tolerance
Diameter of Shank, A	Minus	.0001 to .0005
Length of Shank, L	Plus or Minus	¹⁄₃₂
Dimension, B	Plus or Minus	¹⁄₆₄
Dimension, C	Plus	.002
Dimension, D	Minus	¹⁄₆₄
Dimension, E	Plus or Minus	¹⁄₆₄
Dimension, F	Plus or Minus	.005
Dimension, G	Minus	¹⁄₆₄
Dimension, H	Plus	¹⁄₆₄
Dimension, J	Plus or Minus	.002
Dimension, K	Plus	.003

All dimensions are given in inches.

TABLE 1708

DIMENSIONS FOR END MILL HOLDERS WITH SINGLE SCREW AND AMERICAN NATIONAL STANDARD MILLING MACHINE SHANKS. (TO DRIVE TOOLS WITH WELDON FLATTED SHANKS)

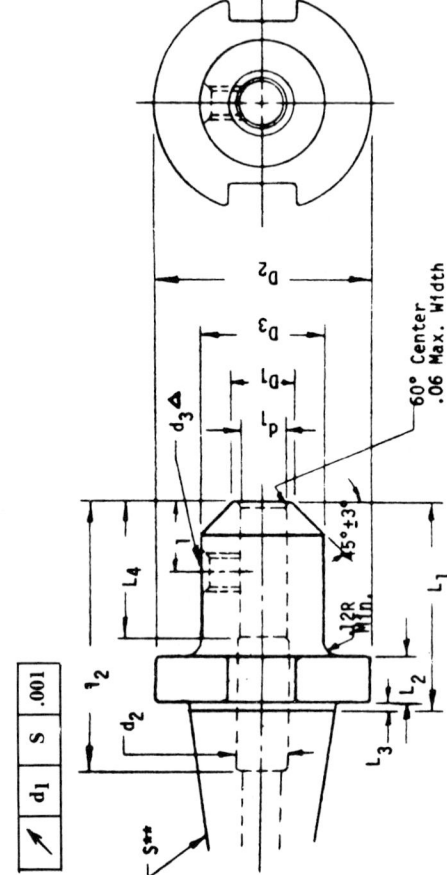

(Continued on following page)

TABLE 1708

DIMENSIONS FOR END MILL HOLDERS WITH SINGLE SCREW AND AMERICAN NATIONAL STANDARD MILLING MACHINE SHANKS.
(TO DRIVE TOOLS WITH WELDON FLATTED SHANKS)

S**	d_1 +.0003 -.0000	D_3 +.00 -.06	L_1 +.06 -.00	D_2 +.00 -.06	L_2 ±.02	L_3 ±.015	d_2 Min.	l_2 Min.	L_4 +.06 -.00	l ±.02	d_3 Δ Class 2B	D_1 ±.03
#30	.1875	1.00	1.75	1.87	.37	.060	.42	Thru	1.00	.56	.250	.44
"	.3750	1.37	1.75	1.87	.37	.060	.42	Thru	1.50	.75	.375	.75
* "	.5000	1.62	1.75	1.87	.37	.060	.52	1.87	1.62	.87	.437	.87
* "	.6250	1.87	1.75	1.87	-	.060	.64	2.06	1.87	.94	.562	1.12
#40	.1875	1.00	2.31	2.50	.37	.060	.53	Thru	1.00	.56	.250	.44
"	.3750	1.37	2.31	2.50	.37	.060	.53	Thru	1.50	.75	.375	.75
"	.5000	1.62	2.31	2.50	.37	.060	.53	Thru	1.62	.87	.437	.87
"	.6250	1.87	2.31	2.50	.37	.060	.64	3.50	1.87	.94	.562	1.12
"	.7500	2.0	2.31	2.50	.37	.060	.77	3.87	2.00	1.00	.625	1.25
#45	.3750	1.37	2.50	3.00	.37	.120	.66	Thru	1.50	.75	.375	.75
"	.5000	1.62	2.50	3.00	.37	.120	.66	Thru	1.62	.87	.437	.87
"	.6250	1.87	2.50	3.00	.37	.120	.66	Thru	1.87	.94	.562	1.12
"	.7500	2.00	2.50	3.00	.37	.120	.77	3.87	2.00	1.00	.625	1.25
#50	.3750	1.37	2.75	3.50	.50	.120	.87	Thru	1.50	.75	.375	.75
"	.5000	1.62	2.75	3.50	.50	.120	.87	Thru	1.62	.87	.437	.87
"	.6250	1.87	2.75	3.50	.50	.120	.87	Thru	1.87	.94	.562	1.12
"	.7500	2.00	2.75	3.50	.50	.120	.87	Thru	2.00	1.00	.625	1.25

All dimensions in inches

*For Single-End End Mills Only
**See ANSI B5 18—Latest Issue
Δ Thread Pitch Optional

TABLE 1709

DIMENSIONS FOR END MILL HOLDERS WITH DOUBLE SCREW
AND AMERICAN NATIONAL STANDARD MILLING MACHINE SHANKS.
(TO DRIVE TOOLS WITH WELDON FLATTED SHANKS)

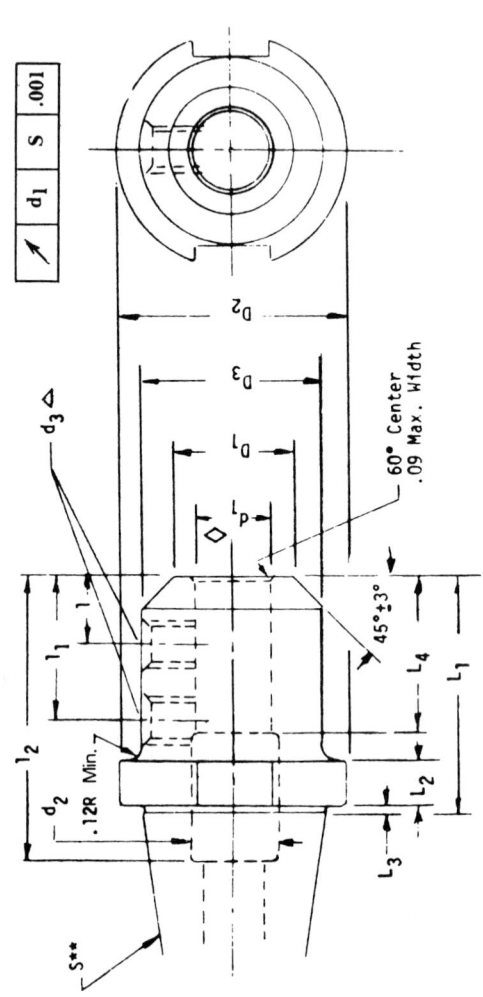

(Continued on following page)

TABLE 1709

DIMENSIONS FOR END MILL HOLDERS WITH DOUBLE SCREW AND AMERICAN NATIONAL STANDARD MILLING MACHINE SHANKS.
(TO DRIVE TOOLS WITH WELDON FLATTED SHANKS)

S**	d₁ ◇	D₃ +.00 -.06	L₁ +.06 -.00	D₂ +.00 -.06	L₂ ±.02	L₃ ±.015	d₂ Min.	l₂ Min.	L₄ +.06 -.00	l ±.02	l₁ ±.02	d₃ Δ Class 2B	D₁ ±.03
#40	.8750	2.25	3.00	2.62	.37	.060	.89	4.19	2.00	1.00	1.81	.625	1.50
* "	1.0000	2.62	3.00	2.62	—	.060	1.02	2.53	2.25	1.12	2.12	.750	1.75
* "	1.2500	2.62	3.00	2.62	—	.060	1.27	2.53	2.25	1.12	2.12	.750	2.00
#45	.8750	2.25	3.12	3.00	.37	.120	.89	4.19	2.00	1.00	1.81	.625	1.50
"	1.0000	2.50	3.12	3.00	.37	.120	1.02	4.44	2.25	1.12	2.12	.750	1.75
* "	1.2500	2.75	3.12	3.00	—	.120	1.27	2.53	2.25	1.12	2.12	.750	2.00
* "	1.5000	3.50	3.50	3.50	—	.120	1.52	2.94	2.62	1.50	2.50	.750	2.50
* "	2.0000	4.00	4.50	4.00	—	.120	2.02	3.50	3.12	1.41	2.91	1.00	3.12
#50	.8750	2.25	3.19	3.50	.50	.120	.89	4.19	2.00	1.00	1.81	.625	1.50
"	1.0000	2.50	3.19	3.50	.50	.120	1.02	4.44	2.25	1.12	2.12	.750	1.75
* "	1.2500	2.75	3.19	3.50	.50	.120	1.27	2.53	2.25	1.12	2.12	.750	2.00
* "	1.5000	3.50	3.50	3.50	—	.120	1.52	2.94	2.62	1.50	2.50	.750	2.50
* "	2.0000	4.00	4.50	4.00	—	.120	2.02	3.50	3.12	1.41	2.91	1.00	3.12
* "	2.5000	5.12	4.50	5.12	—	.120	2.52	3.50	3.12	1.41	2.91	1.00	4.00

All dimensions in inches
Δ Thread Pitch Optional

◇Tolerance: thru 1.2500 dia. +.0003-.0000
over 1.2500 dia. +.0005-.0000

*For Single-End End Mills Only
**See ANSI B5.18—Latest Issue

TABLE 1710

HEXAGON SOCKET SET SCREWS FOR END MILL HOLDERS

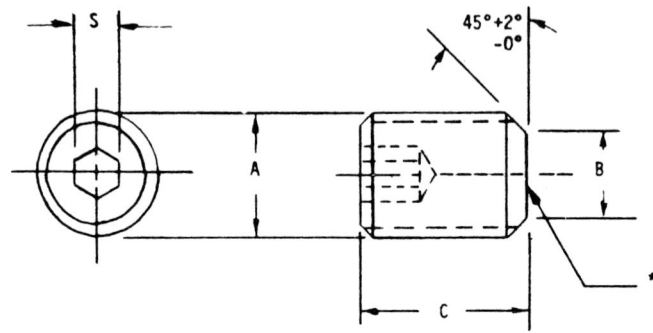

A	B	C	S	
Class 3A Thread Pitch Optional		+.01	Nom.	For Holder Bore Diam.
.250	.118/.132	.37	.125	.1875
.375	.278/.286	.37	.1875	.375
.437	.328/.336	.50	.2187	.500
.562	.398/.406	.50	.250	.625
.625	.453/.461	.50	.3125	.750
.625	.453/.461	.62	.3125	.875
.750	.513/.521	.68	.375	1.000-1.250
.750	.513/.521	.87	.375	1.500
1.000	.698/.706	.87	.500	2.000
1.000	.698/.706	1.12	.500	2.500

All dimensions in inches.
*Surface to be flat and square to thread pitch within .002.
Screws made of oil hardening tool steel-heat treated to 45-53 Rockwell "C"

Stub Taper Shanks
General Dimensions

Stub taper shanks, as shown in the following table, are occasionally used in special applications for drills, reamers, and counterbores.

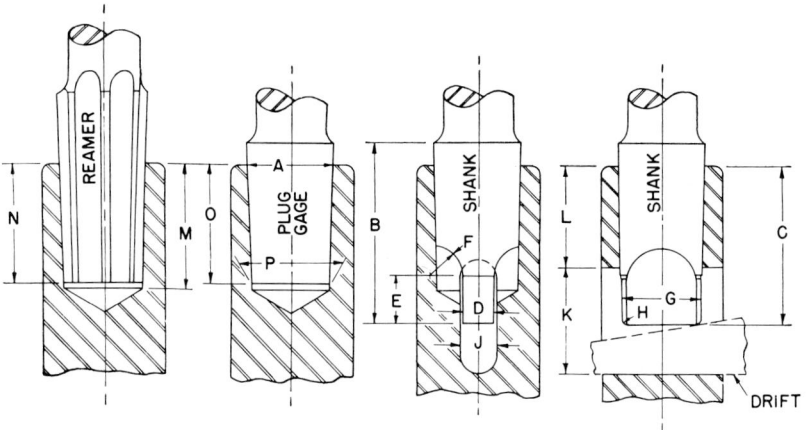

No. of Taper	Diam. at Gage Line	Shank Whole Length	Depth	Tang Thickness	Length	Radius of Mill	Diam.	Radius
	A	B	C	D	E	F	G	H
1	.475	1 5/16	1 1/8	.203	5/16	3/16	13/32	3/64
2	.700	1 11/16	1 1/16	.297	7/16	7/32	39/64	1/16
3	.938	2	1 3/4	.391	9/16	9/32	13/16	5/64
4	1.231	2 3/8	2 1/16	.516	11/16	3/8	1 3/32	3/32
5	1.748	3	2 11/16	.750	15/16	9/16	1 19/32	1/8

No. of Taper	Tang Slot Width	Length	End of Socket to Tang Slot	Depth of Drilled Hole	Depth of Reamed Hole	Std. Plug Depth	Diam. of Plug at Small End	Taper per Inch	Taper per Foot
	J	K	L	M	N	O	P		
1	.219	23/32	25/32	15/16	29/32	7/8	.4314	.049882	.59858
2	.312	15/16	15/16	1 5/32	1 7/64	1 1/16	.6469	.049951	.59941
3	.406	1 1/8	1 1/16	1 3/8	1 5/16	1 1/4	.8753	.050196	.60235
4	.531	1 3/8	1 3/16	1 9/16	1 1/2	1 7/16	1.1563	.051938	.62326
5	.781	1 3/4	1 7/16	1 15/16	1 7/8	1 13/16	1.6526	.052626	.63151

The undercut shown on the tang having diameter G, and length E, may be eliminated at the option of the manufacturer.

TABLE 406
Morse Tapers

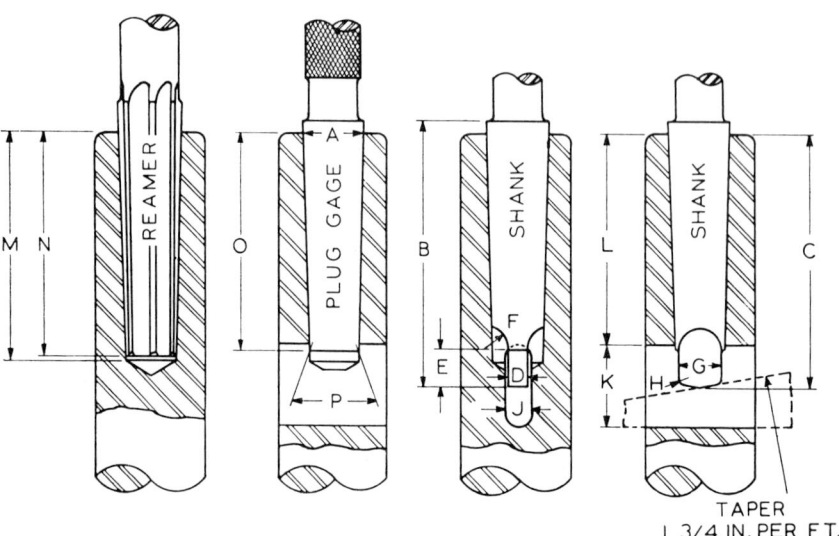

TAPER
1 3/4 IN. PER FT.

Number of Taper	Diam. of Plug at Small End	Diam. at End of Socket	SHANK		SOCKET				TANG					TANG SLOT		Taper per Inch	Taper per Foot
			Whole Length	Depth	End of Socket to Tang Slot	Depth of Drilled Hole	Depth of Reamed Hole	Standard Plug Depth	Thickness	Length	Radius	Diameter	Radius	Width	Length		
	P	A	B	C	L	M	N	O	D	E	F	G	H	J	K		
0	.25200	.35610	2 11/32	2 7/32	1 15/16	2 1/16	2 1/2	2	.156	1/4	5/32	15/64	3/64	.172	9/16	.052050	.62460
1	.36900	.47500	2 9/16	2 7/16	2 1/16	2 3/16	2 5/32	2 1/8	.203	3/8	3/16	11/32	3/64	.218	3/4	.049882	.59858
2	.57200	.70000	3 1/8	2 15/16	2 1/2	2 21/32	2 19/64	2 9/16	.250	7/16	1/4	17/32	1/16	.266	7/8	.049951	.59941
3	.77800	.93800	3 7/8	3 11/16	3 1/16	3 5/16	3 1/4	3 3/16	.312	9/16	9/32	23/32	5/64	.328	1 3/16	.050196	.60235
4	1.02000	1.23100	4 7/8	4 5/8	3 7/8	4 3/16	4 1/8	4 1/16	.469	5/8	5/16	31/32	3/32	.484	1 1/4	.051938	.62326
4 1/2	1.26600	1.50000	5 3/8	5 1/8	4 3/16	4 5/8	4 9/16	4 1/2	.562	11/16	3/8	1 13/64	1/8	.578	1 3/8	.052000	.62400
5	1.47500	1.74800	6 1/8	5 7/8	4 15/16	5 5/16	5 1/4	5 3/16	.625	3/4	3/8	1 13/32	1/8	.656	1 1/2	.052626	.63151
6	2.11600	2.49400	8 9/16	8 1/4	7	7 13/32	7 21/64	7 1/4	.750	1 1/8	1/2	2	5/32	.781	1 3/4	.052138	.62565
7	2.75000	3.27000	11 5/8	11 1/4	9 1/2	10 5/32	10 5/64	10	1.125	1 3/8	3/4	2 5/8	3/16	1.156	2 5/8	.052000	.62400

The undercut shown on the tang having diameter G, and length E, may be eliminated at the option of the manufacturer provided the tang is heat-treated to a minimum Rockwell of C30 with 150 Kg load.

TOLERANCES ON RATE OF TAPER, all sizes 0.002 per foot. This tolerance may be applied on shanks only in the direction which increases the rate of taper and on sockets only in the direction which decreases the rate of taper.

TABLE 407
Brown & Sharpe Tapers

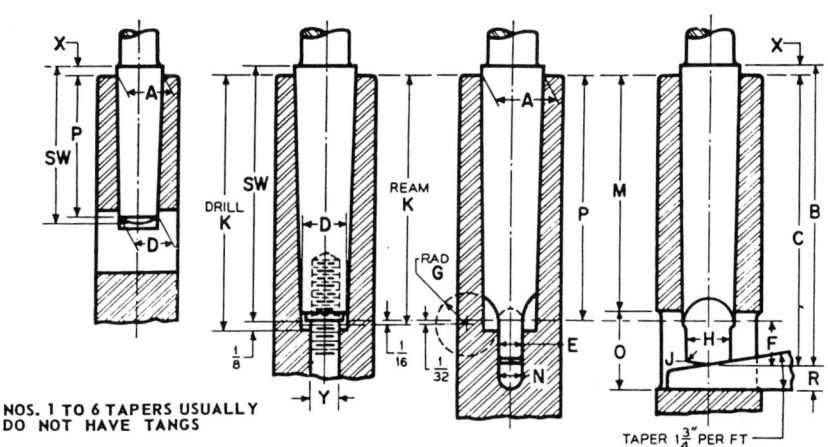

NOS. 1 TO 6 TAPERS USUALLY
DO NOT HAVE TANGS

TAPER 1¾″ PER FT

Y – DIAMETER, NUMBER AND HAND OF THREAD SHOULD BE
SPECIFIED, TO MATCH DRAW-IN BOLT. (WHERE POSSIBLE,
HAND OF THREAD SHOULD MATCH HAND OF CUTTER)

(Continued on following page)

TABLE 407
(Concluded)
Brown & Sharpe Tapers

Number of Taper	D Dia. of Plug At Small End	A Dia. at End of Socket	SHANK B Total Length of Shank With Tang	SHANK SW Shank Length Without Tang	SHANK C Shank Depth	SHANK X Shank Projects From End of Socket	SOCKET M End of Socket To Tang Slot	SOCKET K Drilled Min. Depth	SOCKET K Reamed	P Plug Depth	TANG E Thickness	TANG F Length	TANG G Radius of Mill	TANG H Diameter	TANG J Radius	TANG SLOT N Width	TANG SLOT O Length	TANG SLOT R Shank End To Back of Tang Slot	Taper Per Foot	Taper Per Inch
1	.20000	.23922	1 9/32	1 1/16	1 3/16	3/32	15/16	1 1/16	1 1/16	15/16	.125	3/16	3/16	.170	.030	.141	3/8	1/8	.50200	.041833
2	.25000	.29968	1 9/32	1 9/32	1 7/16	3/32	1 11/64	1 5/16	1 1/4	1 3/16	.156	1/2	3/16	.220	.030	.172	1/2	11/64	.50200	.041833
3	.31250	.37525	1 31/32	1 1/4	1 9/16	3/32	1 15/32	1 5/8	1 9/16	1 1/2	.188	5/16	3/16	.282	.040	.203	5/8	7/32	.50200	.041833
4	.3500	.4023	2 3/32	1 25/32	1 21/32	3/32	1 13/64	1 13/16	1 13/16	1 3/4	.218	3/8	5/16	.320	.050	.228	11/16	15/64	.50240	.041867
5	.4500	.5232	2 9/32	1 29/32	1 15/16	3/32	1 11/16	1 7/8	1 13/16	1 13/16	.250	7/16	5/16	.420	.060	.260	3/4	1/4	.50160	.041800
6	.5000	.5996	2 31/32	2 17/32	2 7/16	3/32	2 19/64	2 1/4	2 7/16	2 7/8	.281	15/32	5/16	.460	.060	.291	7/8	19/64	.50329	.041941
7	.6000	.7254	3 5/8	3 3/32	3 17/32	1/8	2 29/64	3 1/16	3 1/16	3 9/16	.312	1/2	3/8	.560	.070	.322	15/16	5/16	.50147	.041789
8	.7500	.8987	4 1/4	3 11/16	4 1/8	1/8	3 29/64	3 11/16	3 5/8	3 7/8	.343	9/16	3/8	.710	.080	.353	1	21/64	.50100	.041750
9	.9001	1.0671	4 3/4	4 1/8	4 5/8	1/8	3 7/8	4 1/8	4 1/16	4 1/16	.375	21/32	7/16	.860	.100	.385	1 1/8	3/8	.50085	.041738
10	1.04465	1.2893	6 17/32	5 13/16	6 13/32	1/8	5 19/32	5 13/16	5 3/4	5 1/2	.437	3/4	1/2	1.010	.110	.447	1 5/16	7/16	.51612	.043010
11	1.24995	1.5318	7 19/32	6 7/16	7 15/32	1/8	6 19/32	6 7/16	6 13/16	6 3/4	.437	7/8	1/2	1.210	.130	.447	1 5/16	7/16	.50100	.041750
12	1.5001	1.7968	9 3/8	7 1/4	7 15/16	1/8	6 15/16	7 1/4	7 3/16	7 7/8	.500	1 1/32	1/2	1.460	.150	.510	1 1/2	1/2	.49973	.041644
13	1.75005	2.0731	8 11/16	7 7/8	8 9/16	1/8	7 9/16	7 7/8	7 13/16	7 3/4	.500	1 5/16	5/8	1.710	.170	.510	1 1/2	1/2	.50020	.041683
14	2.0000	2.3438	9 9/32	8 3/8	9 9/32	1/8	8 1/32	8 3/8	8 5/16	8 1/4	.562		3/4	1.960	.190	.572	1 11/16	9/16	.50000	.041666
15	2.2500	2.6146	10 3/8	9 3/8	9 21/32	1/8	8 17/32	9 3/8	9 5/16	8 3/4	.562		7/8	2.210	.210	.572	1 11/16	9/16	.50000	.041666
16	2.5000	2.8854		9 7/8	10 1/4	1/8	9	9 7/8	9 13/16	9 3/4	.625		1	2.450	.230	.635	1 7/8	5/8	.50000	.041666
17	2.7500	3.1563		10 3/8		1/8		10 3/8	10 5/16	10 1/4									.50000	.041666
18	3.0000	3.4271																	.5000	.041665

SPINDLE NOSES AND ARBORS FOR MILLING MACHINES

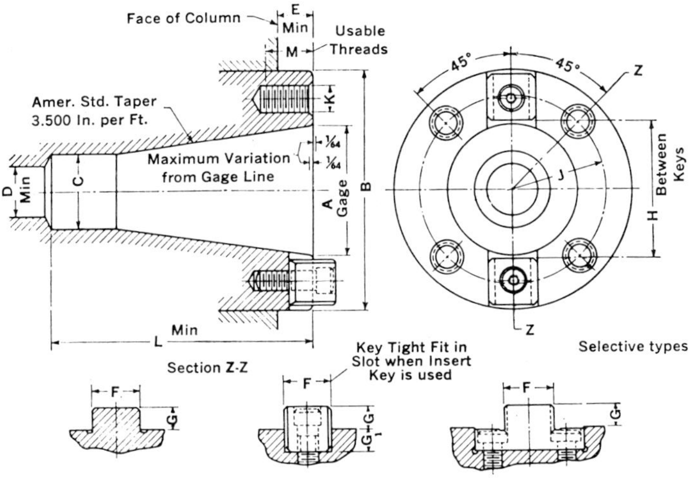

Selective types of Drive Key Construction

ESSENTIAL DIMENSIONS OF SPINDLE NOSE

size number	gage diameter of taper	diameter of spindle flange	pilot diameter	clearance hole for draw-in bolt min	minimum dimension spindle end to column	width of driving key	max height of driving key	min depth of keyway	distance between driving keys	radius of bolt hole circle	size of threads for bolt holes UNC-2B	full depth of arbor hole in spindle min	depth of usable thread for bolt hole
	A	B	C	D	E	F	G	G_1	H	J	K	L	M
30	1¼	2.7493 2.7488	0.692 0.685	21/32	½	0.6255 0.6252	5/16	5/16	1.315 1.285	1.0625 (Note 1)	⅜-16	2⅞	⅝
40	1¾	3.4993 3.4988	1.005 0.997	21/32	⅝	0.6255 0.6252	5/16	5/16	1.819 1.807	1.3125 (Note 1)	½-13	3⅞	13/16
50	2¾	5.0618 5.0613	1.568 1.559	1 1/16	¾	1.0006 1.0002	½	½	2.819 2.807	2.000 (Note 2)	⅝-11	5½	1
60	4¼	8.7180 8.7175	2.381 2.371	1 ⅜	1½	1.0006 1.0002	½	½	4.819 4.807	3.500 (Note 2)	¾-10	8⅝	1 ¼

All dimensions are given in inches.
Note 1: Holes spaced as shown and located within .003R of true position.
Note 2: Holes spaced as shown and located within .005R of true position.

SPINDLE NOSES AND ARBORS FOR MILLING MACHINES—con't

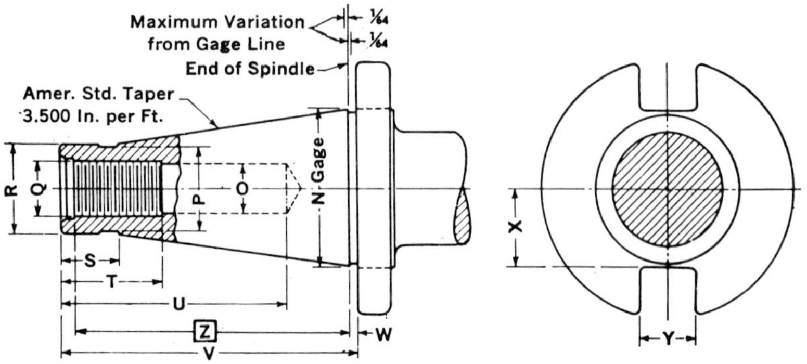

ESSENTIAL DIMENSIONS FOR ENDS OF ARBOR AND ADAPTER

size number	gage diameter of taper	drill clearance for draw-in bolt	diameter of neck	size of thread for draw-in bolt UNC-2B	pilot diameter	length of pilot	length of usable threads	depth of clearance hole	distance from rear of flange to end of arbor	clearance of flange from gage diameter	depth of driving slot	width of driving slot	depth of c'bore
	N	O	P	Q	R	S	T	U	V	W	X	Y	Z
30	1 1/4	27/64	41/64	1/2-13	0.675 0.673	13/16	1	2	2 3/4	1/16	0.640 0.625	0.630 0.640	2 1/2
40	1 3/4	17/32	15/16	5/8-11	0.987 0.985	1	1 1/8	2 1/4	3 3/4	1/16	0.890 0.875	0.630 0.640	3 1/2
50													
50A	2 3/4	7/8	1 1/2	1- 8	1.549 1.547	1	1 3/4	3 1/2	5 1/8	1/8	1.390 1.375	1.008 1.018	4 3/4
60	4 1/4	1 7/64	2 5/32	1 1/4- 7	2.361 2.359	1 3/4	2 1/4	4 1/4	8 1/4	1/8	2.400 2.390	1.008 1.018	7 13/16

All dimensions are given in inches.

TAPERS AND ANGLES

Taper Per Foot	Included Angle		With Center Line		Taper Per Inch	Taper Per Inch from Center Line
	Deg.	Min.	Deg.	Min.		
⅛	0	36	0	18	.010416	.005208
3⁄16	0	54	0	27	.015625	.007812
¼	1	12	0	36	.020833	.010416
5⁄16	1	30	0	45	.026042	.013021
⅜	1	47	0	53	.031250	.015625
7⁄16	2	05	1	02	.036458	.018229
½	2	23	1	11	.041667	.020833
9⁄16	2	42	1	21	.046875	.023438
⅝	3	00	1	30	.052084	.026042
11⁄16	3	18	1	39	.057292	.028646
¾	3	35	1	48	.062500	.031250
13⁄16	3	52	1	56	.067708	.033854
⅞	4	12	2	06	.072917	.036458
15⁄16	4	28	2	14	.078125	.039063
1	4	45	2	23	.083330	.041667
1¼	5	58	2	59	.104166	.052083
1½	7	08	3	34	.125000	.062500
1¾	8	20	4	10	.145833	.072917
2	9	32	4	46	.166666	.083333
2½	11	54	5	57	.208333	.104166
3	14	16	7	08	.250000	.125000
3½	16	36	8	18	.291666	.145833
4	18	56	9	28	.333333	.166666
4½	21	14	10	37	.375000	.187500
5	23	32	11	46	.416666	.208333
6	28	04	14	02	.500000	.250000

Table for Converting Minutes Into Decimals of a Degree

Min.	Dec. of Degree	Min.	Dec. of Degree	Min.	Dec. of Degree	Min.	Dec. of Degree	Min.	Dec. of Degree
¼	0.00416	12¼	0.20416	24¼	0.40416	36¼	0.60416	48¼	0.80416
½	0.00833	12½	0.20833	24½	0.40833	36½	0.60833	48½	0.80833
¾	0.01250	12¾	0.21250	24¾	0.41250	36¾	0.61250	48¾	0.81250
1	0.01666	13	0.21666	25	0.41666	37	0.61666	49	0.81666
1¼	0.02083	13¼	0.22083	25¼	0.42083	37¼	0.62083	49¼	0.82083
1½	0.02500	13½	0.22500	25½	0.42500	37½	0.62500	49½	0.82500
1¾	0.02916	13¾	0.22916	25¾	0.42916	37¾	0.62916	49¾	0.82916
2	0.03333	14	0.23333	26	0.43333	38	0.63333	50	0.83333
2¼	0.03750	14¼	0.23750	26¼	0.43750	38¼	0.63750	50¼	0.83750
2½	0.04166	14½	0.24166	26½	0.44166	38½	0.64166	50½	0.84166
2¾	0.04583	14¾	0.24583	26¾	0.44583	38¾	0.64583	50¾	0.84583
3	0.05000	15	0.25000	27	0.45000	39	0.65000	51	0.85000
3¼	0.05416	15¼	0.25416	27¼	0.45416	39¼	0.65416	51¼	0.85416
3½	0.05833	15½	0.25833	27½	0.45833	39½	0.65833	51½	0.85833
3¾	0.06250	15¾	0.26250	27¾	0.46250	39¾	0.66250	51¾	0.86250
4	0.06666	16	0.26666	28	0.46666	40	0.66666	52	0.86666
4¼	0.07083	16¼	0.27083	28¼	0.47083	40¼	0.67083	52¼	0.87083
4½	0.07500	16½	0.27500	28½	0.47500	40½	0.67500	52½	0.87500
4¾	0.07916	16¾	0.27916	28¾	0.47916	40¾	0.67916	52¾	0.87916
5	0.08333	17	0.28333	29	0.48333	41	0.68333	53	0.88333
5¼	0.08750	17¼	0.28750	29¼	0.48750	41¼	0.68750	53¼	0.88750
5½	0.09166	17½	0.29166	29½	0.49166	41½	0.69166	53½	0.89166
5¾	0.09583	17¾	0.29583	29¾	0.49583	41¾	0.69583	53¾	0.89583
6	0.10000	18	0.30000	30	0.50000	42	0.70000	54	0.90000
6¼	0.10416	18¼	0.30416	30¼	0.50416	42¼	0.70416	54¼	0.90416
6½	0.10833	18½	0.30833	30½	0.50833	42½	0.70833	54½	0.90833
6¾	0.11250	18¾	0.31250	30¾	0.51250	42¾	0.71250	54¾	0.91250
7	0.11666	19	0.31666	31	0.51666	43	0.71666	55	0.91666
7¼	0.12083	19¼	0.32083	31¼	0.52083	43¼	0.72083	55¼	0.92083
7½	0.12500	19½	0.32500	31½	0.52500	43½	0.72500	55½	0.92500
7¾	0.12916	19¾	0.32916	31¾	0.52916	43¾	0.72916	55¾	0.92916
8	0.13333	20	0.33333	32	0.53333	44	0.73333	56	0.93333
8¼	0.13750	20¼	0.33750	32¼	0.53750	44¼	0.73750	56¼	0.93750
8½	0.14166	20½	0.34166	32½	0.54166	44½	0.74166	56½	0.94166
8¾	0.14583	20¾	0.34583	32¾	0.54583	44¾	0.74583	56¾	0.94583
9	0.15000	21	0.35000	33	0.55000	45	0.75000	57	0.95000
9¼	0.15416	21¼	0.35416	33¼	0.55416	45¼	0.75416	57¼	0.95416
9½	0.15833	21½	0.35833	33½	0.55833	45½	0.75833	57½	0.95833
9¾	0.16250	21¾	0.36250	33¾	0.56250	45¾	0.76250	57¾	0.96250
10	0.16666	22	0.36666	34	0.56666	46	0.76666	58	0.96666
10¼	0.17083	22¼	0.37083	34¼	0.57083	46¼	0.77083	58¼	0.96083
10½	0.17500	22½	0.37500	34½	0.57500	46½	0.77500	58½	0.97500
10¾	0.17916	22¾	0.37916	34¾	0.57916	46¾	0.77916	58¾	0.97916
11	0.18333	23	0.38333	35	0.58333	47	0.78333	59	0.98333
11¼	0.18750	23¼	0.38750	35¼	0.58750	47¼	0.78750	59¼	0.98750
11½	0.19166	23½	0.39166	35½	0.59166	47½	0.79166	59½	0.99166
11¾	0.19583	23¾	0.39583	35¾	0.59583	47¾	0.79583	59¾	0.99583
12	0.20000	24	0.40000	36	0.60000	48	0.80000	60	1.00000

Millimeter Equivalents

mm = Inches	mm = Inches	mm = Inches	mm = Inches
.1 = .00394	21. = .82677	48. = 1.88976	75. = 2.95275
.2 = .00787	22. = .86614	49. = 1.92913	76. = 2.99212
.3 = .01181	23. = .90551	50. = 1.96850	76.2 = 3. ins.
.4 = .01575	24. = .94488		77. = 3.03149
.5 = .01968	25. = .98425	50.8 = 2. ins.	78. = 3.07086
		51. = 2.00787	79. = 3.11023
.6 = .02362	25.4 = 1. in.	52. = 2.04724	80. = 3.14960
.7 = .02756	26. = 1.02362	53. = 2.08661	
.8 = .03149	27. = 1.06299	54. = 2.12598	81. = 3.18897
.9 = .03543	28. = 1.10236	55. = 2.16535	82. = 3.22834
	29. = 1.14173		83. = 3.26771
1. = .03937	30. = 1.18110	56. = 2.20472	84. = 3.30708
2. = .07874		57. = 2.24409	85. = 3.34645
3. = .11811	31. = 1.22047	58. = 2.28346	
4. = .15748	32. = 1.25984	59. = 2.32283	86. = 3.38582
5. = .19685	33. = 1.29921	60. = 2.36220	87. = 3.42519
	34. = 1.33858		88. = 3.46456
6. = .23622	35. = 1.37795	61. = 2.40157	88.9 = 3.5 ins.
7. = .27559		62. = 2.44094	
8. = .31496	36. = 1.41732	63. = 2.48031	89. = 3.50393
9. = .35433	37. = 1.45669	63.5 = 2.5 ins.	90. = 3.54330
10. = .39370	38. = 1.49606	64. = 2.51968	
	38.1 = 1.5 ins.	65. = 2.55905	91. = 3.58267
11. = .43307	39. = 1.53543		92. = 3.62204
12. = .47244	40. = 1.57480	66. = 2.59842	93. = 3.66141
12.7 = .5 in.		67. = 2.63779	94. = 3.70078
13. = .51181	41. = 1.61417	68. = 2.67716	95. = 3.74015
14. = .55118	42. = 1.65354	69. = 2.71653	
15. = .59055	43. = 1.69291	70. = 2.75590	96. = 3.77952
	44. = 1.73228		97. = 3.81889
16. = .62992	45. = 1.77165	71. = 2.79527	98. = 3.85826
17. = .66929		72. = 2.83464	99. = 3.89763
18. = .70866	46. = 1.81102	73. = 2.87401	100. = 3.93700
19. = .74803	47. = 1.85039	74. = 2.91338	
20. = .78740			101.6 = 4. ins.

The exact conversion is 1 in. = 25.4 mm.

DATA ON GEARS

COMPARATIVE SIZES OF GEAR TEETH

Chart is intended only as a comparison of gear tooth sizes, and should not be used for actual measurement.

ILLUSTRATION OF GEAR TOOTH TERMS

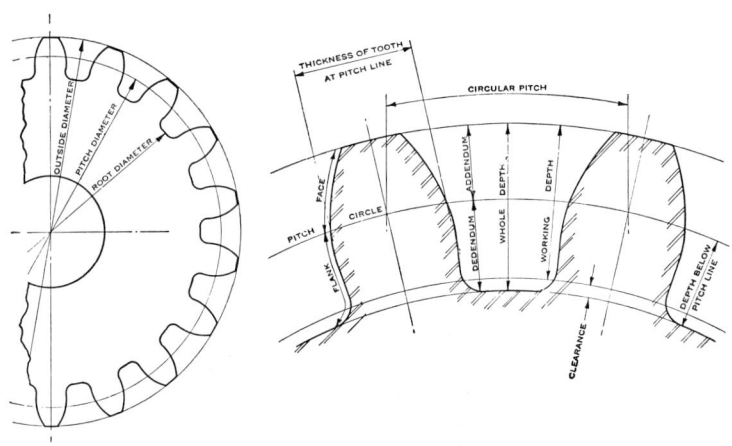

GEAR NOMENCLATURE

The definitions of gear elements are applicable to spur and helical gears.

A *Pitch Cylinder* is the imaginary cylinder in a gear that rolls without slipping on a pitch cylinder or pitch plane of another gear.

The *Axial Plane* of a pair of gears is the plane that contains the two axes. In a single gear, an axial plane may be any plane containing the axis and a given point.

A *Plane of Rotation* is any plane perpendicular to a gear axis.

A *Transverse Plane* is perpendicular to the axial plane and to the pitch plane. In gears with parallel axes, the transverse plane and plane of rotation coincide.

Involute Teeth of spur gears, helical gears, and worms are those in which the active portion of the profile in the transverse plane is the involute of a circle.

A *Pitch Circle* is the curve of intersection of a pitch surface of revolution and a plane of rotation. According to theory, it is the imaginary circle that rolls without slipping with a pitch circle of a mating gear.

The *Pitch Point* is the point of tangency of two pitch circles (or of a pitch circle and pitch line) and is on the line of centers. The pitch point of a tooth profile is at its intersection with the pitch circle.

The Line of Action is the path of contact in involute gears. It is the straight line passing through the pitch point and tangent to the base circles.

Length of Action is the distance on an involute line of action through which the point of contact moves during the action of the tooth profiles.

Pitch Diameter is the diameter of the pitch circle. In parallel-shaft gears, the pitch diameters can be determined directly from the center distance and the number of teeth by proportionality. Operating pitch diameter is the pitch diameter at which the gears operate.

Outside Diameter is the diameter of the addendum (outside) circle.

Circular Thickness is the length of arc between the two sides of a gear tooth, on the pitch circle unless otherwise specified.

Transverse Circular Thickness is the circular thickness in the transverse plane.

Normal Circular Thickness is the circular thickness in the normal plane. In helical gears, it is an arc of the normal helix.

Chordal Thickness is the length of the chord subtending a circular-thickness arc.

Chordal Addendum is the height from the top of the tooth to the chord subtending the circular-thickness arc.

Backlash is the amount by which the width of a tooth space exceeds the thickness of the engaging tooth on the pitch circles.

As actually indicated by measuring devices, backlash may be determined variously in the transverse, normal, or axial planes, and either in the direction of the pitch circles, or on the line of action. Such measurements should be corrected to corresponding values on transverse pitch circles for general comparisons.

Face Width is the length of the teeth in an axial plane.

Lead is the axial advance of a helix for one complete turn, as in the threads of cylindrical worms and teeth of helical gears.

Pressure Angle is the angle between a tooth profile and the line normal to a pitch surface, usually at the pitch point of the profile. This definition is applicable to every type of gear. In involute teeth, *pressure angle* is often described as the angle between the *line of action* and the line tangent to the *pitch circles*.

Circular Pitch is the distance along the pitch circle or pitch line between corresponding profiles of adjacent teeth.

Transverse Circular Pitch is the circular pitch in the transverse plane.

Normal Circular Pitch is the circular pitch in the normal plane at the pitch circle, and also the length of the arc along the normal helix between helical teeth or threads.

Axial Pitch is the circular pitch in the axial plane and in the pitch surface between corresponding sides of adjacent teeth, in helical gears and worms.

The term *axial pitch* is preferred to the term *linear pitch*.

Base Pitch in an involute gear is the pitch on the base circle or along the line of action. Corresponding sides of involute gear teeth are parallel curves, and the base pitch is the constant and fundamental distance between them along a common normal in a plane of rotation.

Addendum is the height by which a tooth projects beyond the pitch circle or pitch line; also, the radial distance between the pitch circle and the addendum circle.

Dedendum is the depth of a tooth space below the pitch circle or pitch line; also, the radial distance between the pitch circle and the root circle.

Clearance is the amount by which the dedendum in a given gear exceeds the addendum of its mating gear.

Working Depth is the depth of engagement of two gears; that is, the sum of their addendums.

Whole Depth is the total depth of a tooth space, equal to addendum plus dedendum, also equal to working depth plus clearance.

Base Diameter is the diameter of the circle from which involute tooth profiles are derived.

Transverse Pressure Angle is the pressure angle in the transverse plane.

Normal Pressure Angle is the pressure angle in the normal plane of a helical or spiral tooth.

Operating Pressure Angle is determined by the center distance at which the gears operate. It is the pressure angle at the operating pitch diameter.

Pitch helix is the curve of intersection of a tooth surface and its pitch cylinder in a helical gear.

Normal helix is a helix on a pitch cylinder, normal to the pitch helix.

Helix Angle is the angle between any helix and an element of its cylinder. In helical gears and worms, it is at the pitch diameter unless otherwise specified.

Diametral Pitch is the ratio of the number of teeth to the number of inches in the pitch diameter. There is a fixed relation between diametral pitch (P_d) and circular pitch (p), namely $p = \pi/P_d$.

Normal Diametral Pitch is the value of diametral pitch as calculated in the normal plane of a helical gear or worm.

Long- and Short-Addendum Teeth are those in which the addendums of two engaging gears are unequal.

Undercut is a condition in generated gear teeth when any part of the fillet curve lies inside of a line drawn tangent to the true involute form at its lowest point. Undercut may be deliberately introduced to facilitate finishing operations.

Equivalent Pitch Radius is the radius of the pitch circle in a cross section of gear teeth in any plane other than a plane of rotation. It is properly the radius of curvature of the pitch surface in the given cross section.

Equivalent Number of Teeth is the number of teeth contained in the whole circumference of a pitch circle corresponding to an equivalent pitch radius.

Involute Contact Ratio is the ratio of the length of action to the base pitch.

Face Contact Ratio is the ratio of the face advance to the circular pitch in the plane of rotation.

Total Contact Ratio is the sum of the involute contact ratio and the face contact ratio, which may be thought of as the average total number of teeth in contact in parallel helical gears.

Active Profile is that part of the gear tooth profile which actually comes in contact with the profile of its mating tooth along the line of action.

Tip Relief is an arbitrary modification of a tooth profile whereby a small amount of material is removed near the tip of the gear tooth.

Basic Rack is a rack that is adopted as the basis for a system of interchangeable gears.

Generating Rack is a rack outline used to indicate tooth details and dimensions for the design of a hob to produce gears of a basic rack system.

Full-Depth Teeth are those in which the working depth equals 2.000 divided by normal diametral pitch.

Stub Teeth are those in which the working depth is less than 2.000 divided by normal diametral pitch.

GEAR INFORMATION

Diametral pitches are numbered according to what part each tooth of a gear represents in the pitch diameter of the gear. Thus a gear of 3 inch pitch diameter having 36 teeth is $36 \div 3 = 12$ pitch. The outside diameter is determined by adding 2 to the number of teeth in the gear and dividing by the diametral pitch. In the above example the outside diameter is $(36 + 2) \div 12 = 3.166$.

The Circular Pitch is the distance from the center of one tooth to the center of the next tooth, measured along the pitch circle. For example:

If the distance from the center of one tooth to the center of the next tooth, measured along the pitch circle, is $\frac{1}{2}$ inch, the gear is $\frac{1}{2}$ inch Circular Pitch.

Having the diametral pitch, to obtain the circular pitch, divide 3.1416 by the diametral pitch. For example:

If the diametral pitch is 4, divide 3.1416 by 4, and the quotient .7854 inch, is the circular pitch.

Having the circular pitch, to obtain the diametral pitch, divide 3.1416 by the circular pitch. For example:

If the circular pitch is 2 inches, divide 3.1416 by 2, and the quotient, 1.5708, is the diametral pitch.

Having the number of teeth and the diameter of the blank, to obtain the diametral pitch, add 2 to the number of teeth, and divide by the diameter of the blank. For example:

If the number of teeth is 40, and the diameter of the blank is $10\frac{1}{2}$ inches, add 2 to the number of teeth, making 42, and divide by $10\frac{1}{2}$. The quotient, 4, is the diametral pitch.

To obtain the distance between the centers of two gears, add the number of teeth together, and divide half the sum by the diametral pitch. For example:

If two gears have 50 and 30 teeth respectively, and are 5 pitch, add 50 and 30, making 80, divide by 2, and then divide this quotient, 40, by the diametral pitch, 5, and the result, 8 inches, is the center distance.

To find the Pitch Diameter of a pair of gears divide the distance between centers by one-half the sum of the teeth and multiply the quotient by the number of teeth in each gear. The product is the Pitch Diameter of each gear.

STANDARD TOOTH DEPTHS

Dia. Pitch	Circular Pitch	Addendum	Full-Depth System		Gear-Shaper System		Pre-Shave System	
			Dedendum	Whole Depth of Tooth	Dedendum	Whole Depth of Tooth	Dedendum	Whole Depth of Tooth
1	3.1415927	1.0000	1.1571	2.1571	1.2500	2.2500	1.3500	2.3500
1¼	2.5132741	.8000	.9257	1.7257	1.0000	1.8000	1.0800	1.8800
1½	2.0943951	.6667	.7714	1.4381	.8333	1.5000	.9000	1.5667
1¾	1.7951958	.5714	.6612	1.2326	.7143	1.2857	.7714	1.3429
2	1.5707963	.5000	.5786	1.0786	.6250	1.1250	.6750	1.1750
2¼	1.3962634	.4444	.5143	.9587	.5556	1.0000	.6000	1.0444
2½	1.2566370	.4000	.4628	.8628	.5000	.9000	.5400	.9400
2¾	1.1423973	.3636	.4208	.7844	.4545	.8182	.4909	.8545
3	1.0471975	.3333	.3857	.7190	.4167	.7500	.4500	.7833
3½	.8975979	.2857	.3306	.6163	.3571	.6429	.3857	.6714
4	.7853981	.2500	.2893	.5393	.3125	.5625	.3375	.5875
5	.6283185	.2000	.2314	.4314	.2500	.4500	.2700	.4700
6	.5235987	.1667	.1929	.3595	.2083	.3750	.2250	.3917
7	.4487989	.1429	.1653	.3082	.1786	.3214	.1929	.3357
8	.3926990	.1250	.1446	.2696	.1563	.2813	.1688	.2938
9	.3490658	.1111	.1286	.2397	.1389	.2500	.1500	.2611
10	.3141593	.1000	.1157	.2157	.1250	.2250	.1350	.2350
11	.2855993	.0909	.1052	.1961	.1136	.2045	.1227	.2136
12	.2617993	.0833	.0964	.1798	.1042	.1875	.1125	.1958
13	.2416609	.0769	.0890	.1659	.0962	.1731	.1038	.1807
14	.2243994	.0714	.0827	.1541	.0893	.1607	.0964	.1679
15	.2094395	.0667	.0771	.1438	.0833	.1500	.0900	.1567
16	.1963495	.0625	.0723	.1348	.0781	.1406	.0844	.1469
17	.1847995	.0588	.0681	.1269	.0735	.1324	.0794	.1382
18	1745329	.0556	.0643	.1198	.0694	.1250	.0750	.1306
19	1653469	.0526	.0609	.1135	.0658	.1184	.0711	.1237
20	.1570796	.0500	.0620	.112	.0625	.1125	.0695	.1195
22	.1427996	.0455	.0565	.1020	.0568	.1023	.0634	.1088
24	.1308996	.0417	.0520	.0937	.0521	.0939	.0583	.0999
26	.1208304	.0385	.0482	.0866	.0481	.0865	.0539	.0924
28	.1121997	.0357	.0449	.0805	.0446	.0804	.0502	.0859
30	.1047197	.0333	.0420	.0753			.0470	.0803
32	.0981747	.0313	.0395	.0708			.0442	.0754
36	.0872664	.0278	.0353	.0631			.0395	.0673
40	.0785398	.0250	.0320	.0570			.0358	.0608
44	.0713998	.0227	.0293	.0520			.0327	.0554
48	.0654498	.0208	.0270	.0478			.0301	.0510
56	.0560998	.0179	.0234	.0413			.0261	.0440
64	.0490873	.0156	.0208	.0364			.0231	.0387
72	.0436332	.0139	.0187	.0326			.0208	.0346
96	.0327249	.0104	.0145	.0249			.0161	.0265

Full-Depth System: $\text{Whole Depth} = \dfrac{2.350}{Pn}$

Pre-Shave System: $\text{Whole Depth} = \dfrac{2.350}{Pn} + .002$

RULES AND FORMULAE

FOR SPUR GEARS
DIAMETRAL PITCH

To Get	Having	Rule	Formula
The diametral pitch	The circular pitch.........	Divide 3.1416 by the circular pitch	$P_d = \dfrac{3.1416}{p}$
The diametral pitch	The pitch diameter and the number of teeth........	Divide number of teeth by pitch diameter......................	$P_d = \dfrac{N}{D}$
The diametral pitch	The outside diameter and the number of teeth.....	Divide number of teeth plus 2 by the outside diameter............	$P_d = \dfrac{N+2}{D_o}$
Pitch diameter	The number of teeth and the diametral pitch......	Divide number of teeth by the diametral pitch................	$D = \dfrac{N}{P_d}$
Pitch diameter	The number of teeth and the outside diameter.....	Divide the product of outside diameter and number of teeth by number of teeth plus 2.......	$D = \dfrac{ND_o}{N+2}$
Pitch diameter	The outside diameter and the diametral pitch......	Subtract from the outside diameter the quotient of 2 divided by the diametral pitch.........	$D = D_o - \dfrac{2}{P_d}$
Pitch diameter	Addendum and the number of teeth..............	Multiply addendum by the number of teeth...................	$D = aN$
Outside diameter	The number of teeth and the diametral pitch.....	Divide number of teeth plus 2 by the diametral pitch............	$D_o = \dfrac{N+2}{P_d}$
Outside diameter	The pitch diameter and the diametral pitch........	Add to the pitch diameter the quotient of 2 divided by the diametral pitch...............	$D_o = D + \dfrac{2}{P_d}$
Outside diameter	The pitch diameter and the number of teeth........	Divide the number of teeth plus 2 by the quotient of number of teeth divided by the pitch diameter......................	$D_o = \dfrac{N+2}{\dfrac{N}{D}}$
Outside diameter	The number of teeth and addendum.............	Multiply the number of teeth plus 2 by addendum...............	$D_o = (N+2)\,a$
Number of teeth	The pitch diameter and the diametral pitch........	Multiply pitch diameter by the diametral pitch................	$N_p = D\,P_d$
Number of teeth	The outside diameter and the diametral pitch	Multiply outside diameter by the diametral pitch and subtract 2	$N_p = D_o\,P_d - 2$
Thickness of tooth	The diametral pitch.......	Divide 1.5708 by the diametral pitch.......................	$t_p = \dfrac{1.5708}{P_d}$
Addendum	The diametral pitch.......	Divide 1 by the diametral pitch	$a = \dfrac{1}{P_d}$
Dedendum	The diametral pitch.......	Divide 1.157 by the diametral pitch.......................	$b = \dfrac{1.157}{P_d}$
Working depth	The diametral pitch.......	Divide 2 by the diametral pitch..	$h_k = \dfrac{2}{P_d}$
Whole depth	The diametral pitch.......	Divide 2.157 by the diametral pitch.......................	$h_t = \dfrac{2.157}{P_d}$
Clearance	The diametral pitch.......	Divide .157 by the diametral pitch	$c = \dfrac{0.157}{P_d}$

RULES AND FORMULAE

FOR SPUR GEARS
Circular Pitch

To Get	Having	Rule	Formula
The circular pitch	The diametral pitch.......	Divide 3.1416 by the diametral pitch........................	$p = \dfrac{3.1416}{P_d}$
The circular pitch	The pitch diameter and the number of teeth........	Divide pitch diameter by the product of .3183 and number of teeth........................	$p = \dfrac{D}{0.3183\,N}$
The circular pitch	The outside diameter and the number of teeth.....	Divide outside diameter by the product of .3183 and number of teeth plus 2.................	$p = \dfrac{D_o}{0.3183\,(N_p + 2)}$
Pitch diameter	The number of teeth and the circular pitch	The continued product of the number of teeth, the circular pitch and .3183.....................	$D = N0.3183p$
Pitch diameter	The number of teeth and the outside diameter....	Divide the product of number of teeth, and outside diameter by number of teeth plus 2.........	$D = \dfrac{ND_o}{N + 2}$
Pitch diameter	The outside diameter and circular pitch...........	Subtract from the outside diameter the product of the circular pitch and .6366...............	$D = D_o - (0.6366p)$
Pitch diameter	Addendum and the number of teeth................	Multiply the number of teeth by the addendum.................	$D = Na$
Outside diameter	The number of teeth and the circular pitch.......	The continued product of the number of teeth plus 2, the circular pitch and .3183..............	$D_o = (N + 2)\,0.3183p$
Outside diameter	The pitch diameter and the circular pitch...........	Add to the pitch diameter the product of the circular pitch and .6366....................	$D_o = D + 0.6366p$
Outside diameter	The number of teeth and the addendum..........	Multiply addendum by number of teeth plus 2..................	$D_o = a\,(N_p + 2)$
Number of teeth	The pitch diameter and the circular pitch...........	Divide the product of pitch diameter and 3.1416 by the circular pitch........................	$N = \dfrac{3.1416D}{p}$
Thickness of tooth	The circular pitch.........	One-half the circular pitch........	$t = \dfrac{p}{2}$
Addendum	The circular pitch.........	Multiply the circular pitch by .3183........................	$a = 0.3183p$
Dedendum	The circular pitch.........	Multiply the circular pitch by .3683........................	$b = 0.3683p$
Working depth	The circular pitch.........	Multiply the circular pitch by .6366........................	$h_k = 0.6366p$
Whole depth	The circular pitch.........	Multiply the circular pitch by .6866........................	$h_t = 0.6866p$
Clearance	The circular pitch.........	Multiply the circular pitch by .05........................	$c = 0.05p$

HELICAL GEAR RULES AND FORMULAE

BASIC RULES FOR HELICAL GEAR CALCULATIONS

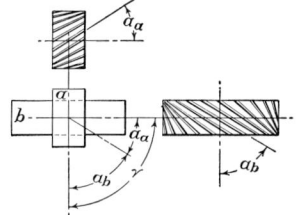

In the formulae, N, a, etc., are the numbers of teeth, helical angle, etc., for *either* gear or pinion; the notations Na, Nb, aa, ab, etc., refer to the teeth or angles in the pinion or gear, respectively, in a pair of gears a and b.

To Find	Rule	Formula
Relation between shaft and tooth angles	The sum of the tooth angles of a pair of mating helical gears is equal to the shaft angle............................	$\gamma = \alpha a + \alpha b$
Pitch diameter	Divide the number of teeth by the product of the normal pitch and the cosine of the tooth angle...................	$D = \dfrac{N}{Pn\cos\alpha}$
Center distance	Add together the pitch diameters of the two gears and divide by 2.........	$C = \dfrac{Da + Db}{2}$
Checking calculations	To prove the calculations for pitch diameters and center distance, multiply the number of teeth in the first gear by the tangent of the tooth angle of that gear, and add the number of teeth in the second gear to the product, the sum should equal twice the product of the center distance multiplied by the normal diametral pitch, multiplied by the sine of the tooth angle of the first gear	$Nb+(Na\times\tan\alpha a)=$ $2CPn\times\sin\alpha a$ for $\gamma = 90°$
Number of teeth for which to select cutter	Divide the number of teeth in the gear by the cube of the cosine of the tooth angle............................	$N' = \dfrac{N}{(\cos\alpha)^3}$ (Approx)
Lead of tooth helix	Multiply the pitch diameter by 3.1416 times the cotangent of the tooth angle	$L = \pi D \times \cot\alpha$
Addendum	Divide 1 by the normal diametral pitch..	$S = \dfrac{1}{Pn}$
Whole depth of tooth	Divide 2.157 by the normal diametral pitch............................	$W = \dfrac{2.157}{Pn}$
Normal tooth thickness at pitch line	Divide 1.571 by the normal diametral pitch............................	$Tn = \dfrac{1.571}{Pn}$
Outside diameter	Add twice the addendum to the pitch diameter..........................	$O = D + 2S$

RULES AND FORMULAE

FOR WORM GEARS

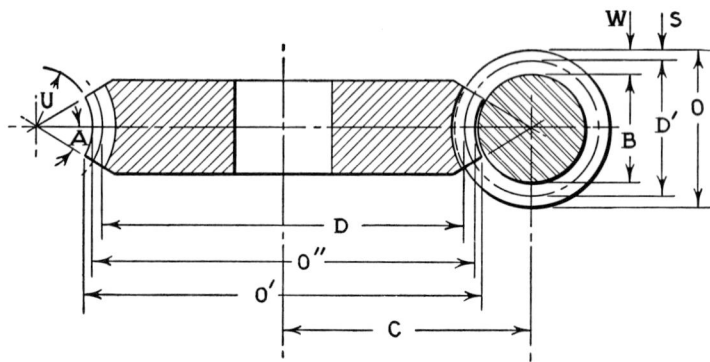

To compute the necessary dimensions for a worm gear drive the following formulae should be used in connection with the above figure.

P   = circular pitch of wheel and axial pitch of worm.

L   = lead of worm.

N′ = number of threads in worm.

S   = addendum

D′ = pitch diameter of worm.

D   = pitch diameter of wormgear.

O   = outside diameter of worm.

O″ = throat diameter of wormgear.

O′ = diameter of wormgear over sharp corners.

B   = bottom diameter of worm.

N   = number of teeth in wormgear.

W = whole depth of worm tooth.

T   = width of thread tool at end.

B′ = helix angle of worm.

90° — B′ = gashing angle of wormgear.

U   = radius of curvature of wormgear throat.

C   = center distance.

A   = face angle of wormgear.

RULES AND FORMULAE

FOR WORM GEARS

To Find	Rule	Formula
Axial Pitch	Divide the lead by the number of threads. It is understood that by the number of threads is meant, not number of threads per inch, but the number of threads in the whole worm—one, if it is single threaded; four if it is quadruple–threaded, etc.	$P = \dfrac{L}{N'}$
Addendum of worm tooth	Multiply the axial pitch by 0.3183...........	$S = 0.3183P,$
Pitch diameter of worm	Subtract twice the addendum from the outside diameter..............................	$D' = O - 2S$
Pitch diameter of wormgear	Multiply the number of teeth in the wheel by the linear pitch of the worm, and divide the product by 3.1416......................	$D = \dfrac{NP}{3.1416}$
Center distance between worm and wormgear	Add together the pitch diameter of the worm and the pitch diameter of the wormgear, and divide the sum by 2.................	$C = \dfrac{D+D'}{2}$
Whole depth of worm tooth	Multiply the axial pitch by 0.6866...........	$W = 0.6866 \times P$
Root diameter of worm	Subtract twice the whole depth of tooth from the outside diameter.....................	$B = O - 2W$
Lead angle of worm	Multiply the pitch diameter of the worm by 3.1416, and divide the product by the lead; the quotient is the cotangent of the lead angle of the worm.......................	$\text{Cotangent } B = \dfrac{3.1416D'}{L}$
Width of thread tool at end	Multiply the axial pitch by 0.31.............	$T = 0.31\ P$
Throat diameter of wormgear	Add twice the addendem of the worm tooth to the pitch diameter of the wormwheel.......	$O'' = D \times 2S$
Radius of wormgear throat	Subtract twice the addendum of the worm tooth from half the outside diameter of the worm.................................	$U = \dfrac{O}{2} - 2S$
Outside diameter of worm	Add together the pitch diameter and twice the addendum...........................	$O = D' + 2S$
Pitch diameter of worm	Subtract the pitch diameter of the wormwheel from twice the center distance.............	$D' = 2C - D$
Diameter of wormgear to sharp corners	Multiply the radius of curvature of the worm-gear throat by the cosine of half the face angle, subtract this quantity from the radius of curvature, multiply the remainder by 2, and add the product to the throat diameter of the wormgear.......................	$O' = 2(U - UX \cos A) + O''$
Gashing angle of wormgear	Divide the lead of the worm by the circumference of the pitch circle. The result will be the tangent of the gashing angle..........	$\text{Tan } (90° - B) = \dfrac{L}{\pi D'}$

GEAR MILLING CUTTERS

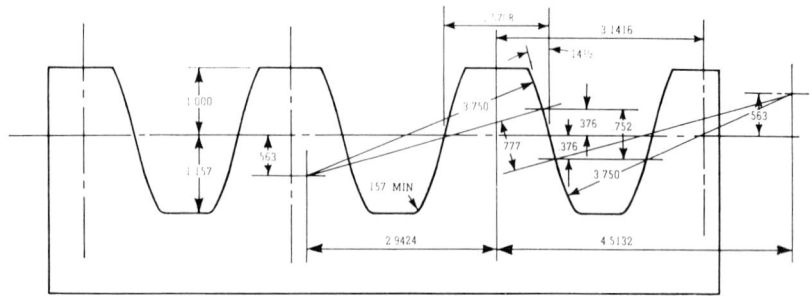

Tooth proportions of the basic rack for standard 1 DP involute gear cutters.

Gear cutters are not universal for a complete range of teeth, as are standard-form hobs. Instead, each cutter will cut only a certain range of teeth within that pitch. The form on the cutter is made correct for the lowest number of teeth in that particular range. Thus, all teeth within the range are provided with sufficient tip relief. The same form is produced on all tooth spaces within that range. Theoretically, the form should change for each different number of teeth, but the change is so small that standard ranges have been established.

If a No. 6 cutter (designed for 17 teeth) is used to cut a gear with 20 teeth, the form will not be correct for 20 teeth. Compared to the true profile for a 20 tooth gear, the No. 6 cutter will produce a tooth which is thin at the top. A close approximation of the form could be obtained from a No. 5½ cutter which has the correct form for 19 teeth. For the utmost accuracy of form, a single-purpose cutter designed for 20 teeth should be ordered.

Gear Milling Cutters for Spur Gears.

Gear cutters have been classified into eight different groups according to the number of teeth in the gear to be cut. The range of teeth for each number of cutter is for spur teeth only.

Number of Cutter	*Range*
No. 1	135 teeth to a rack
No. 2	55 teeth to 134 teeth
No. 3	35 teeth to 54 teeth
No. 4	26 teeth to 34 teeth
No. 5	21 teeth to 25 teeth
No. 6	17 teeth to 20 teeth
No. 7	14 teeth to 16 teeth
No. 8	12 teeth to 13 teeth

These cutters are usually considered as sufficiently accurate for most work. However, the half-size cutters listed below may be used when a more accurate tooth form is required.

Number of Cutter	Range
No. 1½	80 teeth to 134 teeth
No. 2½	42 teeth to 54 teeth
No. 3½	30 teeth to 34 teeth
No. 4½	23 teeth to 25 teeth
No. 5½	19 teeth to 20 teeth
No. 6½	15 teeth to 16 teeth
No. 7½	13 teeth

Gear Milling Cutters for Helical Gears.

To find the range of cutter to be used for helical gears, we must translate the number of helical teeth into an equivalent number of spur teeth. The formula for determining the equivalent number of spur teeth is

$$\text{Equivalent number of teeth} = \frac{N}{\cos^3 \psi}$$

where N = number of teeth in helical gear

ψ = helix angle of gear

See table on page 782.

Gear Milling Cutters for Bevel and Mitre Gears.

Gear cutters for bevel and mitre gears are made in numbers 1 through 8 the same as for spur gears. However, the range of teeth does not correspond with the range listed for spur teeth. The range for bevel gears depends upon the number of teeth in both mating gears rather than a single gear. Since a pair of mitre gears have the same number of teeth, the number of the correct cutter can be determined by multiplying the number of teeth in one gear by 1.41 and selecting the number of the cutter from the table for spur gears. For bevel gears, both the number of teeth in the gear and in the pinion must be considered in determining the number of the required cutters.

SELECTING CUTTERS

FOR MILLING HELICAL GEARS

Angle of Helix	K	Angle of Helix	K	Angle of Helix	K	Angle of Helix	K
0° 0′	1.000	21° 0′	1.228	42° 0′	2.436	63° 0′	10.69
0° 30′	1.000	21° 30′	1.241	42° 30′	2.495	63° 30′	11.27
1° 0′	1.001	22° 0′	1.254	43° 0′	2.557	64° 0′	11.87
1° 30′	1.001	22° 30′	1.268	43° 30′	2.621	64° 30′	12.55
2° 0′	1.002	23° 0′	1.282	44° 0′	2.687	65° 0′	13.25
2° 30′	1.003	23° 30′	1.297	44° 30′	2.756	65° 30′	14.03
3° 0′	1.004	24° 0′	1.312	45° 0′	2.828	66° 0′	14.86
3° 30′	1.005	24° 30′	1.328	45° 30′	2.902	66° 30′	15.80
4° 0′	1.007	25° 0′	1.344	46° 0′	2.983	67° 0′	16.76
4° 30′	1.009	25° 30′	1.360	46° 30′	3.066	67° 30′	17.85
5° 0′	1.011	26° 0′	1.377	47° 0′	3.152	68° 0′	18.98
5° 30′	1.013	26° 30′	1.395	47° 30′	3.242	68° 30′	20.33
6° 0′	1.016	27° 0′	1.414	48° 0′	3.336	69° 0′	21.72
6° 30′	1.019	27° 30′	1.434	48° 30′	3.436	69° 30′	23.33
7° 0′	1.022	28° 0′	1.454	49° 0′	3.540	70° 0′	25.00
7° 30′	1.026	28° 30′	1.474	49° 30′	3.650	70° 30′	26.97
8° 0′	1.030	29° 0′	1.495	50° 0′	3.767	71° 0′	28.97
8° 30′	1.034	29° 30′	1.517	50° 30′	3.887	71° 30′	31.40
9° 0′	1.038	30° 0′	1.540	51° 0′	4.012	72° 0′	33.88
9° 30′	1.042	30° 30′	1.563	51° 30′	4.144	72° 30′	36.92
10° 0′	1.047	31° 0′	1.588	52° 0′	4.284	73° 0′	40.00
10° 30′	1.052	31° 30′	1.613	52° 30′	4.433	73° 30′	43.88
11° 0′	1.057	32° 0′	1.640	53° 0′	4.586	74° 0′	47.79
11° 30′	1.062	32° 30′	1.667	53° 30′	4.752	74° 30′	54.72
12° 0′	1.068	33° 0′	1.695	54° 0′	4.925	75° 0′	57.68
12° 30′	1.074	33° 30′	1.724	54° 30′	5.101	75° 30′	64.15
13° 0′	1.080	34° 0′	1.755	55° 0′	5.295	76° 0′	70.65
13° 30′	1.087	34° 30′	1.787	55° 30′	5.497	76° 30′	79.20
14° 0′	1.094	35° 0′	1.819	56° 0′	5.710	77° 0′	87.78
14° 30′	1.102	35° 30′	1.853	56° 30′	5.940	77° 30′	99.50
15° 0′	1.110	36° 0′	1.889	57° 0′	6.190	78° 0′	111.3
15° 30′	1.118	36° 30′	1.926	57° 30′	6.435	79° 0′	144.0
16° 0′	1.127	37° 0′	1.963	58° 0′	6.720	80° 0′	191.2
16° 30′	1.136	37° 30′	2.003	58° 30′	7.010	81° 0′	261.4
17° 0′	1.145	38° 0′	2.044	59° 0′	7.321	82° 0′	370.6
17° 30′	1.154	38° 30′	2.086	59° 30′	7.650	83° 0′	552.1
18° 0′	1.163	39° 0′	2.130	60° 0′	8.000	84° 0′	876.4
18° 30′	1.172	39° 30′	2.176	60° 30′	8.380	85° 0′	1509.0
19° 0′	1.182	40° 0′	2.225	61° 0′	8.780	86° 0′	2940.0
19° 30′	1.193	40° 30′	2.275	61° 30′	9.209	87° 0′	6990.0
20° 0′	1.204	41° 0′	2.326	62° 0′	9.658	....	....
20° 30′	1.216	41° 30′	2.380	62° 30′	10.160	....	....

Example:—Angle of helix = 30 degrees; number of teeth in helical gear = 18.
Factor *K* for 30 degrees, as found from the table, equals 1.540. Then, number of teeth for which to select the cutter = 18 × 1.540 = 28, approximately. Hence, use spur gear cutter for 28 teeth, or cutter No. 4.